2nd International Conference on Scientific Communication in Fisheries and Marine Sciences (SCiFiMaS 2018)

E3S Web of Conferences Volume 47 (2018)

Purwokerto, Indonesia
7 - 9 May 2018

Editors:

N.A. Prayogo
A. Amron

ISBN: 978-1-5108-6856-4

Printed from e-media with permission by:

Curran Associates, Inc.
57 Morehouse Lane
Red Hook, NY 12571

proceedings.com

Some format issues inherent in the e-media version may also appear in this print version.

Printed by Curran Associates, Inc. (2018)

For additional information, please contact EDP Sciences – Web of Conferences at the address below.

EDP Sciences – Web of Conferences
17, Avenue du Hoggar
Parc d'Activité de Courtabœuf
BP 112
F-91944 Les Ulis Cedex A
France

Phone: +33 (0) 1 69 18 75 75
Fax: +33 (0) 1 69 28 84 91

contact-edps@webofconferences.org

Additional copies of this publication are available from:

Curran Associates, Inc.
57 Morehouse Lane
Red Hook, NY 12571 USA
Phone: 845-758-0400
Fax: 845-758-2633
Email: curran@proceedings.com
Web: www.proceedings.com

TABLE OF CONTENTS

CHITOSAN ADDED AS AN ANTIMICROBIAL IN EDIBLE FILM AND ITS APPLICATION TO THE DODOL OF SEAWEED (EUCHEUMA COTTONII L.) 1
Warkoyo, Harini Noor, Hermawan David, Wulandari Yuliarie

ANTIMICROBIAL COATING ON QUALITY ATTRIBUTES OF SAUSAGE DURING REFRIGERATED STORAGE 9
Naufalin R., Wicaksono R., Arsil P., Salman M.F.

IDENTIFICATION AND EXPRESSION OF VITELLOGENIN GENE IN THE GOURAMY (OSPHRONEMOUS GOURAMMY) UNDER PHOTOPERIODS MANIPULATION 15
Prayogo Norman Arie, Siregar Asrul Sahri, Sukardi Purnama, Nugrayani Dewi, Ekasanti Anandita, Yasumasa Bessho Riviani

GROWTH PERFORMANCE OF CATFISH (CLARIAS GARIEPENUS) CULTURED OF HIGH DENSITY WITH BIOFLOC SYSTEM 25
Hary Tjahja Soedibya Petrus, Listiowati Emyliana, Budhi Pramono Taufik, Arie Prayogo Norman, Taufan Harisam R.

REARING TECHNOLOGY FOR CLIMBING FISH (ANABAS TESTUDINEUS) ON DIFFERENT SALINITY 37
Widiyati Ani, Asyiah B.N., Heru Prihadi Tri, Puspaningsih Dewi

POTENCY OF DIFFERENT SEAWEEDS AS DIETS FOR DEVELOPING ABALONE (H. SQUAMATA) CULTURE IN NUSA PENIDA ISLAND, BALI 42
Heru Prihadi Tri, Rasidi, Ardi Idil, Widiyati Ani, Budi Wiyanto Dwi

POLYCULTURE OF GIANT FRESHWATER PRAWN, MACROBRACHIUM ROSENBERGII AND NILEM CARP, OSTEOCHILUS HASSELTI CULTURED IN RECIRCULATION SYSTEM USING BIOFILTRATION 50
Setiadi Eri, Taufik Imam

WATER QUALITY, SURVIVAL, AND GROWTH OF RED TILAPIA, OREOCHROMIS NILOTICUS CULTURED IN AQUAPONICS SYSTEM 60
Setiadi Eri, Widyastuti Yohana R., Heru Prihadi Tri

GROWTH PERFORMANCE OF SIAMESE GOURAMI TRICHOPODUS PECTORALIS SEED WITH DIFFERENT STOCKING DENSITY TO OPTIMIZE PRODUCTION CAPACITY: OPTIMUM TEMPERATURE AND SALINITY APPLICATION 68
Setijaningsih Lies, Puspaningsih Dewi, Taufik Imam

PRELIMINARY STUDY OF HEAVY METAL (ZN, PB, CR, AS, CU, CD) CONTAMINATIONS ON DIFFERENT SOIL LEVEL FROM POST-MINING BAUXITE PRODUCTION FOR AQUACULTURE 76
Dwirama Putra Risandi, Apriadi Tri, Suryanti Ani, Irawan Henky, Said Raja'I Tengku, Yulianto Try, Kusuma Atmadja Wiwin, Joei Koenawan Chandra

NURSERY I: THE EFFECT OF STOCKING DENSITY ON THE PERFORMANCE OF GLASS EELS, ANGUILLA BICOLOR IN THE BIOFLOC SYSTEM 94
Sukardi Purnama, Prayogo Norman Ari, Winanto Tjahyo, Siregar Asrul Sahri, Harisam Taufan

BIODIVERSITY OF ALGAE POTENTIALLY HABS (HARMFULL ALGAE BLOOMS) IN RESERVOIR MRICA, BANJARNEGARA 99
Andriyani Nuraina, Mahdiana Arif, Dewi Rose, Subagyo Yusuf, sahri siregar Asrul, Junaidi Teuku, harisam Taufan

EFFECT OF MEDIUM AND LIGHT QUALITY ON PINK PIGMENT PRODUCTION OF CYANOBACTERIA OSCILLATORIA SP. BTCC/A0004 111
Karseno, Harada Kazuo, Hirata Kazumasa

POTENTIAL OF SUSTAINABLE AND BIOLOGICAL ASPECT OF STING RAY (DASYATIS SP.) AS CATCH FISHERIES STATUS OVERVIEW IN JAVA SEA 117
Mahdiana Arif, Prayogo Norman Arie, Sulistyo Isdy

SPECIES RICHNESS, ABUNDANCE AND VERTICAL DISTRIBUTION OF COPEPODS IN BANDA SEA, INDONESIA 123
Agus Mulyadi Hanung, Basit Abdul, Yulia Ikhsani Idha, Sudin Abdul Malik

DIVERSITY AND ABUNDANCE OF GIANT CLAMS IN ANAMBAS ISLANDS, INDONESIA 132
Harahap Syawaludin Alisyahbana, Yanuar Yogi, Ilham Yuwanda

SEAWEED SPECIES COMPOSITION, ABUNDANCE AND DIVERSITY IN DRINI AND KONDANG MERAK BEACH, JAVA 141
Ahmad Romdoni Taufiq, Ristiani Ajeng, Dyah Nur Meinita Maria, Marhaeni Bintang, Setijanto

SPATIO-TEMPORAL DISTRIBUTION OF CHLOROPHYLL-A USING MULTITEMPORAL LANDSAT IMAGE AND GROUND CHECK IN SEGARA ANAKAN LAGOON 149
Dewi Rose, Zainuri Muhammad, Anggoro Sutrisno, Winanto Tjahjo, Endrawati Hadi

SUBLETHAL TOXICITY TESTS OF MERCURY (HG) TO NILEM FISH (OSTEOCHILUS HASSELTI) GILLS TISSUE DAMAGE 158
Sahri Siregar Asrul, Arie Prayogo Norman, Listiowati Emyliana, Santoso Mahendro, Gumay Yudha Indra, Wahyuni Sholehah Tri

CLASSIFICATION OF SHALLOW WATER SEABED PROFILE BASED ON LANDSAT 8 IMAGERY AND IN-SITU DATA. CASE STUDY IN GILI MATRA ISLAND LOMBOK, INDONESIA 170
Ayustina Ratih, Aulia Zahra, Mustakin Haji, Alam Fiyesti, Amron Amron, Yuwono Doddy, Ahmad Triguardi, Prayogo Aldo, Sari Fitra

ASSOCIATED OF CORAL REEF WITH REEF FISH IN NORTHERN AND SOUTHERN GILI AIR ISLAND AND GILI TRAWANGAN ISLAND LOMBOK 182
Wahyu A P Trisna, Mubarak Helmi, Lazuardy Rolian Damar, Geraldi Pardede Hanson, Prabowo, Rizka Safitri Dhananta, Zunika Putri Kharisma, Iqbal Kenedi Muhammad, Gita Tiara Shanti Ainia, Nur Afrizal Satria, Larasayu Santoso Niken, Maharani Karina, Widyaningrum Ria, Helmi Muflih Azzam, Nur Faizan Rifki, Safitri Adini, Winanto Tjahjo

THE EFFECT IN THE WASTEWATER TREATMENT AT SOYBEAN CURD OF CONTACT TIME MODIFICATION OF ARTIFICIAL WETLAND USING SSF BY USING SCHOENOPLECTUS CORYMBOSUS TO IMPROVE WATER QUALITY 196
Mahdiana Arif, Sahri Siregar Asrul, Sonny Januar C, Arie Prayogo Norman

ACUTE EFFECTS OF CRUDE OILFOR THREE COMMON MANGROVE SEEDLING IN SEGARA ANAKAN NATURE RESERVE (SANR) CILACAP, INDONESIA 205
Harisam Taufan, Sugiyono, Pribadi Rudhi, Sari Siregar Asrul, Wisudyanti Budi Hastuti Dewi, Arie Prayogo Norman

OCCURRENCE OF BEACH DEBRIS IN TUNDA ISLAND, BANTEN, INDONESIA 214
Maharani Adinda, Purba Noir P, Faizal Ibnu

PHOSPHORUS LOADING FROM FISH FARMING ACTIVITIES TO WADASLINTANG RESERVOIR WATERS 226
Sih Piranti Agatha, RUS Rahayu Diana, Waluyo Gentur

PRESSURES AND STATUS OF SEAGRASS ECOSYSTEM IN THE COASTAL AREAS OF NORTH BINTAN, INDONESIA 237
Karlina Ita, Kurniawan Fery, Idris Fadhliyah

RESPONSES OF HERBIVOROUS FISHES ON CORAL REEF COVER IN OUTER ISLAND INDONESIA (STUDY CASE: NATUNA ISLAND) 243
Dwirama Putra Risandi, Suryanti Ani, Kurniawan Dedi, Pratomo Arif, Irawan Henky, Said Raja'I Tengku, Kurniawan Rika, Pratama Ginanjar, Jumsurizal

POTENTIAL HARMFUL ALGAL BLOOMS (HABS) IN SEGARA ANAKAN LAGOON, CENTRAL JAVA, INDONESIA 261
Dewi Rose, Zainuri Muhammad, Anggoro Sutrisno, Winanto Tjahjo, Endrawati Hadi

TRANSMISSION LOSS ESTIMATION OF UNDERWATER SOUND BASED ON THE NOISE INTENSITY EMMITED BY MV. PENGAYOMAN IV IN TANJUNG INTAN CRUISE LINE, CILACAP 273
Triwahyanti Lesa, Sekar Cyndana Ajeng, Hurly Sefnianti Yessy, Juita Sari Ratna, Amron Amron

THE SLOPE DYNAMIC OF CILACAP BACKSHORE DURING TRANSITION SEASON 281
EkoDwi Haryono Florencius, Nalendra Lazuardy, Trenggono Mukti, Amron Amron, Hartoyo

INTERACTION OF OCEANOGRAPHY PATTERNS TOWARDS THE ABUNDANCE OF PHYTOPLANKTON, ZOOPLANKTON AND ICHTHYOPLANKTON IN TELUK PENYU WATERS OF CILACAP 286
Eko Dwi Haryono Florencius, Dewi Rose, Budhi Pramono Taufik, Ahda Sumantri Rifki, Nur Cahyo Tri, Wisudyanti Dewi

LOCAL FISHERMAN RESISTANCE IN KELAPA KAMPIT COASTAL AREA TOWARD OFFSHORE MINING ACTIVITIES 292
Ibrahim, Haryadi Dwi, Wahyudin Nanang

EFFECTS OF EL NINO ON DISTRIBUTION OF CHLOROPHYLL-A AND SEA SURFACE TEMPERATURE IN NORTHEN TO SOUTHERN SUNDA STRAIT 312
Trenggono Mukti, Amron Amron, Avia Pasha Wanda, Lazuardy Rolian Damar

GENETIC MUTATION IN MANGROVE ACANTHUS ILIFOLICIUS BASE ON DNA BARCODE (RBCL AND MATK GEN) IN THE DIFFERENT ENVIRONMENT CHANGE IN COASTAL CILACAP, CENTRAL JAVA, INDONESIA 323
Harisam R. Taufan, Haryono Florencius Eko Dwi, Marhaeni Bintang, Prayogo Norman Arie, Soedibya Petrus Hary Tjahja

THE PATTERN OF COASTLINE CHANGE BASED ON THE CHARACTERISTICS OF SEDIMENT AND COASTAL SLOPE IN PANGENAN COAST OF CIREBON, WEST JAVA 329
Suhendra, Amron Amron, Hilmi Endang

THE SINKING SHIP POLICY TO THE PERPETRATOR OF ILLEGAL, UNREGULATED AND UNREPORTED FISHING IN CRIMINAL LAW PERSPECTIVE 338
Raharjo Agus, Sudrajat Tedi, Wasi Bintoro Rahadi, Saefudin Yusuf

CORPORATE CRIMINAL LIABILITY TOWARDS THE DOER OF ILLEGAL, UNREGULATED AND UNREPORTED FISHING IN INDONESIA 348
Bintoro Rahadi Wasi, Raharjo Agus, Sudrajat Tedi

HARMONIZATION OF REGULATION IN WATER TERRITORIAL MANAGEMENT BECOMING A FAIR ECONOMIC BENEFIT DISTRIBUTION TOWARDS REGIONAL AUTONOMY 355
Sudrajat Tedi, Raharjo Agus, Wasi Bintoro Rahadi, Saefudin Yusuf

INSTITUTIONAL AND MANAGEMENT ANALYSIS OF THE AUCTION FISH PLACE (AFP) FOR IMPROVING FISHERMEN'S WELFARE IN TEGALSARI AFP, TEGAL CITY, CENTRAL JAVA 361
Junaidi Teuku, Fitri Mardani Irfina, Mahdiana Arif

LAW AND JUSTICE FOR THE OCEANS: STUDY ON ILLEGAL, UNREPORTED AND UNREGULATED FISHING AND RELATED CRIMES 369
Yuliantiningsih Aryuni, Hartiwiningsih, Maman Suherman Ade, Latifah Emmy

THE ROLES OF THE MINISTRY OF MARINE AND FISHERY IN MARINE ENVIRONMENT PROTECTION FROM OIL POLLUTION BY TANKERS IN THE SEA IN INDONESIAN LEGAL SYSTEM 379
Purwendah Elly Kristiani, Triatmodjo Marsudi, Sulistiawati Linda Yanti

FISHING SEASONS OF FISH LANDED AT SUNGAILIAT ARCHIPELAGO FISHING PORT IN BANGKA REGENCY 386
Fauziyah, Agustriani Fitri, Melda Situmorang Desi, Suteja Yuliyanto

PROSPECT ON IMPLEMENTATION OF NATIONAL FISH LOGISTICS SYSTEM: CASE IN INDONESIA 396
Azhar Muhamad, Suhartoyo Suhartoyo, Suharso Putut, Herawati Vivi Endar, Trihastuti Nanik

PRACTICES OF ILLEGAL FISHING IN PEMALANG REGION: A POLICY ANALYSIS 402
Wahyuningrat, Haryanto Tenang, Rosyadi Slamet

SUSTAINABILITY OF SEAHORSES: LESSONS LEARNED OF LOCAL WISDOM FROM BINTAN ISLANDS, RIAU ISLANDS PROVINCE 408
Ulfah Fitria, Afrizal, Pratomo Arief

MAKE PEACE WITH THE DAMAGED COAST(THE EXPERIENCE OF MENTOK COASTAL COMMUNITY IN ADAPTING THE OFFSHORE TIN MINING) 417
Ibrahim, Haryadi Dwi, Wahyudin Nunung

PROTECTION OF TRADITIONAL FISHERMEN IN THE GRANTING OF FISHERY LICENSES IN INDONESIA 427
Azhar Muhamad, Suhartoyo Suhartoyo, ALW Lita Tyesta, Suharso Putut, Herawati Vivi Endar

COASTAL COMMUNITY RESPONSE TO THE MOVEMENT OF LITERACY: A STUDY ON LITERACY CULTURE IN DEMAK PESANTREN'S 433
Suharso Putut, Sarbini Sarbini

THE ROLE OF COMMUNITIES IN STRENGTHENING SOCIAL CAPITAL OF COASTAL COMMUNITIES THROUGH THE LIBRARY 444
Suharso Putut, Sudardi Bani, Widodo Sahid Teguh, Habsari Sri Kusumo

COENZYME Q10 IS POTENTIALLY USEDTO IMPROVE LIPID PROFILE IN DIABETIC HYPERCHOLESTEROLEMIA-INDUCED PERIODONTITIS IN THE COASTAL AREA 450
Saryono Saryono, Devinta Hesti, Widodo Abdul Haris Budi, Hidayat Arif Imam

A DEVELOPMENT POLICY OF NETWORKING-BASED CREATIVE MARINE SMALL AND MEDIUM ENTERPRISES AS A SOLUTION FOR POVERTY ALLEVIATION IN INDONESIA 457
Rosyadi Slamet, Fitrah Elpeni, Sabhita Kusuma Ayusia

GEL ALOE VERA REDUCES MMP-9 IN DIABETIC WOUNDS 464
Sari Yunita, Purnawan Iwan, Wahyu Kurniawan Dhadhang, Sutrisna Eman

Author Index

Foreword

On behalf of Fisheries and Marine Science Faculty, I am thanking all scientists and delegations to attend the 2nd KRIPIK-SciFiMaS 2018. The second International Conference for our newly faculty which is established in 2015.

The 2nd SciFiMas conference 2018 came from the idea of reunification of the vision and commitment of fisheries and marine stakeholders in Indonesia in order to support sustainable development through scientific communication based on the results of basic research, applied and frontier sciences in this field. To reinforce the Indonesian branding, we call this international conference "KRIPIK-SciFiMaS", which comes from the SciFiMaS tranlation in bahasa Indonesia (KRIPIK : Komunikasi Riset Ilmiah Perikanan dan Ilmu Kelautan). The theme of KRIPIK-SciFiMaS 2018 is "Global partnership for sustainability of aquatic biodiversity and environment."

The KRIPIK-SciFiMaS 2018 covers a broad range of topics related to the aquatic resources, aquaculture engineering, biodiversity and bioprospecting, conservation, marine pollution, sea as the world's buffer, social economic aspect of fisheries, coastal community empowerment and environment.

I hope this abstract collection will furnish the participant to follow the conferences. We hope you enjoy 2nd KRIPK-SciFiMaS 2018.

Finally, I would like to thanks the presenter for their willingness to share their latest research and idea, without them, the conference would not be possible.

Dr. Isdy Sulistyo
Dean of Fisheries and Marine Science Faculty

Preface

Dear Author(s), KRIPIK-SciFiMaS participant, esteemed Readers,

First of all, I should like to express my sincere thanks to all of you for participating in the second "Scientific Communication in Fisheries and Marine Sciences" (SciFiMaS) conference, organized by Fisheries and Marine Sciences Faculty - Jenderal Soedirman University in collaboration with the Center for Maritime Biosciences Studies. The venue for which is Java Heritage Hotel Purwokerto, Indonesia.

This abstract book includes all the abstracts of selected papers which are presented at the 2nd KRIPIK-SciFiMaS 2018, 6th-8th May 2018, from our guest speakers, experts speakers and from both type participants in form of oral presentation and speaking poster. All the abstracts published in this issue have been reviewed and accepted by the 2nd Scientific Communication in Fisheries and Marine Sciences (SCiFiMaS) Committee. It was a pleasure for the editorial board to understand that so many research work carried out by leading experts and new scientists in fisheries and marine science field that may be considered, as emerging in the Global partnership for sustainability of aquatic biodiversity and environment.

We are pleased to mention that the event gained significant interest withover 100 National and International speakers from, China, South Korean, Japan, Vietnam, Germany and United States of America. The international participants also come from South Korean, China, Sudan, and New Zeeland. There moreover than 130 oral presenters 7 poster presentations during 8 panel sessions.

We want to thank the contributors for their diligence in preparing their manuscripts. Modifications in the layout of abstracts received from authors have been made to fit with the publication format of abstract book of 2nd KRIPIK-SciFiMaS 2018.

We are looking forward to meeting you during the 2nd KRIPIK-SciFiMaS 2018!

Dr. Norman Arie Prayogo, M.Si
Chairman

Statement of Peer review

In submitting conference proceedings to *Web of Conferences*, the editors of the proceedings certify to the Publisher that

1. They adhere to its **Policy on Publishing Integrity** in order to safeguard good scientific practice in publishing.
2. All articles have been subjected to peer review administered by the proceedings editors.
3. Reviews have been conducted by expert referees, who have been requested to provide unbiased and constructive comments aimed, whenever possible, at improving the work.
4. Proceedings editors have taken all reasonable steps to ensure the quality of the materials they publish and their decision to accept or reject a paper for publication has been based only on the merits of the work and the relevance to the journal.

Title, date and place of the conference

2nd Scientific Communication in Fisheries and Marine Sciences 2018, 7-9 May 2018 in Purwokerto - Indonesia

Proceedings editor(s):

Dr. Norman Arie Prayogo (Jenderal Soedirman University)
Dr. Amron Amron (Jenderal Soedirman University)

Date and editor's signature

17 July 2018

Chitosan Added as an Antimicrobial in Edible Film and Its Application to The Dodol of Seaweed (*Eucheuma cottonii* L.)

Warkoyo[1)], *Noor* Harini[1)], *David* Hermawan[2)], *Yuliarie* Wulandari[1)]
[1)] Department of Food Science and Technology, University of Muhammadiyah Malang, Indonesia
[2)] Departement of Aquaculture, University of Muhammadiyah Malang, Indonesia

Abstract. *Eucheuma cottonii* L. seaweed contains dietary fiber that can cure some degenerative diseases. Dodol seaweed without preservatives can be an alternative as a functional food that many people need to maintain their health. Generally dodol made with the addition of preservatives to enhance the shelf life. Dodol without preservative makes its easily microbes growth and oxidized, thus making it quickly damaged and not accepted by consumers. The use of active edible film on dodol seaweed is expected to inhibit damage and prolong the shelf life. Chitosan is one of the recommended hydrophobic biopolymers to improve the film characteristics based starch while having antimicrobial activity. The purpose of this research is to know the effect of chitosan concentration on character of edible film, and the effect of its application on weight loss, texture and microbial quantity on dodol seaweed in the storage period. The research results showed that the addition of chitosan can increase tensile and transparency, and decrease of elongation and WVTR edible film. Active edible film application has been shown to inhibit the damage of dodol seaweed during the storage period compared without active edible film and inhibiting microbial growth.

1 Introduction

Seaweed contain diaetary fiber. Dodol seaweed without preservatives can be an alternative as a functional food that many people need to maintain their health. Without preservatives, food product becomes good medium for microbial growth, thus making its quickly damaged and not accepted by consumers. The use active edible film to dodol seaweed is expected to inhibit the damage.

Seaweed, in addition to containing nutrients such as carbohydrates, proteins, fats, and minerals, also contains crude fiber and dietary fiber that can provide the function of treating some degenerative diseases, so it can be processed into functional food. One possible form of preparation is dodol. Dodol is generally made by adding inorganic preservatives to increase shelf life, but the presence of preservatives makes its unsafe if consumed in the long time. While dodol without preservatives will be easily damaged by microbes and oxidation, thus making it unacceptable to consumers.

[1]Corresponding author: warkoyo@umm.ac.id

Associated with these problems, the researchers provide solutions with the application of active edible film based taro tuber starch with the addition of chitosan active ingredients on dodol seaweed without inorganic preservatives. The advantages of using active edible film among others can be consumed directly with packaged products, do not pollute the environment, improve the organoleptic properties of packaged products, as flavors, dyes, antimicrobial agents, and antioxidants. Chitosan as a hydrophobic biopolymer is added with the intent as disclosed by Chillo et al. [1] that to improve the film characteristics of hydrophilic starch susceptible to water vapor transmission [2] and simultaneously act as antimicrobials. The aim of this research is to determine the edible character of active films due to the addition of chitosan and to determine the effect of active edible film application on dodol seaweed during storage.

2 Material and Methods

The main ingredients used in this research are taro tuber as an ingredient of edible film, seaweed obtained from seaweed collector of area Sumenep Madura, glutinous rice flour, rice flour obtained from foodstuff and chitosan store obtained from chemical store Makmur Sejati Malang

The study was divided into 2 stages, namely that is making of active edible film with chitosan addition (1, 2, 3%) as antimicrobial material as first stage, then measured its physical character. The physical parameters of edible film include thickness, tensile [3], elongation [3], WVTR [4] and transparency [5]. Further applied to wrapping dodol seaweed to the functional test with a store period of 0, 2, 4, 6 days as second stage. Among the functional test parameters in dodol include weight loss [6], hardness rate, and microbial count (TPC) [7].

3 Result and Discussion

3.1 Characteristics of Edible film

Characteristics of edible film based taro tuber starch due to the addition of chitosan are presented in Table 1.

Table 1. Characteristics of edible film based taro tuber starch due to the addition of chitosan

Chitosan	Thickness (mm)	Tensile Strenght (M.Pa)	Elongation (%)	WVTR ($g.m^{-2}.day^{-1}$)	Transparance (mm^{-1})
1%	0.09 a	1.33 a	87.47 c	8.23 c	4.89 a
2%	0.11 a	1.75 ab	62.78 b	7.77 b	5.27 ab
3%	0.10 a	2.50 b	43.21 a	5.45 a	6.17 b

3.1.1 Thickness

The thickness of edible active film produced ranges from 0.09 to 0.11 mm, this result meets the Japanese industry standards, ie a maximum thickness of 0.25 mm. The thickness obtained slightly increased in the presence of increased chitosan concentration, which increased by 0.005 times the concentration of chitosan (Figure 1). This occurs because the difference in chitosan concentration does not produce the viscosity of different edible film solutions. The addition of the active ingredient in edible film tends to increase its thickness [8,9]

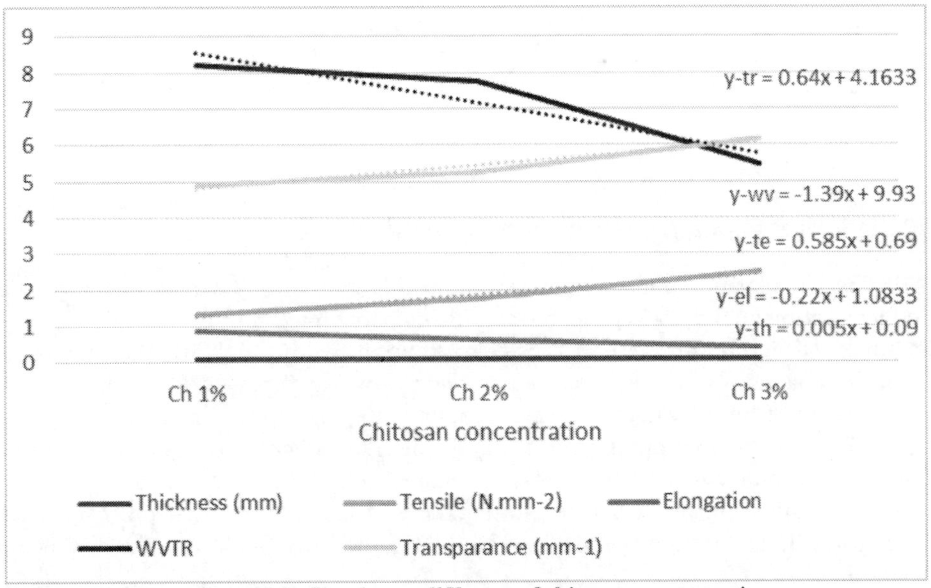

Fig 1. The character of the active edible film due to different of chitosan concentrations

In principle, the more constituent components are added, the thickness of edble film increases. That is supported by the opinion of Nugroho *et al* [10] which states that the increase in the solids amount in solution causes the polymers that make up the edible film matrix more and more. In addition to the total solids in the solution, other things that affect the edible thickness are the viscosity and the constituent polymer content.

3.1.2 Tensile Strength

The tensile strength of the film increases with the added chitosan concentration (Figure 1). The tensile increases with chitosan added, which is 0.585 times, its value ranges from 1.33 to 2.50 MPa, the value available with the Japan Industry Standard (at least 0.39 M.Pa). This shows that the greater chitosan concentration added to edible film production has an effect on the higher tensile strength value. While the research results Warkoyo *et al* and Lim *et al* [8,9] states that the added active ingredients tend to lower the tensile of edible film.

That is happen because chitosan acts as an antimicrobial as well as a film-forming matrix, so the greater the tensile strength value is influenced by the addition of more chitosan, the matrix that is formed will also increase, and the thickness of the film increases, the result is stronger, so the strength which is given to attract bigger. Increased tensile strength shows resistance to damage due to stretching and greater pressure, resulting in improved physical quality of edible film

3.1.3 Elongation

The elongation of edible film increased with higher chitosan concentrations (Figure 1). Based on Japan Industrial Standart, the resulting elongation (43 - 87%) is good (10-50%) to very good (more than 50%.) Elongation is an edible elasticity percentage, the magnitude is inversely proportional to tensile.That is happen because the edible nature of the film is affected by the major constituent components and the amount of active ingredient added, both the type and the level [8]. The higher the concentration of chitosan will affect the viscosity of the edible film solution prior to printing. In addition, it is also influenced by the addition of plasticizer, the increased concentration of dissolved chitosan without being accompanied by an increase in plasticizer concentration will result in a more rigid edible film resulting in lower elongation values associated with the increasing amount of dissolved solids.

3.1.4 Water Vapor Transmission Rate (WVTR)

The water vapor transmission rate of edible films decreases with the greater chitosan concentration added, and compared to JIS standard (maximum 7 $g.m^{-2}.day^{-1}$) the edible film available is edible film with 3% chitosan addition. A low WVTR value indicates the edible film is capable of inhibiting the transmission of water vapor, or is not easy for water vapor to pass. The high value of water vapor transmission is influenced by the ratio of hydrophobic and hydrophilic materials. In addition, The value of the transmission rate is inversely proportional to the thickness value. The thicker the film, the lower the water vapor transmission rate because the thickness is the distance that must be taken by water vapor to diffuse through the film, so the thicker the film, the longer the distance and takes a long time.

3.1.5 Transparency

The greater the concentration of chitosan addition the higher the transparency value with the other meaning the more opaque the edible film produced. The transparency value of edible film may be affected by its thickness and constituent components. The thicker the film, the higher the transparency value. And the color produced by chitosan is yellowish, the greater the concentration of chitosan that is added, the greater the yellowish edible degree. This is because more and more light is absorbed on the sheet of film shot by a spectrophotometer with a certain wavelength. The transparency of a sample will be proportional to the thickness and concentration.

3.2 Functional test

3.2.1 Weight loss of dodol

Weight loss of dodol due to active edible film during storage increased at a low rate of 0.03, while in the dodol without edible film increased at a relatively fast rate of 2 times (Figure 2). The lower the rate of weight loss, the reduction of moisture content during storage is smaller, consequently the freshness of dodol can be maintained. According to Shen et al. (2010) additions to certain limits, the active ingredients will modify the structure of the starch tissue, the film becomes brittle and moisture is easy through it.

Weight loss dodol due to active edible film application is no different from the low rate of increase. That is due to WVTR of the 3 edible films used is very low, ranging from 5-8 $g.m^{-2}.day^{-1}$. The greater the weight loss value means the greater the water vapor coming out of the material, the worse the packaging material is in preventing product damage. Factors that cause increased weights of weight loss during storage are loss of water content and other volatile compounds during storage. In addition, the weight loss correlated positively with the thickness of the edible film

3.2.2 Texture of Dodol

The texture of dodol seaweed due to different active edible film is presented in Figure 3. The harder the product the greater the force required to puncture the product. Dodol with active edible film treatment of is relatively unchanged texture, while dodol texture without edible active film during the store period decreases, the power to suppress the dodol is getting smaller, meaning dodol is getting soft. That indicates that the active film edible treatment is capable of inhibiting dodol damage, so that during the 6-day shelf period the dodol texture has not changed. Warkoyo *et al.* [11]reduced texture rate decreases with the addition of potassium sorbet active ingredients more and more. The texture of the filet with edible coating is harder than the texture of the filet without the edible coating [13].

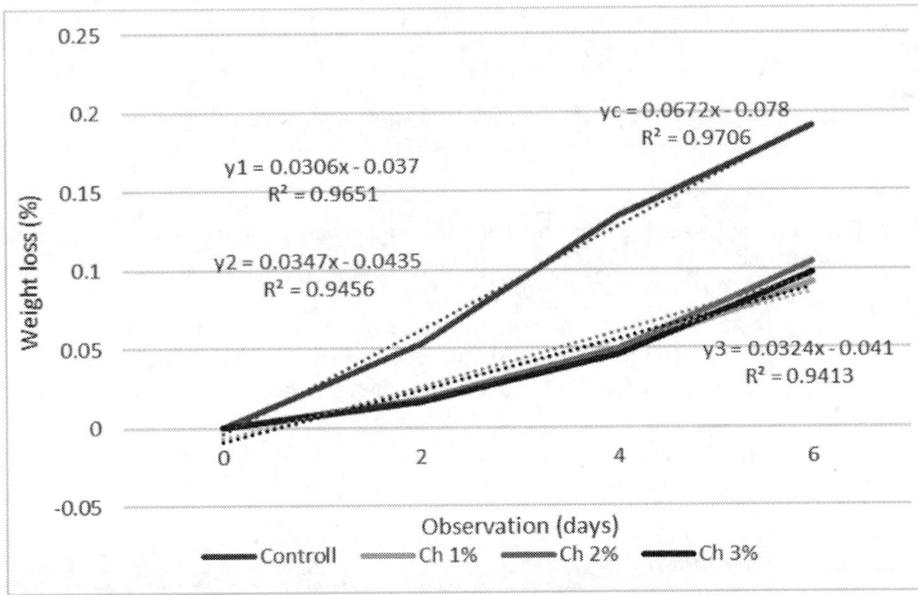

Fig 2. The weght loss of dodol seaweed due to edible film with different chitosan concentrations during the store period

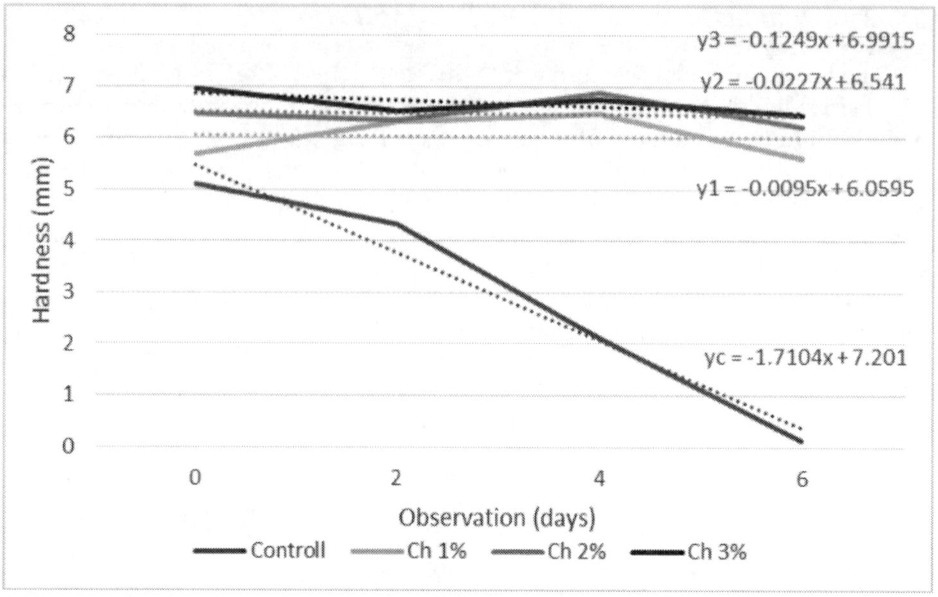

Fig 3. Texture of dodol seaweed due to edible film with different chitosan concentrations during the store period

The main factor affecting dodol texture during storage is the influence of edible film coating, each type of edible film has different physical characteristics and effect on dodol texture value during storage. While texture changes during storage can be caused by retrogradation, enzymatic reactions, and water absorption as well as increased oxidation reactions. The effect of water absorption depends on the level of the product's water content and its texture characteristics, in which water may increase or disappear during storage.

3.2.3 Mikrobes Number (TPC)

The number of microbes in dodol due to the application of active edible films during the store period is presented in Figure 4. The number of microbes in dodol due to active edible film application is changed linearly, while in dodol without active edible film changes exponentially. This is in accordance with the research of Warkoyo *et al.* [11], that the increasing amount of potassium sorbet, resulting in smaller amounts of microbial and increasingly sloping graphics, and its products are increasingly durable. The same results were also produced by Juck *et al* [12] in boiled chicken with edible application of starch-based film that incorporated potassium sorbet.

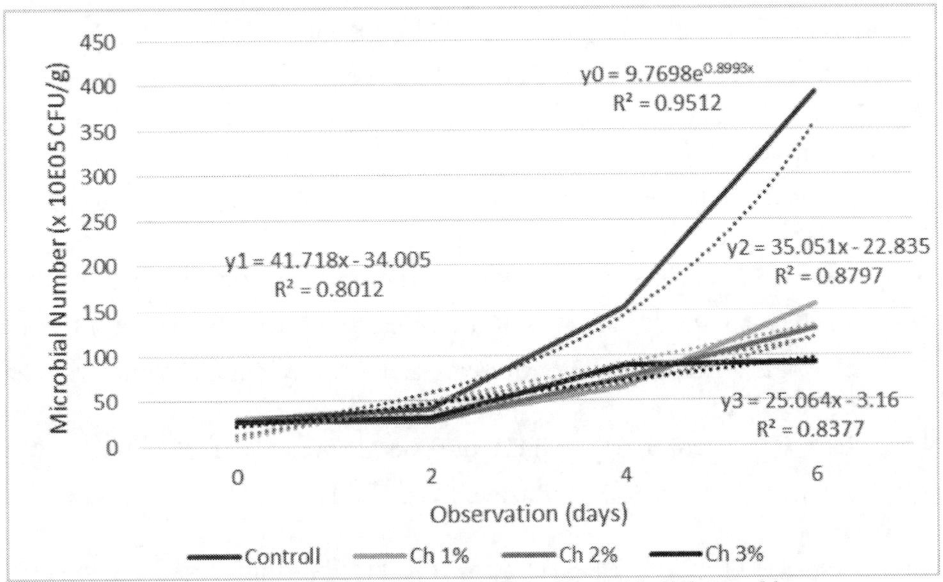

Fig 4. Microbes number (TPC) in dodol seaweed due to edible films with different chitosan concentrations during the store period

Some factors that influence, in addition to the influence of coating type of edible film is also based on the data mentioned above it is proved that edible film is able to inhibit the growth of microorganisms during storage. During the storage process, growth and development of mold and various microbes are inhibited by the presence of chitosan. Chitosan has antimicrobial properties by inhibiting the development of cells that have a negative charge. The concentration of the constituents of the edible film is certainly influential on the inhibition of microbes growth.

4 Conclusion

The research result showed chitosan concentration can effect the character of edible film and dodol seaweed. The addition of chitosan in edible film can increase to thickness, tensile strength, and WVTR, and can decrease to elongation and transparency. The use of active edible film to dodol seaweed can maintain its quality, and weight loss is inhibited to 50%, hardness rate is inhibited up to 10 times compare to without edible film. Microbial number (TPC) can be effectively inhibited until 6th day for 3% chitosan addition, with linear growth while microbial on dodol without edible grow exponentially.

References

1 S. Chillo, S. Flores, M. Mastromatteo, A. Conte, L. Gerschenson, And M.A. Del Nobile. J. Food Engin. **88**,159-168 (2008)

2 M. A. Garcia, M. N. Martino, and N. E. Zaritzky. Journal Of Food Sci. **65**, 94-947 (2011)

3 ASTM. *Standard test method for tensile properties of thin plastic sheeting, Annual books of ASTM standards*. (Designation **E96**-01. Philadelphia: ASTM, 2001a)

4 ASTM. *Standard test method for water vapor transmission of materials. Annual books of ASTM standards*. (Designation **D882**-01. Philadelphia: ASTM, 2001b)

5 A.A. Al-Hasan, and M.H. Norziah. Food hydrocolloids. **26**,108-117 (2012)

6 P. Suppakul, K. Jutakorn, and Y. Bangchokedee. J. Food Engin. **98**, 207-213 (2010)

7 E.A. Yousef, and C. Carlstrom. *Food Microbiology: a Laboratory Manual.* (A John Wiley and Sons., Inc. Publication, USA, 2003)

8 Warkoyo, B. Rahardjo, Dj. W. Marseno, and J. N. W. Karyadi. Jurnal Agritech. **34**, 72-81 (2014)

9 G.O. Lim, S.A. Jang, K.B. Song. Journal of Food Engin. **98**, 585-593 (2010)

10 A. A. Nugroho, Basito, and R. B. Katri. Jurnal Teknosains Pangan. **2**, 73-79 (2013)

11 Warkoyo, B. Rahardjo, Dj. W. Marseno, and J. N. W. Karyadi. Jurnal Agritech. **35**, 61-68 (2015)

12 G. Juck, H. Neetoo, H. Chen.. Int. J. of Food Microbiol. **142**: 302-308 (2010)

13 S.M. Ojagh, M. Rezaei, S.H. Razavi, S.M.H. Hosseini. Food Chemistry. **120**, 193-198 (2010)

14 X.L. Shen, J.M. Wu, Y. Chen, G. Zhao. Food Hydrocolloids. **24**, 285-290 (2010)

Antimicrobial coating on quality attributes of sausage during refrigerated storage

R. Naufalin, R. Wicaksono, P. Arsil, M. Salman. F.

R. Naufalin is with the Food Science and Technology, Jenderal Soedirman University, Purwokerto Indonesia (phone: +6281327135051
R. Wicaksono, is with the Food Science and Technology, Jenderal Soedirman University, Purwokerto Indonesia (e-mail: rumpokowicaksono@gmail.com).
P. Arsil, is with the agricultural Technology, Jenderal Soedirman University, Purwokerto Indonesia (e-mail: poppyarsil@gmail.com).
M.F. Salman. is with the Pharmacy, Jenderal Soedirman University, Purwokerto Indonesia (e-mail: salman.unsoed@gmail.com).

Abstract. Edible coating based on carboxymethyl cellulose (CMC) environmentally friendly. Addition on Kecombrang (*Nicolaia speciosa*) extract used to be antimicrobial and antioxidants coating. CMC-based edible coating added with antimicrobial of kecombrang was used to reduce the oxidative and microbial degradation sausages stored at refrigerator at 10°C for 12 days. The cmc coating reduced malonaldehyde substances and peroxide value by 0.88 mg.kg and 92.29%, respectively, compared with the controls. The moisture barrier effect was significantly better for the CMC coating compared to the control. The CMC coating of sausages inhibits the growth of either the total plate counts of *Bacillus cereus*, *Escherichia coli*, *Staphylococcus aureus*, and *Pseudomonas aeroginosa*. Data show that cmc can effectively be used as a natural antioxidative and moisture barrier coating to extend the quality and shelf life of sausages.

1 Introduction

Gourami fish has the potential to be utilized to be one of the processed food products namely sausage, which is generally made from beef. The high-fat content in sausage cause rancidity due to oxidative damage in the form of lipid oxidation process. Synthetic preservatives such as nitrate/nitrites and / or salts have been widely used to inhibit oxidative damage and inhibit microbial growth in preserving sausages. Consumption of synthetic preservatives continuously causes negative impacts on the health of consumers either directly or indirectly (carcinogenic) [3].

The edible coating is a technology that can minimize the damage and extend the shelf life of food products. Edible coatings enriched with natural ingredients that have antioxidant and antimicrobial compounds can enhance the edible coating ability to extend the shelf life of food products. The results showed that edible coating of teak leaf extract could inhibit microbiological and oxidative damage to the sausage [20], another study also showed that edible coating with the addition of green tea extract could decrease the lipid oxidation rate in the sausage [24].

Several studies have been conducted on bioactive compounds in kecombrang that are able to act as antioxidants and antimicrobials. Phytochemical content contained in flowers, stems, rhizomes and leaves kecombrang on the results of studies of alkaloids, saponins, tannins, phenolic, flavonoids, triterpenoids, steroids, and glycosides that use an active role as antioxidants. The stem and leaves of kecombrang have higher phytochemical compounds than other kecombrang [10].

Edible coating used in this research made of CMC (Carboxy Methyl Cellulose) and glycerol and the addition of concentration of leaf kecombrang at certain concentration. Edible coating base materials of CMC and glycerol are selected to produce stable and plastic edible coatings for easy application to food products. CMC acts as emulsifier and stabilizer, while glycerol acts as a plasticizer which enhances flexibility. The method of edible coating application on gouramy sausage used in this research is spraying because this method has the advantage of producing coating products with thinner and more uniform layers than dyeing techniques. This study aims to determine the effect of

; e-mail: rnaufalin@yahoo.co.id).

addition of kecombrang leaf concentrate to the edible coating on oxidative damage of gouramy sausage during low temperature storage

2 Material and methods

2.1 Place and Time of Research

The research was conducted at the Laboratory of Agricultural Technology, Faculty of Agriculture UNSOED. The research held from January until Juny 2018.

2.2 Materials

Materials needed research consisting of: *Nicolaia speciosa* leaf. Chemicals used among other technical solvents, namely ethanol 96 %; coating (Carboxy Methyl Cellulose), ethanol 99.8 % (PA), linoleic acid, distilled water, ammonium thiocyanate, 0.02 M FeCl2 tetrahydrate, concentrated HCl, N_2 gas. Other materials needed include a coarse filter paper, Whatman No. 41 paper, plastic, paper label, paper towels and aluminum foil.

2.3 Equipment

The tools used in this research that cabinet dryer dryers, blenders, analytical scales, rotary evaporator (Bibby RE 200), shaker, oven (Memmert, Japan), desiccators, pipettes (Pyrex, Germany), pH meter. Tool for the analysis of total phenol and antioxidant activity consists of spectrophotometers, centrifuges, glass tools (Pyrex, Germany), vortex, Fial tubes, micropipette (Gilson), and bath tobacco leaf waste, i.e. shaker, rotary, tube N_2. Tool to extract formulation, a set of tools for testing antimicrobial activity glasses, vortex, micro pipettes, incubator 37°C, desiccator, refrigerator and vacuum pump.

2.4 Undertaking experiment

N. speciosa leaf powder sample preparation, N. *speciosa* flower are selected. Material selection results cleaned with water, then dried in a dryer at a temperature of 50^0C until the moisture content of 8-10%. Furthermore simplicia dry milled to obtain a homogeneous powder.

The *N. speciosa* leaf powder was extracted twice with ethanol (1:4 w / v) [16]. The extraction process is done by maceration at 37°C, with a rotation speed of 150 rpm for 24 hours every level. Filtrat separated from the solvent by evaporation in a rotary evaporator until no solvent dripping again. The solvent was evaporated at a temperature of 50°C. Residual solvent is removed with nitrogen gas to produce an essential oil [16].

Separated *N. speciosa* leaf extract had weighed and added by sterilizing distilled water with a ratio of 1:1 (w/v) respectively. Then blended for 3 min and filtered using a filter clothe to separate the extract and precipitate (residue) to obtain *N. speciosa* leaf extracts with concentration of 50%. The process of *N. speciosa* encapsulation is by adding the CMC with the different concentration. Using hotplate stirrer in 50°C temperature, the filtrate heated and add the coating and stirred on 15 minutes [17].

2.5 Determination procedure

Measurement of Total Phenol was modified. Activity of antioxidant formulas was analyzed using modified iron thiocyanate method. This determination is based on the formation of peroxide as a product of oxidation of linoleic acid. Percent oxidation was calculated by comparing the absorbance value of the test sample (As) with control absorbance value (A ctrl) multiplied by 100%.

2.6 Statistical analysis

The observed variables in this research are antioxidant activity of edible coating pada sosis ikan gurami involved total phenol measurement,. Quantitative data was analyzed used variance test (F test) in 5%, continued in Duncan's Multiple Range Test (DMRT) if it finds the significant difference.

3 Results And Discussions

; e-mail: rnaufalin@yahoo.co.id).

3.1 Total Phenol

The result of variance analysis showed that the kecombrang plant part has no significant effect on the total content of phenol edible coating. The mean total phenol edible coating of kecombrang leaves was 17.97 mg TAE / g samples. The total value of phenol edible coating in this study was higher than 0.13 mg TAE / g sample for edible coating kecombrang and 1.44 mg TAE / g samples for edible coating of kecombrang leaves (with the same concentrate) [4]. This difference in total phenol content can be due to environmental factors for different plants such as plant growth, soil composition, temperature, rainfall, and ultraviolet radiation [2].

The result of variance analysis showed that concentration concentration of carbohydrate have significant effect to total content of phenol edible coating. The average value of total fenol edible coating of leaf of kecombrang on concentration treatment of concentrate kecombrang can be seen in Figure 1.

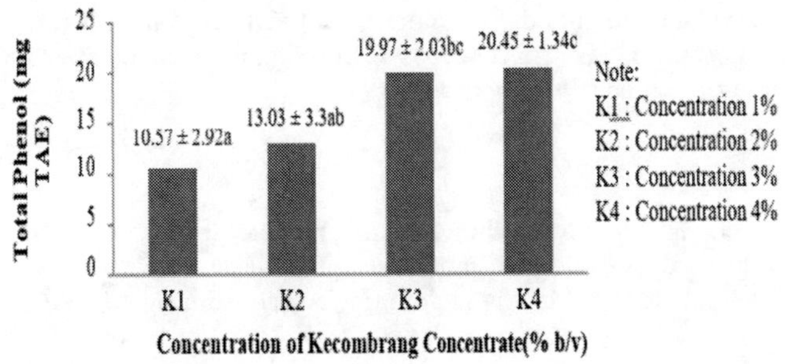

Fig. 1. The mean of total *phenol* edible *coating* on treatment concentration of kecombrang concentrate

Figure 1 shows that the mean total phenol edible coating of leaves of kecombrang increases from addition consentrate. The increase in the mean total phenol edible coatings along with the increased concentration of concentrate due to the total phenol detected in the edible coating derived from phenol compounds contained in the concentrate of leaves kecombrang, so the more concentrated kecombrang will be higher value of total phenol. The phytochemical content of flowers, stems, rhizomes and leaves kecombrang that is composed of alkaloids, saponins, tannins, phenolics, flavonoids, triterpenoids, steroids, and glycosides that use an active role as antioxidants [10].

3.2 Antioxidant activities

Generally, the antioxidant activity is directly proportional to the total phenol because the total phenol is quantitative of all antioxidant compounds of phenol group in a plant, including flavonoids (Widyastuti, 2010). This study showed the opposite result was suspected because the flavonoid compound on the leaves kecombrang is a class of phenol compounds that play the most antioxidant role, so in this study the value of flavonoids is directly proportional to the antioxidant activity. The flavonoids act as antioxidants by giving their hydrogen atoms or through their ability to metal clatting [23].

The result of variance analysis showed that kecombrang concentration very significant effect on edible coating antioxidant activity. The average value of antioxidant activity of edible coating of leaves of kecombrang on concentration treatment of kecombrang concentrate can be seen in Figure 2.

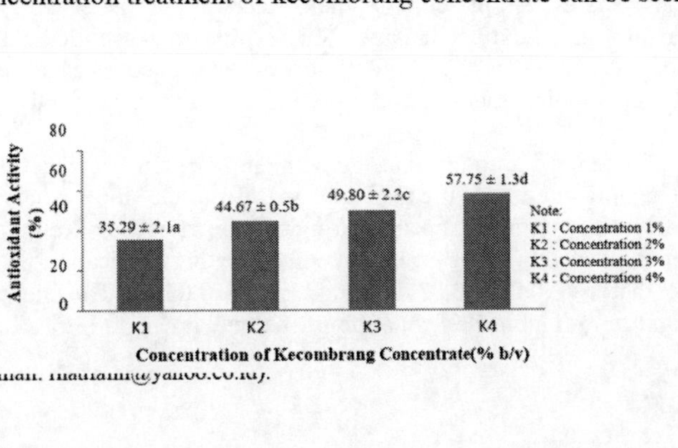

; e-mail: [illegible]

Fig. 2. The mean of antioxidant activity *edible coating* on treatment concentration of kecombrang concentrate

Figure 2 shows the antioxidant activity of leaves increased as the concentration of concentrate increased. This is because the mount of bioactive compounds in the edible coatings increases as the concentration of added ingredients increases. Several studies have shown that the concentration of extracts and plant concentrates has a linear correlation with the result of antioxidant activity. This is reinforced by the results of research Cahyaningtiyas [4] that the more concentrated kecombrang added to the edible coating can increase antioxidant activity.

The results of analysis of antioxidant activity in this study indicate a linkage with the total value of flavonoids is the higher the value of flavonoids then the value of antioxidant activity produced. The correlation coefficient between the total flavonoid with IC50 cape fruit with a variety of solvent extract is equal to 0.9993 [2]. That is, increased antioxidant activity is closely related to the increase in the total value of flavonoids. There are several factors that influence the correlation between antioxidant activity and the type of antioxidant compound in a material. The first factor is the type of plant. The antioxidant compound content is scattered throughout the plant with types and concentrations according to the type of plant [5]. This type of antioxidant compound can cause synergistic and antagonistic effects of these antioxidant compounds. The antioxidant activity is not always correlated with both phenol and flavonoid levels [6]. This is due to the difference of active components in plants that cause synergistic and antagonistic effects that affect the antioxidant activity in plants.

3.3 Oxidative Damage

Malondialdehyde levels are one indicator of oxidative damage levels in foodstuffs. Malondialdehyde is a lipid peroxidation end product, usually used as biological biomarker of lipid peroxidation and represents the degree of oxidative stress. The value of Malondialdehyde levels in gouramy sausage during storage at plant part treatment can be seen in Figure 3.

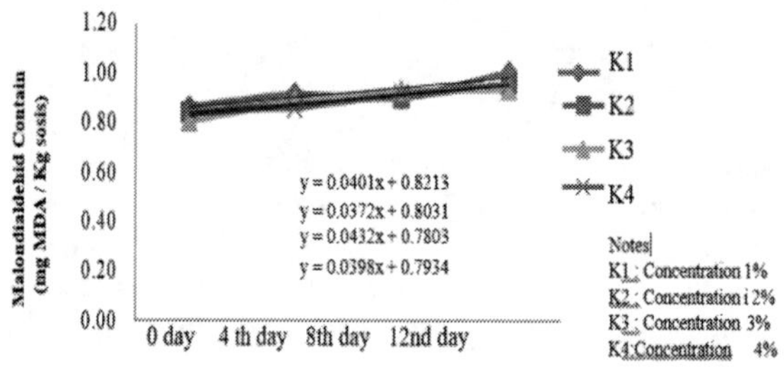

Fig. 3. Malondialdehyde levels have increased during storage

Figure 3 shows that malondialdehyde levels have increased during storage. This is because the free fatty acids present in the sausage undergo further oxidation so that it can produce malondialdehyde. The malondialdehyde is an end product of lipid peroxidation, which levels are influenced by the high number of free radicals [29]. The presence of oxidation rates in foodstuffs may lead to increased levels of malondialdehyde [25].

Based on the result of the analysis of the variety, the treatment of plant part on the application of edible coating kecombrang significantly to the malondialdehyde content of gouramy sausage in this study. Based on Figure 7, malondialdehyde content of gouramy sausage with edible coating application of stems and leaves of kecombrang were 0.039 and 0.0408 respectively. The slope value of the edible coating of the kecombrang rod is lower than the slope value of the edible coating of the leaves. This suggests that edible coating sticks kecombrang better able to suppress malondialdehyde levels in gouramy sausage due to oxidative damage. This is due to the antioxidant activity of edible coating sticks kecombrang higher than the antioxidant activity of edible coating leaves kecombrang in this study. The consumption of antioxidant supplements cause the presence of antioxidant activity can suppress levels of malondialdehyde in the body [22].

Based on the analysis of variance, malondialdehyde content of gouramy sausage with edible coating application of kecombrang on concentration treatment did not have significant effect. It can also be seen from the slope value of the average regression equation of malondialdehyde content of gouramy sausage during storage at each concentrate which is fluctuating. Slope value of malondialdehyde content of gouramy sausage on edible coating application with concentration of 1%, 2%, 3% and 4% respectively, that is 0,0401; 0.0372; 0.0432; and 0.0398. The higher concentrations of antioxidants are proportional to the ability to inhibit the formation of malondialdehyde [26]. This

; e-mail: maufalin@yahoo.co.id).

mismatch is caused by changes in oxidative stress levels that occur in gouramy sausages. This indicates that the levels of malondialdehyde are unstable in a material. The situation with high levels of oxidative stress can significantly increase malondialdehyde levels, but if the oxidative stress condition is resolved, the malondialdehyde level will decrease again [21].

4. Conclusion

Leaves part of kecombrang effect on flavonoid and edible coating antioxidant activity, while concentration variation effect to total phenol, flavonoid, and edible coating antioxidant activity. Application of edible coating kecombrang can be effect on free fatty acid content and malondialdehyde gouramy sausage during storage.

Reference

1. Alfiani, S.,Liling T., dan Malikhatun N. *Jurnal Pharmascience*, **1(1)**: 7-13 (2014)
2. Borges, L., Alves S., Sampaio B., Conceicao E., Bara M., and J. Paula. *Brazilian Journal of Pharmacognosy*, **23(2)**: 230-238 (2013)
3. Cahyadi, W. *Edisi Kedua*. Bumi Aksara, Jakarta (2009)
4. Cahyaningtiyas, N. Skripsi. Universitas Jenderal Soedirman, Purwokerto (2018)
5. Chan, E. W.C., Chew, Y. K., and Y. Y. Lim. *Food Chemistry*, **102**: 1214-1222 (2007)
6. Gengaihi, S. E., Faten M. A. E., Emad M. H., Emad S., and H. Doha. *Food Process and Technol*, **5(2)** : 1-5 (2014)
7. Hendromartono, S. *Majalah Penyakit Dalam Udayana*, **1**:89-92 (2000)
8. Naufalin, R. dan S. R. Herastuti. 2012. *Jurnal Agricola*, **2(2)**: 124-147.
9. Naufalin, R., dan Tri Y. *Artikel ilmiah*. Universitas Jenderal Soedirman, Purwokerto (2009)
10. Naufalin, R., Jenie, B. S. L., Kusnandar, F., Mirnawati, dan S. R. Herastuti. *Jurnal Teknologi dan Industri Pangan*, 17(3): 197-203 (2006)
11. Naufalin, R., Herastuti SR dan Yanto T. 2010. Formulasi dan Produksi pengawet alami estrak kecombrang. Laporan Hibah Kompetensi Tahun II. DP2M Dikti.
12. Naufalin R and Herastuti S.R. *Prosiding Pengembangan sumber daya pedesaan dan kearifan lokal berkelanjutan I.* (2011)
13. Naufalin R and Herastuti S.R. Proseeding Pengembangan sumber daya pedesaan dan kearifan lokal berkelanjutan II (2009)
14. Naufalin R and Herastuti S.R. *Proceeding International Conference: Future of Food Factors,* Jakarta (2012b)
15. Naufalin R. and Herastuti S.R. UPT Percetakan dan Penerbitan Unsoed (2012c)
16. Naufalin, R. Prosiding Seminar Nasional PATPI, Jember 26-29 Agustus 2013 : 243-252 (2009)
17. Naufalin, R. and Herastuti, S.R. Journal animal production. **1(15)**: 8-14 (2009)
18. Naufalin, R and Herastuti, S.R. Proceeding Asean Food Conference, Singapure (2013b)
19. Naufalin, R., Herastuti, SR and R. Wicaksono. 2nd International Food Safety Conference (IFSAC2013). Malaysia 2-3 Desember 2013 : 269-276 (2012)
20. Pahlevi, Y. R. *Skripsi*. Universitas Sebelas Maret, Surakarta (2011)
21. Papas, A. M. *Chem. Prog*, **7(1):** 34-39 (2011)
22. Ramatina, Leily A., dan I. Ekayanti. *Jurnal Gizi dan Pangan*, **9(1)**: 35-45 (2011)
23. Redha, A. *Jurnal Belian*, **9(2)**: 196-202 (2011)
24. Shokraneh, N., Arlai P., Ghahroudi F. R., Hasannia F. and S. Sabaghpour. *Journal of Food Biosciences and Technology*, 7(1): 23-30.

; e-mail: rnaufalin@yahoo.co.id).

25. Sreejai, R. dan D. S. Jaya. 2010. Studies on the Changes in Lipid Peroxidation and Antioxidants in Fishes Exposed to Hydrogen Asulfide. *Toxicol Int*, **17(2)**: 71-77 (2011)

26. Ulilalbab, A., Bambang W., dan M. Adriani. *Jurnal Ilmu Kefarmasian Indonesia*, **13(2):** 215-220 (2011)

27. Wahdaningsih, S., Erna P. S., dan S. Wahyuono. *Majalah Obat Tradisional*, **16(3)**: 156-160 (2011)

28. Widyastuti, N. *Skripsi*. Insitut Pertanian Bogor, Bogor (2011)

29. Wresdiati, T. *Biota*, **8(1):** 107-112 (2003)

; e-mail: maufalin@yahoo.co.id).

Identification andExpression ofvitellogenin gene in the Gouramy (Osphronemous Gourammy) under photoperiods manipulation

Norman Arie Prayogo[1,2], Asrul Sahri Siregar[1], Purnama Sukardi[1], *Dewi* Nugrayani[1], *Anandita* Ekasanti[1], *Riviani*, Yasumasa Bessho[3]

[1]Fishery and Marine Faculty, Jenderal Soedirman University
[2]Center of Marine And Bioscience Jenderal Soedirman University
[3]Laboratory of Gene Expression Research, Nara Institue Science and Technology

Abstract .Vitellogenin was protein hormone where controlled gonad development in fish. Expression of this gene regulated from the external and internal factor.Photoperiods are the external factor that regulated endocrine gland activities in gonadal development, gametogenesis and reproductive cycles. Aim of the research to find out the effect of photoperiod on gouramy reproductive performance by manipulating photoperiod. Design experimental with three photoperiod treatments, namely 14L: 10D (control), 8L: 16D (short photoperiods) and 18L: 6D (long photoperiods). Four aquaria consisting of nine fishes each were served as replicates. Fishes were kept under these photoperiods for 8 weeks. The observed variable was the liver activities. Liver activity was evaluated by measuring gene expression of Vitellogenin. The normalized data were subjected to ANOVA followed by Tukey's multiple-comparison tests. The length of Vitellogenin cDNA was 1136 bp. The vitellogenin precursors encoded cDNA consisted of 378 amino acids. The average of vitellogenin gene in each experimental group significantly increased according to longer photoperiods ($P<0.05$). These results indicated that photoperiods had a stimulatory effect in improving gouramy reproductive performance

1 Introduction

Vitellogenin (Vg) is the precursor of vitellin (Vn), the major egg yolk protein of oviparous animals. The vg gene is generally expressed in a female-specific manner. In vertebrates, Vg is synthesized in the liver, secreted into blood, and accumulated in growing oocytes in the ovaries (3). The synthesis of Vg is induced by estrogens in females and is inducible in immature females by a synthetic estrogen, estradiol-17β (E_2). In fishes, Vgs are produced only in females, not in males due to the absence of endogenous estrogen; however, males begin synthesizing Vgs when they are exposed to exogenous estrogens (29). The expression of vitellogenin gene in fish regulated from environmental factor like photoperiods.

Photoperiod was external factor that controlled fish reproduction. In many study found that photoperiods in primary environmental factor that regulated pituitary activities via melatonin (5,7,16). Study for the effect of photoperiods timing of spawning in many fish species as European sea bass, *Dicentrarchus labrax* (17), Atlantic cod, *Gadus morhua L* (19), dan Chinook salmon, Oncorhynchus *tshawytscha* (6). The majority of studies were conducted on temperate-zone fishes in which photoperiod strictly differ between seasons. Studies on influence of photoperiod on tropical fishes are still limited. We had already research with hard-lipped barb (*Osteochilus hasselti* C.V) tropical fishes. We found that manipulations with longer photoperiods 18L:6D (18 hour light and 6 hour dark) were significantly decreased melatonin level and increased gene expression cGnRH-II and also incread the vitelogenin gene (Prayogo et al., 2012; Prayogo et al., 2017). Conduct in that study in this research, we wants to evaluated another economically tropical fish like gouramy (Osphronemous Gouramy) in Indonesia. This research was to examined that long photoperiods will increasing gene expression of vitellogenin altered ovarium development in gouramy.

Gouramy (*Osphronemusgourami*) is one of the local freshwater fish species with very potential for propagation by fish farmers, especially in the area around Banyumas District, Central Java, Indonesia. Gourami have high economic value, easily cultivated, and favored by consumers in Indonesia. Many studies were conduct for increasing and mantain fish production in Indonesia (Prayogo et al 2012; Prayogo et al 2016a; Prayogo et al 2016b). Osphronemous Gouramy is a synchronous batch spawner fish (Prayogo et al., 2011) capable of spawning several time during the peak of the spawning period. Osphronemous Gouramy has been adapted to a photoperiod of 12L: 12D to 14L:10D. The present study was examined the effect of different photoperiods on gene expression vitellogenin of the Gouramy. We also identified of vitellogenin genes in the gourami for the first time. This study will inform photoperiods effect to increasing gouramy reproduction.

2 Material and Methods

2.1 Treatment and sampling of fish

In this research we used144 sexually female hard-lipped barb weighing of 100 g in average and maintained inResearch Laboratory of Fisheries and Marine, Jenderal Soedirman University. The female'sgourami was divided into 3 groups photoperiods treatment. Each group consisted of 4 aquaria with 9 fish/50 L water.

In this study, three types of photoperiods namely 6L: 18D, 12L:12D and 18L:6D have been tested toward vitelogenin gene expression. The aquaria were covered with lightproof black polybag. The light source provided from 25-Watt (*Phillips*) bulb regulated by automatic timers 24 hours cycles which were placed at the top of each aquarium. In photoperiod of 14L: 10D, light was turned on since 06.00 am until 08.00 pm and in 8L:16D light was turned on since 06.00 am until 02.00 pm and in 18L:6D light was turned on since 06.00 am until 12.00 pm local time.

The fish were reared photoperiods treatment for four months.During the research, fish were fed on commercial pellet (protein 37% and fat 10%) as much as 3% of total body weight daily. The water was siphoned regularly to maintain water quality. The water temperature, dissolved oxygen, pH and carbon dioxide were monitored every month. Every sampling time,thepituitary was collected from 3 fish of each group and was snap-freeze on liquid nitrogen for vitelogenin expression study. The expression of vitelogenin geneswas evaluated using Real Time PCR applying primers derived from vitelogenin genes. Real Time PCRwas conducted at the Research Laboratory Jenderal Soedirman University.

https://doi.org/10.1051/e3sconf/20184702001

2.2. mRNA Isolation and DNAse Treatment

Total mRNA was extracted from whole liver using blue Sepasol R-RNA super-1 reagent, based on Ethanol-phenol-chloroform extraction method. The integrity of the RNA was verified in a denaturing agarose gel, stained with ethidium bromide. The RNA samples were treatment with DNAse free RNAse (Takara). The quality and concentrations of total RNA were determined by agarose gel electrophoresis and optical density reading at 260 and 280 nm (Figure 1), the RNA were aliquoted in batches and frozen at -70°C.

2.3 RT-PCR

Total mRNA samples (1µl) were reverse transcript using cDNA synthesis kit (PrimeScriptTM Reverse Transcriptase, Takara Bio.Inc) using Random 6mers (sequence pd (N)6, 50µM) primers and prime script R-tase with manufacture instruction.

2.4 cDNA Amplification

The Degenerated primer pairs, vitelogenin gene were designed from cDNA *cyprinidae* like *cyprinus carpio* and *carrasius auratus*. The forward primer was 5-CGTGGATCHYTGMARTACGAGTT-3 and reverse was 5-ATGGTGGCRTCATTGAT-3. All sequence was aligned with multalin to identify the conserve region in ORF (Open Reading Frame) region. The primers to amplify the vitelogenin gene were designed using primer 3 software.

Thirty five cycles of PCR for hard lipped barb vitelogenin were carried out using a thermal cycler (Robocycler, Stratagene) according to the following cycle, 95°C for 2 min, 35 cycles to 95°C for 30 s, 55°C for 30 s, 72°C for 60 s, followed by a 5 min extension at 72°C . After amplification, the PCR products was electrophoretic ally separated on a 1.5% agarose gel and stained with ethidium bromide.

2.5 Cloning and Sequencing of PCR Products.

PCR products amplified from cDNA were separated by agarose gel electrophoresis, and the incised gels were purified using the DNA gel extraction procedure. The desired DNA fragments from mRNA vitelogenin were sub cloned into T vector (10 ng) (Takara) and ligated with T4 ligase. The plasmid were transfected into *E. coli* and were spread into LB medium. The recombinant positive colonies were screened using ampicilin. Positive colonies were treated with mini scale plasmid preparation for sequencing. DNA sequences of these fragments were determined using the Big Dye version 3.1 sequencing method with specific primers (Table 1). The data were automatically collected on the ABI PRISM 3100 Genetic Analyzer (PE Applied Bio-systems).

2.6 Sequence analysis

The cDNA sequences for vitelogenin gene were checked using BLASTN searches (http://www.ncbi.nlm.nih.gov/ BLAST/) were performed with default settings on the complete, non redundant GenBank database nucleotide sequences.

2.7 Phylogenetic analysis

For phylogenetic analyses, gourammy cDNA vitelogenin genes were compared to cDNA vitelogenin sequences from ten fish species. All sequences were retrieved from NCBI GenBank. The relationship between gourammyvitelogenin and other teleost vitelogenin was generated with CLUSTAL W with scoring method percent and the unrooted tree was generated using Treeview version 1.5.2.

2.8 Quantitative Real Time Analysis

The primers were designed based on vitelogenin(submit number: 1995561), using the Primer 3.0 software. For the actin used from gouramy actin, used as endogenous control, was amplified by the following primers-actin forward 5-GAGCTATGAGCTCCCTGACGG- 3, actin reverse 5- AAACGCTCATTGCCAATGGT-3-and were used to normalize variations in RNA. After optimization, PCR reactions were performed in a 10 µl volume containing 2 µl cDNA, 5 µl SYBR mix (Applied Biosystem), 0.3 µl forward primer, 0.3 µl reverse primer and 2.4 µl DDW using the following condition: 95°C for 45 s, (45 cycles of 95°C for 15 s and 60°C for 1min), then95°Cfor15s, 60°Cfor15sand95°Cfor15s. The results were analyzed using the standard curve mode, according to the manufacturer's recommendations (Applied Biosystems).

3 RESULTS

3.1 Identification of vitelogenin genes in gourammy

The vitelogenin genes of the gourammy were successfully amplified from cDNA. The agarose gel electrophoresis of the cDNA vitelogenin showed a specific band, approximately 360 bp (submit number:1995561). The corresponding cDNA sequences were called Vitelogenin. The cDNA sequences were check with BLAST and we found there wasnt 100% identity with another vitelogenin gene. The nucleotide sequence identity of vitelogeninmRNA was 82% with mud crap(cirrhinus moliterella, *GU324313.1*), 81 % with grasscarp (*ctenophyringodon idella*, KT984759.1), 81 % chinesse minnow (Gobiocypris rarus, EU623080.1), and 81% with fathead minnow (Pimephales promelas ,AF130354.1).

3.2 Gene and amino acid Structure vitelogenin

Vitellogenin genes in gouramy successfully amplified and contain 361 basepair nucleotide. Based on alignment to another teleost, this region includesopening reading frame in vitellogenin gene. (Figure. 1)

CATTCAACTCATGAGATCAGTGATGCACCAGCCCAGAGATTATTGAAGTTCTTACGCAC
CTTGTTACAAACCATGTGGCCATGGTCGATGATGATGCTCCTCTCATGTTTATTCAGCTCA
TTCAACTCCTGCCTGTTGCCACCCTTGAGAATGATGAGTCTATCTGGGCTCACTACAAGGA
CAAACCAGTTTACAGGGCGCTGGCTTCTGGATGCTCTTCCTGCTGTGGGCACACCAGTAAT
GGTGAAAATGTACAAGGAGAAATGCCTGGCTGTCGATCTTACCCTTTCTGAGAGTCATTCT
GACTGTTGTGGGGCTCTGCAAATGCTTTTGCTGACATCGAAAACATCCGGTGACCGCT

Fig 1. Vitelogenin nucleotide and amino acid sequence of gouramy

The connection vitellogenin sequence in gourami with anotherfishcan be seen as the distance at the phylogenetic tree (Figure.2, 3,4). The vitellogenin sequences in gourami had conserved area with another fish and The striking contrast between conservation of the vitelogenin coding sequence and lack thereof in the GAP coding sequence is evidence of differential selective pressure within the gene. This is evident in cases where the identity and similarity of vitelogenin and GAP coding sequences have been compared for mRNAs of different vitelogenin genes within a species.

3.3 Phylogenetic analyses

Phylogenetic analyses were performed to establish an evolutionary context for the vitellogenin gene. Genetic distances (measured as substitutions per site) showed moderate low values, and the topology was well supported by strong bootstrap values. As expected, vitellogenin in gouramy was included within a sub-cluster of the *carp* (*ctenopharynodon idella)* and minnowwith high bootstrap values (Figure 5).

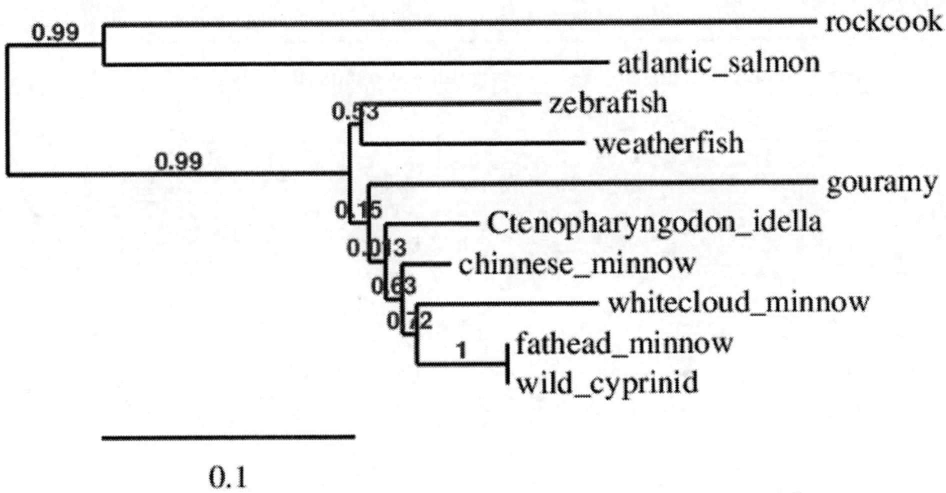

Fig 4. Phylogenetic relationship of precursors derived from known nucleotide encoding vitelogenin. The relationship was generated with CLUSTAL W and the unrooted tree was generated using Treeview version 1.5.2. The scale bar represents the estimated evolutionary distance as 0.1 amino acid substitutions per site.

4 DISCUSSION

This paper reports for the first time the clone of differing vitelogenin genes from pituitary tissues of gouramy. Comparison the vitelogenin gene structure with previously reported gene structures of other fish species shows a high conservation. The nucleotide sequence identity of vitelogenin mRNA also very similar with another variant vitelogenin, from BLAST result showed was 82% with mud crap (cirrhinus moliterella, *GU324313.1*), 81 % with grasscarp (*ctenophyringodon idella*, KT984759.1), 81 % chinesse minnow (Gobiocypris rarus, EU623080.1), and 81% with fathead minnow (Pimephales promelas , AF130354.1) (Figure 2).

https://doi.org/10.1051/e3sconf/20184702001

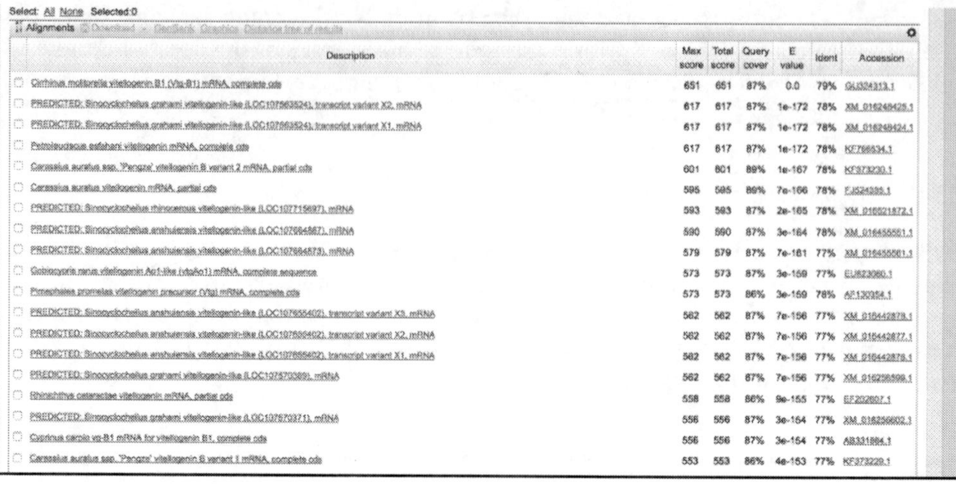

Fig 2. Comparison of vitelogenin nucleotide between gouramyand another species.

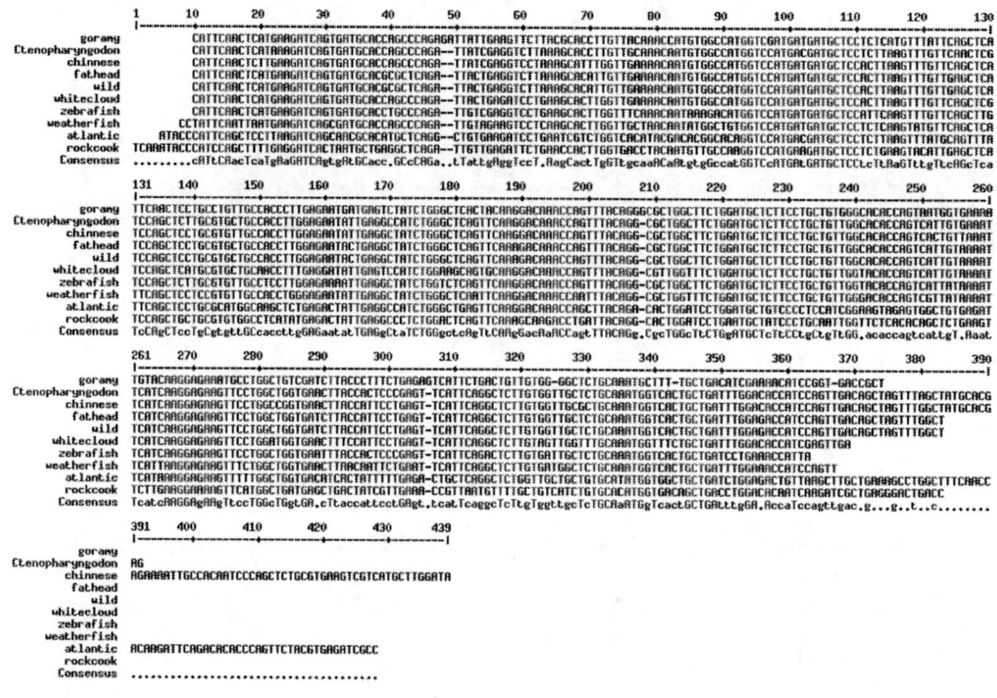

Fig 3. Multalin of Vitelogenin Sequence of gouramy with another fish

These results showed that gouramy vitelogenin genes had a high similarity with other teleost. The highest similarity was identified between gouramy and *grasscrap* (82%). Based on these results we suggested that goramy had structure of vitellogenin gene similar to teleost (figure 3).

The present study is the first to describe the vitelogenin genes in gouramy, providing new information on this gene family in the gouramy. The vitelogenin in gouramy were grouped together with other teleost in the phylogenetic tree, suggesting a common ancestor for both groups of genes. Phylogenetic analysis shows that vitellogenin can be separated into 2 major groups. Subgroup I contains vitelogenin from rockcook and salmon, sub group II fromzebrafish until minnow. (Figure.4).

4.1Expression of vitelogenin mRNA gouramy under photoperiods manipulation

In gouramy, relative vitellogenin mRNA expression levels in eight weeks were 0.5-2.35 (Figure 5). The highest vitellogenin mRNA expression (2.35) was observed LP group in eight month significantly different ($P<0.05$). mRNA expression for 18L:6D group increased with post spawning periods. The vitelogenin mRNA expression for other treatment photoperiods in second weeks, fourth weeks and sixted weeks had non significantly different ($P>0.05$), but in eighted month long photoperiods (LP) had higher gene vitellogenin than short photoperiods (SP) and control ($P<0.05$).

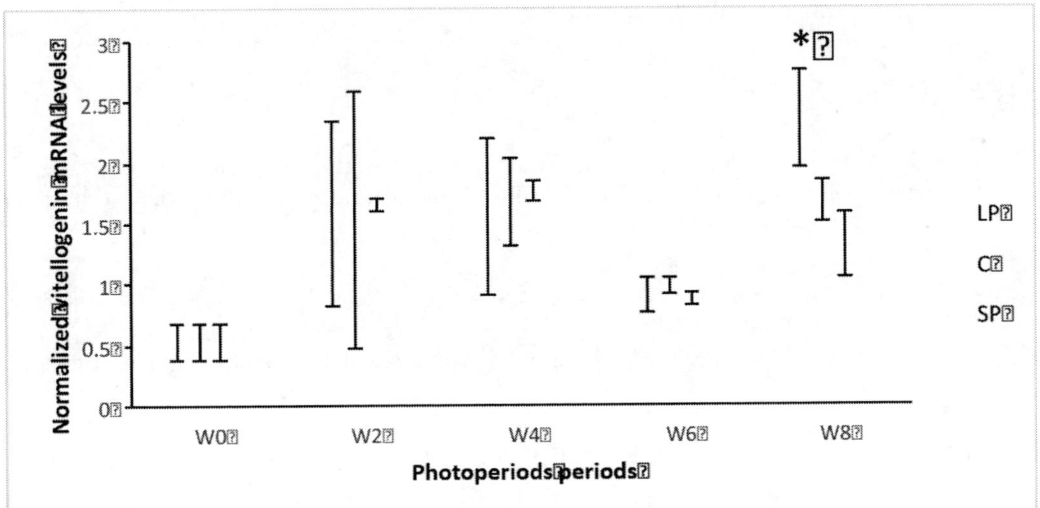

Fig 5. Vitelogenin gene expression of gouramy kept under different photoperiod for 8 weeks. C=control, LP=18L:6D, SP=6L:8D). (*: Significantly different)

In this study, changes in, gene expression of vitellogenin in gouramy was analyzed to characterize the role of neuropeptides in the control of reproduction under photoperiods manipulation. This study confirms previous results from fisheries laboratory showing increased in vitelogeningenes levels during photoperiods manipulation in gouramy. In addition, we report for the first time, changes in the gene expression levels of vitellogenin genesin correlation with photoperiods manipulation. Although we are aware that mRNA levels do not always match with protein levels and/or the physiological effects of the protein products, the regulation of mRNA levels provides an indication of the activity of a particular peptide neuronal system.

In this study showed that vitelogenin level increased equivalent with the long photoperiod increased. This is proved that photoperiod exert it role on reproduction through hipotamaus-pituitary-gonad, that integrates and conveys input from external and internal cues to the pituitary organs [4,5]. Photoperiods regulated melatonin production and melatonin mediated cyclical regulation of GnRH mRNA expression involve the protein kinase C and the extracellular signal-regulated kinase 1 and 2 pathways. Melatonin regulated act through membrane receptors to trigger the protein kinase C pathway and 12-O- tetradecanoyl phorbol-13-acetate (TPA), a modulator of this pathway, has been shown to suppress GnRH gene expression through the promoter (16). GnRH binds to GnRH receptor and active G protein mediated phosphorilation to protein kinase C and shyntezis Gonadotrophin (GtH-I and GtH-II). GtH-II secreted into blood vessel, to reseptor in theca cell actived G protein and adenilat cyclase to phosphorilation cAMP and activation staR protein. staR protein regulated cholesterol. Steroid hormone will produce vitelogenin in liver.

5 CONCLUSIONS

In summary, the present work has reported for the first time sequence of vitelogenin an gouramy, the phylogenetic results presented in this work support the idea that vitellogenin genes share the same basic structure. That was meaning vitellogenin in gouramy very conserve, that assumed had a same function with another teleost. Photoperiod affected regulation of gene expression of vitellogenin, in the gouramy. The longer photoperiod increased gene expression of vitellogenin, via HPG axis.

REFERENCES

1. M.J. Bayyari, L. Rodrıguez, S. Zanuy, J.A. Madrid, F.J. Sanchez-vazquez, H. Kagawa, K. Okuzawa, and M. Carrillo. General and comparative endocrinology. **136**, 72–81 (2004)

2. S. Bhattacaraya. Current science. **63**, 3 (1992)

3. A.K. Biswas, T. Morita, G. Yoshizaki, M. Maita and T. Takeuchi. Aquaculture. **243**, 229-239(2005)

4. N.R. Bromage, M.J.R. Porter, and C.F. Randall. Aquaculture.**197**, 63–98 (2001)

5. V.M.Cassone. Chronobiology. International. **15**, 457-473 (1998)

6. J. Chyb, Mikolajczyk, B.J endocrinol.**63**, 87-97 (1999)

7. P. Ekstrom, and H. Meissl. Fisher. 7, 199–284 (1997)

8. M.R Flack, J. Froehlich, A.P Bennet. J biol chem. **269**,14–20 (1994)

9. J.E. Juell, and J. E. Fosseidengen. Aquaculture. **233**, 269–282 (2004)

10. P. Melamed, G. Gur, H. Rosenfeld, H. Elizur, A. Yaron. Biochem. **17**, 93–98 (1997)

11. F. Minniti, M. Maisano,A. Giannito,A. Mauceri, and F. Sasuolo. Italian journal of zoology. **76**, 269-278 (2007)

12. L.A. Miranda, C. A. Strussmann, and G. M. Somoza.Fish physiology biochemical. **35**, 101–108 (2008)

13. Prayogo.n.a, g.e. wijayanti, murwantoko, m. Kawaichi and p. Astuti. (2012) effect of photoperiods on melatonin levels, estradiols level and the expression of cgnrh-ii and sgnrh genes, in hard-lipped barb (*osteochilus hasselti* c.v.).*Journal of global veterinaria*, 8(6) : 591-597.

14. N.A. Prayogo, G.E. Wijayanti, I. Sulistyo,P. Sukardi. Biodiversitas.**17**, 523-530 (2016)

15. N.A. Prayogo, A. Siregar, P. Sukardi. Turkish journal fishery and aquatic science. **16**, 1003-1009 (2016)

16. T. Qingbo, M. Mazur, and P. L. Mellon. Molecular endocrinology.**19**, 2769–2779 (2005)

17. L. Rodriguez, M. Carrillo, L. Sorbera, Y. Zohar, and S. Zanuy. General and comparative endocrinol. 136: 37-48. (2004)

18. H. Rosenfeld, B. Levavi-sivan, P. Melamed, Z. Yaron, and A. Elizur.Biochem. **17**, 85– 92 (1997)

19. J.E. Skjæraasen, T. Nilsen, and O. S. Kjesbu. Fish aquatic science. **63**, 310–320 (2006)

20. I. Sulistyo, J. Rinchard, P. Fontaine, J.N. Gardeur, B. Capdeville, and P. Kestemont. Aquatic living resource. **11**, 101-110 (1998)

21. I. Sulistyo,P. Fontaine, J. Rinchard, J.N. Gardeur, H. Migaud, B. Capdeville, and P. Kestemont. Aquatic living resource. **11**, 101-110 (1998)

22. V.L. Trudeau, B.D. Sloley, and R.E. Peter. Possible role for taurine in regulating pituitary hormone release in fish *taurine minisymposium, fourth international congress on amino acids* (Vienna, Austria, 1995)

23. A.A. Ulloa, C. Timossi, M.P. Damian. **11**, 205–215 (1999)

24. T.N. Utoh,A. Horie, Y. Okamura, S. Yamada, N. Tanaka, A. Mikawa, H. Akazawa, and P. Oka. Fisheries science.**69**, 181–188 (2003)

25. V.L.A. Vieira, S.J.S. Johansen, R. Bickerdike, and I.A. Aquaculture.**246**, 197– 208 (2005)

26. H. Volkoff, and R.E. Peter. Gen comp endocrinol. **116**, 347–355 (1999)

27. Q. Wang, K.W. Sham, S. Ogawa, S. Li, I.S. Parhar, C.H. Cheng, X. Liu, and H. Lin. Cell. Endocrinol. (2013)

28. F. Xiong, and C.L. Hew. J. Zool. **69**, 2572–2578 (1991)

29. Z. Yaron. Aquaculture. 129, 49-73(1995)

Abbreviations

1. GnRH: Gonadotrophin Releasing Hormone
2. GtH : Gonadotrophin Hormone
3. mRNA: Messenger RiboNucleoAcid

https://doi.org/10.1051/e3sconf/20184702001

4. DNA : Deoxy NucleoAcid
5. cDNA: complementary Deoxy Nucle Acid
6. GnRH: Gonadotrophin Releasing Hormone
7. TPA : tetradecanoyl phorbol-13-acetate

GROWTH PERFORMANCE OF CATFISH (*Clarias gariepenus*) CULTURED OF HIGH DENSITY WITH BIOFLOC SYSTEM

Petrus Hary Tjahja Soedibya[1], *Emyliana* Listiowati[1], *Taufik* Budhi Pramono[1], *Norman* Arie Prayogo[2], *R.* Taufan Harisam[2]

[1]Aquaculture, Faculty of Fisheries and Marine Science, Jenderal Soedirman University, Indonesia
[2]Center of Maritime Bioscience, Jenderal Soedirman University, Indonesia

Abstract. Catfish is a species of freshwater fish that contains a source of protein and it has economic value and has become one of the famous commodities in Indonesia. Biofloc can be one alternative waste fixers intensive cultivation for reduce the waste of inorganic nitrogen and also provide additional protein for increase growth and feed efficiency. This study was to determine the effect of high stocking densities on the growth performance of african catfish in biofloc system. This study used completely randomized design (CRD) with different stocking densities in the system biofloc T1.(1000/m^3), T2.(1500/m^3), T3.(2000/m^3), and T4.(2500/m^3). Animal trials were using jouvenile african catfish with an average weight of 1.06±0.3g, which maintained in a pool tarpoulin cage with water volume ±2000L for 50 days with feeding ratio 3% of the weight biomass. The results showed a significantly different effect ($P<0.05$) against the value of hepatosomatic index, absolute growth and daily growth rate, but the result not significant at spesific growth rate. The second treatment (T2) showed the best results than others in the growth rate and daily growth rate with a value of 6.45±3.1g for absolute growth and 0.13±0.06g for daily growth rate. The four treatment showed the best results than others in the hepatosomatic index with value 4.7 ± 1.8%. These findings demonstrate a role of biofloc technology in catfish aquaculture.

1 INTRODUCTION

Catfish is a species of freshwater fish that contains a source of protein and it has economic value. Catfish has become one of the commodities that serve as food ingredients into the food menu like in Indonesia. fishery commodities continues to increase each year so that the need for innovation in order to increase production [1]. Catfish production progress over the last few years show results very significantly by 21.82% per year from 69386 tonnes in 2005 to 145 099 tonnes in 2009. During the 2010-2014 rise in catfish fish production increased by 450%, or an average increase by 35% per year which in 2010 amounted to 270.6 thousand tons increased to 900.000 tons in 2014.

[1]Corresponding author: hary_tjahja@yahoo.co.id

Limited natural resources such as water and land, making intensification as the option most likely to increase aquaculture production [6]. Many studies have been carried out to develop reproduction and culture techniques for *C. Gariepinus* [2,3,4]. Intensive aquaculture systems are characterized by an increase in the density of fish and additional food from outside. On the cultivation environment is good and sufficient feed, an increase in density will result in increased production of aquaculture [5]. Moreover, the intensification of aquaculture, especially stocking rate increase unfavorable impact on the preservation of health and the environment in the form of environmental degradation cultivation. Environmental degradation caused by organic waste from feed residue and dirt, the waste is generally dominated by inorganic nitrogen compounds are toxic.

Biofloc technology can be one alternative waste fixers in intensive cultivation, this technology is the most profitable to reduce the waste of inorganic nitrogen, this technology can also provide additional protein for animal feeding to increase growth and feed efficiency. Biofloc is a combination of macro-and microorganisms including bacteria, microalgae, fungi, protozoa, metazoan, and nematodes. Floc biomass formed in bodies of water may be consumed by fish as additional sources of feed and as water purifier.

Several studies have shown that the application of biofloc technology plays a role to improve water quality, biosecurity, productivity, feed efficiency, and reduce production costs through lowered feed expenses. Theoretically and practically, the application of biofloc technology improves water quality by controlling ammonia concentrations and improving nutrient consumption as it is consumed by the organism cultured [8]. Floc biomass formed in bodies of water may be consumed by fish as additional sources of feed and as water purifier. Bacteria as main component of biofloc is capable to produce polyhydroxybutyrate (PHB) which is useful in aquaculture. The advantages of PHB are an energy reserve for fish, digestible in intestine, increasing unsaturated fatty acid, and increasing growth of fish [7]. Therefore, the aim of this study was to evaluate growth rates of African catfish, *Clarias gariepenus* juveniles treated with high density bioflocs system.

2 MATERIAL AND METHODS

2.1 Experimental Fish

Catfish (*C. gariepinus*) with size approximately 1.06±0.3g/ fish was used as tested material with different stocking density. For seven days prior to treatment, catfish was cultured in a pool terpaulin cage. Volume of each pool was 2,3 m^3, filled with 2000 L well water, equipped with aerator system with minimum water exchange. The experiments divided into 4 groups, which were 1.(1000/m^3), 2.(1500/m^3), 3.(2000/m^3), and 4.(2500/m^3). The fish was treatment for 50 days. Sampling to measure fish growth was conducted once everyday. During the research, fish were fed on commercial pellet (protein 37% and fat 10%) as much as 3% of total body weight daily. The water temperature, dissolved oxygen, pH , Ammonia, nitrite and nitrate were monitored every weeks. The sample of water was taken every week to calculate the amount of phytoplankton. Every sampling time, 6 fish captured for the replicant samples.

2.2 Specific growth rate

This parameter was calculated as the weight of the fish after 50 days, calculation was conducted using equation by Abdel-Wahab *et al.,* [9]:

$$SGR = (\text{Ln } Wt - \text{Ln } Wo)/t \times 100\% \quad (1)$$

SGR expressed as %/day, Wo and Wt were the weight of the fish during the day of measurement (t) and initial stage (o) expressed as gram (g), while experiment period was expressed as day (t).

2.3 Daily growth rate

This parameter was calculated as the weight of the fish after 50 days, calculation was conducted using equation by Abdel-Wahab *et al.*, [9]:

$$DGR = \frac{(Wt-W0)}{t} \quad (2)$$

DGR expressed as g/day, W0 and Wt were the weight of the fish during the day of measurement (t) and initial stage (o) expressed as gram (g), while experiment period was expressed as day (t).

2.4 Absolute Growth

This parameter was calculated as the weight of the fish after 50 days, calculation was conducted using equation by Abdel-Wahab *et al.*, [9] :

$$AG = Wt - W0 \quad (3)$$

AG expressed as g, W0 and Wt were the weight of the fish during the day of measurement (t) and initial stage (o) expressed as gram (g).

2.5 Hepatosomatic Index

This parameter was calculated as the weight of the fish after 50 days, calculation was conducted using equation by Abdel-Wahab *et al.*, [9]:

$$\%HSI = WL / Wb*100 \quad (12)$$

HSI expressed as %, WL were the weight of the liver during the day of measurement and Wt were the weight of the fish during the day of measurement expressed as gram (g).

2.6 Data analysis

Data from specific growth rate, daily growth rate and absolute growth was statistically measured using one-way analysis of variance using SPSS software (versi 17.00) at confidence interval of 95% ($p<0.05$). LSD test was following when significant difference present. Data of water quality was analyzed in description.

3 RESULTS AND DISCUSSION

3.1 Specific growth rate

In catfish, specific growth rate (SGR) on 50 days biofloc treatment were 0.87-1.15 (Figure 1). The peak level of SGR (1.15) was recorded at 50 days treatment on group 4 (2500/m^3). SGR in catfish was increased based on biofloc density treatment. Growth of fish depends on several factors: the type of fish, genetic nature and the ability to use food, disease resistance and supported by environmental factors such as water quality, feed and space or stocking density [10]. Although the quality of the water can affect the growth rate, but in general the quality of the water treatment biofloc in this study showed good results so that the water quality is not a factor for the low growth rate of fish biofloc treatment. Fish growth can occur if the amount of feed nutrients are digested and absorbed by fish greater than the amount required for maintenance of the body [11]

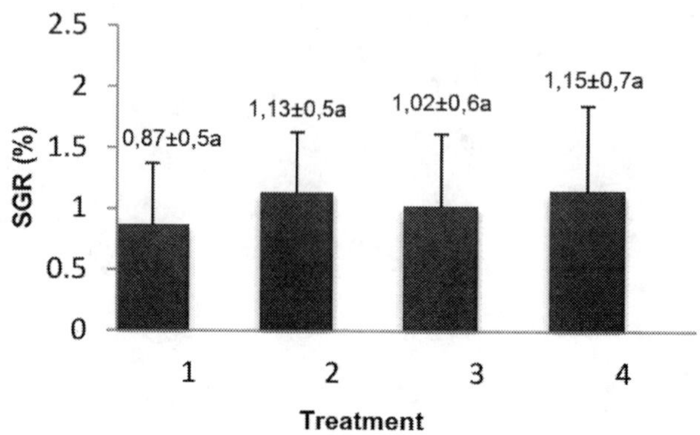

Fig 1. Specific growth rate of catfish (*C. gariepinus*) Are Cultured Of High Density With Biofloc System. Different letters indicate Significantly different ($p<0.05$). 1). Treatment density $1000/m^3$ in biofloc system, 2 Treatment density $1500/m^3$ in biofloc system 3. Treatment density $2000/m^3$ in biofloc system and 4.Treatment density $2500/m^3$ in biofloc system.

The Specific Growth Rate (SGR) of all treatment were not significantly. In this study, showed that the value of SGR of all treatment were high. The addition of Bacillus sp., cells into the feed had positive impact on fish digestive system due to exogenous enzymes released by the cells, thus expenditure energy was more effective during digestion [12]. Energy difference resulting from this process can be used for the growth. Ekasari *et al* (2016) mentioned that the increasing growth rate of aquatic organisms provided with probiotic in their feed was related to higher enzymatic digestion activity together with vitamin synthesis, therefore, digestibility and organism weight were also increased. Spesific growth rate value in all treatment of the obtained values average of 1.04±0.12%/day. one treatment shows that at low stocking density generate value growth also lower. it is suspected because the space that is wide enough to make the fish a lot of activities so that a lot of energy used for activity and metabolic processes than for growth . In line with the statement Soedibya [14, energy used for metabolic processes very high, so the energy portion reserved growth to be reduced.

3.2 Daily growth rate

The average daily growth rate in the one treatment of 0.09 ± 0.02 g/day, two treatments of 0.13±0.06g/day, three treatments of 0.10±0.04g / day and the treatment of four of 0,10±0,03 g / day. ANOVA test based on the value of daily growth rate of catfish (Clarias gariepinus) showed a significantly ($P < 0.05$), followed by further test of LSD. Average daily growth value based Least Significant Difference test showed that the first treatment was significantly different to two treatments. Treatment of two significantly different with the other treatment. Two treatment showed best growth results seen from the average absolute growth was 0.13 ± 0.03 g. The average value of daily growth rate in the treatment of two highest than the other treatments. It is believed to be related to the carrying capacity (carrying capacity of the environment), the higher stocking densities used, the carrying

 https://doi.org/10.1051/e3sconf/20184702002

capacity decreases [15]. Competition in obtaining food is thought to be one cause of declining daily growth rate in the treatment of the three, although food is available but the fish can not reach due to space limitations. This can lead to stress which results in decreased appetite [12]. Reduction appetite resulted in low nutrients entering and absorbed in the body of the fish. As a result, the energy obtained from the overhaul of these nutrients is only used for fishing activities, whereas the energy for growth is reduced so that the growth of fish decreased.

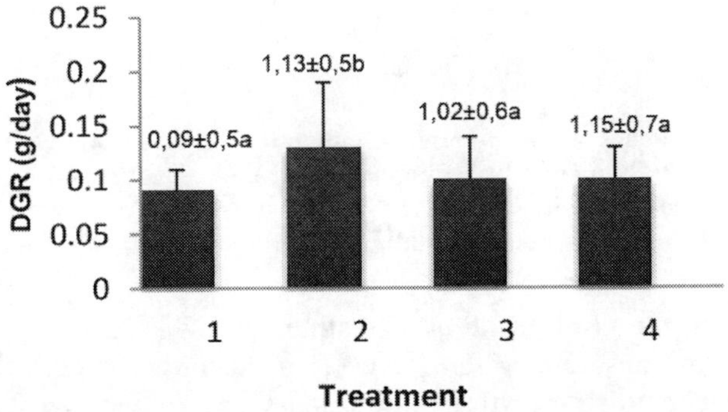

Fig 2. Daily growth rate of catfish (*C. gariepinus*) Are Cultured Of High Density With Biofloc System. Different letters indicate Significantly different ($p<0.05$). (1. Treatment density 1000/m^3 in biofloc system, 2 Treatment density 1500/m^3 in biofloc system 3. Treatment density 2000/m^3 in biofloc system and 4.Treatment density 2500/m^3 in biofloc system)

Treatment two with a stocking density of 1500/m3 able to utilize the feed is well evidenced by the value of daily growth rate higher than the one treatment, the treatment of three and four treatments. The feed given to each of equal treatment that is equal to 3% of the body weight of fish. Fish body weight at the two higher treatment so that the amount of feed given to more than one treatment, treatment of three and four treatments. Therefore, the energy produced was greater, the energy obtained from the feed is used for the body's activity and growth. Contributions floc is available for 24 hours also affect growth. Flocs contains nutrients needed by fish, one of which is a protein. Protein floc in the treatment of two higher at 22.24%, while on the treatment, the treatment of three and four treatment floknya protein content of 19.25%, 15.83% and 21.60% with a protein content of feed (pellets) which given the same on each treatment which amounted to 28.75%. The protein content of different floc this is one cause of fish growth rate is different for each treatment.

3.3 Absolute Growth

Absolute growth value in the treatment of the obtained values of 4.76±1.3 g, while the treatment of the two obtained values of 6.45±3.1 g, and the treatment was 5.14±2.2 g three and four treatments of 5.14±2.2 g. Based on ANOVA test, the value of the absolute growth of african catfish (Clarias gariepinus) were cultivated at high stocking density with biofloc system showed significantly different results. Absolute growth value based Least Significant Difference test showed that the first treatment was significantly different to two treatments. Treatment of two significantly different with the other treatment. Two treatment showed best growth results seen from the average absolute growth is 6.45 ± 3.1 g.

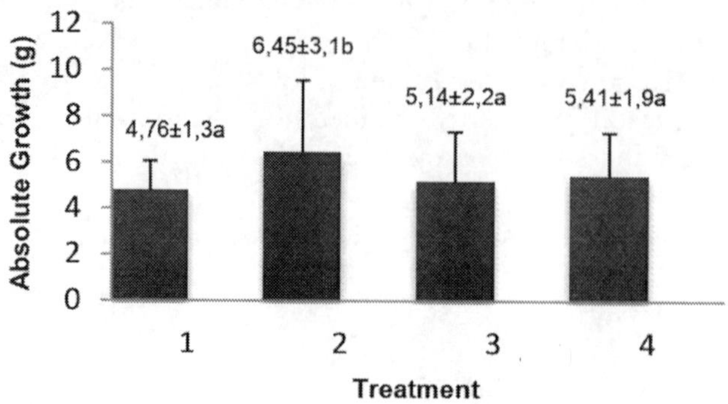

Fig 3. Absolute Growth of catfish (*C. gariepinus*) are cultured of high density with biofloc system. Different letters indicate Significantly different ($p<0.05$). (1. Treatment density 1000/m^3 in biofloc system, 2 Treatment density 1500/m^3 in biofloc system 3. Treatment density 2000/m^3 in biofloc system and 4.Treatment density 2500/m^3 in biofloc system)

Growth performance in the third and fourth treatment showed that the higher density, growth was even more low absolute weight. This is in accordance with the opinion of Hickling [17] which states that fish growth will be faster when kept at low stocking density which would otherwise be slow when a solid high stock density. Increased density will be followed by a decrease in growth (critical standing crop) and if it has to some extent (carrying capacity) growth is to be stopped altogether. Factors that affect the carrying capacity include water quality, feed, and the size of the fish. Space and food supply are factors that also affect the growth of fish, where the fish will grow better if both of these factors can be met, and otherwise growth will slow if one or both are lacking. This result indicated that Biofloc was able to increase growth. Similar result also indicated by another research using inoculating heterotrophic bacteria addition into biofloc-based culture media for catfish was able to increase the survival rate and growth performance [12].

Biofloc which also serves as a source of natural food fish available 24 hours in the maintenance of an effect on the growth of the body. The nutritional content of different flocks on each pool resulted in an absolute growth of fish each different treatment. Protein is available in all four treatment tends to be higher than other treatments that is equal to 28.75% of feed pellets and floc protein analysis results in the amount of 21.60%. This causes excess protein, so there is no storage repository for amino acids. Therefore, foods containing excess amino acids will be used immediately to meresintesis body proteins and other nitrogen compounds. The carbon skeleton oxidized or converted to glucose or fats, and amino group is converted into ammonia [13].

3.4 Hepatosomatic Index (%)

Hepatosomatic index (HSI) value in the treatment of the obtained value of 3.3 ± 1.5%, while the treatment of the two obtained a value of 3.7 ± 1.8%, and the treatment of the three at 3.5 ± 1.4% and the treatment of four of 4.7 ± 1.8%. Based on ANOVA test, hepatosomatic index value of african catfish (Clarias gariepinus) were cultivated at high stocking density with biofloc system showed significantly different results. Furthermore, the test results Least Significant Difference, the index value hepatosomatic showed that the treatment of the significantly different from the treatment of four and is equal to the treatment of two and three, the treatment of two significantly different from the treatment of four and the same as the treatment of one and three, and treatment three significantly different treatments four and equal treatment with one and two. The four treatments significantly different from the treatment of one, two and three.

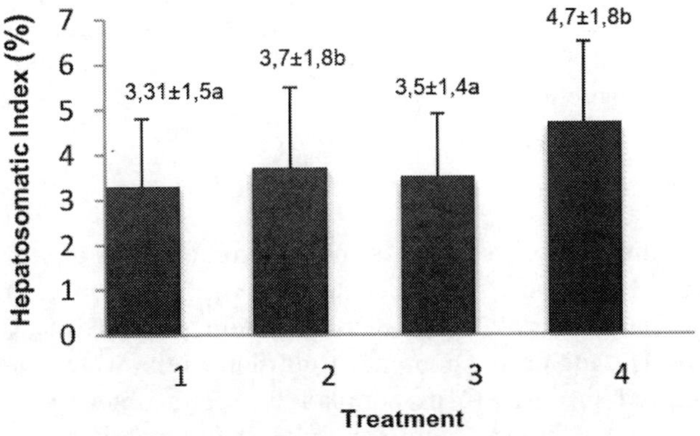

Fig 4. Hepatosomatic Index of catfish (*C. gariepinus*) are cultured of high density with biofloc system. Different letters indicate Significantly different ($p<0.05$). (1. Treatment density $1000/m^3$ in biofloc system, 2 Treatment density $1500/m^3$ in biofloc system 3. Treatment density $2000/m^3$ in biofloc system and 4.Treatment density $2500/m^3$ in biofloc system)

The liver is an important organ which secretes material for the process of digestion. Material reserve nutrients commonly found in liver cells are fat and glycogen granules. In general, the liver serves as the metabolism of carbohydrates, fats and proteins as well as a place to produce bile [18]. One of the sources of energy used in the fish while maintaining its development comes from the heart. The fish feed supply shortage in addition to its metabolism is reduced, the fish did not get enough energy source, so it will use an energy source that comes from the heart. Hepatosomatic index value in the treatment of the first, second and third tend to be lower than the fourth treatment. It is suspected that more fish using existing reserves of energy in the body compared to the fourth treatment. This is consistent with the statement Lenhardt *et al.* [19], the index hepatosomatic used to describe the distribution of energy in fish, namely the decrease in the index value hepatosomatic. This indicates that the presence of energy reserves in the liver is used to maintain the body's metabolism.

The fourth treatment showed the best hepatosomatic index value is 4.7 ± 1.8%. It is suspected that the fish on the fourth treatment at a density of 2500 head/m3 tend not many activities due to limited space for stocking density is very high, so much energy is stored in the liver than used to perform the activity. Furthermore, the results of the proximate analysis of protein content in the pellet given was quite high at 28.75%. In addition, the fish in the treatment of the fourth gain additional nutrients from the floc that act as natural food sources are available for 24 hours. Based on the results of the proximate analysis of the nutrient content of floc on media pisciculture fourth treatment showed a fairly high value protein content reached 21.60 %. This shows that the four treatments, nutrition fish have been met so that the metabolism of many stored in the liver. Hepatosomatic index is usually used to indicate a problem in nutrition, because of the relative size of the liver correlated with the state of nutrition given to the quality of the fish [9]. Increased value HSI showed increased amounts of nutrients are absorbed and subsequently cause the amount of nutrients accumulated in the liver increased [20].

The results are consistent with research that has been done by the Soedibya *et al.* [21] showed that the index value hepatosomatic resulting from maintenance on the system biofloc with a stocking density of fish catfish $500/m^3$ significantly different from the stocking density $250/m^3$ and $375/m^3$, dense stocking higher yield index value hepatosomatic better for allegedly fishing save more energy than is used to move because of the limited space due to high stocking density. Sadekarpawar and Parikh [22] states that the liver is a metabolic organ in fish that HSI values can also be used as a biomarker in detecting environmental conditions of test animals. Hepatosomatic high index value does not always show high growth anyway, since fish growth is also affected by other factors such as environmental and genetic.

3.5 Abundance of phytoplankton

The abundance of plankton in a biofloc is more common than phytoplankton groups. Where, plankton from Microcystis and Coelosphaerium genus belongs to a group of phytoplankton that is Cyanophyceae class. Phytoplankton growth was supported by the availability of nutrients. The higher the content of nutrient in the water, the more abundant phytoplankton that spurred the growth of zooplankton. The phytoplankton of the class Cyanophyceae or blue-green algae was highly abundant because this type of phytoplankton was easy to grow in water with high nutrients in the form of N, P and K. Besides, this type of algae is normally used by the trophic structure or organism as a source of food (Bernardi and Giussani, 1990).

Table 1. Abundance of phytoplankton in catfish cultivation with biofloc system

Number	Class	Abundance of phytoplankton (ind/l)			
		T1	T2	T3	T4
	Chlorophyceae				
1	*Actinastrum* sp.	392	1124	1400	865
2	*Scenedesmus* sp.	589	1463	1186	2122
3	*Pediastrum* sp.	107	250	375	811
4	*Diatoma* sp.	9	18	0	0
5	*Characium* sp.	0	0	36	0
	Chyanophyceae				
6	*Oscillatoria* sp.	999	829	1828	936
7	*Coelosphaerium* sp.	2961	2550	3380	2809
8	*Spirulina* sp.	0	214	178	0
9	*Anabaena* sp.	0	0	45	0
	Sum	5057	6448	8428	7543

Biofloc technology, forming a floc from organic material life, is fused into clumps consisting of various water microorganisms including bacteria, algae, fungi, protozoa, metazoan, rotifers, nematodes, Gastrotricha, and other organisms that are suspended by detritus. Giving molasses in a biofloc could supply organic C in continuity or in accordance with the content of nitrogen in the water. Thus, the content of nutrients like N, P and K will be abundant in the biofloc. The abundance of the nutrient content can lead to the growth of plankton in the form of Phy (plants) that will make the growth of zooplankton sustainable. The research conducted by Andrian (2010) about the growth of plankton in various types of different animal manure that contains nutrients such as nitrogen, phosphorus, and potassium shows that plankton will be easy to grow in a pool that has much nutrient content of 1.00 ppm N, P 0, 80 ppm and 0.4 ppm K.

According Avnimelech [8], biofloc is generally can be formed from the residue of feed, metabolism, and feces from aquaculture. Residual feed and fecal waste water will produce inorganic nitrogen. The inorganic nitrogen can be converted into single cell protein with the addition of carbon material water and can be used to feed fish, shrimp and plankton. The condition of C:N ratio is balanced in cultivation media. Heterotrophic bacteria will utilize N, both in organic and inorganic form for the growth of plankton so that the concentration of N in the water is reduced. Comparison between the carbon (C) and nitrogen (N) (C:N ratio) is very important to note in biofloc system so that bacteria can grow well and it will affect the growth of phytoplankton.

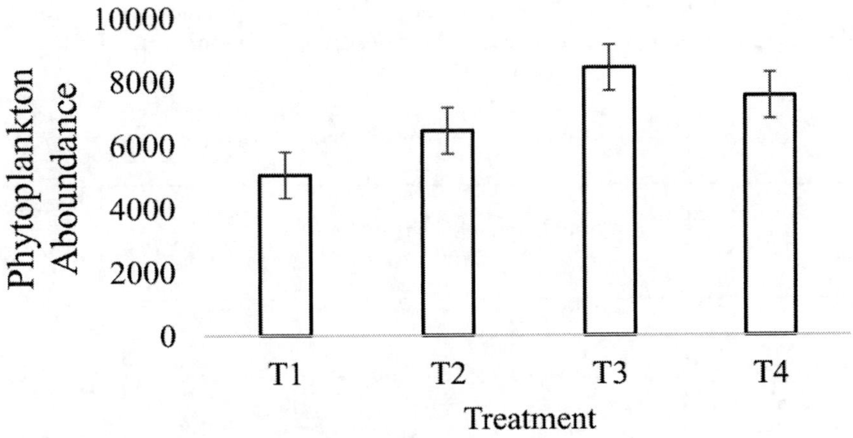

Fig 5. Phytoplankton abundance in catfish cultivation with biofloc system

Plankton may experience changes in community composition (succession) as a result of changes in physical (light intensity and temperature), chemical (nutrients, water quality, and toxins), and biological (competition and predation) condition. This opinion is justified by Reynolds (1990), who stated that phytoplankton abundance change is a response to environmental variables such as temperature, light, nutrient availability, and the abundance of predator. Odum (1971) in Pratiwi *et al.* (2010), states that an ecosystem is developed from time to time. It is occurred by stimulus in the form of nutrients and it is characterized by the change of the genus that is dominant.

The dominance of plankton in a biofloc was mostly composed of groups of phytoplankton especially Cyanophyceae class of Microcystis genus and the difference in the dynamics of plankton that occurred during the treatment period. However, not all plankton could be found in significant amount in each pool per week. In addition, differences in the dynamics of plankton in biofloc pool were quite diverse as seen from the difference in the amount of plankton abundance on a weekly basis for each pool. In each week of observation, the phytoplankton of Cyanophyceae class showed the highest abundance thatwas varied in each pond.

3.6 Water quality

Water quality in every treatment were analyzed. Ammoniac on 50 days biofloc treatment were 0.70-13.01 ppm, Nitrat were 2.17-12.17 ppm, nitrit 0.03-3.66 ppm, DO were 7.4-8 ppm, Temperature were 25-29^0, and pH were 5-7 (Table 2). The peak level of Ammoniac (13.01) was recorded at group 4 (2500/m^3), the peak level of Nitrat (12.17) was recorded at group 1 (1500/m^3), and the peak level of nitrit (3.66) was recorded at group 2 (2500/m^3). For the DO, temperature and pH, every treatment group had a same condition.

Ammonia is a major end product of protein breakdown in fish. The fish will digest the protein in the feed and excrete ammonia through their gills and urine. Ammonia in the cultivation environment is also derived from the decomposition of organic material such as food remains, dead algae, and aquatic plants [28]. At the beginning of the study of ammonia ranged 8.63-13.005mg/L. High levels of ammonia caused by the accumulation of the metabolism of the fish and the unconsumed feed consumed. Once that happens a gradual decline in each treatment until the end of the study of ammonia ranged 0.7488-7.6123mg/L. According to Green and Schrader (2015) that the increased levels of ammonia in the culture system can be affected by pH and temperature. At low water pH (acidic) ammonia tend to be more in the form of NH4+, while the high water pH (alkaline) ammonia tend to be more in the form of NH3. At low water temperatures tend to be more ammonia in the form of NH4+, while the high water temperatures tend to be more ammonia in the form of NH3.

Table 2. Water quality during maintenance

Treatment	Ammoniac (ppm)	Nitrat (ppm)	Nitrit (ppm)	DO (ppm)	Temperatur (°C)	pH
1	0.70-10.98	5.49-12.17	0.03- 2.11	7.6-8	25-29	5-7
2	7.61-8.64	4.32-11.96	0.06-3.66	7.6-8	25-29	5-7
3	0.83-8.77	5.48-10.90	0.08-2.11	7.6-8	25-29	5-7
4	0.75-13.01	2.17-11.03	0.03-2.00	7.4-8	25-29	5-7

Nitrite is a result of the oxidation of ammonia in a nitrification process which is then converted into nitrate [27[. The measurement results nitrite levels were obtained at the start of the maintenance ranges 0.0323-0.0792 mg/L. The concentration of nitrite tend to fluctuate to indicate the occurrence of nitrification in maintenance media. The occurrence of nitrification in maintenance media can be seen by comparing the value of ammonia, nitrite, and nitrate, which the current value of high ammonia nitrite then the value is low and vice versa. According to Mosha *et al.* [16], the main effect of nitrite compounds in the body of the fish is a change in the transport of oxygen and the oxidation of compounds in tissue. Ferric ion oxidizes nitrite which converts hemoglobin into methemoglobin. Methemoglobin low ability to bind and transport oxygen from aquatic environments throughout the body causing the fish to experience hypoxia (lack of oxygen) and cyanosis (blue skin bruising and because of the blood on hold). The high increase in the concentration of nitrite nitrogen in the first treatment that stocking density 2.000 head per pond, thought to be one of the factors that cause high levels of mortality in the treatment.

Nitrate (NO3) is the main form of nitrogen in natural waters and is a major nutrient for plant growth and algae. Nitrates are highly soluble in water and are stable. This compound is produced from the oxidation of nitrogen compounds in the water is perfect. Nitrification which is the process of oxidation of ammonia to nitrite and nitrate is an important process in the nitrogen cycle and takes place in aerobic conditions. The oxidation of ammonia to nitrite made by the bacteria Nitrosomonas, while the oxidation of nitrite to nitrate made by the bacteria Nitrobacter. Both types of bacteria are a chemotrophic bacteria, ie bacteria that obtain energy from a chemical process . Nitrate is an end product of the nitrification process. Nitrates are not toxic for fish except in very high concentrations (>100 mg/L). Under anaerobic conditions, nitrate can be removed with the aid of denitrifying bacteria will convert nitrates into nitrogen gas. The observation of the nitrate content of water pisciculture media during the study in Figure 8 shows that the concentration of nitrate nitrogen all treatments tend to fluctuate with the patterns and values are almost the same. Nitrates have 1.000-3.000 ppm limit value so it is rarely the cause of death in fish[27].

Dissolved Oxygen plays an important role in the culture system, especially in intensive aquaculture systems that employ biofloc. This is because the metabolic activity of microbes to decompose the organic material requires sufficient amount of oxygen continuously (Green and Matthew., 2017). The results of the measurement of dissolved oxygen (DO) in the water used for pisciculture *C. gariepenus* at 7.4 to 8 ppm which is still appropriate quality standards applied cultivation Government Regulation No.28 Year 2001. Oxygen levels are not only important for the metabolic activity cells but also can affect the structure of floc. According to Wilen and Balmer [24], the dissolved oxygen concentration is higher, floc size will be larger and more compact so that it will more easily be used as feed by fish.

The pH value of the water reflects a balance between acids and bases in water. The observation of the pH value of maintenance media showed that at the beginning of the maintenance of the pH value is quite low, but in the middle of maintenance, normal pH value in the range of 6.8 to 7.0. According Boyd [25], a pH below 6.5 or greater than 9.0 can reduce the ability of fish reproduction and growth. While the pH value ranging from 6-9 maintenance media, it is still in line with the range of water quality standard for cultivation in Government Regulation No. 82 Year 2001.

Temperature is a factor that affects the growth and survival of fish. The water temperature will affect the rate of growth, the rate of metabolism and appetite of fish [25]. The difference in temperature that occurs during pisciculture looks quite volatile, which in the early morning just 25°C temperature maintenance, then increased fairly high at 28°C, and then stabilized until the end of the study. According Green and Schrader [23], optimal temperature range for catfish farming is 25-30°C. So that the temperature fluctuations that occur during maintenance is still in normal condition with a range that can be tolerated by fish. Temperature affects the various chemical reactions in the body of water, including the effect on the solubility of oxygen in water and metabolism of fish, so it will affect fish growth [26]. Fluctuations in temperature also affects the activity of bacteria in media biofloc maintenance. The high activity of bacteria characterized by decreasing the amount of oxygen in the afternoon, when the temperature reached the peak or the maximum value. The high activity of these bacteria is offset by the addition of molasses as a carbon source to be converted into cellular proteins of bacteria in the afternoon after feeding. At the time of low temperature, high oxygen solubility, and vice versa. In addition, the temperature also affects the strength and morphology floc. The optimal temperature to maintain the stability of flocks ranged between 20-25°C, while the temperature is too high (> 35 ° C) may cause a decrease in the stability of the floc [26].

4 Conclusions

In summary, the bioflocs system can enhance growth of juvenile catfish *C. batrachus*. Density of biofloc have played an important role in promoting growth in catfish, however this deserve further investigations

The first author would like thank to Ministry of Research, Technology and Higher Education of the Republic of Indonesia for providing research funding of the study.

References

1 Hermawan, A., Sudaryono, and S. Prayitno. Journal of Aquaculture Management and Technology. **3**, 35-42 (2014)
2 Fagbenro, O. and Jauncey, K. Biores. Tech. **51,** 29-35 (1995)
3 Richter, C.J.J., Eding, E.H., Verreth, J.A.J. and Fleuren, W.L.G. Blackwell Science, Cambridge, pp. 242-277. (1995).
4 ÇEK, Fiehriban and Erdal Yilmaz. Turk J Zool. **31**, 35-46 (2007)
5 Shafrudin, D.,Yuniarti., and M, Setiawati. Jurnal Akuakultur Indonesia. **5**, 137-147 (2006)

6 Asaduzzaman, M., M,A, Wahab., M,C,J, Verdegem., S, Huque., M,A, Salam., and M,E, Azim. Aquaculture. **280:** 117–123 (2008)
7 De Schryver., P. R. Crab., T, Defoirdt, N. Boon, and W. Verstraete. Aquaculture**. 277**, 125–137 (2008)
8 Avnimelech, Y. Aquaculture. 176, 227-235 (1999)
9 Abdel-Wahab A. Abdel-Warith, El-Sayed M.I. Younis, Nasser A. Al-Asgah. Saudi Journal of Biological Sciences 23, 404–409 (2016)
10 Hepher. B. and Y. Pruginin. *Commercial fish farming: With special reference to fish culture in Israel.* (John Wiley and Sons, New York, 1981)
11 Amon P. Shoko, Samwel M. L., and Yunus D. Mgaya. Effect of stocking density on growth performance, survival, production, and financial benefits of African sharptooth catfish (Clarias gariepinus) monoculture in earthen ponds. Journal Journal of Applied Aquaculture Volume 28, Issue 3. 2016.
12 Yusuf, M. W., N. B. P. Utomo., M. Yuhana., Widanarni. Journal of Fisheries and Aquatic Science. **10**, 523-532 (2015)
13 Julie Ekasaria, Muhammad A.S., Wiyoto W., Ratih F. H., Gilang S. L., Rosi S., Muhammad A. and Muhammad Z. J. Aquaculture Volume **464**, 1 2016
14 Soedibya PHT. *Variasi Fisiologis Ikan Gurami (Oshpronemus gouramy Lac.) dalam Menghadapi Ketersediaan Sumber Pakan [Disertasi]*. (Bogor: Institut Pertanian Bogor. 1999)
15 Onada OA, Ogunola OS Effects of Catfish (*Clarias gariepinus*) Brood-stocks Egg Combination on Hatchability and Survival of Fish Larvae. J Aquac Res Development **2**, (2017)
16 Mosha S.S., Kang'ombe J., Jere W., Madalla N. J Aquac Res Development **7**, 441 (2016)
17 Hickling, C.F. *Fish Culture*. (London : Faber and Faber. P : 348. 1971)
18 Farhat and Mukhtar A.K. Journal of Applied Aquaculture. **23**, (2011)
19 Lenhardt, Mirjana., I. Jarić., P. Cakić., G. Cvijanović., Z. Gačić., J. Kolarević. Turk. J. Vet. Anim. Sci. **33**, 209-214 (2009)
20 Bekeh A.F., Olatunji A.E., Bassey W.K. J Aquac Res Development. 6, 378 (2015).
21 Soedibya, P. H. T., E. Listiowati. *Prosiding Seminar Nasional Percepatan Desa Berdikari Melalui Pemberdayaan Masyarakat Dan Inovasi Teknologi.* (LPPM Unsoed Purwokerto. 2014)
22 Sadekarpawar, S., and P. Parikh. World Journal of Zoology **8** ,110-118 (2013)
23 Green B.W., and Schrader KK. J Aquac Res Development. **6,** 314, (2015)
24 Wilen, B. M. and P. Balmer. Water Research **33**,391-400 (1999)
25 Boyd, C. E. and Likhtykopper. *Water Quality Management in Pond* (Fish Culture. Auburn: University Alabama, 1981)
26 Kuhn, D.D., Lawrence, A.L., Boardman, G.D., Patnaik, S., Marsh, L., and Flick G.J. Aquaculture. **303**, 28-33 (2010)
27 Green B. W. and Matthew E. M. Journal Journal of Applied Aquacultur. **29**, (2017)

Rearing technology for climbing fish (*Anabas testudineus*) on different salinity

Ani Widiyati[1,*], *B.N.* Asyiah[2], *Tri* Heru Prihadi[1], and *Dewi* Puspaningsih[1]

[1]Research Institute for freshwater Aquaculture and Fisheries Extension, Jl.Sempur No.1, Bogor, Indonesia
[2]Student of Department of Biology, Faculty of Mathematic and Science, University of Indonesia, Jakarta, Indonesia

Abstract. Optimal salinity on rearing media is needed to get the best production performance of climbing fish culture. The aim of this experiment is to get an optimal salinity on rearing media of climbing fish. This experiment was conducted in wet lab of Research Institute for freshwater Aquaculture and Fisheries Extension. The treatments were different salinities on rearing media; 0 ppt, 3 ppt, 6 ppt and 9 ppt. Each treatment has 3 replicates. Experiment design use completely randomized design. The initial weight of fish was 1 ± 0.11g and initial length of fish was 2 ± 0.3 cm. Aquaria of 40x22x20 cm was used for rearing media. Each aquarium with 15 lt of water was stocked 350 fish. Fish was fed frozen blood worm (protein contains of 62.5%) at ad libitum. Duration time of rearing was 40 days. Sampling was conducted every 10 day for fish weight, lenght and water quality. The result showed that optimal salinity of survivale rate, specific growth rate and total length were on salinity of 1.8 ppt, 2.5 ppt and 3.3 ppt. The value of water quality parameter in aquaria such as temperature, pH, Oxygen, nitrite, nitrate an amonia is still feasible for climbing fish rearing.

1 Introduction

Climbing fish (*Anabas testudineus*) is a kind of fresh water and brackish water specific local fish [1] and has high economic value in Sumatra and Kalimantan. Demand of climbing fish as food increase significantly, however population of climbing fish is decreasing in their habtitat [10, 7]. Culture technology is needed to develop a climbing fish production, any efforts of culture was conducted but production is still low. This case predicted caused of unsuitable water quality. Water quality condition is related to capability to get sucsesfully aquaculture.

*Corresponding author: ani_widiyati@yahoo.co.id

Water quality condition affects fish physiology proccess so it can be a limited factor. Salinity one of water quality parameter affects directly to fish metabolism. Too high, too low and big different fluctuated salinity would cause fish death. Salinity will affect directly by osmotic pressure dan osmoregulation proccess. Kaneko and Hiroi [11] said omoregulation process is needed to control water balancing and ion between in the fish body and environment so physiology process in fish body works normally.

Climbing fish was found on swamp area in South kalimantan. Swamp was area which watering affected by tidal of sea water, salinity fluctuated from 0 – 20 ppm [16]. Fluctuated salinity was susceptible from survival rate of fry because it has big osmotic pressure. Osmoregulation process was needed to control water balance of ions between fish body and environment so that physioligy process in fish body works normally [8, 11]. And hence, salinity was an urgent limiting factor to fish survival in aquaculture.

Many researchers conducted culture experiment of climbing fish, however no one researcher did an optimal salinity on rearing media to get maximum productivity from rearing technology of climbing fish fry.

2 Materials and methods

Experimental design used was Completely random Design with 4 treatments and 3 replicates. The treatments of rearing media salinity were 0 ppt; 3 ppt; 6 ppt; 9 ppt. Fish fry was got from naturally breeding, with initial length was 2 – 3 cm, density of fry was 2 per litre. Fish was adapted before treatment as long as 60 days. Fish was reared 40 days. Sampling was conducted every 10 day to get weigh and length gain. Size of aquarium was 40x22x20 cm with water volume of 15 litre. Feed for feeding fish was blood worm with protein contains was 62.5%. feed fed et ad libitum, 3 times a day, at 08.00 am, 12.00 am and 17.00 pm.

Parameters evaluated were Total growth weight, spesific growth rate, feed conversion, water quality and survival rate. Salinity on media was checked everyday while water temperature, disolve oxygen, pH and amonia were checked every 10 days. Data were analyzed using ANOVA for growth, survival rate and feed conversion while water quality was analyzed by deskriftive analysis. Polinomial orthogonal was used with tolerant significance of 95 %.

3 Result and Discussion

Specific growth rate (SGR), total weight gain, total lenght gain, feed efficiency, and survival rate of climbing fish fry from each treatment on 40 days rearing time could be seen on table 1. Treatment of 0 ppt, 3 ppt, 6 ppt and 9 ppt showed significant different on survival rate ($P > 0.05$). Rearing on media 0 ppt with 67.78 % was the best survival rate, while media on 9 ppt with 46.67% was the worse survival.

Reriang media of 0 ppt (0 ppt treatment), 3 ppt, 6 ppt and 9 ppt gave significant difference on specific growth rate ($P<0.05$). rearing media of 0 ppt gave the best specific groth rate with 5.61 %, however 9ppt gave the worse with 4.53 %.

All treatments of media rearing (0 ppt, 3 ppt, 6 ppt and 9 ppt) gave significant difference on total weight gain and length gain ($P<0.05$). rearing media of 0 ppt gave the best total weight gain with 2.73 g, however 9 ppt gave the worse with 1.83.

Feed eficiency gave significant difference among tratments (0, 3, 6 and 9 ppt). The best feed efficiency was from treatment on rearing media of 3 ppt with 64.70 % however the worse from 9 ppt with 49.26 %.

Table 1 showed the value of water quality parameters on all treatments during rearing.

Table 1. Value of specific growth rate (SGR), total weigh gain, total length gain, feed efficiency and survival rate for climbing fish fry during rearing period (40 days)

Parameter tests	Treatments (rearing media)			
	0 ppt	3 ppt	6 ppt	9 ppt
SGR (%)	5.61±0.20[a]	5.01±0,15[ab]	4.61±0.62[b]	4.53±0.48[b]
Total weigh gain (g)	2.73 ±0.10[a]	2.38±0.24[ab]	2.02±0.40[ab]	1.83±0.30[b]
Feed efficiency (%)	53.01±4.58[ab]	64.70±9.00[a]	56.52±4.50[ab]	49.2±4.29[b]
Survival rate (%)	67.78±6.94[a]	60.00±5.77[a]	55.56±1.92[ab]	46.67±6.67[b]

Note : Value followed the same Superscript word showed no significant difference among treatments (P>0,05).

Table 2 showed water quality parameter value on rearing media for climbing fish fry from all treatments (0, 3, 6 and 9 ppt).

Table 2. Value of water quality parameters of rearing media from all treatments

Parameters	Treatments			
	0 ppt	3 ppt	6 ppt	9 ppt
Temperatur (^{0}C)	25,5—26,4	25,8—26,3	25,6—26,6	25,8—26,2
pH	6,0—6,7	5,8—7,1	6,0-7,20	6,2—7,2
Salinity (ppt)	0	3—3,5	6—6,5	9—9,5
DO (mg/l O_2)	5,58—9,41	5,47—8,69	5,48—8,66	5,41—8,93
Amonia (mg/l NH_3)	0,90—1,30	0,60—1,30	0,83—1,50	1,0—1,7

Value of SGR decreased if salinity increased, this happened was occured because climbing fish fry tended more stress cause of salinity, so that fish would use more energi to balance lose salinity by non active diffusion. Energy got from feed was foccused to control capability fish adapted with environment so that the growth became slow and decreasing [22]. Garcia [9] added, salinity on rearing media can give bad effect on fish physiology mecanism, such as decreasing on growth. According to Ye [21] and De Boeck [5] decreasing on growth was predicted lack of input nutrient caused no availabity of suitable salinity.

Relathionship between salinity (x) and specific growth rate (SGR) has coefficien value (R^2) was 79.3%. it means any changing 79.3 % of SGR was affected by salinity. Based on formula $Y= 73.04 + 1,03\ x + -0.20\ x^2$, was found optimum salinity of 2.5 ppt for specific growth rate.

Relathionship between salinity (x) and total growth gain has coefficien value (R^2) was 79.3%. it means any changing 79.3 % of total growth gain was affected by salinity. Based on formula $Y= 3.19 + 0.4\ x + -0.14\ x^2$, was found optimum salinity of 2.8 ppt for total growth gain. Relathionship between salinity (x) and total length gain has coefficien value (R^2) was 80.9%. it means any changing 80.9 % of total length gain was affected by salinity. Based on formula $Y= 3.49 + 1.00\ x - 0.15\ x^2$, was found optimum salinity of 3.33 ppt for total length gain.

Salinity of 9 ppt was the worse feed efficiency with value of 49.26 %, this happenned was predicted more stress with increasing salinity. Energy got from feed in advance was used for osmoregulation procces, so that feed was in efficiency. According Andrianto (2005) high and low salinity on rearing media affect directly to feed efficiency and feed consumed. As well as Bardach *et al.* (1972) *in* Rahmadani [17] said feed efficiency was affected by stress and so feed consumed decreasing.

Salinity of 3 ppt was the highest feed efficiency with value of 64.70 %, but if compared to growth gain on salinity of 3 ppt was 2.38 gram and still under treatment salinity of 0 ppt with 2.73 gram. This showed that feed was consumed by climbing fish fry on 3 ppt treatment was in efficiency for growth because part of energy gor feed consumed was used for osmoregulation procces and for balancing too high salinity in environment condition. The more feed efficiency the more energy produced for osmoregulation procces and part of it was used for growth.

Relathionship between salinity (x) and feed effiiency having coefficien value (R^2) was 80.9%. it means any changing 80.9 % of feed efficiency was affected by salinity. Based on formula Y= 37.04 + 21.75 x + -4.73 x^2, was found an optimum salinity of 2.29 ppt for feed efficiency.

The highest suvival rate was got from treatment of 0 ppt with 67.78 %, rearing media of 0 ppt close to body fuid of climbing fish. So that the fish use energy minimal for osmoregulation proccess. According to Stickney [20], if environment osmotic pressure close to fish osmotic pressure, energi produced by metabolism would not be used for adapted to environment osmotic pressure. So that energy was more used for growth and survival. As well Chotipuntu dan Avakul [4] in their experiment showed climbing fish with size of 3 – 4 cm will ablle to adapt more high salinity compare to egg and larvae.

Relathionship between salinity (x) and survival rate having coefficien value (R^2) was 72.0%. it means any changing 72.0 % of survival was affected by salinity. Based on formula y= 73.04 + 0.54x + -0.30x2, was found an optimum salinity of 1.8 ppt for survival rate.

Water quality measurement during experiment showed feasible for climbing fish. According to Rahman [18] water temperature of 26,06°-31,97°C was suitable for fish culture. While ph on rearing media was 5.8 – 7.2. According to Boyd (1982) in Rasmawan [19] pH would die fish if the pH less than 4 and more than 11. pH 5 – 10 was the can survive, while the pH ini this experimen was till normal condition. Salinity in this experiment was control as the treatment. Salinity was fluktuated no more than 0.5 ppt. Disolved oxygen on rearing media was 5.41 – 9.41 mg/l. According Peterson *et al.*(1999) disolved oxygen more than 4 mg/l was good for fish culture. Amonia contain ini this experiment was 0.60 – 1.7 mg/l. This condition still feasible to fish for survive however if the amonia contain more than 2.4 mg/l will affect fish for death [3].

4 Conclusion

The optimum salinity for survival, specific growth rate, total weight gain and feed efficiency were 1.8 ppt, 2.5 ppt, 2,8 ppt and 2,29 ppt.

References

1. H. Akbar. Bogor. Agr. Univ. 35 (2008)
2. J. Akbar, M. Adriani, S. Aisiah. J. Bionatura, **13**, 9 (2011)

3. I.M. Chotiba. Padjadjaran Univ. 40 (2014)
4. P. Chotipuntu, P. Avakul. Walailak J. **7**, 7 (2010)
5. G. De Boeck, H. Smet, R. Blust. Physiol. Biochem. Zool. **73**, 9 (2000)
6. S. Djarijah. *Natural fish food*. 87 (1995)
7. Y. Ernawati, M.M. Kamal, N.A.Y. Pellokila. J. Ikh. Ind. **9**, 25 (2009)
8. Y.Fujaya. *Fish physiology*. 179 (2004)
9. C. Garcia, F. Pozet, C. Michel. Dis. Aquat. Organ**.** **42**, 7 (2000)
10. Isriansyah, K. Sukarti. Mulawarman University, **3,** (2007)
11. T. Kaneko, J. Hiroi. *Fish Larval Physiology*. **20,** (2008)
12. E. B. Kuncoro. The popular encyclopedia of freshwater fish. **2**, (2009)
13. H. Kurata. Mexican Journal of Agricultural Science. **20**, 21 (1959)
14. Mudjiman. Fish food. 190 (1998)
15. F. Nahar. Dhaka university press. 72 (2014)
16. S. Najiyati, L. Muslihat. Know the Type of Peat Land. **4**, (2015)
17. D. Rahmadani. Bogor. Agr. Univ. 80 (2007)
18. M.S. Rahman, M.Y. Chowdhyry, A.K.M.A. Fiaque, M.S. Haque. Bangl. J. Fish. Res. **5**, 10 (1982)
19. Rasmawan. Bogor. Agr. Univ. 67 (2009)
20. R. R. Stickney. *Principle of Warmwater Aquaculture*. 375 (1979)
21. W. Ye, D. Cheng, L. Ma. J. Envirn. Bio. **1, 7 (**1990)
22. R. Woo, M. Kelly. J. Fish. Phys. **78**, 76 (1995)

POTENCY OF DIFFERENT SEAWEEDS AS DIETS FOR DEVELOPING ABALONE (*H. squamata*) CULTURE IN NUSA PENIDA ISLAND, BALI

Tri Heru Prihadi[1], *Rasidi*[2], *Idil* Ardi[3], *Ani* Widiyati[1] and *Dwi* Budi Wiyanto[4]

[1]Research Institute for freswater Aquacultuire and Fisheries Extension, Jl. Sempur 1 Bogor, Indonesia; [2]Research Center for Fisheries, Jl. Ragunan 20 Pasar Minggu South Jakarta, Indonesia; [3]Institute for Research of Ornamental Fish Aquaculture, Depok West Java, Indonesia; [4] Marine and Fisheries Faculty of Udayana University, Denpasar, Bali, Indonesia

ABSTRACT. Nusa Penida island is one of center for seaweed culture in Bali province. Seaweed that cultured is *Euchema spinosum* and *Euchema cottoni* spread around coastal. There also found wild seaweed, such as *Ulva* sp and *Gracillaria* sp. Abalone (*H. squamata*) is one of kind Mollusca which high economic value, but abalone cultured is not yet develop in Nusa Penida island, although this island have high potency for developing abalone culture. Abalone culture needs seaweeds as a diet. The abundance of seaweeds as abalon diet is important for developing abalon culture in Nusa Penida Island in the future. The aim of this research was to asses the potency of seaweeds as diet for developing abalone culture in Nusa Penida island. Research was done in Batununggul village, Nusa Penida Island. Experimental design was used Completely Randomized Design (CRD) with 3 treatments and 3 replications. The treatments were used different seaweeds as diet for abalone culture: (DG=Diet *Gracillaria* sp, DS=Diet *Spinosum* sp. and DU=Diet *Ulva sp.*). Abalon were cultured for 4 months by feeding the diet at satiation. Data analysis was used analysis varian (Anova) with SPSS. 16. The result showed abalon fed with *Gracillaria* sp (DG) at 4.73 g was the highest growth by 4.73 g, followed DU by 3.93 g and DS by 3.43 g. Meanwhile the abalon shell length fed with *Gracillaria* sp (DG) was the highest growth by 6.55 mm, followed DU by 5.97 mm and DS by 5.60 mm. Based on variant analysis showed growth performance (length shell and weight) abalon, all treatments were not significantly different (P>0.05). The conclusion this research, the three species of seaweed can be used as diet for abalone culture. These seaweeds have same potency as diet for developing abalone culture in Nusa Penida Island

Corresponding author: rsdrasidi@gmail.comtriheru_p@yahoo.com

1 INTRODUCTION

Abalone is one kind of mollusca which high economic value, the price depends on size and species, between IDR 300.000 – 400.000/kg. Abalone production from legal catch was 8.650 mt and illegal 5.300 mt in 2008 [1]. Now on production from catch was decreasing so culture of abalone is needed al least to maintaine production of abalone. In Nusa Penida Island was found abalone in the coastal when low tied many last year. According to local people in Nusa Penida the local name of abalon is Seven Eye Shell. They Could catch the Seven eye Shell easily 10 years ago how ever currently it was difficult to get. This information showed in Nusa Penida area was abalone habitat.

Nusa Penida Island Bali is wellknown as center for seaweed culture in Bali province. They are many kinds of seaweed found in Nusa Penida, such as *E.spinosum*, *E.cottoni* and as well found seaweed, *Gracillaria* sp. and *Ulva* sp. Because so many kinds of seaweed found in that area and we know that seaweed is also as feed resources for abalone And hopefully developing aquaculture will be easier for development in the future. Therefore, this study tried to assess potency the use of many kind of seaweeds as abalone diet in the Nusa Penida island. Food source selection for abalone is influenced by local conditions and seasons in an area development. Seaweed on Nusa Penida growth in the rocks, nets and pole. Wild seaweed can be found easily in the region at low tide. *Gracillaria* sp.can be found throughout the year, but for *Ulva* sp. found in the rainy season.

Food is one of the factors that determine the successfull of abalone aquaculture. According Oktaviany [2] abalone requires macro algae such as seaweed which used as an energy source for growth and survival. *Gracillaria* sp. and *Ulva* sp. can be use as food for abalone culture [3]. Ungson [4] also reported have been used seaweed *Sargassum* sp. as abalone food. The aim this research was to know the influence of seaweed as food to abalon growth culture in Nusa Penida Island.

2 MATERIALS AND METHODS

Research was conducted in the village Batununggul, Nusa Penida island, Klungkung regency, province Bali (Figure1). Location area of the research was in coastal area with sandy and rocky bottom texture, this location usually was used for seaweed culture. Water depth level was about 0.5–1 m at the lowest tide some time depends on tidal. Following the conditions existing land abalone culture methods are applied using off-bottom technology. Abalone cultured were used basket containers made from polyethylene, with height of 80 cm and diameter of 40 cm. The baskets were placed on the bottom and bound with wooden stakes.

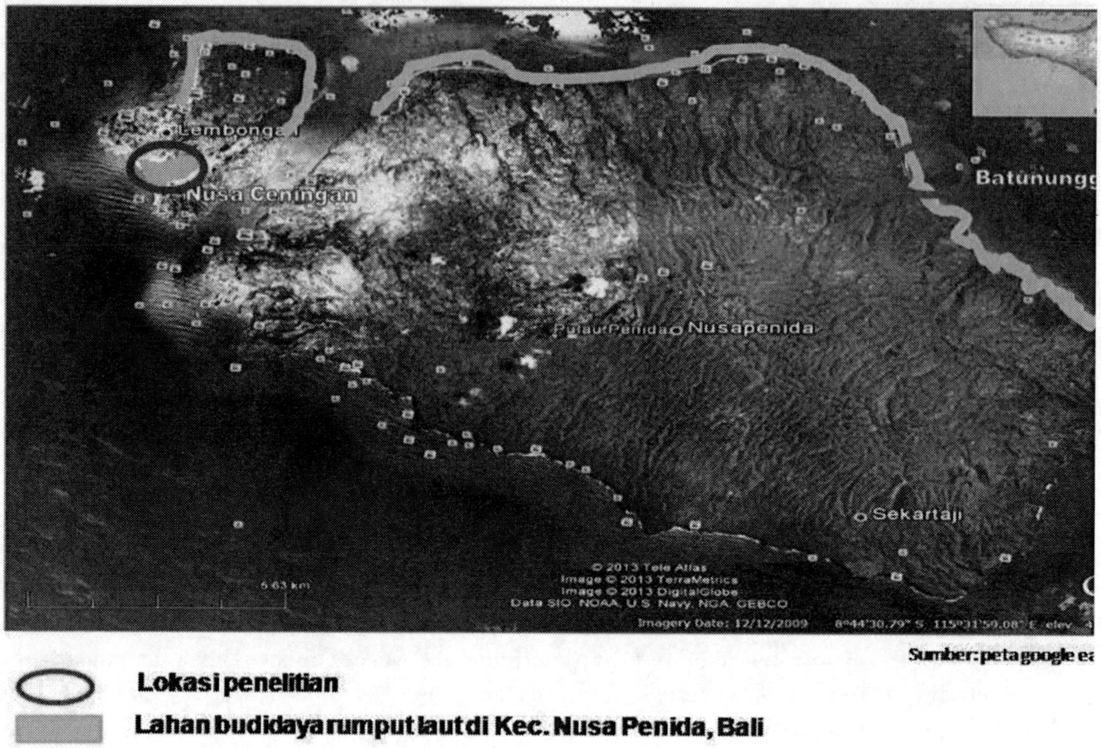

Fig 1. Research location in Batununggul village Nusa Penida Island.

Abalone (*Haliotis squamata*) were carried out from the Institute for Research and Development of Marine Aquaculture Gondol, North Bali. The research was conducted with stocking density of 100 pcs/basket. The research design was used a Completely Randomized Design (CRD) with 3 treatments x 3 replicates. The abalone was fed different seaweed, as treatments. there are (DG) Diet Gracillaria, (DS) Diet Spinosum and (DU) Diet Ulva. Feeding rate used 15% of the biomass abalone were given every 3 days. According Ungson [4], measurement of shell length and width were done every 15 days using a caliper and weight used a digital scale with a precision of 0.01 g. Measuring the shell length, width, and weight of abalone by random sampling, the number of abalone are measured by 20 pcs/basket.

Length and weight of abalone were used to calculate an absolute growth, Daily Increasing of body weight (DIBW) = (Wt–W0)/time length, growth rate of shell in length = mm, Daily increasing of shell length (DISL)=(L_t–L_0)/t. Where, W_0 = initial average weight (g), W_t= average weight in the end of study (g), L_0=initial shell length, L_t= length of the shell the end of study, t= long time culture. Data were analyzed using ANOVA test if there is a real difference followed by a further test using SPSS software version 16. Seaweed and abalone samples were analyzed proximate in the laboratory IPB Bogor, to know the nutritional content.

3 RESULT AND DISCUSSION

3.1 Seaweed for abalone diet

Seaweed is a natural food for abalone in the their habitat [5]. We found many kind seaweed culture and wild seaweed in the coastal near experiment location. Seaweed cultured such us *E.spinosum dan E. Cottoni,* and wild seaweed are *Gracilaria* sp. and *Ulva* sp. Wild seaweed growth up in the rock, net, and the other place near location seaweed farming. *Gracillaria* sp seaweed. and *Ulva* sp. commonly found growing attached to the net/peg seaweed cultivation.

The research used three kind of seaweed are *Gracilaria* sp. *Ulva* sp. and *E. spinosum* sp. We analyze proximate sample seaweeds to know nutrient content in the laboratorium IPB Bogor. Nutrients content of seaweed are presented in Table 1. Based on Table 1 obtained protein content in seaweed *Gracilaria* sp is higher than the other seaweeds. According Handy (2006) protein levels vary depending on the species. The protein content of the proximate analysis results in this study is not different from the protein content of sea *Gracillaria* used in previous studies [3].

In addition to nutrient content in abalone diets need to be considered such as texture and colour. *Gracillaria* sp and *E.spinosum colour* is brown while *Ulva* sp. Is bright green. Culture development of Abalone continuously as sea food should also be considered. Seaweed *Gracilaria* sp and *E.spinosum* sp available throughout the year In Nusa Penida, but *Ulva* sp. availability depending on the season. *Ulva* sp. usually appear during the rainy season.

Table 1. Proximate analyzed from different seaweeds used on this experiment

Parameters	Unit	Species of seaweed			*Technique Analysis*
		Gracillaria sp./DG	*E. spinosum/DS*	*Ulva sp./DU*	
Water Content	%w/w	45.55	48.02	38.08	Gravimetry
Ash Content	%w/w	24.8	25.49	34.49	Gravimetry
Protein	%w/w	8.01	3.48	6.77	Kjeldhal (Titrimetry)
Fat	%w/w	0.1	0.14	0.19	Soxhlet
Carbohydrate	%w/w	22.56	22.49	21.3	Phenol Sulphate (spektrometry)
Crude Fiber	%w/w	5.15	2.76	3.71	Gravimetry

Table 1. showed proxymate analyzed from different seaweed used in this experiment. Spinosum looked the highest of water content with 45.55 %w/w, while Ulva was the lowest water containt with 38.08 %w/w. The highest 0f Ash was Ulva with 34.49. the highest of protein, carbohydrat and crude fiber was belonging of Gacilaria with 8.01; 22.56 and 5.15 %w/w, however gracilaria has the lowest in fat with 0.1 %w/w. Data in detail of proxymate analyzed could be read in table 1.

3.2 Abalone Growth

Feed or Diets is one factor that affect growth of abalone, biomass growth is very important in the cultivation of abalone [6]. Based on Figure 2 and 3, shows absolute growth (shell length and weight) obtained in abalone with treatment DG is the higher followed by DU and DS is the lowest. Based on the results of analysis variance absolte growth shows that treatment give effect in growth (shell length and weight) of abalone did not differ significantly between all treatments (P>0.05).

Three species of seaweed which are used as food or diets for abalone culture as *Gracilaria* sp, *E.Spinosum* and *Ulva* sp. *Gracillaria* sp. had growth response as the highest compare to other treatments (Figure 2). This happens was caused by nutrient content of *Gracillaria* sp. is the highest compared to others (Table 1). Abalone need nutrient such us protein, carbohydrat, fat and the other for their growth. Protein content in *Gracillaria* sp. is the higher than the other seaweed. Protein are important nutrient for growth abalone [7].

The result in this experiment statistically produced growth abalone (shell length and weight) did not differ significantly among treatment. These results experiment indicate three different seaweed are equally can be uses as a source of food for abalon culture. Compare than previous studies, *Gracillaria* sp. from the sea was used as food for abalone resulted growth is lowest compared *Ulva* sp. and *Gracilaria* sp. from brachiswater in the laboratory experiment [3]. Different seaweed for abalone fed increased growth, it can be seen from the length and weight of abalone shells during the study. Daily increase body weight and shell length of abalone all treatments are presented in Table 1.

Table 2. Size, Daily Increasing of Body Weight (DIBW) and Daily Increasing of Shell Length of abalone during experiment..

Treatment	Initial experiment		End experiment		DIBW	DISL
	Weight (g)	Shell Length (mm)	Weight (g)	Shell Length (mm)	(g/day)	mm/day
DG	10.91	40.32	15.64	46.87	0.045	0.063
DS	10.73	40.50	14.22	46.10	0.034	0.054
DU	10.22	38.90	14.19	44.87	0.038	0.057

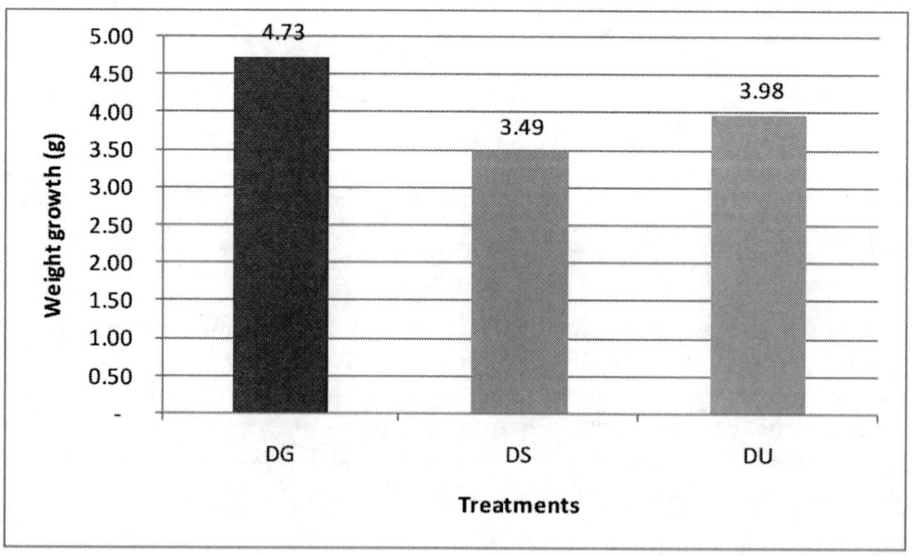

Fig2. Total growth (weight) of abalone among treatment during 4 months rearing(DG = Diet Gracilaria, DS = Diet Diet Spinosum and DU = Diet Ulva)

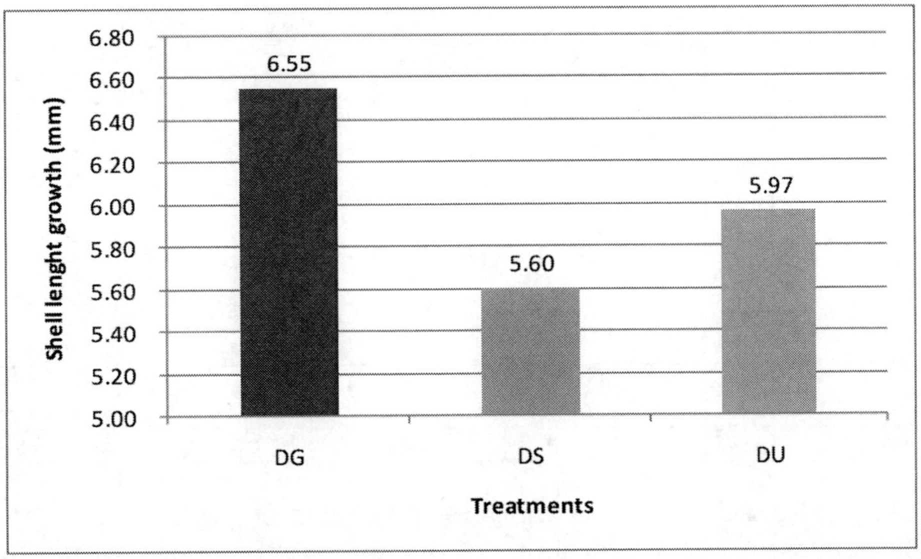

Fig 3. Total growth (shell length) of abalone among treatment during 4 months rearing (DG = Diet Gracilaria, DS = Diet Spinosum and DU = Diet Ulva)

Abalone were spread has increased shell length during cultivation. According by Figure 3, abalone shell length in DG is higher than the other treatments. Seaweed *Gracillaria* sp. have high nutrition so abalone shell length also increased like their weight. This result experiment is higher than result previous studied [3].

Abalone shell is formed from mineral and calcium contained in the waters. Shell color also is influenced by the food, abalone fed seaweed Ulva color will seen bright green. Likewise abalone fed Gracillaria and spinosum seaweed would brownish colour. These results are consistent with studies using laboratory scale Gracillaria marine origin and Ulva [3]. Yong Ju, *et al.*, [8] also reported the use of brown seaweed and produce red abalone with shell color matches the color of seaweed are given.

The results of the proximate analysis of abalone at the end of the experiment are presented in Table 3. Base on Table 3 showed protein content abalone with food seaweed Ulva (DU) is highest followed by treatment abalone food *Gracilaria* sp. (DG) and the lowest in abalon with diet seaweed *Spinosum* sp. (DS). It is inversely proportional to the protein content of seaweed that are used as food abalone. This condition is the same as the results of research Ungson [4] reported protein content two kinds of seaweed *Sargassum* sp is higher, it used food for abalone but have growth did not differ significantly. In addition to nutrient content, seaweed texture also affect the outcome of the study, the texture of *Ulva* sp. is softer than *Gracillaria* sp. and *Spinosum* sp., so more easily digested by the abalone.

Table 3. The proximate analysis of abalone at the end of research

Parameters	Unit	Treatment			Technique Analysis
		DG/Gracillaria sp.	*DS/Spinosum* sp.	*DU/Ulva* sp.	
Water Content	%w/w	23.15	29.26	24.58	Gravimetry
Ash Content	%w/w	8.99	9.7	8.97	Gravimetry
Protein	%w/w	42.68	33.22	46.63	Kjeldhal (Titrimetry)
Fat	%w/w	3	2.19	1.85	Soxhlet
Carbohydrate	%w/w	22.14	25.36	17.41	Phenol Sulphate (spektrometry)
Crude Fiber	%w/w	0.3	0.38	0.45	Gravimetry

Table 3 showed proximate analyisis (water content, ass content, protein, fat, carbohydrate and crude fiber) of abalone which were feed from three species was not significant different.

4 CONCLUSIONS

Three species of seaweed *Gracillaria* sp, *E. spinosum* and *Ulva* sp. which are used as feed for abalone produced the same growth respon among treatments. Based on these data, three species of seaweed can be used as feed in abalone culture in Nusa Penida Island Bali.

The authors would like to thank Mr I Nyoman Landep, The Dean of Faculty of Fisheries and Marine, Udayana University for the collaboration and cooperated research. This research were supported by government fund budget from Research Center for Aquaculture Jakarta. Indonesia

Refference

1 Cook P.A.,& H. R Gordon. 2010. World Abalon Supply, Markets, and Pricing. Journal Shellfish Research 29 (3): 569-571.

2 Oktaviany, M.J. 2007. Beberapa catatan tentang aspek biologi dan perikanan abalon. Oseana XXXII (4): 9 hlm.

3 Susanto, B., I. Rusdi & I.N. Adiasmara Giri. 2012. Optimalisasi pemeliharaan yuwana abalon (*H. squamata*) dengan kepadatan dan jenis pakan yang berbeda. Jurnal Riset Akuakultur 7(1): 21 – 31.

4 Ungson, J.R. 2003. Fooding of Abalon Juveniles with Two Species of Sargassum (*Sargassum cristaefolium* and *Sargassum polycystum*). Philippine Journal of Science 132 (1): 33-38

5 Body, A. G. C.,1987. Abalone Culture in Japan. Marine Fisheries Review 49(4): 75–76.

6 Francis T.L. 2008. The Effects of Differences in Fooding Regime and of Export Simulations on the Growth of the abalon *Haliotis midae* Linnaeus. Thesis. University of the Western Cape. South Africa.104 hal

7 Fleming, A.E.,Van Barneveldt, R.J & Hone, P.W. 1996. The development of artificial diets for abalone: A review and future directions. Aquaculture, 140: 5–53.

8 Yong Ju, Z., Dong-Fang Deng, W. Dominy, C. Viljoen, P. Hutchinson. 2013. Remarkable Affect of Diets on Abalon Meat Nutrient Composition and Shell Pigmentation. http://www.ctsa.org/files/publications/AbalonDiets_Ju_OI.pdf. dowlod at 24 Desember 2013.

Polyculture of Giant Freshwater Prawn, *Macrobrachium rosenbergii* and Nilem Carp, *Osteochilus hasselti* Cultured in Recirculation System Using Biofiltration

Eri Setiadi[1] and *Imam* Taufik[1]

[1] Research Institute for Freshwater Aquaculture and Fishery Extension, Bogor, West Java, Indonesia

Abstract. Giant Freshwater Prawn is one of the freshwater commodity which have an economy value where Nilem carp has also potentially to be cultured to produce fingerling size as a "baby fish". The purpose of this experiment is to determine the optimal survival, growth, FCR, and productivity of Giant Freshwater Prawn and Nilem carp cultured on polyculture system and also biomass productivity of bio filter (snail, water spinach, and Cabomba). The different stocking density of Nilem carp as a treatments were as followed: A) 25 fish m^{-3} B) 50 fish m^{-3}, and C) 75 fish m^{-3}. Density of Giant Freshwater Prawn was 20 individuals m^{-3} for all treatments. The result showed that the optimal of survival, growth (length and weight), productivity and FCR for Giant Freshwater Prawn was found at density of 25 fish m^{-3} (A) than that of density 50 fish m^{-3} (B) and 75 fish m^{-3} (C) ($P<0.05$). The highest of biomass productivity of bio filter (snail, water spinach, and Cabomba) was found at density of 25 fish m^{-3} (A) than that of density of 50 fish m^{-3} (B) and 75 fish m^{-3} ($P<0.05$). Water quality such as DO, temperature and pH showed almost the same at all treatments.

1 Introduction

Giant Freshwater Prawn (*Macrobrachium rosenbergii*) is the one of commodity species that has an economy value. In Indonesia, Giant Freshwater Prawn has been initially cultured since 1970. Most of the culture technique on the earthen pond in Tasikmalaya and Ciamis areas are intensive system where rice field have been used for culture in Lamongan [1]. Thereafter, Giant Freshwater Prawn have been distribute to Bali and Borneo Islands [2]. Although Giant Freshwater Prawn have been cultured for 40 years but the productivity is low and unstable caused by traditional and semi-intensive system where stocking density is low (3-5 individuals m^{-3}) which have resulted in low of survival (12.30-35.70%) [2].

[1]Corresponding Author: erisetiadi@yahoo.com

On the other hand, disease outbreak has also contributed to low productivity [3]. Some effort have been conducted to improve the production such as mono sex culture, polyculture, and using shelter in an extensive culture [1, 4]. Polyculture of Giant Freshwater Prawn to another fish species in Indonesia is clearly lacking. For instance, Polyculture Giant Freshwater Prawn with Gouramy [5, 4] and Red Tilapia [6]. However, the production is still not significant increase due to semi intensive system. Therefore, innovation to improve productivity via polyculture with intensive system is needed.

Nilem carp have a potential to be cultured, especially, to produce of fingerling 'baby fish''. Market demand of this fish have increased in year by year Thus, Nilem carp has an economy value as a local fish [7]. Nilem carp is herbivore species because the main food of this species is detritus, phytoplankton and pheriphyton. Based on such condition, Nilem carp is called bio-cleaning agent [8].

Moreover, Nilem carp is easy to culture and resistance on environmental and disease [9]. Therefore, Nilem carp is suitable for culture with Giant Freshwater Prawn in a polyculture system.

Intensive culture is the most system that applied by fish farmer in Indonesia. This system believe that can increase the fish production. However, the negative impact of this system has a seriously problem even in the fish pond and environment. [10] reported that waste of the aquaculture activity have resulted in eutrophication and sedimentation in water bodies (e.g. lake and river) and salinization or acidification of soils. Richness nutrients from aquaculture activity had been increased in ammonia, nitrite, nitrate, and phosphate concentrations [11]. Ammonia and nitrite are toxic to aquatic organisms including fish [12]. To solve such problems, the recirculation system using bio filtration is known effective to reduce of waste come from aquaculture activity and water conserve. Recirculation system is efficient in term of water utility due to during the operation no need water exchange [13]. Microorganisms (bacteria and plankton), mollusk, and plants have been commonly used as a bio filtration to reduce of water pollutants from aquaculture effluents [14]. Mollusk have a function as filter feeder and very efficient in reducing of total suspended solid [15] where plants can absorb pollutants such as heavy metals, organic matters, nitrogen and phosphor [11].

The purpose of this experiment is to determine the optimal survival, growth, and biomass of Giant Freshwater Prawn and Nilem carp and also the biofiltration biomass such as water spinach (vegetable), snail, and Cabomba (ornamental aquatic plant).

2 Materials and Methods

The experiment was conducted from July to September 2016 at Research Station for Freshwater Aquaculture Environment Technology and Toxicology, Cibalagung, Bogor, Research Institute for Freshwater Aquaculture and Fishery Extension, Bogor, West Java, Indonesia.

Nine of concrete tanks with size of 4 m^2 (2.3 length x 1.9 wide x 1 depth m) with completed by water pump and aeration were used for Giant Freshwater Prawn and Nilem carp culture. Giant Freshwater Prawn with 7.73±0.56 cm in total length and 4.31±0.98 g in body weight was used. Feeding rate was 5% a day $biomass^{-1}$ and artificial diet with containing 35% protein was fed three times a day. Stocking density was 20 prawns for each tank. Newly hatched larvae (D1) of Nilem carp with 0.31±0.0316 cm in total length and 0.0012±0.00007 g in body weight were used. No feeding was fed during the experimental period. Culture period of this experiment was 60 days

Bio filtration box was made from wooden board with size of 2 length x 0.6 wide x 0.4 depth m for water spinach (*Ipomoea aquatic*) while wooden boards with size of 1 length x 0.6 wide x 0.4 depth m were used for snail (*Pila ampulacea*) and ornamental aquatic plant of Cabomba (*Hydrilla* sp). Initially of wet weight (biomass) of snail, water spinach, and Cabomba were 500 g, 540g, and 216 g, respectively. Lime stone with 4-5 cm in diameter and gravel with 2-3 cm in diameter were used as a substrate for snail where limestone and pumice stone with 2-3 cm in diameter were used for water spinach. Gravel and sand were used for Cabomba. All boxes were covered using plastic sheet inside with completed by inlet and outlet. Net with size of 0.8 length x 0.5 wide x 0.4 depth m was placed inside the box then the snail were introduce into the net which completed by zippers to avoid of escape.

All concrete tank were set up in recirculation system. Water from concrete tank was transferred using water pump into the snail box then the water through into the water spinach box and then into the Cabomba box. Finally, water flow back through the concrete tank. The layout of recirculation system and implementation of the system in the field were shown in Figure 1.

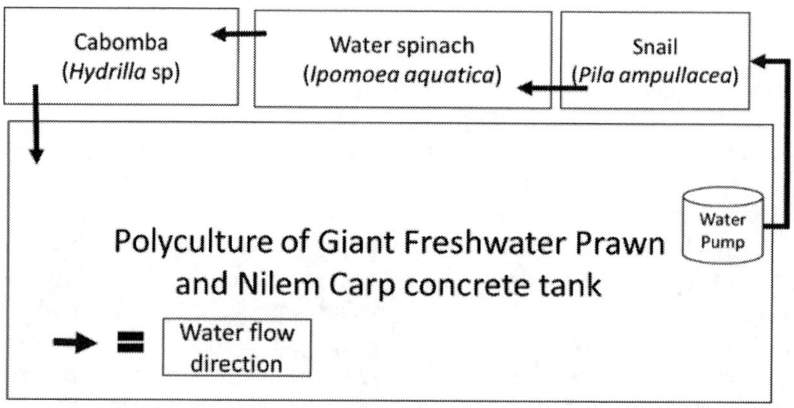

a

b

Fig 1. Recirculation system of polyculture Giant Freshwater Prawn (*Macrobrachium rosenbergii*) and Nilem Carp (*Ossteochilus hasselti*) using biofiltration. a) The layout of recirculation system and b) Implementation of the recirculation system in the field.

Different stocking density of Nilem carp as a treatment were as followed A) 25 fish m^{-2}, B) 50 fish m^{-2}, and C) 75 fish m^{-2}. Parameters observed were survival, growth, and biomass. Data of the total length and weight, 35 Giant Freshwater Prawn were sampled for initial and final weight while 10 newly hatched larvae and 10 fish of Nilem carp. The survival and biomass data were obtained from the population and total weight of population both of Giant freshwater Prawn and Nilem Carp at the times of harvested. All data were analyzed using formula [16] were as followed:

- Weight gain (g):

$$Weight\ gain = Mean\ final\ weight - Mean\ initial\ weight \tag{1}$$

- Percent weight gain (%):

$$\%weight\ gain = \frac{Mean\ final\ weight - mean\ initial\ weight}{mean\ initial\ weight} x\ 100 \tag{2}$$

- Survival rate (%):

$$Survival\ Rate\ (\%) = \frac{No.of\ total\ live\ prawn}{Total\ no.of\ prawn\ stocked}\ x\ 100 \tag{3}$$

- FCR (Feed Conversion Ratio):

$$FCR = \frac{Total\ feed\ consumed\ (kg)}{Total\ yield\ (kg)} \tag{4}$$

- Biomass/net Production:
  The net production of prawn (Kg/decimal) was calculated by the following formula:

$$Biomass\ (kg/decimal) = \frac{survival\ rate\ x\ stocking\ density\ x\ weight\ gain\ (gm)}{100x1000} \tag{5}$$

All data of survival, growth (length and weight), and biomass were statistically analyzed using completed randomized design with three treatments and three replicates. If there was significant different among the treatments, Tukey's test was performed. Water quality parameters such as temperature, pH, and dissolved oxygen (DO) were analyzed using descriptive.

3 Results and Discussion

Result of polyculture of Giant Freshwater Prawn and Nilem carp on survival, growth (length and weight), and biomass are shown in Figure 2. Over all data showed that the survival, total length, weight, and biomass both of Gian Freshwater Prawn and Nilem Carp decreased led to increasing of stocking density of Nilem carp (Fig. 2). The highest of survival, total length, weight, and biomass whether Giant Freshwater Prawn or Nilem carp were found at density of 25 fish m^{-2} (A), followed by density of 50 fish m^{-2} (B), and 75 fish m^{-2} (C) ($P<0.05$). This indicated that different stocking density of Nilem carp affect the survival, growth (total length and weight), and biomass where at high stocking density have resulted in low of survival, growth, and biomass. At high stoking density in polyculture have resulted more competitive in term of food, locomotion, and space.

https://doi.org/10.1051/e3sconf/20184702005

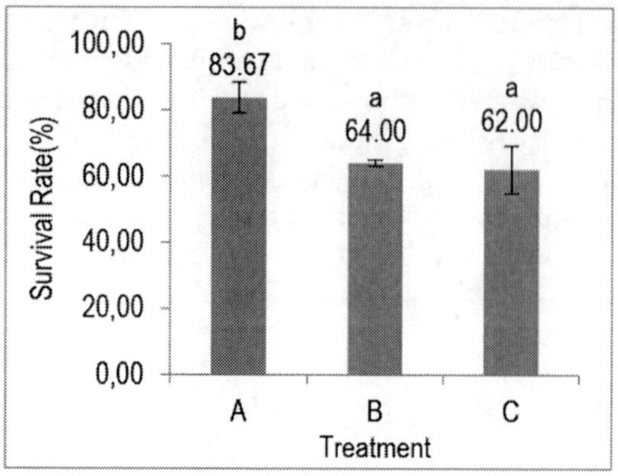

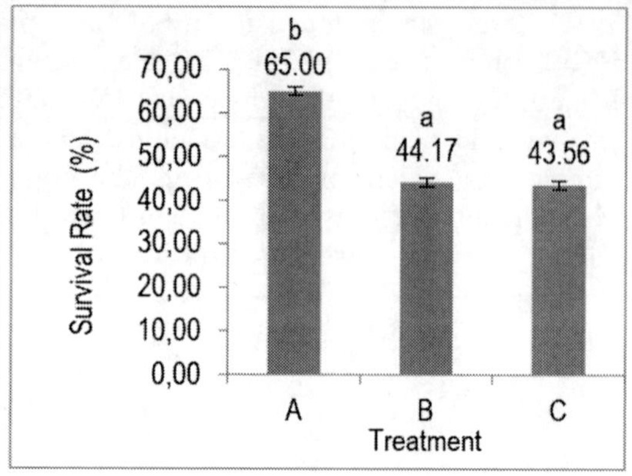

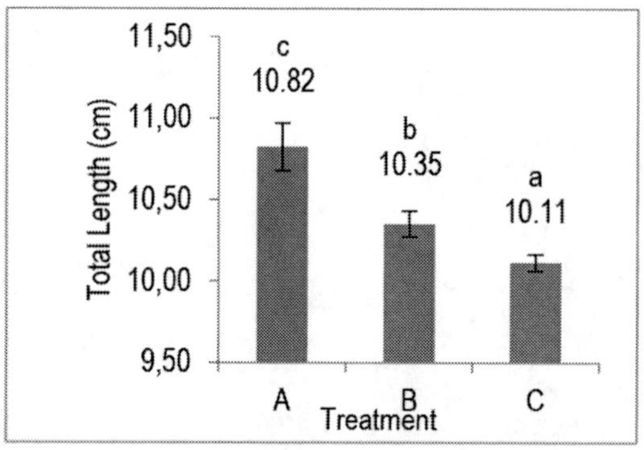

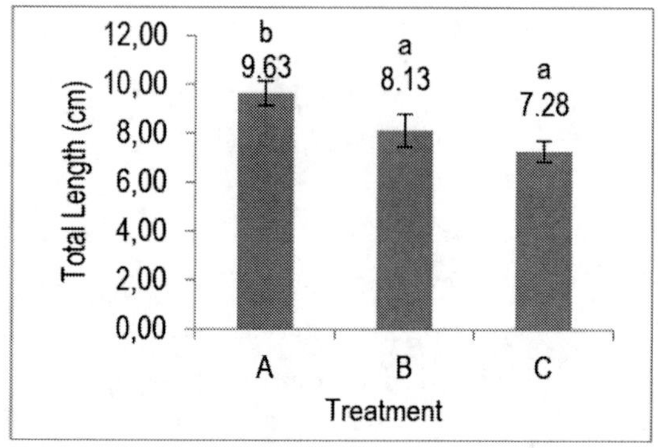

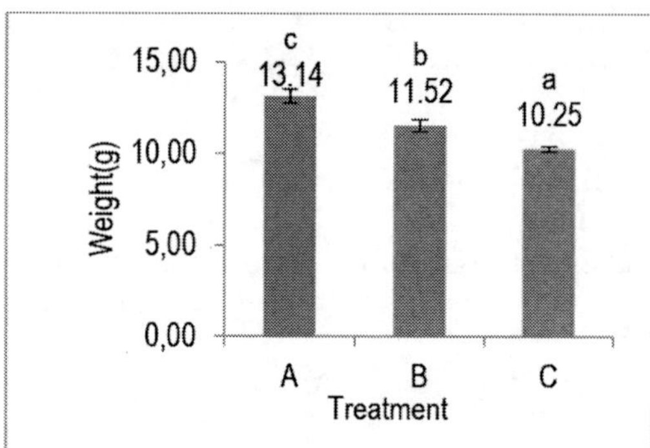

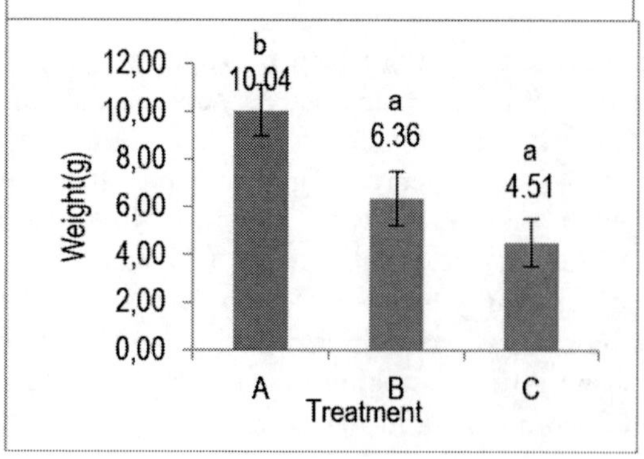

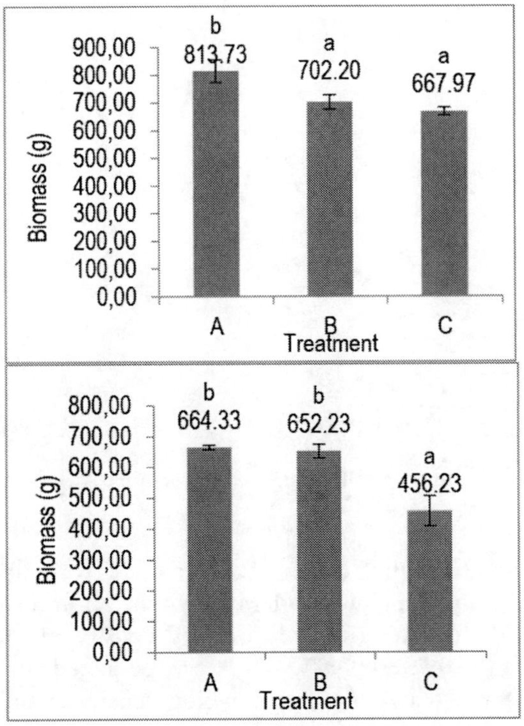

Fig 2. The survival, total length, weight, and biomass of Giant Freshwater Prawn (*Macrobrachium rosenbergii*) and Nilem carp (*Ostechilus hasselti*) cultured in polyculture system

[17] stated that inappropriate ratio or number between one species to the other species in polyculture affect the survival, growth, and biomass which is caused by high competitive food sources, space, and locomotion. The present experiment, although Giant freshwater Prawn have feeding habit and niches different from Nilem carp but at high stocking density of Nilem carp shows clearly that food and space competitive occurs. High stocking density is also impact on physiological, mainly, stress that affect fish growth and survival [18]. The different stage and fish species selection for introducing into the pond with polyculture system is very important in order to obtain optimal production [19]. [4] reported that high stoking density of Gouramy is followed by high mortality and low of biomass.

Feeding conversion ratio (FCR) of Giant Freshwater Prawn and Nilem Carp cultured on polyculture system is shown in Figure 3.

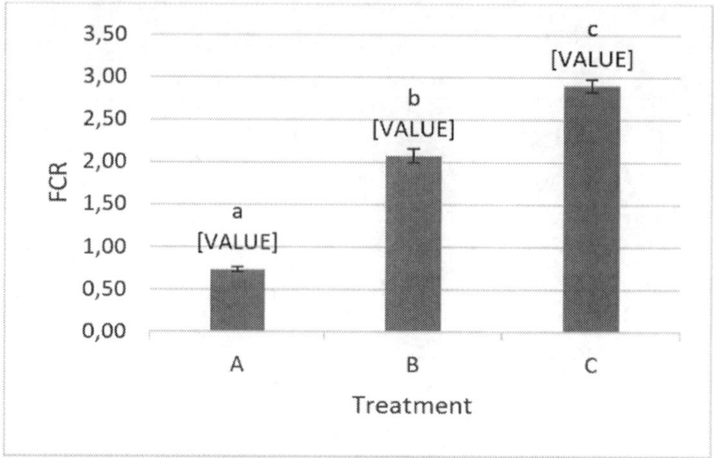

Fig 3. Feeding conversion ratio of Polyculture of Giant Freshwater Prawn, *Macrobtachium rosenbergii* and Nilem Carp, *Osteochilus hasselti*

Feeding conversion ratio of polyculture of Giant Freshwater Prawn and Nilem carp (Fig. 3) showed increase with increasing in stocking density of Nilem carp. The highest of FCR (2.90±0.07%) was found at density of 75 fish m^{-2} (C), followed by density of 50 fishm^{-2} (2.07±0.08%) and density of 25 fish m^{-2} (0.73±0.03%) (A). It can be stated that stocking density affect the feeding conversion ratio where the high stocking density would be led to increase FCR. Similar result had been reported that increasing on stocking density of Gouramy was led to increase in FCR [4]. The present experiment can be explained that food source such as natural food that exist in the pond insufficient for supplying Nilem carp. Thus, artificial diet that mainly feed to Giant Freshwater Prawn only is also taken by Nilem carp. [20] reported that inadequate the number of stocking density have influenced on food supply where the main food become used for consuming all fish.

Biomass productivity of bio filtration such as snail, water spinach, and cabomba on polyculture of Giant Freshwater Prawn and Nilem carp is shown in Figure 4.

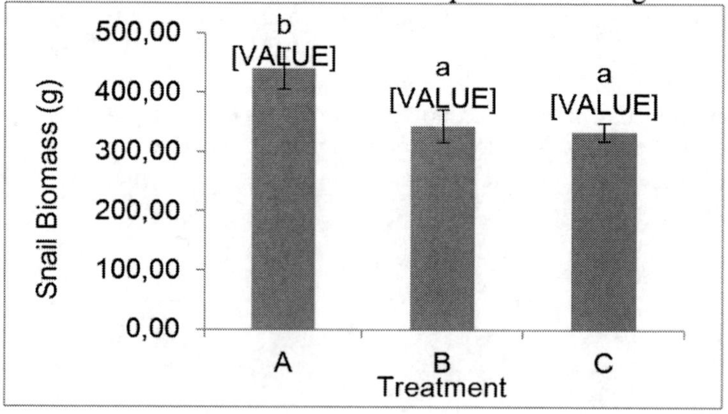

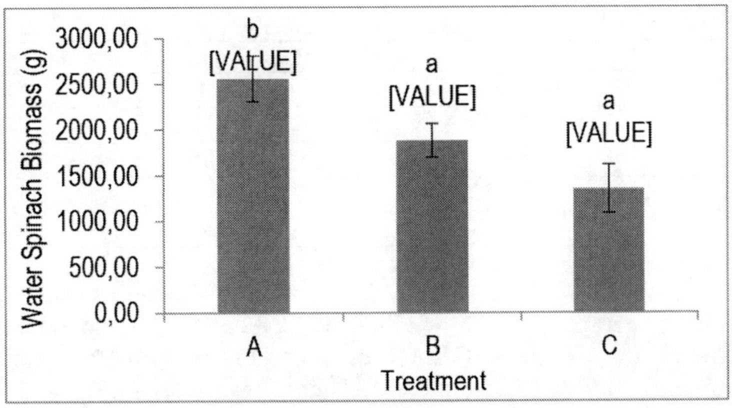

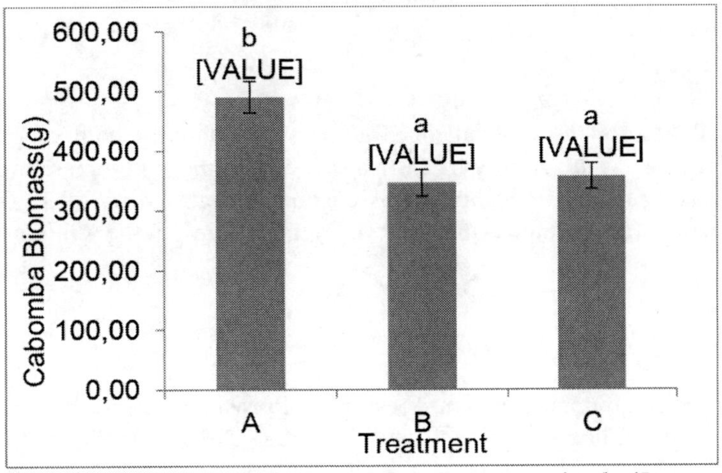

Fig 4. Biomass biofiltration of snail (*Pila ampulacea*), water spinach (*Ipomoea aquatic*), and cabomba (*Hydrilla* sp)

The highest of biomass productivity of bio filtration such as snail, water spinach, and cabomba (Fig. 4) was encountered at density of fish 25 m-2 (A), followed by density 50 fish m-2 (B), and 75 fish m-2 (C) ($P<0.05$). This suggested that biomass productivity of bio filtration might be related to the waste product that obtained by aquaculture activity. Waste aquaculture increase with increasing the biomass of fish and time period of culture [11]. The present experiment shows that the highest of Giant freshwater Prawn and Nilem carp biomass was found at density of 25 fish m^{-2}. Therefore, high waste of aquaculture have also high containing nutrients that consume by snail and to be absorbed by plants. [15] reported that total suspended solid concentration have correlated to mussel grows. High nutrients concentration in the pond water column that derived from feces, urine, and uneaten food. These nutrients containing nitrogen and phosphor elements that important for plant growth [21, 22].

Water quality such as temperature, pH, and dissolved oxygen (DO) during culture period of Giant Freshwater Prawn and Nilem Carp is presented at Table 1.

Table 1. Water quality

Treatments	Parameters		
	DO (mg L)	Temperature (° C)	pH
A	3.3 – 3.6	27.5 – 27.8	6.7 – 7.1
B	3.1 – 3.9	27.0 – 27.6	6.3 – 6.9
C	3.2 – 3.8	27.2 – 28.0	6.4 – 6.8
Optimal Range	≥ 3[23)]	26-30[23)]	6-9[23)]

Remaks: [23)] *government regulation No. 82 (2001)*

All water quality parameters such as dissolved oxygen (DO), temperature, and pH were within the optimal range both of Giant Freshwater Prawn or Nilem Carp culture. The present experiment imply that Polyculture of Ginat Freshwater Prawn and Nilem Carp in the recirculation system using biofiltration can keep water quality..

4 Conclusions

The optimal of growth, survival, and biomass productivity both of giant freshwater prawn and nilem carp is the density of 25 fish m^{-3}. The higher of biomass production of bio filter (snail, water spinach, and cabomba) is encountered at density of 25 fish m-3. Water quality parameters (temperature, pH, and DO) is within in the optimal range for all treatments.

References

1. F. Ali. Growth of Indonesian Giant Freshwater Prawn (*Macrobrachium rensenbergii*) in a closed aquaculture system with artificial shelter. LIMNOTEK, **XIV**(1): 29-36 (2007)
2. L. E. Hadie, W. Hadie, Imron, I. Khasani, N. Listyanto. *Strategi pengembangan budidaya udang galah GIMacro*. Forum Inovasi Teknologi Akuakultur. p.67-77 (2010)
3. I. Khasani. Upaya peningkatan produktivitas dalam usaha pembesaran udang galah (*Macrobrachium rosenbergii* De Man). Media Akuakulture, **3** (1): 25-30 (2008)
4. D. Rohmana, C.H. Adi, A. J. Pamungkas, S. Roselia. *Polikultur udang galah (Macrobrachium rosenbegii) dan gurami (Osphronemus goramy) sistem esktensif*. Seminar Nasional Hari Pangan Sedunia XXVII. p. 222-226 (2007)
5. A. Yustiati, T. Herawati, W. Lili, A. Nurhayati, Rosidah, I. B. B. Suryadi. Budidaya Polikultur Ikan Gurame (*Osphronemus gouramy*) dengan Udang Galah (*Macrobrachium rosenbergii*). Jurnal Pengabdian kepada Masyarakat. **2**(1):44-46 (2018).
6. DKPDIY (Dinas Kelautan dan Perikanan Daerah Istimewa Yogjakarta). Polikultur udang galah dan nila merah. (2016)
7. D. T. Mulyasari, Soelistyowati, A. H. Kristanto, and I. I. Kusmini. Karakteristik Genetik enam populasi ikan Nilem (*Osteochilus hasselti*) di Jawa Barat. Jurnal Riset Akuakulture, **5**(2): 175-182 (2010)
8. F. Cholik, A. G. Jagatraya, R. P. Poernomo, and A. Jauzi. *Akuakultur: Tumpuan Harapan Masa Depan Bangsa*. Masyarakat Perikanan Nusantara dan Taman Akuarium Air Tawar TMII, Jakarta p. 415. (2005)
9. Y. Hermawan, Rosmawati, and Mulyana. Gorwth and Survival Rate of Nilem (Osteochillus hasselti) with Different Feeding Rate. Jurnal Mina Sains, 1(1): 19-23 (2015)

10. M. Martinez-Porchas and L. R. Martinez-Cordova. World Aquaculture: Environmental Impacts and troubleshooting alternatives. The Scientific World Journal (2012)
11. E. Setiadi, & L. Setijaningsih. Improving water quality and productivity of tilapia *Oreochromis niloticus* using constructed wetland. Indonesian Aquaculture Journal, **6:** 107-122. (2011)
12. O. F. Joel and C. A Amajuoyi. Determintation of Concentration of Ammonia that could have Lethal Effect on Fish Pond. Asian Research Publishing Network Journal of Engineering and Applied Sciences, **5** (2): 1-5 (2010)
13. FAO. *A Guide to Recirculation Aquaculture an Introduction to Environmentally friendly and Highly Productive Closed Fish Farming System*. Edited by Jacob Bregenballe. EUROFISH International Organisation. pp 100. (2015)
14. A. E. Turcios and J. Papenbrock. Sustainable Treatment of Aquacuaculture Effluents-What Can We Learn from the Past for the Future. Sustainability, **6**: 836-856 (2014)
15. D. Puspaningsih and E. Setiadi. *Potensi kijing local (Pilsbryoconcha exillis) sebagai biofilter untuk mengurangi TSS pada budidaya ikan nila*. Prosiding Forum Inovasi Teknologi Akuakultur. Pusat Penelitian dan Pengembangan Perikanan, Badan Penelitian dan Pengembangan Kelautan dan Perikanan, Jakarta. (2013)
16. P. Paul, A. Rahman, M. Hossain, S. Islam, S. Mondal, and M. Haq. Effect of Stocking Density on the Growth and Production of Freshwater Prawn (*Macrobrachium rosenbergii*). International Journal of Fisheries and Aquaculture Sciences. **6** (1): 77-86 (2016)
17. I. Taufik, Sutrisno, and E. Setiadi. The Optimal Ratio of Nle Tilapia *(Oreochromis nilotica)* and Common Carp *(Cyprinus carpio)* for Improving on Deep Water Pond. Indonesian Aquaculture Journal, **8** (1): (2013)
18. I. Lupatsch, G. A. Santos, J. W. Schrama, and A. J. Verreth. Effect of stocking density and feeding level on energy expenditure and stress responsive in European sea bass Dicentrarchus labrax. Aquaculture, **298**:245-250 (2010)
19. B. C. Mohapatra, S. K. Singh, B. Sarkar, D. Majhi, and N. Sarangi. Observation of Carp poluculture with Giant Freshwater Prawn in Solar Heated Fish Pond. Journal of Fisheries and Aquatic Sciense. 2(2):149-155 (2007).
20. M. J. Alam, D. A. Jahan, W. A. Pramanikand, and M. E. Hoq. Poly culture of freshwater prawn, Macrobrachium rosenbergii, de Man with carps: effects of prawn stocking density. Bangladesh]. Fish. Res., **5**(2): 135-144 (2001)
21. Nuryadi and E. Setiadi. Produksi benih ikan lele (Clarias sp.) menggunakan sistem firoremdiasi. Prosiding Forum Inovasi Teknologi Akuakulture. Pusat Penelitian dan Pengembangan Perikanan, Badan Penelitian dan Pengembangan Kelautan Perikanan, Jakarta. (2012)
22. E. Setiadi. Kleinschlige aquaponics voor voedselzekerheid in Indonesia. Aquaculture, 3:34-37. (2017).
23. Government Regulation. Government Regulation of Republic of Indonesia Number 82. Water quality management and Water pollution protection. Jakarta (ID): Sekretariat Negara. 46. (2001).

Water Quality, Survival, and Growth of Red Tilapia, *Oreochromis niloticus* Cultured In Aquaponics System

Eri Setiadi[1], *Yohana* R. Widyastuti[1], and *Tri* Heru Prihadi[1]

[1] Research Institute for Freshwater Aquaculture and Fishery Extension, Bogor, West Java, Indonesia

Abstract. Red tilapia, *Oreochromis niloticus* is the main commodity of freshwater fish in Indonesia. Red tilapia have a good price compared to tilapia. The aims of this experiment to determine of survival, growth, and water quality of red tilapia cultured in aquaponic system. The treatment of this experiment, namely A) Red tilapia cultured without aquaponic (control), B) Red tilapia cultured with pakcoy, and C) Red tilapia cultured with caisin. The result showed that the highest of survival rate, weight, and length absolute found at Red tilapia cultured with pakcoy (96.00±1.73%, 32.31±0.74g, and 7.57±0.21 cm) and Caisin (96.00±1.73%, 32.31±0.74g, and 7.57±0.21 cm) than that of without aquaponic (86.67±1.15%, 25.77±1.05g, and 6.43±0.31 cm) ($P<0.05$). Vegetable leaf production of pakcoy was 6.57±0.16 Kg and Caisin was 6.17±0.11 Kg. The water quality parameters such as DO, TAN, nitrite, and nitrate of Red tilapia cultured using aquaponics was better than that of without aquaponic.

1 Introduction

The main cultivated freshwater species in the world belong to carps and tilapia. The Nile tilapia, Oreochromis niloticus is one of the most important freshwater fish in world aquaculture. It is widely cultured in many tropical and subtropical countries. Some of advantages of Nile tilapia are rapid growth, high tolerance to environment conditions, easy to spawn, resistance to disease, and good consumer acceptance make it a suitable fish for culture [1]. This species is the most familiar and popular fishes in Indonesia, especially, in Sumatera, Java, Bali, and Borneo Islands.

The production of aquaculture technique is usually extensive, semi-intensive and intensive culture systems, but the commonest technique in Indonesia is semi-intensive and intensive systems. The intensive culture has usually high level of management input such as feed and fertilizers are intensively applied following appropriate recommended rates. Most commercial farm adopts this approach due to the fish grow is very faster compare to the others [2]. However, intensive culture system has encountered many problems such as high cost due to commercial food utility during the operation where commercial food constitutes 50 – 70% of the total production cost [3].

Corresponding Author: erisetiadi@yahoo.com

On the other hand, intensive culture system by product such as uneaten feed, feces, and urine have resulted in poor of water quality and eutrophication which negative effect to fish production [4]. Feed leachate into aquaculture systems results in the release of additional nitrogen (N), phosphorus (P), organic matter and trace elements into the water environment [5]. Annual nutrient discharge of about 5100kg N, 2900kg P, and 29500kg organic matter from a fish farm due to high level of feed, feces, and excreta in the water [6]. This condition affects water quality such as depletes of dissolved oxygen and increase concentration of nitrite and ammonia, which are toxic for fish [7, 8, 9]. Thus, poor of water quality has resulted in low production due to high mortality [4].

Some efforts of Nile tilapia culture have been carried out to increase productivity in order to fulfill fish demand such as polyculture using deep pond technology [10], constructed wetland [4], stock density [11], polyculture [12], probiotic utility [13], and aquaponic system [14]. Although these techniques can increase fish production, but base on economy value is still unprofitable due to high price of commercial food, unless fish culture in aquaponic system [15]. Therefore, technology innovation of aquaculture in term of increasing fish production and eco-friendly is needed. Aquaponics is an integrated farming concept that combines fish and vegetable (hydroponic). This integration aims to convert the normally wasted nutrients excreted by fish into valuable plant biomass [16]. The purpose of this experiment is to determine water quality, survival, growth, and biomass of Red Tilapia cultured in an aquaponics system.

2 Materials and Methods

The experiment was conducted from June to August 2017 at Research Station for Freshwater Aquaculture Environment Technology and Toxicology, Cibalagung, Bogor, Research Institute for Freshwater Aquaculture and Fishery Extension, Bogor, West Java, Indonesia.

Nine of concrete pond with size of 3 length m x 3 wide m x 0.7depth m was used. 11 bucket with volume of 12 L was used for aquaponics where the bucket with volume of 25 L was used for physical filter. All pond were installed water pump and completed by aeration. Clearly, the layout of aquaponics design construction was shown in Figure 1.

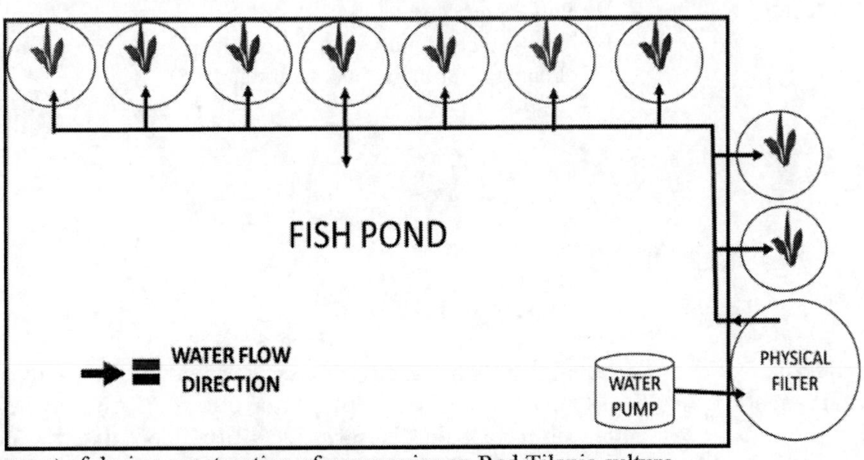

Fig. 1. Layout of design construction of aquaponics on Red Tilapia culture

Red Tilapia with average of 3.37 cm in total length and 2.67 g in body weight was used. Stocking density was 50 fish m^{-3}. Feeding rate was 5% a day biomass-1. Culture period was 50 days. 20 fish were sampled at the beginning and at the end of the experiment. Vegetable plant species was Pakcoi and Caisin. The substrate of aquaponics were lime stone with diameter of 4-5 cm, gravel with diameter of 2-3 cm, and fern root. One bucket was planted by three plants. Parameters observed for fish was survival, growth (absolute length and absolute weight), and biomass where vegetable parameter observed was biomass (wet weight). The treatment of this experiment were as followed: A) Red Tilapia cultured without aquaponic, B) Red Tilapia cultured with Pakcoi, and C) Red Tilapia cultured with Caisin

Water quality parameter such as dissolved oxygen (DO), pH temperature, total ammonia (TAN), nitrite, nitrate, and alkalinity were measured using water checker and analyzed in the laboratory. Water sample was taken at the beginning, in the middle, and at the end of the experiment.

Data of survival, absolute weight, absolute length, and biomass was calculated using formula [17, 18] were ass followed:

Survival Rate

$$SR = \left(\frac{Nt}{No}\right) \text{ x } 100\% \tag{1}$$

where: *SR* = Survival Rate (%)
Nt = Number of fish at the end of experiment
No = Number of fish at the beginning of the experiment

Absolute Length

$$Pm = Lt - Lo \tag{2}$$

where : Pm = absolute length growth
Lt = length average of individual at day t (cm)
Lo = length average individual at initial day/day 0 (cm)

Absolute Weight Growth

$$\Delta W = Wt - Wo \tag{3}$$

where: ΔW = absolute weight growth
Wt = Weight average of individual at day t (g)
Wo = weight averageof individual at initial day/0 (g)

Biomass:

$$B = W \text{ X } N \tag{4}$$

Where: B : Biomass
W : weight average (g)
N : number of population (fish)

All data such as survival, growth, and biomass were analyzed using statistic (ANOVA), except water quality parameters (DO, pH, temperature, TAN, nitrite, nitrate, and alkalinity) and vegetable biomass using descriptive only. Completed randomized design with three treatments and three replicates were performed. If there was significant different among the treatments, Duncan test was performed.

3 Results and Discussions

Water quality parameters such as dissolved oxygen (DO), pH, temperature, total ammonia (TAN), nitrite, nitrate, and alkalinity during the Red Tilapia cultured is shown in Table 1.

Table 1. The range of water quality parameters (Dissolved oxygen, pH, temperature, total ammonia (TAN), nitrite, nitrate, and alkalinity) during culture period of Red tilapia in aquaponics system

Parameters	Unit	Treatments			Optimal Range
		A	B	C	
Dissolved Oxygen (DO)	mg¯L	3.05-3.87	3.12-4.67	3.08-4.61	≥ 3[19,20]
pH		6.71-7.34	6.73-7.45	6.72-7.41	6-9[19]
Temperature	°C	25.8-27.6	25.6-27.2	25.5-27.7	26-30[19,20]
Total Ammonia (TAN)	mg¯L	0.267-0.582	0.052-0.192	0.055-0.185	0.17-3.87[21]
Nitrite	mg¯L	0.065-0.183	0.011-0.078	0.015-0.072	0.02-0.12[19,20]
Nitrate	mg¯L	0.924-5.763	0.478-3.924	0.517-3.586	0.2-219[21]
Alkalinity	mg¯L	116.14-158.71	117.22-149.84	116.10-159.87	20-200 [20]

Remaks: [19] *Government regulation No. 82 (2001),* [20] *Cavalcante et al. (2014),* [21] *Caldini et al. (2015)*

All water quality parameters were within the threshold, except for nitrite was slightly beyond the optimal range for Tilapia culture (Table 1). However, the water quality parameters at Red Tilapia cultured using aquaponics whether Pakcoy or Caisin showed the lowest than that of Red Tilapia cultured without aquaponics. This imply that Pakcoy and Caisin are great efficient for reducing the waste, especially total ammonia, nitrate, and nitrite. Some of studies have been reported that the aquaponics system can reduce the total nitrogen was 30,6% [22], 25% [23], 40,32 % [24], 78,3% [25], and 41,5% [26]. The different result that had been reported caused by different fish species, feeding management, design construction, retention times, and plant species affect the different removal efficient [4]. The optimum range such as temperature (16-30 °C), pH (5.5-7.5), ammonia (<30). Nitrite (<1), nitrate (-), and dissolved oxygen (>3) for plant growth have been reported [27]. The present experiment shows that all water quality parameters are within the optimal range for plant grows.

The survival, absolute weight, absolute length, and biomass of Red tilapia cultured in without aquaponincs and aquaponics system and also biomass of Pakcoy and Caisin is shown in Figure 2.

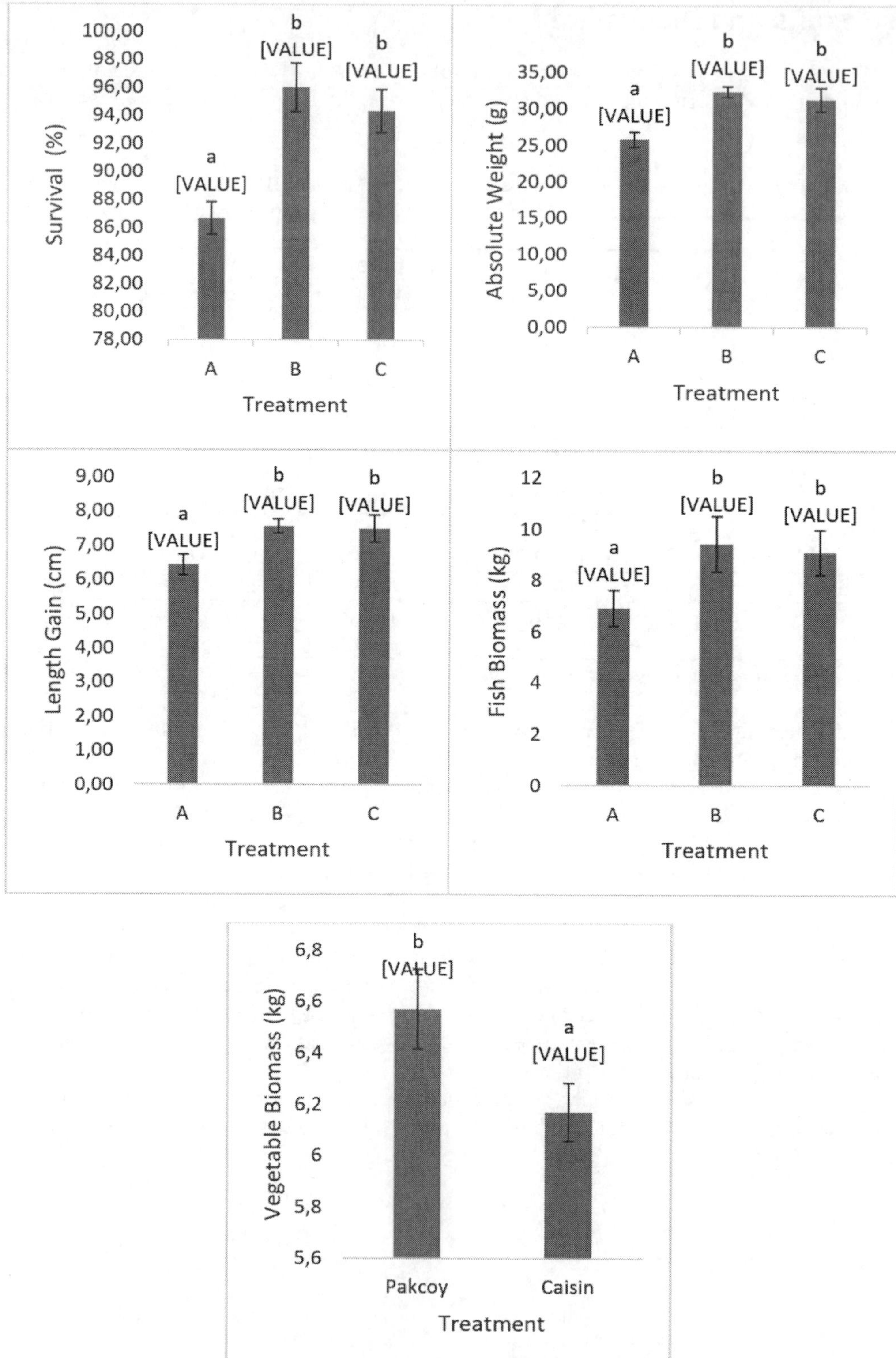

Fig 2. Survival, Absolute length, absolute weight, and biomass of Red Tilapia cultured without aquaponics and with aquaponics system and biomass of Pakcoy and Caisin

Survival, absolute weight, absolute length, and biomass of Red Tilapia was the higher than that of Red tilapia cultured without aquaponics system ($P<0.05$) but no significantly different ($P>0.05$) Red Tilapia cultured in aquaponics using Pakcoy and Caisin (Figure 2). This indicated that aquaponics system affect the survival, absolute weight, absolute length, and biomass. Low in survival, growth, and biomass might be caused by water quality. As mentioned above (Table 1) that water quality of Red Tilapia cultured without aquaponic system showed poor water, especially nitrite concentration. Water quality is very important in aquaculture as a limiting factor. Nitrite are toxic to fish and influence on fish growth in sub-lethal concentration [28, 29].

4 Conclusion

Red tilapia reared in aquaponics system was better than that of without aquaponics in term of survival, absolute weight, absolute length, biomass, and water quality (total ammonia, nitrite, and nitrate). The biomass both of Pakcoy and caisin are the same.

Reference

1. D. M. S. D. El-Saidy, and M. M. A. Gaber,. Effects of dietary protein levels and feeding rates on growth performance, production traits and body composition of Nile tilapia, *Oreochromis niloticus* (L) culture in concrete tanks. Aquaculture Research, **36**: 163-171 (2005)

2. E. D. P. Maia, G. A. Modesto, L. O. Brito, A. O. Galves, T. C. V. Gesteira. Intensive culture system of *Litopenaeus vannamei* in commercial ponds with zero water exchange and addition of molasses and probiotics. Revista de Biologia Marina y Oceanografia, **51**(1): 61-67 (2016)

3. C. C. Alceste. Alternative protein sources in tilapia feed formulation. Aquaculture Magazine, **26** (4) : 3-6 (2000)

4. E. Setiadi, L. Setijaningsih. Improving water quality and productivity of Nile tilapia (*Oreochromis niloticus*) using constructed wetland. Indonesian Aquaculture Journal, **6**(2): 107-122 (2011)

5. L. C. Nwana. Risk management in aquaculture by controlled feeding regime. Pakistan Journal of Nutrition, **2**(6): 324-328 (2003)

6. E. Kerepeczki, D. Gal, F. Pekar. Studies on the utilization of discharged nutrients from an intensive fish production plant. In Proceedings of the 10th International Symposium on Nutrition and Feeding in Fish, Rhodes Greece, 236 (2002)

7. V. Voslarova, V. Pistecova, Z. Svobodova, I. Bedanova. Nitrite toxicity to Danio rerio; Effects of subchronic exposure on fish growth. Acta Vet. BRNO, **77**:445-460 (2008)

8. H. H. Abbas. Acute toxicity of ammonia to common carp fingerlings (*Cyprinus carpio*) at different pH levels. Pakistan J. Bio. Sci. **9** (12): 2215- 2221 (2006)

9. A. C. Karasubenlu, G. I. K. Koksal. The Acute Toxicity of Ammonia on Tilapia (*Oreochromis niloticus* L.) Larvae and Fingerlings. Turk J. Vet. Anim. Sci. **29**: 339-344 (2005)

10. I. Taufik, Sutrisno , E. Setiadi. Intensive culture of BEST Nile tilapia in deep pond. Prosiding Forum Inovasi Teknologi Akuakultur, Research and Development Center for Aquaculture, Jakarta, **2**: 531-538 (2011)

11. J. J. Danaher, J. H. Tidwell, S. D. Coyle, S. Dasguta. Effects of Two Densities of Caged Monosex Nile tilapia, *Oreochromis niloticus*, on Water Quality, Phytoplankton Populations, and Production When Polycultured with Macrobrachium rosenbergii in Temperate Ponds. Journal of the World Aquaculture Society, **38**(3): 367-382 (2007)

12. M. Marceli, D. Santos, W. C. Valenti. Production of Nile tilapia *Oreochromis niloticus* and Freshwater Prawn *Macrobrachium rosenbergii* Stocked at Different Densities in Polyculture Systems in Brazil. Journal of the World Aquaculture, **33** (3): 369-376 (2002)

13. M. H. Mohamed, N. A. G. A. Refat. Pathological Evaluation of Probiotic, *Bacillus Subtilis*, against *Flavobacterium columnare* in Tilapia Nilotica (*Oreochromis Niloticus*) Fish in Sharkia Governorate, Egypt. Journal of American Science, **7**(2): 244-256 (2011)

14. I. Taufik. Culture of Nile tilapia (*Oreochromis niloticus*) fingerling with aquaponic system at different locations. Forum Inovasi Teknologi Akuakultur, Ujung Pandang. (2012)

15. Nuryadi, E. Setiadi. Catfish (*Clarias* sp) seed production Using Phytoremediation System. Forum Inovasi Teknologi Akuakultur, Ujung Pandang. (2012)

16. E. Setiadi. Kleinschlige aquaponics voor voedselzekerheid in Indonesia. Aquaculture, 3:34-37. (2017)

17. S. Goddard. *Feed Management in Intensive Aquaculture*. New York (UK): Chapman and Hall. 174 (1996)

18. H. Effendi. Telaah Kualitas Air Bagi Pengelola Sumber Daya dan Lingkungan Perairan. Kanasius: Yogyakarta (2003)

19. Government Regulation. Government Regulation of Republic of Indonesia Number 82. Water quality management and Water pollution protection. Jakarta (ID): Sekretariat Negara. 46 (2001)

20. D. D. H. Cavalcante, N. N. Caldini, J. L. S. da Silva, F. R. D. S. Lima, M. V. Docarmo. Imbalance in the hardness/alkalinity ratio of water and Nila tilapia's growth performance. Acta Scientiarum, Technology, **36** (1):49-54 (2014)

21. N.N. Caldini, D. D. H. Cavalcante, P. R. N. R. Filho, M. V. Docarmo. Feding Nile Tilapia with artificial diet and dried bioflocs biomass. Acta Scientiarium. Animal Sciences, **37**(4):335-341 (2015).

22. I. Wenxiang, L. Zhongjie. In situ nutrient removal from aquaculture wastewater by aquatic vegetable *Ipomoea aquatica* on floating beds. Water science and technology, **59**,1937-1943 (2009)

23. Y. Shimoda, M. Takagaki, P. Thongbai, K. Ohyama, and K. Ozawa. Improving water quality by using plants with water convolvulus (*Ipomoea aquatica* FORSK.) as a model. International Society for Horticultural Science (2005)

24. X. Zhou, J. Wang, L. Xue, X. Xu, L. Yang. N and P removal characters of eutrophic water body under planted float. Water Res. 16(11); 219-223 (2005)

25. A. Enduta, A. Jusohb, N. Alib, W. N. S. Wan Nikc, A. Hassand. Effect of flow rate on water quality parameters and plant growth of water spinach (*Ipomoea aquatica*) in an aquaponic recirculating system. Desalination and Water Treatment, **5**: 19–28 (2009)

26. H. Hum, S. Aoy, X. E. Yang, Q. Lit. Treating eutrophic water for nutrient reduction using an aquatic macrophyte (*Ipomoea aquatica* Forsskal) in a deep flow technique sistim. Aquaculture, 95, 607-615 (2008)

27. H. Y. Yildiz, L. Robaina, J. Pirhonen, E. Mente, D. Dominguez, G. Parisi. Fish Welfare in Aquaponic System: its Relation to Water Quality with an Emphasis on feed and Faeces. Water. **9**(13): 1-17 (2017)

28. P. Dolezelova, S. Macova, V. Pistekova, Z. Svobodova, I. Bedanova, E. Voslarova. Nitrite toxicity assessment in *Danio rerio* and *Poecilia reticulata*. ACTA VET. BRNO, **80**:309-312 (2011).

29. W. Yanbo, Z. Wenju, L. Weifen, X. Ziong. Acute toxicity of nitrite on tilapia (*Oreochromis niloticus*) at different external chloride concentrations. Fish Physiol. Biochem. **32**(1):49-54 (2006)

Growth performance of siamese gourami *Trichopodus pectoralis* seed with different stocking density to optimize production capacity: optimum temperature and salinity application

Lies Setijaningsih[1,*], *Dewi* Puspaningsih[1], and *Imam* Taufik[1]

[1]Research Institute for Freshwater Aquaculture and Fisheries Extention, Jl. Sempur No.1, Bogor, West Java, Indonesia

Abstract. Fulfill needs of Siamese gourami still rely on the catch from nature, so the population in nature has become decreased. The increase of Siamese gourami fish production from culture is still constrained by the survival rate and slow growth rate in the seed phase. The purpose of this research was to determine the effect of Siamese gourami stocking density on the application of optimum temperature and salinity to growth performance for optimization of production capacity. Application of optimum temperature and salinity refers to the results of previous research, i.e., temperature of 28 °C and salinity 3 ‰. Siamese gourami seed with length of 2.96 ± 0.42 cm and weight of 5.27 ± 0.10 g used in the research. The experimental design used completely randomized design with four treatments, i.e. 1 fish L^{-1}, 2 fish L^{-1}, 3 fish L^{-1} and 4 fish L^{-1}. The results showed that there were an effect of different stocking density on absolute growth weight and length, survival rate, specific growth rate and productivity. This is shown in the C treatment which growth was 4.95 ± 0.037 g, length 3.51 ± 0.01 cm, survival rate 93.47%, specific growth rate 1.22% and fish productivity of 625.76 g.

1 Introduction

Siamese fish commodities are endemic but very potential to be culture, because in addition to being sold as fresh fish in the market, Siamese fish is often preserved in the form of salted fish. Fulfill the needs of Siamese fish still rely on the catch from nature, so the population in nature has decreased. The population decline in nature, indicated by the smaller size of individual Siamese fish caught by the community (Ath-thar et al., 2014). Efforts to increase Siamese fish production from culture are still constrained by the survival rate and slow growth in the seed phase. Problem-solving attempts have been made through the study of aspects of reproductive biology [10]; aspect of feed [14], growth and development performance (Prakoso et al., 2014), the influence of water temperature on the culture of Siamese fish [11] and salinity media on fish culture [9].

In the development of siamese gourami culture, approach on physiological factors such as temperature and salinity can improve the performance of fish growth. Temperature may affect fish activity, where temperature fluctuations are closely related to energy use, and feed consumption affects on the growth. Optimal water temperature will increase the activity of fish feeding, thus accelerating of growth [8]. While salinity has an effect on digestibility and growth [13]. The salinity study on growth performance in freshwater fish has been done, but information on optimum temperature and salinity application to support the growth performance of Siamese gourami seed has not existed. The purpose of this research is to determine the growth performance of Siamese gourami *Trichopodus pectoralis* with the difference of stocking density to optimize production capacity with optimum temperature and salinity application.

2 Materials and methods

The research was conducted at Research Installation of Environmental Technology and Toxicology of Freshwater Aquaculture, Research Institute of Freshwater Aquaculture and Fisheries Extension, Bogor, Indonesia. The test fish came from the fish farmer in Jambi with the initial weights of 5.27 ± 0.10 g and the length of 2.96 ± 0.42 cm. This experiment used an experimental laboratories method. The design pattern used was Completely Randomized Design (RAL) consisting of 4 treatments and 3 replications, can be seen in Table 1.

Table 1. Treatments of difference stocking density on culture of Siamese gourami seed with optimum temperature and salinity application

Parameters	Treatments of difference stocking density (fish)			
	A	B	C	D
Temperature (28^0C) Salinity 3‰	**400**	**800**	**1200**	**1600**

Trial container is a fiberglass tanks with size of 140 × 50 × 50 cm. Restricted feeding is based on FR 5% biomass and given three times daily. Application of water temperature 28^0C refers from the result of the research that the seeds with size of the initial weight of 0.89 ± 0.06 g has a 100% survival rate [11]. Each fiberglass tanks is equipped with heather and for salinity referring to Setijaningsih [9], the optimum salinity 3 ppm gives 97.33% of survival rate.

The parameters observed during the experiment were the survival rate according to Effendi (2004), the specific growth rate according to Godart (1996), the increase in absolute weight and total length according to Zonneveld [16]. Measurements of water quality parameters include temperature, dissolved oxygen, pH and ammonia were done every ten days. Data analyzed with variance (ANOVA) to know the effect of interaction between treatment given, if there is difference then Duncan advanced test were done to know which treatment gave highest and lowest result. Water quality parameters data were analyzed descriptively.

3 Results and Discussion

3.1 Survival rate and specific growth rate

After passing culture period for 50 days research of optimum temperature and salinity application on Siamese gourami seeds achieved 93.67% survival rate. The highest survival rate is in the C treatment with stocking density of 1200 fish (Figure 1). The results of the statistical test, the survival rate value of treatment C showed significantly different ($P < 0.05$) with treatment A, B and D, while treatment A and B were not significantly different ($P > 0.05$). The more the amount of food consumed, the growth of fish will increase. Growth will occur when there is excess energy after the energy produced is reduced by energy used for all activities including energy lost through feces and urine. The excess energy is used to build a new tissue that results in growth rate. On the other hand, the low survival rate in the D treatment with density of 1600 fish is assumed to be due to the lack of equilibrium between the energy for movement activity and the energy for growth. Survival rate related to fish comfort will be followed by a specific growth rate (LPS) (Figure 2). The highest LPS value is shown in the C treatment with 1.22%, with that condition or with Salinity 3 ‰ represents optimal growth because the energy for the ion adjustment process is diverted for metabolism and growth of fish with a value of 0.005 osmol kg^{-1} H_2O^{-1} [9], so that it can optimize the growth weight of Siamese gourami seed. But if the salinity is not optimum, then the fish will spend more energy to regulate the osmotic balance [6].

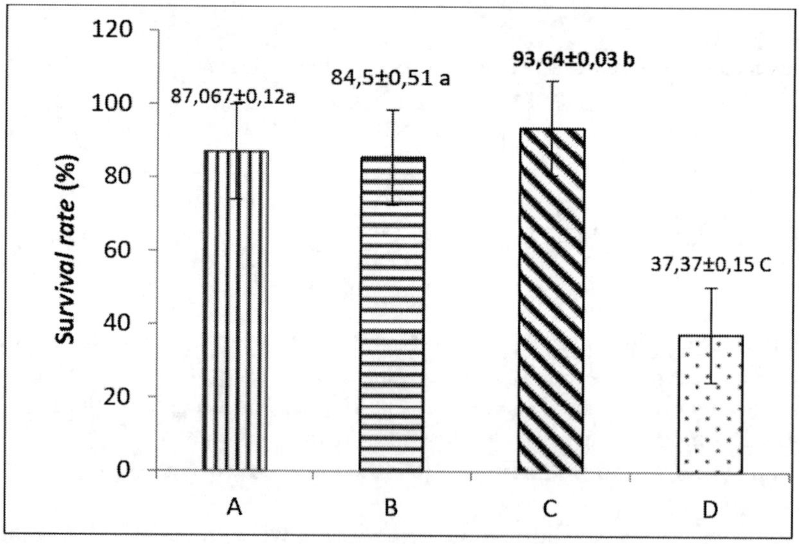

Fig 1. Survival rate of *Trichopodus pectoralis*

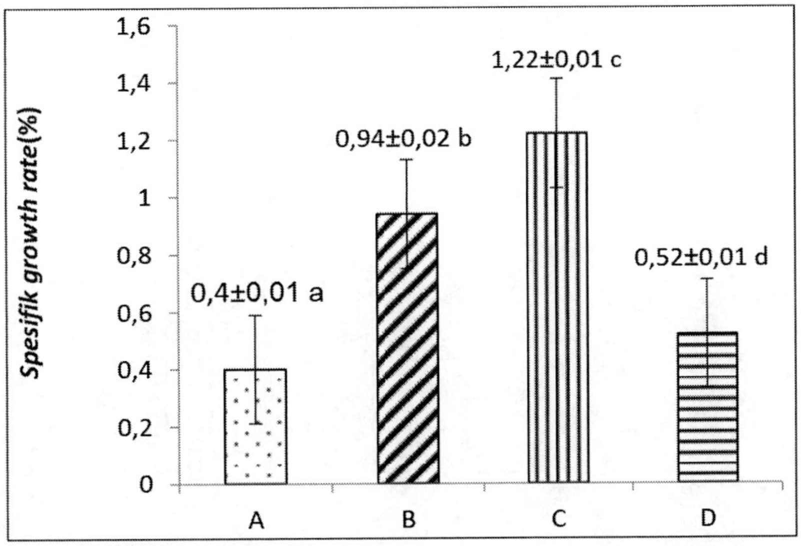

Fig 2. Spesific growth rate of *Trichopodus pectoralis*

3.2 Weight and length gain

The measurement result of weight and length of Siamese gourami seeds during 50 days of culture gave the best result on the C treatment (stocking density 1200 fish) of 4.95 ± 0.037 g and 3.51 ± 0.01 cm (Figures 3 and 4). The result of the analysis of variance showed that the C treatment was significantly different ($P < 0.05$) with treatment of A, B and D. This means that the increase of Siamese gourami seed weight in treatment C, with 1200 stocking density was influenced by optimum temperature and salinity application, which is better than other treatments. The existence of optimum temperature and salinity application hence no temperature fluctuation which is very influential to the availability of dissolved oxygen and pH value. While the low weight increase in treatment D, the high of stocking density is suspected to result in the movement of fish so that it can cause inconvenience or fish to be stressful and resulted in disruption of metabolism and growth of fish. The active movement of fish requires stored energy in the body that is used for self-adjustment to adverse environmental conditions, consequently disrupting the metabolic or biological processes [15]. While at treatment A with stocking density 400 fish / tank has low weight gain. When viewed from water quality data, the alkalinity value of treatment A especially at the beginning of measurement and its range is also low enough to cause a decrease in pH and dissolved oxygen values.

The highest productivity of Siamese fish seeds was achieved at the C treatment of 1371.04 g, followed by treatment A, B and D of 320.16 g; 642.96 g and 309.50 g. The result of the analysis of the variation of the productivity of Siamese fish seeds on treatment C was significantly different ($P < 0.05$) with treatment A, B and D, but treatment A and D were not significantly different ($P > 0.05$).

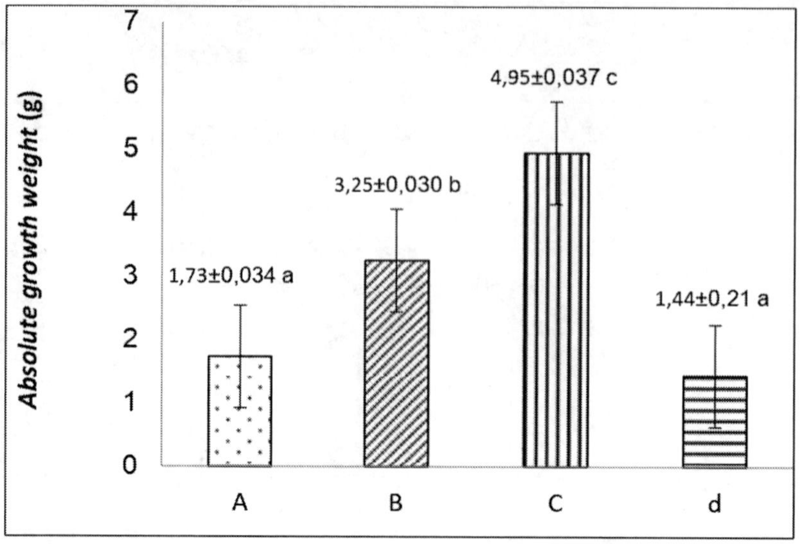

Fig 3. Absolute growth weight of *Trichopodus pectoralis*

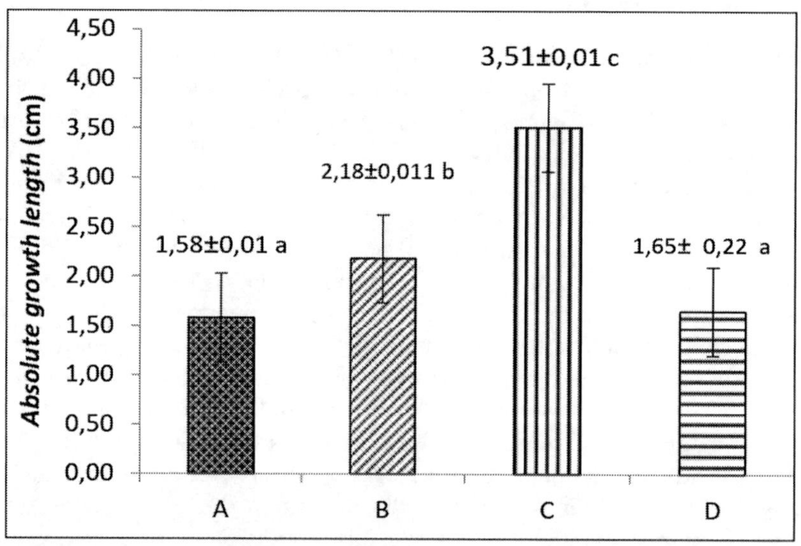

Fig 4. Absolute growth length of *Trichopodus pectoralis*

3.3 Blood figure

The osmotic gradient obtained at all treatments at the beginning of the study achieved at about 0.316 Osmol kg H_2O^{-1} and at the end of the study showed a decrease in treatments A, B and C, while an increase in gradient value was achieved in treatment D (Figure 5). Osmotic gradients and blood glucose are indicators of fish in comfortable conditions or not in the medium of life. Siamese fish seeds maintained on culture media with optimum temperature and salinity applications are in optimum or non-stressful conditions, whereas fish in stress conditions will cause physiological processes to disrupt and even cause death [12]. In the treatment of A, B and C, the Siamese Siamese fish live with sufficient energy. This is supported by the results of the measurement of dissolved oxygen and pH value as well as the results of biological index measurements such as survival rate, specific growth rate, weight gain and absolute length. The optimum application of temperature and salinity to the culture of fish seeds is shown by the effectiveness of its performance, where the activity of Siamese fish is influenced by the optimum temperature so that the use of energy obtained from feed consumption spur on the growth of fish. In addition, the improved performance of Siamese fish seed growth is also influenced by the optimum salinity application that is by feed utilization which affects the digestibility of feed.

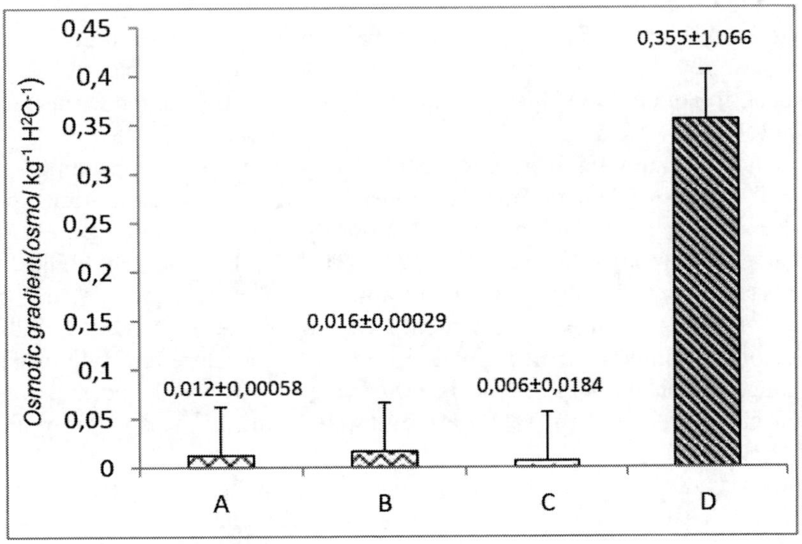

Fig 5. Osmotic gradient *of Trichopodus pectoralis*

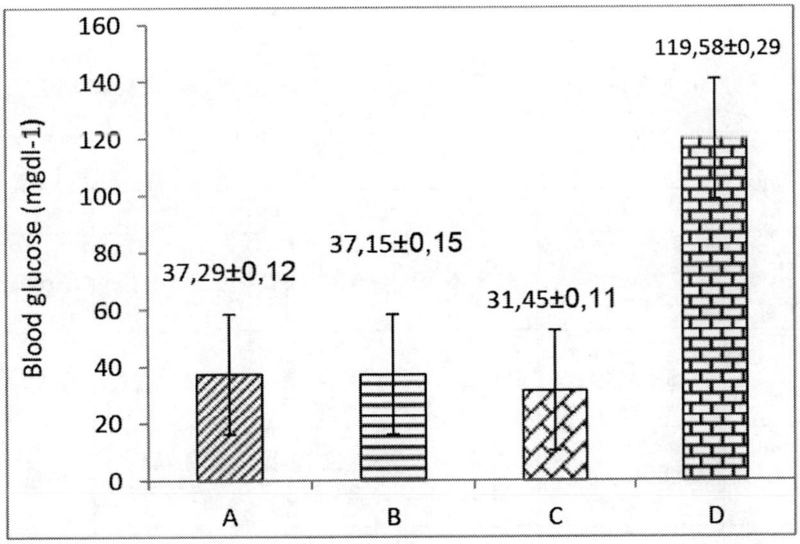

Fig 6. Blood Glucose *of Trichopodus pectoralis*

3.4 Water quality

The role of water quality parameters becomes one of the supporting factors as well as to increase the growth and survival of fish such as temperature [1]. The optimum temperature and salinity application have an effect on the endurance level of fish. In this research, the highest concentration of oxygen was found in the treatment of C, which was 1200 fish/container, followed by treatment of B (800 fish/container), treatment A (400 fish/container), and D treatment (1600 fish/container) (Table 2). The results of the temperature study on the performance of the growth and survival of Siamese fish seeds showed that the dissolved oxygen content obtained was 4.97 - 5.19 at 28°C better than at temperatures of 26°C, 30°C and 32°C [11]. The pH value at treatment A (400 fish/container), treatment B (800 fish/container) and C (1200 fish/container) ranged from 7.04 - 7.51 (Table 2).

The amount of ammonia excretion stimulated from feed consumption, ammonia excretion will increase when the media temperature to the optimum limit and ammonia excretion decreases when the temperature is more than the optimum limit. At treatment C was reached the ammonia range of 0.0012-0.0213 mg L^{-1} and the highest range was achieved at treatment D of 0.1541-0.269 mg L^{-1}. The data in Table 2 shows the physiological processes of each treatment, which in the treatment C illustrates the convenience of fish on culture media with optimum temperature and salinity applications. While the range of fish ammonia values is 2.0-6.0 mg $L-^{1}$ [4] and the normal concentrations of ammonia in uncontaminated waters are below 0.25 mg L^{-1} and in the polluted waters concentrate below 1,0 mg L^{-1} [2].

Alkalinity is often referred to as a quantity showing the buffer capacity of the bicarbonate ion (HCO_3^-), at a particular stage to carbonate ions (CO_3^{2-}) and hydroxide (OH^-) in water. The value of alkalinity during culture ranged from 35.16 to 116.02 mg L^{-1} $CaCO_3$ (Table 2) and the value was still within the range that could support the survival of Siamese fish seeds. The ideal alkalinity range for pond water culture is 20-120 mg L^{-1} $CaCO_3$ [3].

Table 2. The value of water quality on culture media of Siamese fish

Parameters	Treatment				Reference
	A	B	C	D	
Dissolved oxygen (mg L^{-1})	3,03- 5,67	3,26-6,59	5,92- 6,94	2,96-4,23	23-28 (Froese&Pauly, 2014)
pH	5,74-6,02	7,07-7,51	7,12-7,84	5,65-6,21	6,0-8,3 (Froese&Pauly, 2014)
Ammonia (mg L^{-1})	0,009-0,0143	0,013-0,0245	0,0012-0,0213	0,1541-0,269	2,0-6,0 (Froese&Pauly, 2014)
Alkalinity (mg L^{-1} $CaCO_3$)	20,16-40,74	35,14-67,52	35,34-62,36	80,21-116,62	(Boyd & Tucker, 1998)

4 Conclusions

The application of temperature 28°C and salinity of 3 ppm gives the optimum value range for the growth of Siamese fish seeds in the stocking density of 1200 fish/container. Optimum growth is supported by optimum values on water quality parameters and blood image.

Our thanks go to Research Institute for Freshwater Aquaculture and Fisheries Extention, Bogor, Indonesia which has funded this research through APBN of Ministry of Marine Affairs and Fisheries of the Republic of Indonesia 2016.

References

1. F.J. Arjona, L.V. Chacoff, I.R. Jarabo, O. Gonçalves, I. Páscoa, P. María, M.D. Río, J.M. Mancera. Aqua. **87**. (2009)
2. C.E. Boyd. Water quality. doi:10.1007/978-3-319-17446-4. (2015)
3. C.E. Boyd, C.S. Tucker. doi:10.1007/978-1-4615-5407-3. (1998)
4. R. Froese, D. Pauly. Fish Base. www.fishbase.org. (2014)
5. S. Goddard. *Feed Management in Intensive Aquaculture*. **174** (1996)
6. S. Kucuk. Af. J. Biotech. **12**, 5 (2013)
7. D.M. Kelabora. Fish.Terrest.J. **38**, 10 (2010)
8. N.S. Madinawati, Yoel. Media Litbang Sulteng. **4**, 5 (2011)
9. L. Setijaningsih, I. Taufik, T. Prihadi, D. Radona, A. Saputra. Aq. J. Ind. (to be published).
10. D. Oktafia, R.M. Putra, D. Efizon. Reproductive Biology of *Trichogaster pectoralis* From Flood Plane captured in the Tangkerang Barat District and Delima District. (2014)
11. L. Setijaningsih. *Prosiding Simposium Nasional Ikan dan Perikanan*. 9 (2017)
12. M. M. Porchas, L. R. M. Cordova, R. R. Enriquez. Pan-Amer. J. Aq. Sci. **4**, 20 (2009)
13. D. Rachmawati, J. Hutabarat, S. Anggora. Marine Science. **17**, 7 (2012)
14. Siswanto, U. Bijaksana, I. Fitriliyani. J. Fish. Sci. **6**, 2 (2016)
15. P.M. Schulte. Ency. Fish. Phys. Volume 3. *Energetics, interactions with the Environment, Lifestyles, and Applications*. Inggris (GB): Elsevier. 6 (2011)
16. N. Zonneveld, E.A. Huisman, J.H. Boon. *The Principles of Fish Cultivation*. 318 (1991)

PRELIMINARY STUDY OF HEAVY METAL (Zn, Pb, Cr, As, Cu, Cd) CONTAMINATIONS ON DIFFERENT SOIL LEVEL FROM POST-MINING BAUXITE PRODUCTION FOR AQUACULTURE

Risandi Dwirama Putra[1*], *Tri* Apriadi[2], *Ani* Suryanti[3], *Henky* Irawan[4], *Tengku* Said Raja'I[4], *Try* Yulianto[4], *Wiwin* Kusuma Atmadja[4] and *Chandra* Joei Koenawan[1]

[1] Marine Science Departement, Faculty of Marine Science and Fisheries, Raja Ali Haji Maritime University, Jl. Politeknik, Senggarang 29111, Tanjungpinang, Indonesia
[2] Aquatic Resources Management Departement, Faculty of Marine Science and Fisheries, Raja Ali Haji Maritime University, Jl. Politeknik, Senggarang 29111, Tanjungpinang, Indonesia
[3] Center/Division/Laboratory, Institution, City with Postal Code, Country
[3] Sosio-Economic Departement, Faculty of Marine Science and Fisheries, Raja Ali Haji Maritime University, Jl. Politeknik, Senggarang 29111, Tanjungpinang, Indonesia
[4] Aquaculture Departement, Faculty of Marine Science and Fisheries, Raja Ali Haji Maritime University, Jl. Politeknik, Senggarang 29111, Tanjungpinang, Indonesia

Abstract Bauxite production and exploration give negative impact environmental modification and increasing concern pollution of heavy metals. One of an alternative to reduce the negative impact of bauxite post-mining production is by re-utilizing the abandoned bauxite land and necessary exploratory study to obtain the current environmental conditions primarily related to heavy metals in bauxite mining area. Soil and water quality samples from 5 (five) sites based on post-mining activities were used to determine concentration heavy metals of As, Cd, Pb, Zn, Cr, Cu from different soil level. They were also collected for grain size analysis including those of sand and mud percentages along with the mean, and sorting analysis. The mean grain size post-mining bauxite ranged between very fine sand to medium sand and had heterogeneous texture. The coarse grain size percentage increased towards the bottom of the soil. Total heavy metal contents for post-mining bauxite soils are 0,081 ppm, 0,245 to 0,471 ppm and 0,007 ppm for As, Pb and Cd respectively. Heavy metals for Cd, uniformly at every depth from soil level. On the other hand, Pb showed significance pattern, it was indicated every depth from soil level, the concentration of Pb was different.

*Corresponding author, e-mail: yanti.ajb@umrah.ac.id

1. INTRODUCTION

Many pollutants are accumulated in the sediments and have harmful effects on the aquatic environments [1]. The sediments of the aquatic environment act as a major reservoir of heavy metals and source of contaminants [2]. Heavy metals are natural constituents of sediment and their concentration varies depending on parental materials [3]. Heavy metals can be present in the soil as a product of the weathering of the natural rocks, or because they come as part of pollution loads generated by human activities [3].

In many cases of miner exploration, the action of the water over tailings and other miner wastes causes a phenomenon known as acid mine drainage (AMD) [4]. The activities developed by bauxite mining are, from the very start of operations, capable of degrading the surrounding environment and causing major environmental modification [4]. Several areas which are former bauxite mining are damaged and give negative impact environmental modification. The process of dredging bauxite in some mining sites resulting in the emergence of puddles as well as leaving a barren and unproductive land.
Some research related to bauxite mining has been studied by researcher and almost researcher generally investigated on the impact of bauxite tailing/red mud. Inappropriate storage of red mud, a by-product of alumina production, is of increasing concern, as illustrated by a major pollution incident [5].

The average metal concentration from the largest to the smallest was Fe, Ni, Mn, Zn, Pb, Cu, Cd, and Cr. The value from results of metal concentration in bauxite sample was still below from the specified standard quality threshold. Bauxite has a variety of important minerals, including iron, silica, titanium, calcium, and magnesium. Further complementary minerals contained in bauxite were Na, K, P, Cr, V, Ga, Zr, Zn, Pb, Cu, Ni, Mn and Co [6] It is further explained that the main chemical content of red mud from bauxite tailing consists of Fe_2O_3 (48.50%), SiO_2 (11.53%) and Al_2O_3 (14.14%) and additives of TiO_2 (5.42%) and V_2O_5 (up to 0.116%). Zr element has the largest proportion element that contained in bauxite samples ranging from 460-1790 ppm and abundant element found in next is V element with a range of 210-416 ppm then Cr with a range of 108-326 ppm. Ni, Cu, Zn, Rb, Sr, Y, Nb, Ce, La, and Pb are found in small amounts (<100 ppm) [7]. Inappropriate storage of red mud, a by-product of alumina production, is of increasing concern, as illustrated by a major pollution incident[6]. Red mud can also contain problematic concentrations of potentially toxic metals and metalloids, including As, Cr, Ni, Pb, Mo and V (Brunori et al., 2005; Grafe et al., 2011; Klebercz et al., 2012) and contains several potentially toxic elements, including arsenic [8].

The most common characteristic of exploration and production mining on bauxite give the impact destruction of soil which allows the erosion and causing the topsoil to become thinner or even lost and bauxite extraction clearly disturbs the upper regolith and eliminates any indurated layer present [9]. The former or primary soil of bauxite mines usually dense and difficult to process because of the structure, texture, porosity, and bulk density of primary soil that affects root system development and disrupts plant growth [10].

One of an alternative to reduce the negative impact of bauxite mining production is by re-utilizing the abandoned bauxite land. Excavations of bauxite mining will generally be filled by water (generally rainwater) that formed a puddle of water. Puddles on the former bauxite excavations are thought to have the potential for fish farming. Potential land for bauxite excavation for fish cultivation has not been studied before Therefore, it is necessary exploratory study to obtain the current environmental conditions primarily related to heavy metals in bauxite mining area and concentration of macro and micro minerals concentrate that function as nutrients in the process of fish farming. Based on information on the concentration of minerals and related metals contained in the bauxite land will be known potential development of future utilization of bauxite land.

2 MATERIALS AND METHODS

2.1 Sampling and sediment collection

Each sampling location was determined using a GPS and located from 5 (five) sites based on post-mining activities (Table 1) were used to determine concentration heavy metals of As, Cd, Pb, Zn, Cu, Cr from different soil level (0-30 cm), (30-60 cm) and (60-100 cm). They were also collected for grain size analysis including those of sand and mud percentages along with the mean, and sorting analysis [11]. The concentration of total heavy metals was measured using Flame-Atomic Absorption Spectrophotometer using AA-6200 Shimadzu, 10 mA currents by adjusting the specified wavelengths; acetylene air gas 21/min; O_2 15 L /min, temperature 2000°Cc.

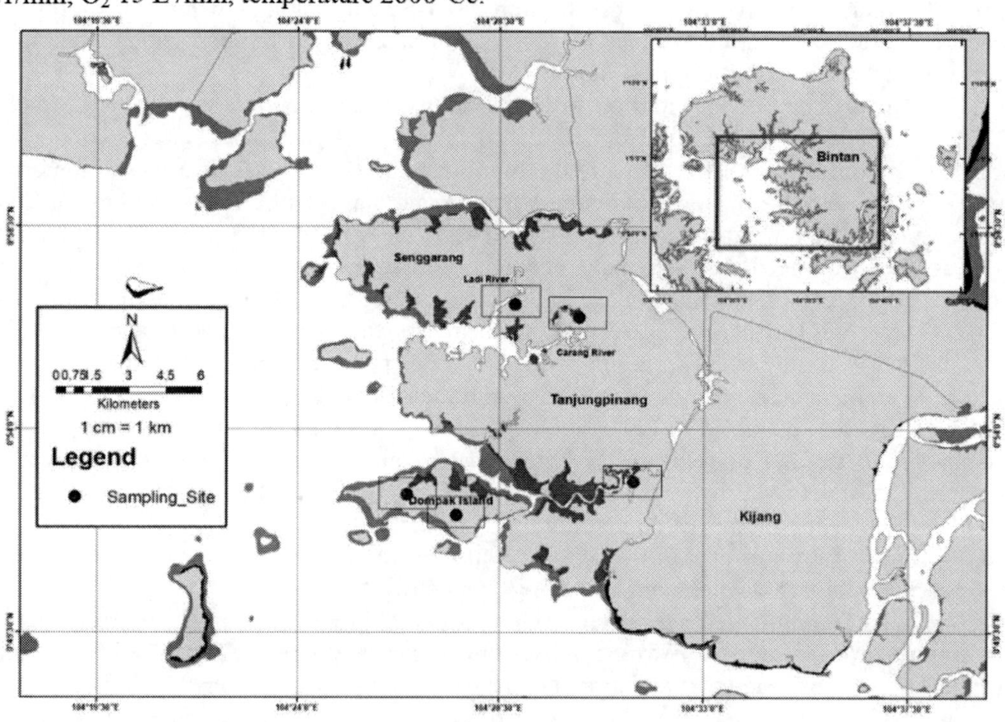

Fig 1. Map of Study area

Table 1. Geographic locations of sampling site and data of collecting cores samples

Sampling Site	Location		Soil depth (cm)	Water Depth (cm)
	Latitude	Longitude		
Sei. Ladi	0.94636	104.4807	100	37
Sei. Carang	0.94169	104.5041	100	28
Simpang Dompak	0.87582	104.4406	100	10
Dompak	0.86830	104.4591	100	21
Wacopek	0.88071	104.5240	100	8

2.2 Sediment core and water quality parameter instrument

15 sediment cores for the laboratory experiment were collected within an area of post-mining bauxite production. The cores had an internal diameter 65 mm and consisted of 100 cm depth of sediment [5]. DO concentration. conductivity. pH and water temperature were measured in *situ* at the sample site and in the collected sediment cores using a SensoDirect 150 Lovibond® Portable Multi-Parameter Meter and Probes.

2.3 Dilution of Sediment for trace metals analysis

Chemical Acid (70% HNO_3). Hydrofluoric acid (HF) and deionized water were used for dilution. Blank samples and quality control standard for trace metals analysis [12].

2.4. Sample Preparation

Sediment samples were prepared according to an adapted USEPA 3050B method to determine the total heavy metals for Cr. Mn. Fe. Cu. Zn. and Pb in core sediments [35]. The method was altered by using approximately 0.20 g of the sample as described by [13]. Briefly. Samples were dried in an oven at 105°C and ground into a powder using a mortar and pestle. Nitric acid (70% HNO3) and hydrofluoric acid (HF) were used for the total metal dissolution. Sediment samples were digested in heavy acid with 3:1 mixture of HNO3 and HF acid. Nine ml of HNO_3 and three ml of HF were added to the sediment samples inside a fume hood. Sediment samples plus HNO3 and HF acid were digested at 100°C on a hot plate until dryness [12].

2.5 Enrichment Factor (EF)

The Enrichment Factor (EF) in metals and Geo accumulation Index (I_{geo}) are indicators used to assess the presence and intensity of anthropogenic contaminant deposition on surface soil [3]. The EF calculation seeks to reduce the metal variability associated with variations in mud/sand ratios and is a convenient tool for plotting geochemical trends across large geographic areas. Which may have substantial variations in the mud (i.e. clay-rich) to sand ratios [14]. For assess the possibility of anthropogenic influence in the study area. EF was calculated for the investigated three total heavy metals. The essential reason for using geochemical studies is to determine differences of metals from anthropogenic activities and those from natural sources [12]. The EF method normalizes the measured heavy metal content with respect to a sample reference metal using Fe based on [14]. Fe as an acceptable normalization element to be used in the calculation of the enrichment factor since they considered the Fe distribution was not related to other heavy metals. The EF is calculated according to the following equation:

$$EF = M_x \text{ x } Fe_y / M_b \text{ x } Fe_b \quad (1)$$

Where M_x and Fe_x are the sediment sample concentration of the heavy metal and Fe [14]. while M_b and Fe_b are their concentrations in a suitable background or baseline reference material [15]. Hence. for quality classification. EF results are applied to a qualitative scale of metal concentration based on [16] (*reference levels; <2 deficiency to minimal enrichment ; 2-5 moderated enrichment; 5-20 significant enrichment.; 20-40 very high enrichment ;> 40 extremely high enrichment*).

$$I_{geo} = \log_2 C_n / 1.5 B_n \quad (2)$$

Where Cn is the concentration of the element in the enriched samples. and the Bn is the background or pristine value of the element. Factor 1.5 is introduced to minimize the effect of possible variations in the background values, which may be attributed to lithologic variation in the sediments [1,14]

Hence. for quality classification. I_{geo} results are applied to a qualitative scale of enrichment intensity based on [17] (*reference levels; 0-1 uncontaminated; 1-2 unpolluted to moderately contaminated; 2-3 moderately contaminated; 3-4 moderately to strongly contaminated; 4-5 strongly contaminated;>5 very strongly contaminated*) and. according to this scale. samples with I_{geo} above 1 show signs of enrichment/contamination. [1,2,18]

2.6 Statistical Analysis

The statistical analysis measurements using Bivariate-Plot sediment analysis to calculate mean grain size, degree of scatter of sorting, kurtosis and the degree of skewness as describe by [11, 19, 20, 21) using packages rysgran for R statistical analysis [22]. Sorting or standard deviation measures of post-mining bauxite sediment indicates the fluctuations in anthropogenic conditions of the tailing and excavating condition on bauxite mining. Graphical Anova to calculate variance based on post-mining bauxite sediment with One-way ANOVA analysis using packages granova for R statistical analysis [23], Effect of heavy metal contamination for aquaculture water quality measure using Principle Component Analysis (PCA) with multivariate data from pH, dissolved oxygen, temperature water column and all heavy metals tracer (Zn, Pb, Cr, As, Cu, Cd) based on confidence ellipse around the categories of site location [35]. Linear regression analysis from heavy metal tracer calculated using CCM (Coefficient Correlation Matrix) based on person method every sediment layer.

3 RESULTS AND DISCUSSION

3.1 Textural characteristics of sediment

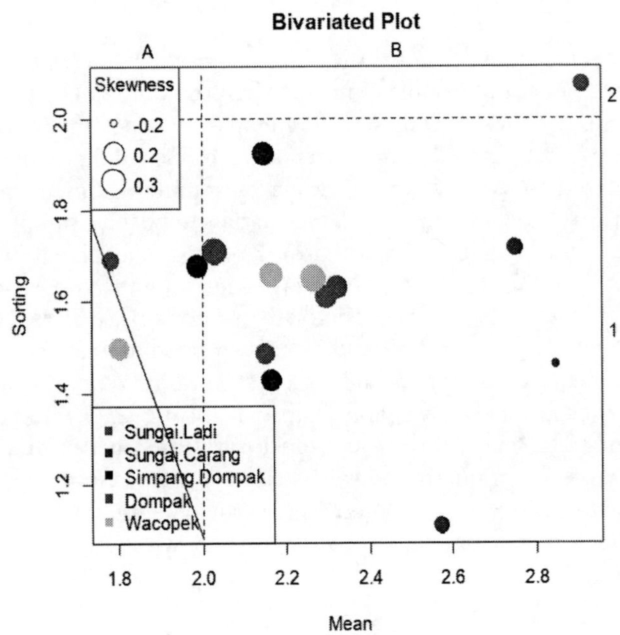

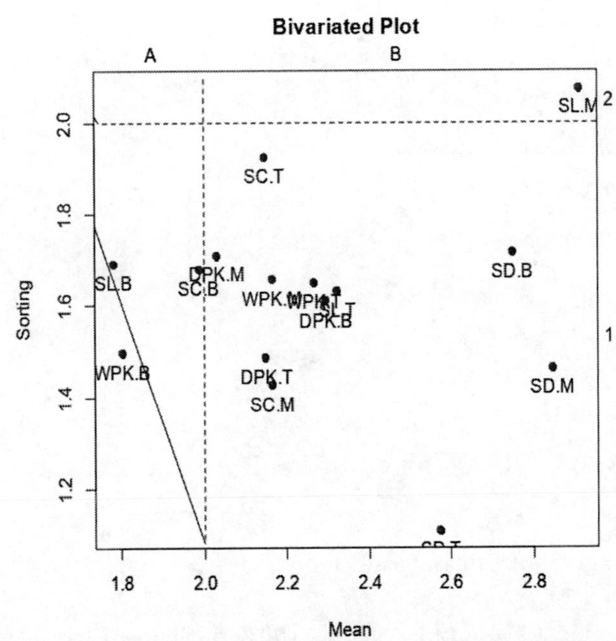

Description

Mean	Verbal.Mean	Sorting	Verbal.Sorting
A	Medium sand	1	Poorly sorted
B	Fine sand	2	Very poorly sorted

Fig.2. Bivariate-Plot of sediment between Mean Size and Sorting Post-mine Bauxite Bintan Island.

The statically analysis reported cumulative probability curve provide skewness, kurtosis, sorting and mean size for each sediment sample (Fig. 2). The post-mining bauxite production in Bintan Island show textural facies of sediment dominated by sand (medium sand and fine Sand). Overall, most of every layer sediment enriched with fine sand and has relatively high sorting values categorized by poorly sorted. Some of sample sediment in bottom layer with medium sand (Sungai Ladi, Wacopek and Sei.Carang) and one sample categorizes in very poorly sorted (medium layer in Sei Ladi site). Kurtosis analysis show that all sample are better sorted at the tail than the central portions (approximately symmetrical), hence the mesokurtic curves. The sample obtained in surface layer show variation in the mean grain size between ϕ 2.14 to ϕ 2.57 and are categorized as fine sand. The sediment is poorly sorted, with standard deviation as high as 1.93. Bottom layer on bed material sediment on the hand has larger grain size, varying from ϕ 1.78 to ϕ 2.75 falling categorized medium sand to find sand. The mean size in bottom layer from post-mining bauxite sediment show consistent decreasing grain size toward sediment layer. Grain size distribution of post-bauxite sediment are generally associated with vertical sediment layer. In view of this, the small particles of sediment dominated in surface layer while the larger particles concentration in bottom layer.

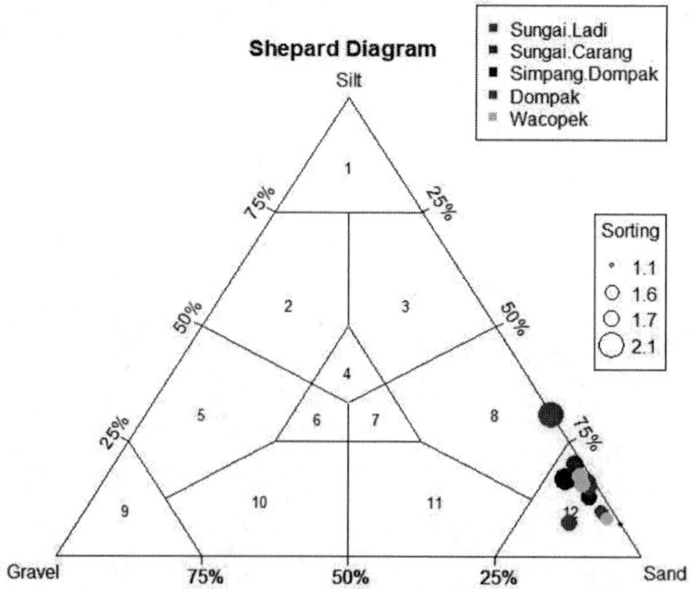

Fig 3. Shepard post-mining Bauxite diagram of Bintan Island.

The sum of these three basic component should be equal to 100% in Shepard triangle classification system. In study are (Fig 3), two kinds of sediment types were found, they were dominated categorized sand and only one site sampling categorized sandy silt which was in accordance with field survey result. The average percentage of sand was 89.92% and Silt was 14.97%.

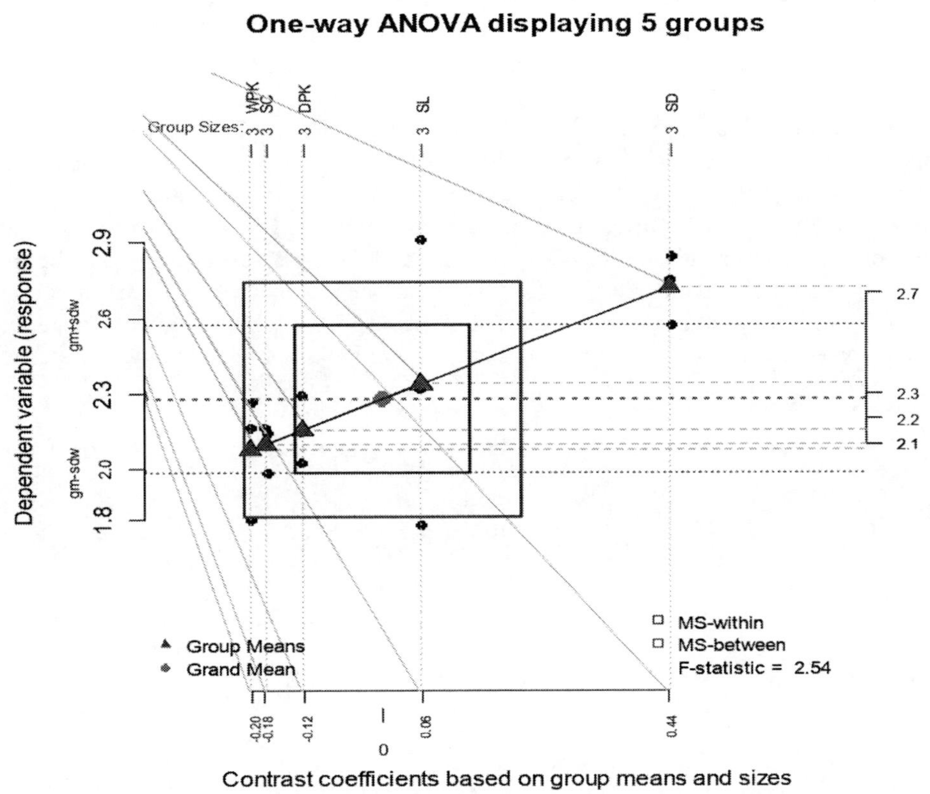

Fig 4. Graphical Anova post-mining Bauxite soil diagram of Bintan Island.

Comparison of 5 site sediment sampling in post-mining bauxite production, summary measure of mean ϕ and percent sediment resulted in same conclusion, there were sand category sediment. The estimated mean ϕ was equivalent across sediment layer for all sample (Fig 4). In addition, estimates of percent grain size ϕ non-significantly different among site and sediment layer (f-test 2.54 and p-value <0.05). The measurement variance non-differed over the range of half-ϕ intervals for all sediment site and layer. Summary on variance analysis show the highest grand-mean of sediment grain size from Simpang Dompak Site with mean ϕ value 2.7. In other hand, the lowest grand-mean of sediment grain size from Wacopek site with mean ϕ value 2.1. The highest and lowest contrast coefficient based on site group mean and size was 0.44 and 0.20 respectively.

3.2. Enrichment Factor (EF)

The result of Enrichment Factor of Heavy metals in Post-mining Bauxite from different soil level are represented in **Fig.5**. In the present study, Fe has been used as a conservative tracer to differentiate natural from the anthropogenic component. The results are indicated that the EF of Zn ranged from 6.93 to 42.51, which is significant to very high enrichment, respectively. The EF for Cd ranged between 6.15 and 6.26, which is significant enrichment. The EF of Pb varied from 32.55 to 63.25, which is very high enrichment to extremely high enrichment. The EF of Cu varied from 2.25 to 14.49, from 0.53 to 16.03, and from 107.61 to 206.29 for As. These result indicate that the EF for all of the studied metals is in moderated enrichment to significant enrichment except for Arsenic (As), which ranged extremely high enrichment. These heavy metals are classified in the sediment of post-mining bauxite by their source into a natural source (<1.5) and anthropogenic source (>1.5). Most of these metals come from the surrounding anthropogenic activities and exploration bauxite mining. The sequence of EF for heavy metals in the sediment of post-mining bauxite is in the following order: As > Pb > Zn > Cr > Cu > Cd. This indicates that arsenic was more abundant when compared with the others metals, where cadmium had the lowest appearance. An average enrichment factor of different soil level post-mining bauxite are relatively similar for surface level (0-30) cm, middle level (30-60) cm, and bottom level (60-100) cm. The obtained result of EF calculations showed Zn, Cd, Mn, Cu, and Cr to have EF values lowest in middle level of soil except for Pb which is lowest in surface level. Almost average enrichment factor from heavy metal had the highest value in bottom level.

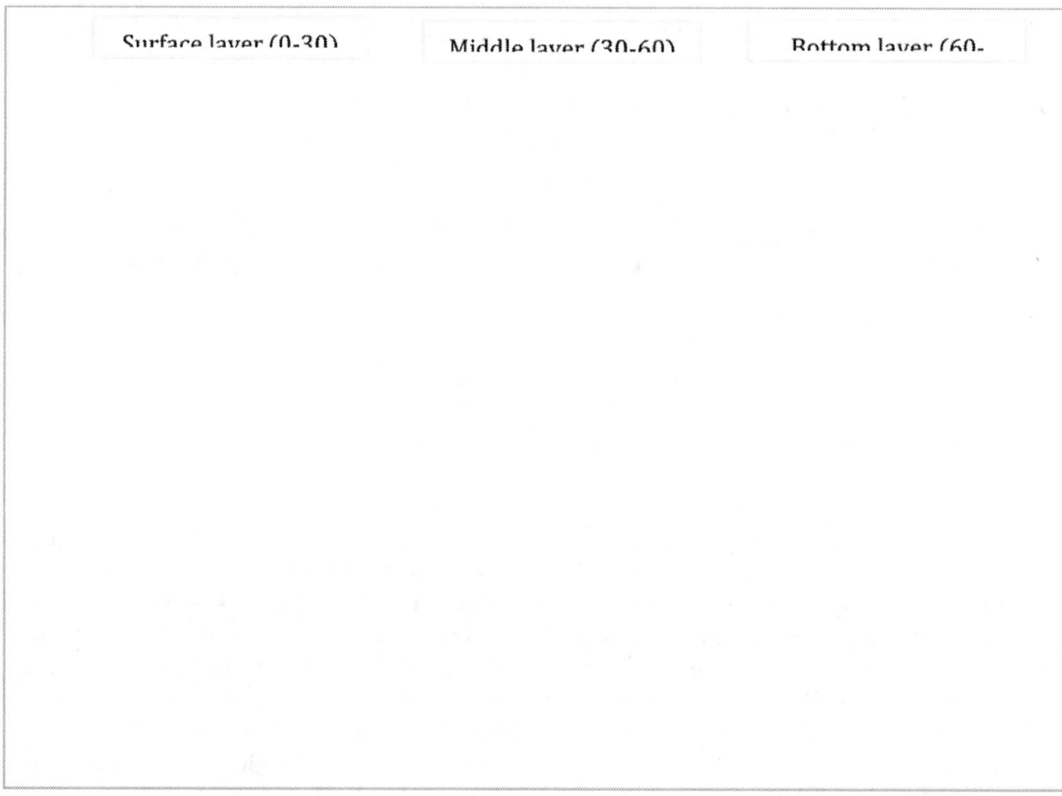

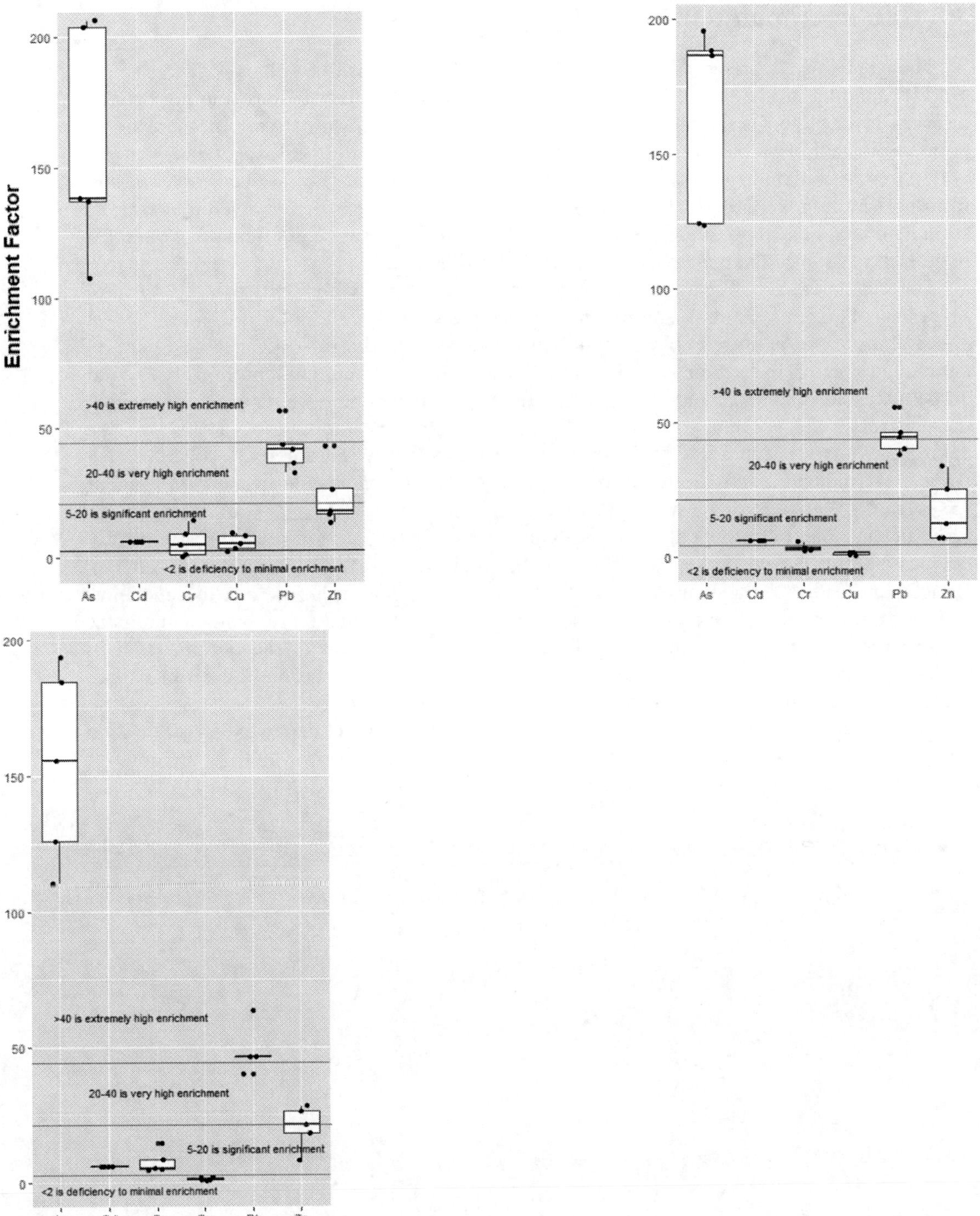

Fig 5. Enrichment Factor of heavy metal on post-mining bauxite from different soil level.

3.3 Geo-accumulation index (I_{geo})

The results of I_{geo} are as shown illustrated in Fig. 6. The positive values of heavy metal > 1.5 depending on the classification based on [17] indicated that the post-mining bauxite in every level of soil is polluted with this metal and used to quantitatively measure metal pollution in sediments based on a pollution intensity classification [16]. I_{geo} values showed very strongly contaminated pollution with arsenic (As) and lead (Pb) at every level of soil from post-mining bauxite production. For zinc (Zn), the values of I_{geo} showed moderate to strongly contaminated pollution degree. On the basis of above said classification, Cr and Cu in both showed uncontaminated to moderately pollution load.

The characteristic of sediment in bauxite mining is by having a red color [24,25,26] Presence of arsenic natural source in the sediment with red color provides lowest concentration of As, where black sediment produces mostly As. [27]. The result of enrichment factor of arsenic which ranged extremely high enrichment EF (107.61 to 206.29) was described as a mining exploration which contains huge amounts of wastes that may increase the amounts of Arsenic [2]. The high concentration of arsenic in the environment is suspected due to bauxite mining activities. High arsenic levels can also come from industrial activities such mining also contribute to arsenic in the environment [28]. The total content of Pb was indicative of enrichment at every level soil. Enrichment Pb indicating that the sediments were polluted [29] and allegedly originated consequently may cause significant ecological effect on the sediment surface (Desert et al., 2017). Enrichment lead (Pb) on sediment indicates the anthropogenic influence due to deposition and absorption by dredging on tailing process and disposal gas and oil on heavy equipment bauxite mining [30]. The enrichment of Zn appear to be present in the bottom level in amounts, which is explained by the high adsorption capacity of ground because Zn had correlation mainly in smaller grain size [31].

In summary, concentrations of Zn, Pb and As in sediment depended largely on the distance away from the mining area, or more specifically, the farther away from the mining area, the lower the concentrations of heavy metals [32].

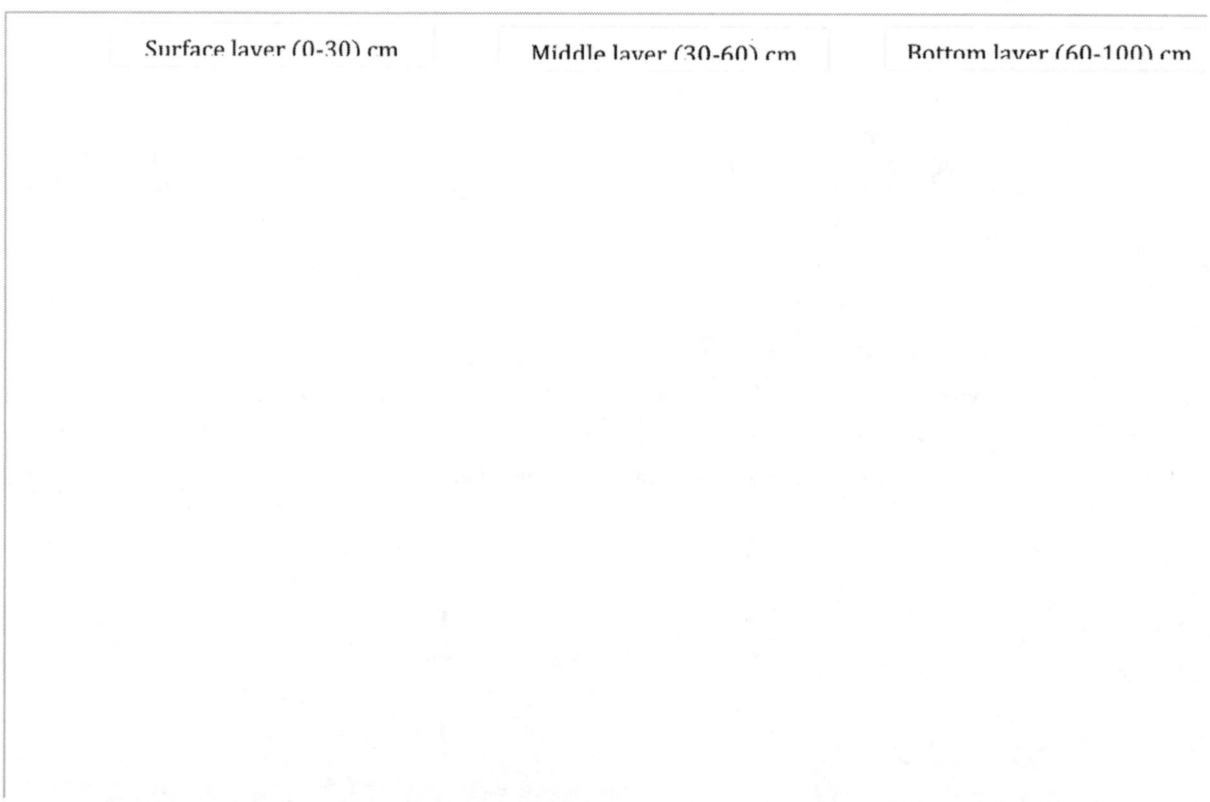

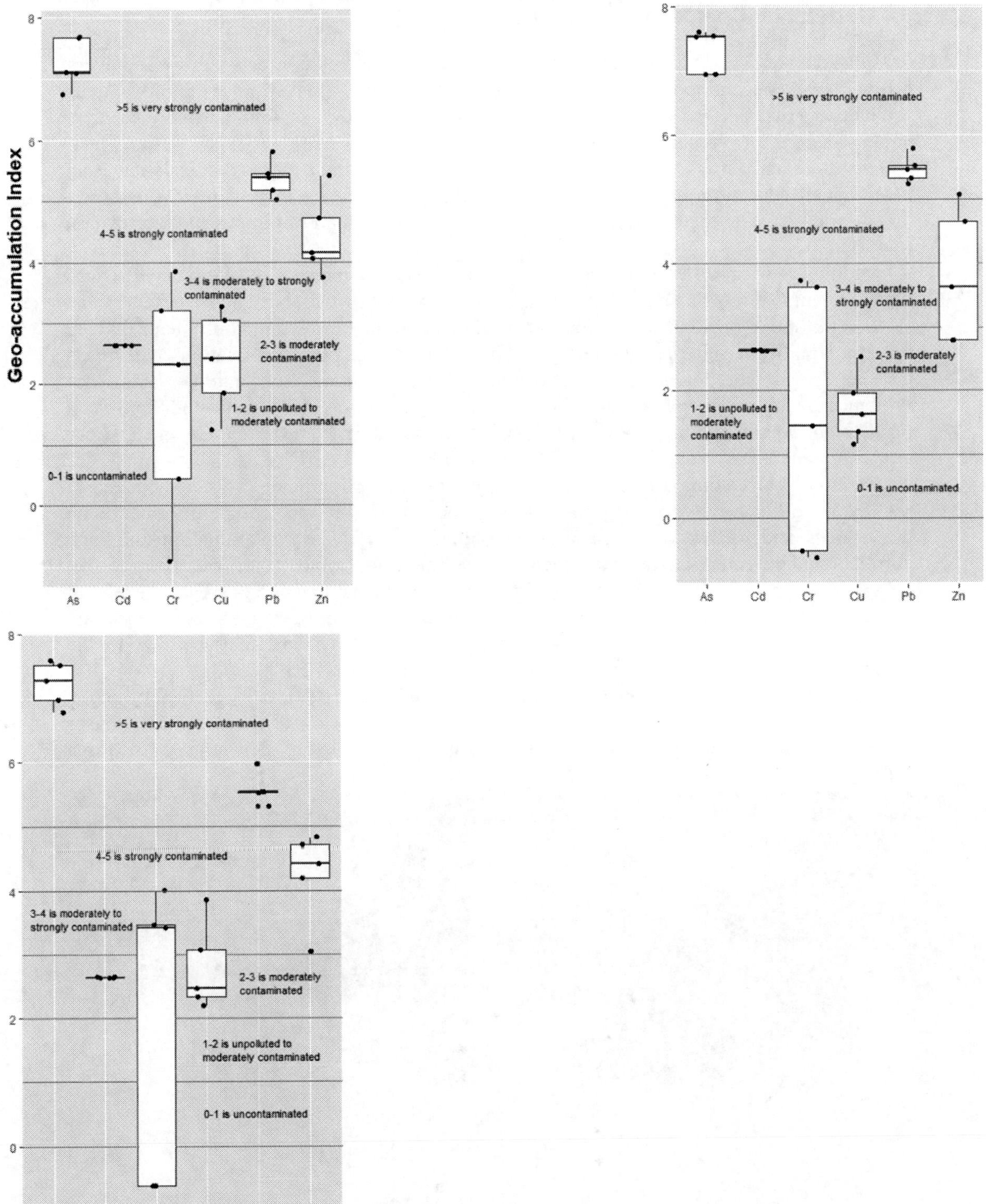

Fig 6. Geo-accumulation Index (Igeo) of heavy metal on post-mining bauxite from different soil level

3.4 Correlation Coefficient Matrix (r^2)

Linear regression analysis showed the strong positive correlation between Fe and every heavy metal concentration (Cr, Cu, As, Pb, Zn) on surface layer post-mining bauxite production (**Fig.7**). The strong significant positive correlation is found in surface layer among Fe and Cr (r^2 = 0.906), Fe and Cu (r^2 = 0.853), Fe and As (r^2 = 0.831), Fe and Pb (r^2 = 0.691), Fe and Zn (r^2 = 0.798). Furthermore the significantly positive correlation from heavy metal concentration in surface layer of Cr (r^2 = 0.803), Cu (r^2 = 0.892), Pb (r^2 = 0.736) with Zn. The toxic pollutant Pb and As have significantly high positive correlation (r^2 = 0.827).

Middle layer had different pattern from surface layer (0-30 cm). Correlation coefficients between pairs of metals show that Cr and As (r^2 = 0.995), Cu and Cd (r^2 = 0.992) that showed the strong positive correlation, others showed of the correlation coefficients that are not at a significant value in middle layer. The values of Pearson´s correlation coefficients that confirmed that bottom layer almost have similar pattern from middle layer. The influence of anthropogenic activities from bauxite production result indicated significant positive correlation among Fe and Cr (r^2 = 0.923), As and Cr (r^2 = 0.772), Cu and Zn (r^2 = 0.951), Pb and Fe (r^2 = 0.823).

Pearson's correlation (PC) matrix for analyzed sediment parameters was calculated to see if some of the heavy metal interrelated with each other [33] and provide information on the sources and pathways of heavy metal pollution [34]. The contamination and high correlations between specific heavy metals in sediment in surface layer, this showed that heavy metal pollutants were most likely derived from same source similar levels of contamination and/or release from the same sources of pollution, mutual dependence and identical behavior during their accumulation in sediment system, may be due to the contamination from production and tailing bauxite production. [12,33].

The results of correlation matrix of each core sample indicate that a significant fraction of the trace metal is found adsorbed on to Fe (surface level), geochemical phases controlling the trace metals in sediments which may be attributed to their large surface area [16]

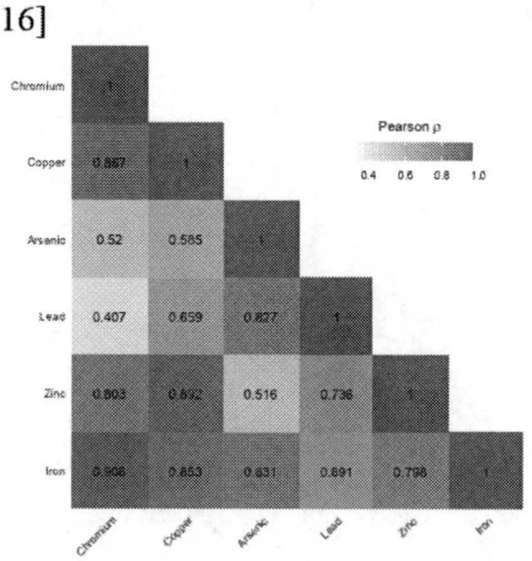

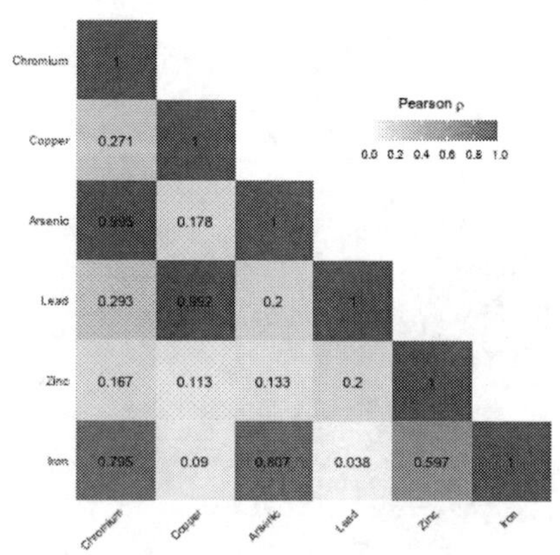

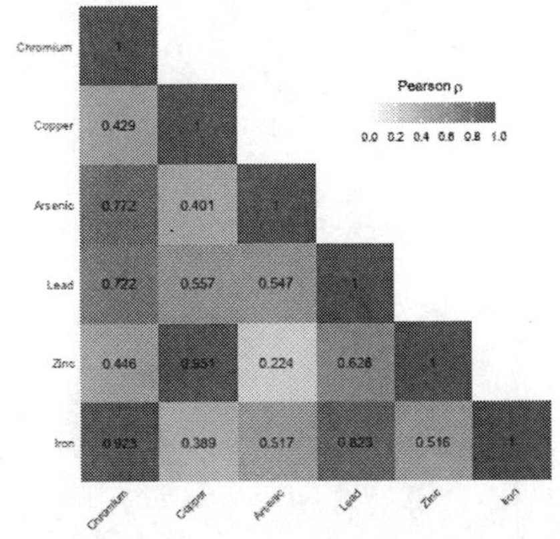

Fig 7. CCM (r^2) for heavy metal concentration in three level core sediment (n=105) from post-bauxite mining

3.5 Effect of heavy metal contamination for aquaculture water quality

Tabel 2. Water quality parameters of post-mining bauxite Bintan Island

Sites	pH	Dissolved Oxygen	Temperature
Sei.Ladi	4.94 ± 0.006	6.43 ± 0.208	30.2 ± 0.115
Sei.Carang	5.56 ± 0.027	6.37 ± 0.513	30.4 ± 0.231
Dompak	4.45 ± 0.058	7.33 ± 0.058	29.7 ± 0.001
Simpang Dompak	6.36 ± 0.046	6.37 ± 0.058	29.1 ± 0.001
Wacopek	6.13 ± 0.057	5.93 ± 0.058	28.7 ± 0.058

The physicochemical parameters of the water column such as pH, temperature and dissolved oxygen are presented in Table 2. The physicochemical parameters are very important because they have significant effect on the water quality [33]. Furthermore, aquatic life also suffer due to degradation of water quality. Among the external factors temperature, pH and dissolved oxygen are important factors which influence the aquaculture ecology.

The average values pH water column post-mining bauxite were ranged from 4.45 with SD ±0.058 to 6.36 with SD ±0.046. Dissolved oxygen (DO) refers to the oxygen gas that is dissolved in the water and made available to aquaculture. The average dissolved oxygen was 6.43 with SD ±0.208 to 7.33 with SD ±0.058. Temperature was measured in water column and the values average of temperature were ranged 28.7 with SD ±0.058 to 30.4 with SD ±0.231.

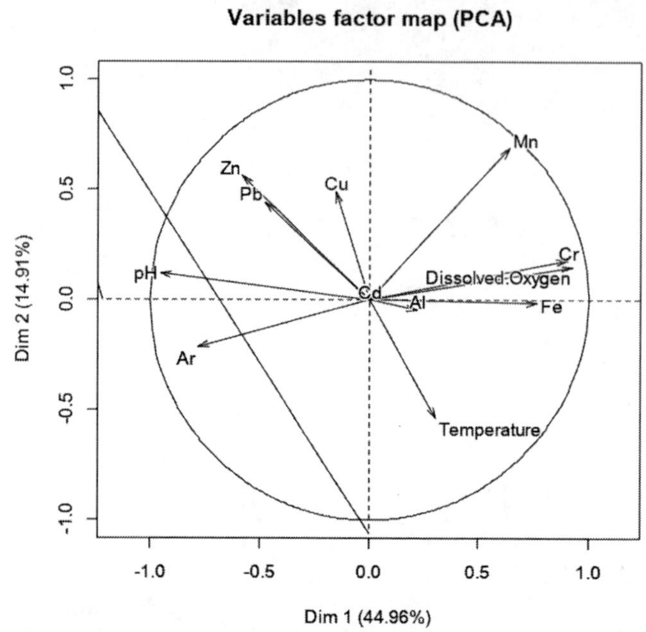

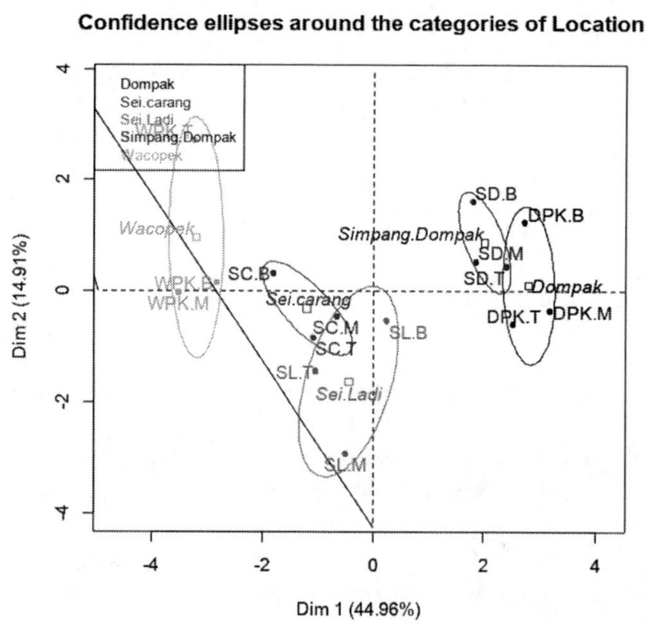

Fig 8. Principle Component Analysis of Effect of heavy metal contamination for water quality post-mining bauxite (a) and confidence ellipse correlation matrix around the categories of site location post-mining bauxite. In both figures the factor loadings of the PCA-axes 1 and 2 are shown, which explain both figure have 44.96% and 14.91% of the variation. Data standardized to mean zero and unit deviance prior to analysis.

The first two axes of the PCA explain 59.87% of the variation between effects of heavy metal contamination for water quality in post-mining bauxite. Dissolved oxygen (DO) were clustered along the first axis with Manganese (Mn) and Chromium (Cr). Temperature was clustered with iron (Fe) and Aluminum (Al) since their arrow on the x-axis pointed in the same direction (Fig 8a). pH was clustered with copper (Cu), Zinc (Zn) and lead (Pb) and was negatively correlated with temperature water column and their exact opposite scores on the x-axis. Dissolved oxygen (DO), Mn and Cr varied independently from temperature and pH, since their lines in the PCA plot are positioned perpendicular to each other (Fig. 8a.) site clustering show similar condition water quality parameter and heavy metal concentration with Dompak and Simpang Dompak in same clustering with intersect on confidence ellipse around (Fig 8b.). Sei Carang and Sei Ladi intersect on confidence ellipse around that show their water quality parameter and heavy metal concentration have similar condition but not significant in the path analysis. Wacopek have different condition from other station

4 CONCLUSIONS

Free-ranging herbivores of interest in study were fishes in the family Acanthuridae (Surgeonfish). Scaridae (parrotfishes). and Siganidae (Rabbitfishes) was found 41 species herbivore fish; Scaridae (25 Species). Siganidae (10 Species) and Acanthuridae (six species). The highest abundance and biomass of herbivorous fish were site RS15 at Penagi on east Natuna. The environmental condition in Natuna Island has good condition with the less of ecological stresses although in several site (at the western Natuna) have been function as destructive fishing area). The post-mining bauxite production in Bintan Island show textural facies of sediment dominated by sand (medium sand and fine Sand). Overall, most of every layer sediment enriched with fine sand and has relatively high sorting values categorized by poorly sorted. Kurtosis analysis show that all sample are better sorted at the tail than the central portions (approximately symmetrical), hence the mesokurtic curves. Grain size distribution of post-bauxite sediment are generally associated with vertical sediment layer. In view of this, the small particles of sediment dominated in surface layer while the larger particles concentration in bottom layer. The sequence of EF for heavy metals in the sediment of post-mining bauxite is in the following order: As > Pb > Zn > Cr > Cu > Cd. This indicates that arsenic was more abundant when compared with the others metals, where cadmium had the lowest appearance. Igeo values showed very strongly contaminated pollution with arsenic (As) and lead (Pb) at every level of soil from post-mining bauxite production. For zinc (Zn), the values of Igeo showed moderate to strongly contaminated pollution degree. On the basis of above said classification, Cr and Cu in both showed uncontaminated to moderately pollution load. The average values pH, dissolved oxygen and temperature water column post-mining bauxite were ranged from 4.45 with SD ±0.058 to 6.36 with SD ±0.046, 6.43 with SD ±0.208 to 7.33 with SD ±0.058 and 28.7 with SD ±0.058 to 30.4 with SD ±0.231 respectively.

The authors wish to thanks Dr. Syakti. A.D for reading manuscript and suggestion. We also thanks to (COREMAP – CTI – LIPI) Indonesia provide opportunities in conducting research activities and the anonymous reviewers for their time to provide us their constructive comments.

REFERENCES

1 J. Nilin, L. B. Moreira, J. E. Aguiar, R. Marins, D. Moledo de Souza Abessa, T. Monteiro da Cruz Lotufo, & L. V. Costa-Lotufo. Marine Env Res, **91**, 89–96. (2013)

2 Y. A. El-amier, A. A. Elnaggar, & M. A. El-alfy. Egyptian J. Basic & Applied Sciences. **4**, 55–66 (2017)

3 M. Barbieri. J Geol & Geophysics. **5**, 1–4 (2016)

4 V. M. G. Oliveira, L. F. T. de, Ribeiro, & M. C. R. da. Silva. Geofisica Internacional. **53**, 135–151 (2014)

5 J. P. Olszewska, K. V. Heal, I. J. Winfield, L. J. Eades, & B. M. Spears, Water Research. **123**, 569–577 (2017)

6 V. Cabllk. *The British Library " the World's Knowlege,"* 27–38 (2007)

7 A. O. Ayorinde. *Mineralogy and Geochemistry of the Bauxite Deposits (Cretaceous), Wilkinson County* (2011)

8 C. L. Lockwood, R. J. G. Mortimer, D. I. Stewart, W. M. Mayes, C. L. Peacock, D. A. Polya, I. T. Burke. Appli Geo. **51**, 268–277. (2014)

9 A. H. Grigg. J of Hydrology: Regional Studies. **12**, 150–164. (2017)

10 Sembirirng, S. Info Hutan. **5**, 123–134 (2008).

11 R. L. Folk, & W. C. Ward. J Sedimentary Petrology. **27**, 3–26 (1957)

12 B. A. Al-Mur, A. N. Quicksall, & A. M. A. Al-Ansari. Oceanologia. **59**, 262–270 (2017)

13 N. B. E. Badr, A. A. El-Fiky, A. R. Mostafa, & B. A. Al-Mur. Environmental Monitoring and Assessment. **155**, 509–526 (2009)

14 G. M. S. Abrahim, & R. J. Parker. Environ Monitoring and Assessment. 136, 227–238 (2008)

15 W. Solomons, & U. Forstner. *Metals in the Hydrocycle* (Springer, Berlin, 1984)

16 R. Kalaivanan, M. Jayaprakash, S. Nethaji, V. Arya, & L. 8, 1–10. (2017)

17 G. Muller. Geojournal. **2**, 108-118 (1969).

18 W. Desert, M. I. I. Mohamaden, M. K. Khalil, S. E. O. Draz, & A. Z. M. Hamoda. The Egyptian J of Aquatic Res. **43**, 31–43 (2017).

19 Otto, G. H. Journal of Sedimentary Petrology. **9**, 62–76 (1939)

20 Trask, P. D. Economic Geology. **25,** 581– 599 (1930)

22 E. R. Gilbert, M. G. deCamargo, L. *Package "rysgran", R Staticaly Analysis*. Version 2.1.0 (2015)

23 R. M. Pruzek, J. E. *Package "granov", R Statisticaly Analysis*. Version 2.1. (2015)

24 D. Dodoo-Arhin, D. S. Konadu, E. Annan, F. P. Buabeng, & A. Yaya. Fabrication and Characterisation of Ghanaian Bauxite Red Mud-Clay Composite Bricks for

Construction Applications. American J of Mat Sci. **3**, 110–119 (2013)

25 D. Dodoo-Arhin, R. A. Nuamah, B. Agyei-Tuffour, D. O. Obada, & A. Yaya. Case Studies in Construction Materials. 7, 45–55 (2017)

26 X. Hao, K. Leung, R. Wang, W. Sun, & Y. Li. Geoscience Frontiers. **1**, (2010).

27 M. Hossain, P. Bhattacharya, S. K. Frape, G. Jacks, M. M. Islam, M. M. Rahman, & K. Matin Science of the Total Environment, The. **493**, 615–625 (2014)

28 S. Pradit, M. S. Pattarathomrong, & S. Panutrakul. Procedia - Social and Behavioral Sciences. **91**, 573–580 (2013)

29 M. A. M. Abdallah. Arabian Jo of Chem. **10**, 68–75 (2017)

30 I. V. Galitskaya, K. R. Mohan, A. K. Krishna, G. I. Batrak, O. N. Eremina, V. S. Putilina, & T. I. Yuganova. Procedia Earth and Planetary Science. **17**, 674–677 (2017)

31 S. Ni, Y. Ju, Q. Hou, S. Wang, Q. Liu, Y. Wu, & L. Xiao. Progress in Natural Science, **19**, 1133–1139 (2009)

32 Q. Yi, X. D. Dou, Q. R. Huang, & X. Q. Zhao. Procedia Environmental Sciences, **13**, 43–52 (2012)

33 M. M. Ali, M. L. Ali, M. S. Islam, & M. Z. Rahman. Environmental Nanotechnology, Monitoring & Management, **5**, 27–35 (2016).

34 H. Effendi, M. Kawaroe, Mursalin, & D. F. Lestari. Procedia Environmental Sciences. **33**, 574–582 (2016)

35 A. Kassambara, F. Mundt. *Pakage "factoextra", R Statiticaly Analysis*. Version 1.0.5 (2017)

Nursery I: The effect of stocking density on the performance of glass eels, Anguilla bicolor in the biofloc system

Purnama Sukardi[1]*, Norman Ari Prayogo[3], Tjahyo Winanto[2], Asrul Sahri Siregar[3], Taufan Harisam[2]

[1]Department of Aquaculture, Universitas Jenderal Soedirman, Indonesia
[2]Department of Marine Science, Universitas Jenderal Soedirman, Indonesia
[3]Department of Water Resource Management, Universitas Jenderal Soedirman.

Abstract. Glass eels of Anguilla bicolor is an expensive and still abundant commodity in Laguna Segara Anakan, Central Java, Indonesia. However, the growth of glass eels to elver is still a problem because of high mortality in nursery I and II. The objective of the study was to evaluate the result of stocking density on the performance of glass eels in the biofloc system during nursery I. Glass eels were stocked at densities of 54.95, 109.89 and 164.84 fish/m^3, respectively, with three replicate ponds for each density. Eels were fed a formulated pasta-diet containing 40% crude protein and , 4% crude lipid, 5% crude fiber, 11.5% ash and 12% moisture, respectively. The water quality were maintained at levels of for fish culture througout the experiements: water temperature was 27.1°C (ranged from 26.3 to 28.2°C), pH (7.6, ranged from 6.8 to 7.8) and DO (7.2 mg.L^{-1} , ranged 6.9 to 7.5 mg.L^{-1}). The results showed that the stocking density did not significantly affect the final weight, weight gain, AGR, SGR, FCR and survival, however this had a significant effect on the yield. The biofloc system was suitable for raising glass eels.

1 Introduction

Eels, Anguilla bicolor, Gouramy, Osphronemus gouramy, Tawes, Barbonymus goniono, and Nilem, Osteochilus hasselti have high economic value, and favored by consumers in Indonesia(23,24,25). Eels, Anguilla sp. is one of a high-priced export commodities, glass eels are still abundant, there is a ban on export glass eel, it could be stunting, and many investors are interested, but the availability of elver for growth up to the size of consumption the number is still not fulfilled [1, 2, 3]. In general, cultivation stage of eels in Java are nursery 1, 2 and then growth-out stage. Nursery 1, rearing of glass eel weighing between 0.1 g to elver weighing 1.5 -2 g within 2-3 months. Nursery 2, rearing of elver weighing between 1.5-2 g to fingerling weighing 10-15 g within 2-3 months. Growth-out stage, fingerling size eels weighing 10-15 g reared up to 250-500 g (consumption size) within 5-7 months [1, 2].

[1]Corresponding author: purnamas@unsoed.ac.id; purnamaskd@gmail.com

There were at least six species including Anguilla marmorata, A. celebensis, A. ancentralis, A. borneensis, A. bicolor bicolor and A. bicolor pacifica in Indonesian waters [4,5]. Drastic decline of eel populations in Japan, Korea, China and Taiwan, even in Japan eel is included indangered category [1,6]. This status becomes an opportunity for Indonesia to be able to increase production and export.

Biofloc technology is an environment-friendly aquaculture system which is considered as a sustainable approach system for future aquatic organism production [7,8,9]. The system is based on a live microbial which beneficially affects the host animal by improving its microbial balance [10]. The role of microorganisms is increasing culture feasibility by reducing FCR and depreciating feed cost [8,9, 11]. Thus, high stocking density can be applied in this system that is biologically safe [12].

In aquaculture, fish stocking density directly affects water quality, behavioral, health, feeding, growth, survival and yield. Increased density leads to stress that causes increased use of energy and feed and reduces growth [13,14]. Optimum stocking densities required to be discovered for each species to modify efficient management and to maximize production and profitability [15]. The present study was to evaluate the result of stocking density on the performance of glass eels in the biofloc system during nursery I.

2 Materials and Methods

Glass eels were obtain from fisherman and were transported to 5000 L indoor tank in Aquaculture Laboratory, Faculty of Fisheries and Marine Science, for two days. They were randomly selected, counted and stocked into 1815 L $\approx$ 1.82 m^3 cylinder plastic pond {(3.14)x(85x85)x(80)}, each tank contained 100, 200 and 300 fish, respectively (equivalent to 54.95, 109.89 and 164.84 fish/m^3), respectively) with three replicate ponds for each density. Pond frames were made of steel covered with a plastic HDPE (300μm). Glass eels were reared for 60 days and fed four times a day with a pasta feed with 40% crude protein, crude lipid, crude fiber, ash and moisture (40%, 4%, 5%, 11.5% and 12%, respectively). Fish were fed at rates of 3% body weight/day until termination. The pasta were applied to fish twice daily at 07:00 h and 16:00 h.

2.1 Water quality

Temperature and dissolved oxygen were monitored twice daily at 08:00 and 17:00 h with a digital oxygen-meter (YSI 55, Yellow Springs, OH, USA). The pH was measured daily at 12:00 using a digital pH meter (Hanna HI98128 pH meter). CaCO3 added when the pH drops below 7. The theoretical adding quantity of carbohydrate sources calculated according to Avnimelech[16]. In this study, molasses as carbohydrate sources were set at level of 25% of the theoretical quantity. Molasses were mixed well and spread out to the water tank in the morning. Fish were harvested after 60 days and absolute growth rate (AGR), specific growth rate (SGR), survival rate (SR), feed conversion ratio (FCR) and yield were calculated using the following equations: AGR = g/fish/day; SGR (%/day)= ln final weight −ln initial weight/ days× 100; SR (%) = 100 × (final fish count/initial fish count), FCR = total dry weight of feed offered/total fish wet weight gained, Yield (heads/m^3) = total fish (heads)/volume of fish- pond (m^{-3}).

2.2 Statistical analysis

One way analysis of variants (ANOVA) by the SPSS (version 14) statistical package used to compare growth, survival, yield, and efficiency parameters. The ANOVA followed by Tukey's multiple range tests to identify differences among experimental groups. Prior to analysis, arcsine-transformation applied to the percentage.

3 Results and Discussion

3.1 Water quality parameters

During the study, parameters of water quality were maintained within the tolerance range for most freshwater species used in aquaculture. The water characteristics did not present any significant difference among fish densities. Average water temperature was 27.1°C (ranged from 26.3 to 28.2°C), pH (7.6, ranged from 6.8 to 7.8) and DO (7.2 $mg.L^{-1}$, ranged 6.9 to 7.5$mg.L^{-1}$).

Tabel 1. Performance of glass eels, Anguilla bicolor, in the BFT at the end of experiment

Parameter	Stocking densities (54.95 fish/m3)	Stocking densities (109.89 fish/m3)	Stocking densities (164.84 fish/m3)
Initial weight (g)	0.09 ± 0.01^a	0.09 ± 0.01^a	0.09 ± 0.01^a
Final weight (g)	1.41 ± 0.13^a	1.42 ± 0.12^a	1.41 ± 0.12^a
Weight gain (g)	1.33 ± 0.13^a	1.34 ± 0.11^a	1.32 ± 0.12^a
AGR (g/day)	0.022 ± 0.002^a	0.022 ± 0.002^a	0.022 ± 0.002^a
SGR (%/day)	4.62 ± 0.23^a	4.62 ± 0.21^a	4.60 ± 0.21^a
FCR	1.26 ± 0.01^a	1.24 ± 0.02^a	1.26 ± 0.04^a
Survival (%)	51.33 ± 2.52^a	51.33 ± 1.53^a	50.33 ± 1.53^a
Yield (heads/m3)	32.3 ± 1.6^a	64.7 ± 1.9^b	92.6 ± 1.9^c

Means (±SD) of the initial and final weight (g), weight gain (g), AGR (g), FCR, survival and yield (kg m−3). Means within a row followed by different superscript letters were significantly different ($P < 0.05$)

Our finding showed that stocking density was not significantly influenced final weight, weight gain, AGR, SGR, FCR and survival, it was significantly influenced production. Glass eels growth was not significantly influenced by density. Mean final weight were 1.41, 1.42, 1.41 g and mean weight gained were 1.33, 1.34, 1.32 g at 54.95, 109.89 and 164.84 fish/m3, respectively, which were in agreement with Björnsson and Ólafsdóttir [17] study in juvenile cod (Gandus morhua L.). AGR and SGR was not significantly influenced ($P<0.05$) by density. Our study showed that survival of glass eels was not significantly affected by stocking density. Average of survival rates were 51.33, 51.33, 50.33% at 54.95, 109.89 and 164.84 fish/m3, respectively. Similarly, stocking density of fingerlings of silver perch, Bidyanus bidyanus, stocked at 100 or 200 fish/m3 was not influenced survival. Several studies have also shown that the density have not significantly effect on survival of rainbow trout, Oncorhynchus mykiss [18], tambaqui, Colossoma macropomum [19] and Oreochromis niloticus [20]. Conversely with these study, it found in the endangered teleost species piabanha, Brycon insignis was higher survival at low stocking density [13]. In this experiment found that higher productivity for glass eels, Anguila bicolor at higher stocking density. It shown that a productivity of 92.6 heads/m3 was obtained at a stocking density of 164.84 fish/m3. High density in the pond also obtained high yields on the matrinxã larvae, Cephalus characidae [21] and it was directly proportional to production, which could be used to increase profits [22]. The yield of this study could probably be improved with the addition of glass eel density and better seed quality.

4 Conclusions

The data obtained suggest that stocking density was not significantly influenced final weight, weight gain, AGR, SGR, FCR and survival, however it was significantly influenced yield. Further studies could be conducted to determine optimal feeding rate of glass eels in the biofloc system. The continuation of the study with glass eels is important to provide data and to optimize the production, as well as to improve the management of glass eels in the biofloc system.

We are indebted to the Directorate General of Higher Education, Ministry of Research, Technology and Higher Education of the Republic of Indonesia for the financial support of this work, which was made available through Hibah Strategis Nasional (STRANAS)

Reference

1 Murjani, M. *Discussion forum of eel aquaculture national III.* (Yogyakarta, September, 24, 2014)
2 Sukardi, P., Winanto, T. and Harisman, T. *Proceeding International Conference of Aquaculture Indonesia (ICAI).* (2012)
3 Sukardi, P., Hana, Prayogo,N.A., Sulistyo,I.,Soedibyo, P.H.T., Harisam,T.,Winanto,T.. Acta Scientiarum Animal Sciences. **40**, (2018)
4 Affandi, R. Jurnal lktiologi Indonesia. **5**, 77-81 (2005)
5 Shen, K.N. and Tseng, W.N. Zoological Studies. **46**, 441-453 (2007)
6 Nakai, T. *Discussion forum of eel aquaculture national III.* (Yogyakarta, 24, 2014)
7 Hargreaves, J.A. *Southern Regional Aquaculture Center (SRAC)* (2013)
8 Schveitzer, R., R. Arantes, M.F. Baloi, P.F.S. Costodio, L.V.Arana, W.Q. Seiffert, E.R. Andreatta. Aquacultural Engineering. **54**, 93– 103 (2013)
9 Xu, W.J., L.Q. Pan, and D.H. Zhao. Aquaculture. 147–153 (2012)
10 Perez-Sanchez,T., Ruiz-Zarzuela, I., de Blas,I. and Balcazar,JL. Reviews in Aquaculture. **5**, 1–14 (2013)
11Martínez-Córdova, L. R., M. Martínez-Porchas, M.G.C. Emerenciano, A. Miranda-Baeza, T. Gollas-Galván. Critical Reviews in Biotechnology. **37**, 1-9 (2017)
12 Goodall, J.D., Wade, N.M., Merritt, D.J., Sellars, M.J., Salee, K., Coman, G.J. Aquaculture. **450**, 206–212, 2016)
13 Tolussi, C.E., A.W.S. Hilsdorf, D. Caneppele, R.G.Moreira. Aquaculture. **310**, 221-228 (2010)
14 Schram, E., J. W. Van Der Heul, A. Kamstra, M. C J Verdegem. Aquaculture. **252**, 339-347 (2006)
15 Huertas, M., J. Cerda. Biological Bulletin. **211,** 286-296 (2006)
16 Avnimelech, Y. Aquaculture . **176**, 227–235 (1999)
17 Björnsson, B., and S.R., Ólafsdóttir. ICES Journal of Marine Science. **63**, 326-334 (2006)
18 Wallat, G.K., Tiu, L.G., Rapp, J.D., Moore, R. Journal of Applied Aquaculture. **15**, 73–82 (2004)

19 Gomes, L, E. Chagas, H. Martins-Junior, R. Roubach, E.A. Ono, J.N. De Paula Lourenço. Aquaculture. **253**, 374-384 (2006)

20 Gibtan, A., A. Getahun and S. Mengistou. Aquaculture Research. **39**, 1450-1460. (2008)

21 Gomes, L. C., B. Baldisserotto and J.A. Senhorini. Aquaculture. **183**, 73-81(2000)

22 Rahman, M., M. Islam, G.C. Halder and M. Tanaka. Aquaculture Research. **37,** (2006)

23. N.A Prayogo, g.e. wijayanti, murwantoko, m. Kawaichi and p. Astuti. 8(6) : 591-597. (2012)

24. N.A Prayogo, G.E. Wijayanti, I. Sulistyo, P. Sukardi. Biodiversitas. **17**, 523-530 (2016)

25. N.A. Prayogo, A. Siregar, P. Sukardi. Turkish journal fishery and aquatic science. **16**, 1003-1009 (2016)

Biodiversity of algae potentially HABS (Harmfull Algae Blooms) in reservoir Mrica, Banjarnegara

[1]*Nuraina* Andriyani, [1]*Arif* Mahdiana, [1]*Rose* Dewi, [1]*Yusuf* Subagyo, [2]*Asrul* sahri siregar, [2]*Teuku* Junaidi, [2]*Taufan* harisam

[1]Department of Management Waters Resources, Faculty of Fisheries and Marine Resources, University of Jenderal Soedirman
[2]Center of Maritime and Bio science, University of Jenderal Soedirman

Abstract. It has been reported that the waters of the reservoir Mrica-Banjarnegara were mesotropik-hypereutrophic conditions, potential algal bloom is of Division: Cyanophyta is Microcystis; Chlorophyta is pediastrum; Chrysophyceae is Pyrophyta nitzchia and is Ceratium. Research the types of algae that Potentially HABS (harmfull Algae Blooms) has been Carried out as a continuation of the kinds of research that have the potential of algae blooming. Survey methods was used. Sampling was done by purposive sampling, through 5 (five) sampling sites. Data were Analyzed descriptively. The results Showed that the algae are Potentially HABS of the division Cyanophyta are Microcystis, Anabaena, Oscillatoria and phormidium.

1 INTRODUCTION

Reservoir Mrica located in District Onion, Banjarnegara district, Central Java Province established on1988. The main objective is the construction of reservoirs for hydropower Mrica with an installed capacity of 180.93 MW, in addition reservoirs used for irrigation Mrica DI Banjarcahyana area of 6,550 ha and DI Penaruban area of 900 ha, fish cage system by the local community and tourist attraction. Reservoir catchment area is 957 km^2 Mrica. At the reservoir water level +231 m asl initial reservoir capacity is 148.287 million m^3 [1].

The main river that acts as a medium influx of water and sediment into the reservoir Mrica are Serayu and Merawu River, which flow region is an area of volcanoes (Sumbing, Sindoro and Dieng Mountains). Serayu River is the largest river that goes into Mrica Reservoir. Serayu through an area that largely has undergone further weathered so that they are loose and easily eroded. Serayu watershed morphology has steep cliffs and steep valleys. Vegetation that are often found in the form of crop plants and shrubs, perennials are very rare. Merawu river passing through the area with loose rocks and easily eroded. The rocks on the cliffs of the river is also very prone to erosion because of the soft and easily crushed. Geological conditions in the watershed Merawu unstable so easily happen avalanches. And land-use patterns that there is support in terms of preventing erosion. Land cover in the catchment area (DTA) in the form of forest is less than at the close of land for agricultural activities such as plantation, growing vegetables, rice paddy, Polowijo and garden plants. The practice of planting potatoes and cassava less attention to soil and water conservation, thus increasing the rate of erosion. In a hilly area that has a fairly steep slope found many farms with dry land farming practices. Mechanical erosion control have been carried out in the form of a terrace DTA reservoir. In addition to the geological reservoir DTA is a potential area for the occurrence of avalanche danger. From the description above can be concluded DTA Mrica Reservoir is a fertile area with a large enough potential erosion,

Fishing activities by floating net (KJA) have now grown very rapidly and is thought to have passed through the carrying capacity for example is KJA in Cirata. KJA waste crop activities may result in remnants of feed accumulate in the bottom of the reservoir and its decomposition produces nutrients that fertilize the waters often have a negative impact on the marine environment so that water quality continues to decline. One of the main causes of water quality degradation in Cirata is the number of waste from fish farming activities in the KJA. Feed the fish that is not utilized by the fish as much as 27 -31%, then of feed consumed by 20% of the fish will eventually end up being a stool and fell into the water. The rest of the feed and the fish metabolic waste is what is assumed to increase the nutrients in the water such as phosphate and nitrogen [2]. Fish farming activities in the KJA is the largest contributor of domestic waste on Cirata, which is about 80%. The increase in these nutrients can cause *bloom* toxic algaethat could result in mass death of fish. [3,4,5,6].Toxic algal bloom known as harmfull algae blooms (HABs).

Algae is a plant organism, chlorophyll a role in photosynthesis that produce organic matter and oxygen in waters that are used as a basic link in the feed chain cycle*(foodchains)*.[7,8]. However, these algae can be very dangerous when it becomes blooming (growing very quickly or there is a population explosion) as a result of the carrying capacity of water is very high, among others, for example due to the high content of nutrients nitrogen (N) and Phosphorus (P) [9].

Microcystis sp and *sp*Ceratium,is an example of algae that often bloming in stagnant waters (reservoirs) in Indonesia. One example is the impact of blooming algae bloom *of Microcystis sp.* , Resulting in a natural ecosystem becomes impaired waters since the algae produce toxins microcystin stable in the water and are hepatotoxic [10, 11]. The toxin inhibits the metabolic processes of plants macrophytes, causing the death of the fish and organisms waters through the chain of feed, also cause liver disease in cattle due to drinking water contaminated with microcystin.

The formulation of the problem of this research is the growth of algae in the waters affected by physical, chemical and biological. Mrica dam water has been polluted by sediment originating from the mountainous Dieng through Serayu and Merawu. Sediment besides containing soil particles also contain nutrients N and P, so the effect on fertility waters in this case is the growth of algae. Therefore, Blooming algae can occur due to changes in pH, light penetration, temperature, TSS, turbidity, CO2, ortho phosphate, nitrate resulting from sediments earlier. The purpose of this study was to determine the types of algae that potentially HABs in reservoir Mrica Banjarnegara.

2 MATERIAL AND METHODS

The research used descriptive method and the data were analyzed comparatively with Cyanophyta division and table toxic compounds from the Public Health Advisory Guideline 2010. Algae which studied is the nature of planktonic algae, so the sampling and the data analyse together with how to analyze plankton (Table 1).

Table 1. Sampling Station Location

Stations	Point Coordinates	Research Location	Description Location
I	°23'25,1S 07 " E 10939'59,4 "	Merawu	Rice Fieldriver,moor, sand mining
II	S 07°22'48,5 " E 10938'23,1 "	Serayu.	Rice paddies, fields, mines pasr
III	S 07°23 ', 15.2' 'E10936'56,7'	'Middlereservoir	dam water Amid
IV	S 07°22', 48.8 " E 10936'40,9 "	Areas of KJA (Coral Jambe)	KJA Activities
V	S 07'23 ', 29.1' 'E10936'30,8'	'Outletdam	reservoir At the outet

2.1 Water sampling

Decisionand preservation of water samples is only done for the parameters that are *eksitu* (measured in laboratory), which includes TSS, nitrate, orthophosphate, BOD and turbidity by means of a water sample is cooled in the ice box (Table 2). Parameters that are *in situ* (directly measured in the field) include measurements of temperature, pH, salinity, and carbon dioxide-free.

Table 2. Methods for analysis and standard quality water for phytoplankton (Yuliana, 2012)

No.	Parameter	Unit	Method / Tools	Source	Source
1	Temperature	^0C	Expansion / thermometer	APHA (2005)	(Yuliana, 2012)
3	Turbidity	NTU	Nephelometri / turbidimeter	APHA (2005)	<25

 https://doi.org/10.1051/e3sconf/20184703001

4	TSS	mg / L	Grafimetri / Whatman filter paper No. 41	APHA (2005)	-
6	CO2	mg / L	Erlen Meyer	APHA (2005)	-
7	pH	-	Colorimetric / pH paper	APHA (2005)	-
8	BOD	mg / L	Winkler	APHA (2005)	<25
10	Nitrates	mg / L	brusin / spectrophotometer	APHA (2005)	0.9-35
11	orthophosphate	mg / L	Stannous chloride / spectrophotometer	APHA (2005)	0.27 - 5.51

Monitoring of water quality and plankton found identification conducted in the laboratory of Water Quality Management at Fisheries and Marine Faculty, Laboratory of Soil Science- Faculty of Agriculture Unsoed., And Purbalingga Health Laboratory.

2.2 Retrieval and Preservation Phytoplankton

Samples were filtered water as much as 200 L with a volume of 10 L plastic bucket and poured into a plankton net no. 25. Water samples were deposited in a plankton net reservoir bottle is transferred into a 30 mL sample bottle and added with 3 mL of 40% formalin so that the water in the bottle normality to 4% as well as a few drops of Lugol three drops. The samples are stored in the ice box and made observations in the laboratory.

2.3 Identification and Calculation of abundance of algae

Phytoplankton samples of 30 mL were taken observed with a binocular microscope. The samples were homogenized with a pipette and taken as many as three drops, then placed on glass slide. The slides are put in place binocular microscope object. Observations were carried out with repeat 5 times to 30 times the observation point. Phytoplankton had been identified by using the book Davis (1955), Edmonson [2], Sachlan (1982) and the APHA [12]. The calculation of the amount of the abundance of phytoplankton per liter using the formula Microtransect Lackey Drop Counting (APHA, 1989), with the following formula:

 https://doi.org/10.1051/e3sconf/20184703001

$$N = nx\frac{A}{B}x\frac{C}{D}x\frac{1}{E} \tag{1}$$

Where:

N	= the number of phytoplankton species per Liter (individual / Liter)
n	= number of phytoplankton species averaging per field of view (1/30)
A	= area of the glass cover (mm2) (22 mm x 22 mm = 484 mm2)
B	= broad one field of view (mm2) (2,405 mm2)
C	= volume of concentrated water (30 mL)
D	= volume drops below a glass cover (0.05 mL)
E	= volume of filtered water (200 L)

3 RESULTS AND DISCUSSION

3.1 Reservoir water quality Mrica

Reservoir water quality Mrica compared with the entry in the dam water level mesotropik- Mrica hypereutropik (Table 3). This means that the water conditions are already in the stage of high fertility. High fertility rates will affect the growth of algae, because algae is in dire need of light penetration, phosphorus as well as nitrogen. TSS Serayu and Merawu very high (> 100 NTU) indicate that the water contains organic particles and organic carried from upstream or Dieng. Turbidity Serayu and Merawu also high when compared to the central area, KJA and outlets. BOD indicates that the water has been polluted organic waste, as has been widely known that a farm in Dieng as the area above the reservoir Mrica used manure or organic in maintaining soil fertility. However, due to the use of manure is contaminating bottom areas, reservoirs consequently increased nitrate and phosphate nutrients, it will result in the enrichment of the algae [13,14].

Table 3. Physical parameters -kimia reservoir waters Mrica

PARAMETER	MERAWU	SERAYU	TENGAH	KJA (Kr.Jambe)	OUTLET
	I	1	1	1	1
Jam	15.01	10.25	11.22	11.53	12.57
Lokasi	S 07° 23'25,1"	S 07° 22'48,5"	S 07° 23',15,2"	S 07° 22',48,8"	S 07° 23',29,1"
	E 109°39'59,4"	E 109°38'23,1"	E 109°36'56,7"	E 109°36'40,9"	E 109°36'30,8"
	Cerah	Cerah	Cerah	Cerah	Cerah
PH	7,00	7,00	7,00	7,00	7,00
Temperatur (oC)	28,00	20.55	28,50	28,50	28,00
Penet.Cahaya (cm)	25,50	17.5	76,25	111,25	100,00
TSS (ppm)	101,00	127,00	57,00	33,00	16,00
Kekeruhan (NTU)	39,00	50,00	5,00	2,00	4,00
O2 terlarut (ppm)	6,15	4,80	6,10	6,40	6,60
CO2 (ppm)	0,66	3,08	tt	tt	tt
BOD (ppm)	8,33	16,67	8,89	26,67	14,87
N Total (ppm)	1,79	4,50	1,45	0,18	1,28
P Total (ppm)	2,81	3,16	3,10	1,09	4,45
Ortho Pospat (ppm)	0,53	0,52	0,65	0,55	0,42
Nitrat (ppm)	0,08	0,07	0,10	0,31	0,39
Klorofil (mg/m3)	0,18	0,18	3,92	2,67	10,28

Increased nitrogen and phosphate, in addition to coming from Dieng also from KJA (Karamba Floating Net). The use of artificial feed or excessive pellet will leave a lot of garbage, litter or feed residue will settle to the bottom waters[15]. The amount of feed residue that settles causing aerobic and anaerobic bacteria are not able to degrade, so that the rest of the food are not degraded perfectly able to form ammonia that are harmful to the organism.

3.2 The diversity and abundance of algae in the reservoir Mrica

The diversity and abundance of algae in the reservoir Mrica, the highest of the division Chlorophyta and chrysopyta (Figure 1.). Genus of division Chlorophyta represented pediastrum; Cyanophyta genus of division is represented by Microcystis; genus of divisions represented Chrysophyta nitszchia; and genus of pyrophyta represented by ceratium (Figure 2,3,4,5). Genus Microcystis, although the number of individuals less compared to the number of individuals of other genera Microcystis others in fact have a larger size and more because of the size of the colony (Table 4) [16] Algae that live in colonies had greater biomass than the phytoplankton that live solitary (Sachlan, 1982). In quantitative terms Cyanophyta biomass will be greater than the biomass divisio others. So that the state of Microcystis bloom, surface waters can be fully closed. , The Role of the waters of this genus as pemfiksasi nitrogen-free under normal circumstances, but if it happens blooming will remove toxins that will kill the organisms in it, such as toxic by the cyanobacterium Anabaena sp., Microsystis sp., And Nodularia sp.(Table 5) [17,18,7].

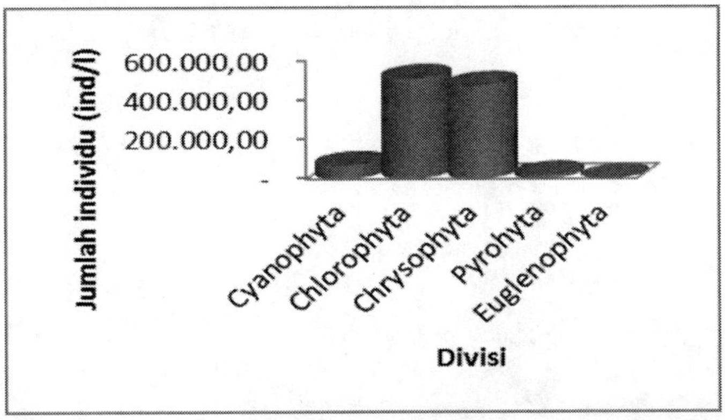

Fig 1. Number of individuals in the genus division of phytoplankton

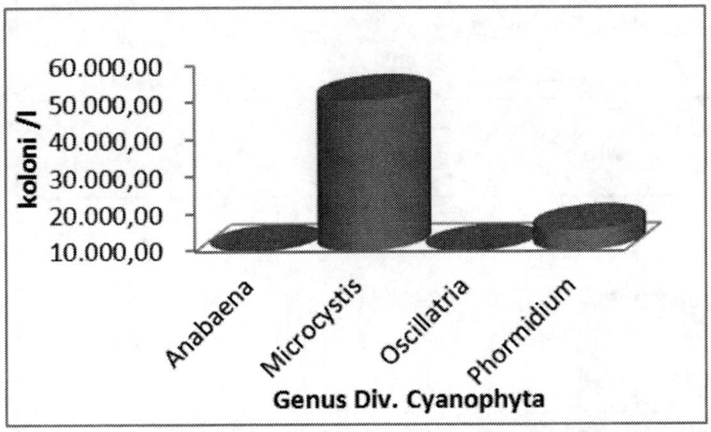

Fig 2. Number of individuals in the genus of div. Cyanophyta

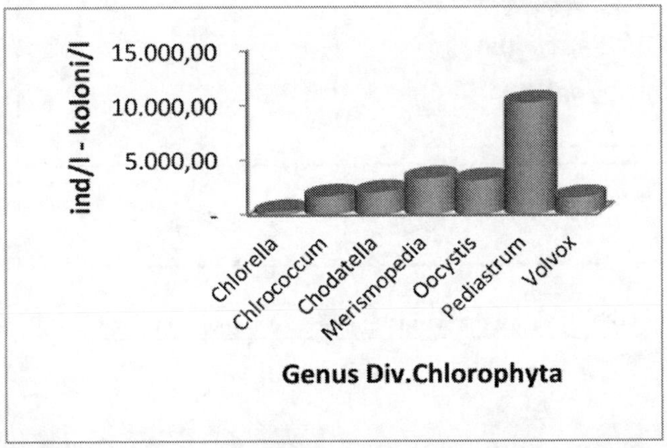

Fig 3. Number of individuals in the genus of div. Chlorophyta

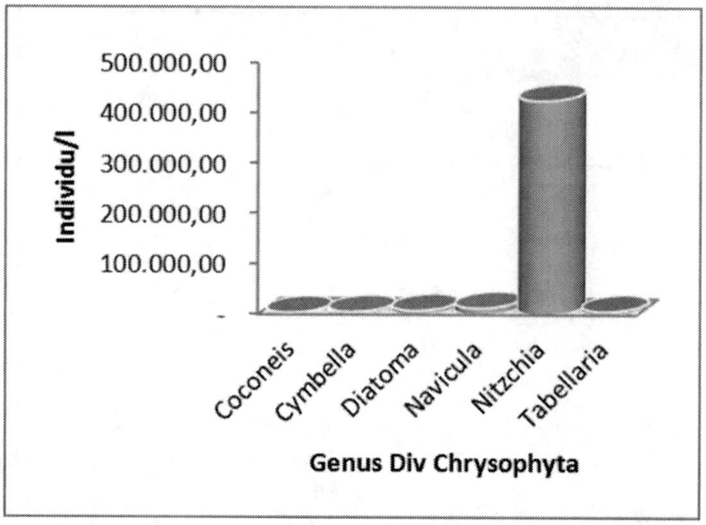

Fig 4. The number of individuals in the genus of div. Chrysophyta

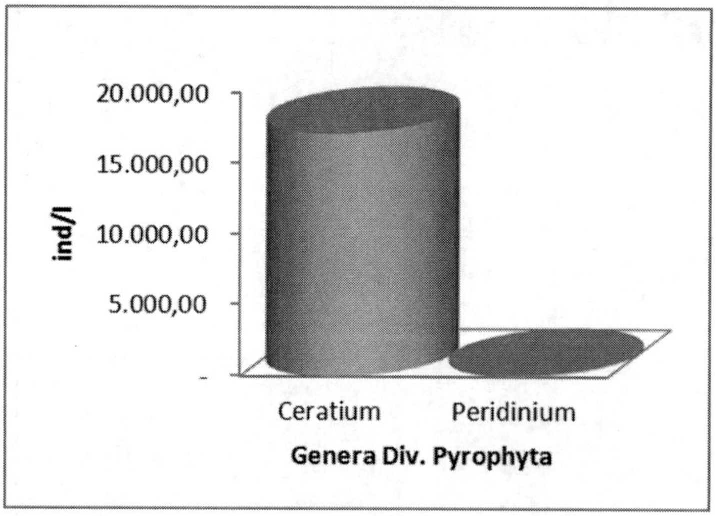

Fig 5. Number of individuals in the genus of div. Pyrophyta

Table 4. Standard rate of eutrophication of waters by the Minister of Environment no. 28 of 2009.

Parameters	**Fertility classificationwaters**		
	oligotrophic	**Mesotrofik**	**eutrophic**

light penetration (cm)	≥400 ≥250		≥1000
total phosphorus (mg / L)	<0.01	<0.03	<0.10
total nitrogen (mg / L)	≤0	≤0,65, 75	≤1,9

Table 5. Compound toxin of Division Cyanophyta (Public Health Advisory Guidelines, 2010)

Hepatotoxins			Neurotoxins	
microcystin	Nodularin	Cylindro-spermopsin	Anatoxin-a	saxitoxin
Anabaena	+	+	+	+
Anabaenopsis			+	
Aphanizomenon(Except A. flos-aquae)	+		+	+
Arthrospira			+	
Cyanobium			+	
Cylindrospermopsis		+		+
Gloeotrichia			+	
Hapalosiphon			+	
Limnothrix			+	

Lyngba			+	
Microcystis		+		+
Nodularia			+	
Nostoc			+	
Oscillatoria		+		+
Phormidium		+		+
Planktothrix	+		+	+
Raphidiopsis		+		+
Schizothrix				
Synechocystis			+	
Umezakia			+	

3. 3 The types of algae that could potentially lead HABS in waters Mrica

Algae potentially HABS based on the types of potentially blooming algae were: division Cyanophyta is Microcystis. Microcystis is recommended to be avoided if a water containing up to 40,000 individuals / l because it can cause illness to humans (Public Health Program) in addition to Microcystis of Cyanophyta division also found Anabaena, Oscillatoria, and Phormidium. The four genera containing toxins: hepatotoxin and neurotoxins (Table 5). Anabaena containing hepatotoxins (nodularian and cylindrospermin) and neuotoksin (anatoksin and saxitoksin) whereas Microcystis, Oscillatoria and phormidium containing hepatotoxins (cylindrospermin) and neurotoxins (saxitoksin).

4 CONCLUSIONS

From the discussion above it can be concluded that the waters of the reservoir Mrica on mesotropik-hipereutropik conditions. Algae are blooming potential of divisions: Cyanophyta is Microcystis; Chlorophyta is pediastrum; Chrysophyta is nitzchia and Pyrophyta is ceratium. Genera which can cause HABs is a division Cyanophyta: Anabaena, Microcystis, Oscillatoria and phormidium.

Many thanks to DIPA funds PKA (BOPTN) Number: DIPA-023.04.2.189899 / 2013 dated December 5, 2012 Revision to 02 on 1 May 2013 on funds provided sehigga this study can be held. Thanks also we would like to Typhoon Harisam SPi, SPi Ramsie Genda, Fa Rara Yuniarti Anggraeni and Melissa for their assistance and cooperation.

REFERENCE

1 Andriyani, N; US Siregar; and A. Mahdiana. 2013. The types of algae blooming in the reservoir potential Yag Mrica-Banjarnegara. Paper presented at the National Seminar on "Rural Resource Development and Local Wisdom sustained in LPPM Unsoed Purwokerto. 26 to 27 November 2013.

2 Edmondson, WT 1959. *Freshwater Biology.* John Willey and Sons Inc. New York.

3 Mizuno, T. 1978. Illustration of The Fresh Water Plankton. Japan: Heikusha Publishing, co. Ltd Osaka.

4 Odum, E.P. 1988. Fundamental of Ecology.Phidelphia: W.B. Sounders Company.

5 Soylu, E.N., and A. Gönülol. 2003. Phytoplankton and seasonal variations of the River Ye ilırmak, Amasya, Turkey. Turkish Journal of Fisheries and Aquatic Sciences 3: 17-24.

6 Stoddard, A., J.B. Harcum, J.T. Simpson, J.R. Pagenk opf, and R.K. Bastian. 2003, Municipal Wastewater Treatment: Evaluating Improvements in National Water Quality. Published by John Wiley and Sons, Inc.

7 Oren, A. 2011. Naming Cyanophyta / Cyanobacteria-A Bacteriologist's View. Fottea. 11 (1): 9-11.

8 Barange, M and B. Campos. 1991. Model of spesies abundance: a critique of and an alternative to the dynamic model. Marine Ecology Progress Series 69: 293-298.

9 Chapman, VJ and DJ Chapman.1973. *Algae Freshwater EcologyThe 2nd Edition.*The Mac Millan Press, London Hutabarat, S. 2000. *Marine Productivity and*Plankton.Publisher Agency Dipenogoro University, Semarang.

10 Magurran, A. E. 1988. Ecological Diversity and Its Measurement. Princeton, N.J.: Princeton University Press.

11 Zalocar de Domitrovic, Z.Y., A.S.G. Poi de Neiff, and S.L. Casco. 2007.Abundance and diversity of phytoplankton in the Paraná River (Argentina) 220 km downstream of the Yacyretá reservoir. Brazil

12 APHA (American Public Health Association). 2005. *Standard methods for the examination of water and waste water.* 21th edition. Washington DC

13 Onyema, I.C. 2007. The phytoplankton composition, abundance and temporal variation of a polluted estuarine creek in Lagos, Nigeria, Turkish Journal of Fisheries and Aquatic Sciences 7: 89-96.

14 Piirsoo, K., P. Peeter, A. Tuvikene, and V. Malle. 2008. Temporal and spatial patterns of phytoplankton in a temperate lowland river (Emajõgi, Estonia). Journal of Plankton Research 30 (11): 1285-1295.

15 Bledsoe, E., E.J. Phlips, C.E. Jett, and K.A. Donnelly. 2004. The relationships among phytoplankton biomass, nutrient loading and hydrodinamics in an inner shelf estuary. Ophelia 58 (1):20-47.

16 Lacerda, S R., M.L. Koening, S. Neum ann-Leitão, and M.J. Flores-Montes. 2004. Phytoplankton Nyctemeral variation at a tropical river estuary (Itamaracá-Pernambuco-Brazil). Brazilian Journal of Biology 64 (1): 81-94.

17 Tuvikene, A., K. Piirsoo, and Pall. 2005.Effect of nutrient load on the planktonic biota in the River Narva drainage area. InRusso, R. C. (ed.), 2005. Modelling Nutrient

Loads and Responses in River and Estuary Systems. Report No. 271. Brussels: Committee on the Challenges at Modern Society, NATO.

18 Ekwu, A.O. and F.D. Sikoki. 2006. Phytoplankton diversity in the cross river estuary of Nigeria, Journal of Applied Sciences & Environmental Management10 (1): 89-95.

Effect of medium and light quality on pink pigment production of cyanobacteria *Oscillatoria* sp. BTCC/A0004

Karseno[1*], *Kazuo* Harada[2], and *Kazumasa* Hirata[2]

[1)]Department of Agriculture Technology, Universitas Jenderal Soedirman
Kampus Karangwangkal, Purwokerto 53123 Indonesia
[2)] Department of Applied and Environmental Biology, Graduate School of Pharmaceutical Sciences, Osaka University, 1-6 Yamada-oka, Suita, Osaka 565-0871, Japan

Abstract. Cyanobacteria are well known as promising source of valuable chemicals for human usage. Especially, cyanobacteria in tropical area are very wide in diversity and they are potent producers of unique metabolites which exhibit interesting bioactivities. *Oscilatoria* sp. BTCC/A0004 produce pink pigments extracellularly (OsPP). The effects of various environmental factors on the production of cyanobacteria metabolites were well documented. In this research, the effect of medium and light quality on cell growth and OsPP production were investigated. In case, three different culture media, named No 18, C, and modified C media, in which nutrient compositions are different, and light quality (white, blue, green, pink) were tested. The highest cell growth and OsPP production were obtained in modified C medium. The nitrogen concentration in modified C medium is higher (5 g/L) than in No 18 medium (1.5 g/L) or C medium (1 g/L). In addition, cell growth and OsPP production were significantly stimulated by pink light radiation.

1 Introduction

Cyanobacteria are well known as promising source of valuable chemicals for human usage. They have been identified as one of the most attractive group of organisms for novel bioactive natural products [1]. Especially, cyanobacteria in tropical area are very wide in diversity and they are potent producers of unique metabolites which exhibit interesting bioactivities. However, most of them remain unexplored. In this study, we found tropical freshwater cyanobacterial strains, *Oscillatoria* sp.

[1]*Corresponding author*: karseno@unsoed.ac.id

BTCC/A0004 produce pink pigments extracellularly (OsPP) and it was identified as phycoerytrin like pigment [2].

In previous study, OsPP exhibited physical properties similar to phycoerytrin (PE). PE has gained tremendous interest due to excellent spectroscopic properties. It has been extensively investigated and exploited commercially for usage such as for fluorescent dyes in bio-assay [3, 4]. In this regards, OsPP might be substituted chemicals of PE. In addition, OsPP has potent growth inhibitory activity against green algae and cancer cells with function as photosensitizer. These results indicated that OsPP is promising substance for application in various fields.

In many papers, the effects of nutrient, light, pH, temperature and co-cultivation with competitive organisms on the synthesis of cyanobacterial metabolites were well documented [5-7]. It was expected that those factors might stimulate the production of OsPP. Therefore, in this research the effects of nutrient and light quality on stimulate cell growth and OsPP production were investigated.

2 Materials and methods

2.1. Algal cultivation condition

Oscillatoria sp. was sub cultured in test tubes (3 cm i.d. x 20 cm) containing 50 ml of modified C medium at pH 7.5 with the following composition: (per liter) 5 g KNO_3, 0.1 g KH_2PO_4, 0.05 g $MgSO_4.7H_2O$, 0.005 g $FeCl_2$, 2.86 mg H_3BO_3, 1.81 mg $MnCl_2.4H_2O$, 0.22 mg $ZnSO_4.7H_2O$, 0.018 mg $(NH_4)_6Mo_7O_{24}.4H_2O$, and 0.075 mg $CuSO_4.5H_2O$. The cells were cultivated with aeration (10 ml min^{-1}, 1 % CO2) at 25°C under 50 µmol photons m^{-2} s^{-1} of continuous illumination using white fluorescence lamps. For measurement of cell growth and pigment production, 4-day-old seed cultures of *Oscillatoria* sp. was transferred to fresh media and adjusted to 0.1 of OD_{680}. Incubation conditions was the same as for seed cultures except for light intensity was 100 µmol photons m^{-2} s^{-1}.

2.2. Effect of different media

Three different culture media named C, modified C and No 18 media, in which nutrient compositions are different were tested (Table 1). The cultivation methods were same to basal cultivation condition.

Table 1. Chemical composition of three culture media (L^{-1})

Components	No 18	C	Modified C
NaCl (mg)	70	-	-
$MgSO_4$ $7H_2O$ (mg)	380	40	50
$CaCl_2$ H_2O (mg)	106	-	-
$Fe_2(SO_4)_3$ nH_2O (mg)	10	-	-
Na_2EDTA $2H_2O$ (mg)	27	1000	-
K_2HPO_4 (mg)	600	-	100

H_3BO_3 (mg)	3	-	2.86
$MnSO_4$ $4H_2O$ (mg)	2	-	-
Na_2MoO_4 $2H_2O$ (mg)	8	2.5	-
$ZnSO_4$ $7H_2O$ (mg)	0.3	22	0.22
$CuSO_4$ $5H_2O$ (mg)	0.08	-	0.075
$CoCl_2$ $6H_2O$ (mg)	0.03	4	-
Titriplex III (g)	0.07	-	-
$Ca(NO_3)$ $4H_2O$ (mg)	-	150	-
KNO_3 (g)	-	0.1	5
$FeCl_3$ (mg)	-	196	-
$MnCl_2$ $4H_2O$ (mg)	-	36	1.81
Tris (hydroxymethyl) aminomethane (mg)	-	500	-
NaH_2PO_4 $2H_2O$ (mg)	-	50	-
Vitamin B_{12} (μg)	-	0.1	-
Biotin (μg)	-	0.1	-

2.3. Effect of light quality

To investigate the effects of light quality on cell growth and pink pigment production, the cells were cultivated in 100 ml of modified C medium in Petri dishes (6 and 9 cm in depth and diameter, respectively). The dish was placed in a transparent cultivation box and illuminated continuously with 30 μmol photons m^{-2} s^{-1} fluorescence light quality and aeration with 1 % CO_2 in air, 25°C. White, blue and green light were supplied with three 10 W fluorescence lamps (National, Japan). The pink light was supplied with three 10 W fluorescence lamps (NEC, Japan).

3 Results

3. 1. Effect of different culture media

The effects of different media on cell growth and OsPP production were presented in Figure 1. The highest cell growth and OsPP production were obtained in modified C medium. Modified C medium contains more nitrogen (KNO_3) than two other media evaluated (Table 1). It is well known that nitrogen is an essential element required for the growth and synthesis of various metabolites [5]. This might be one of the reasons why cell growth and OsPP production in modified C medium were higher than those in the other media.

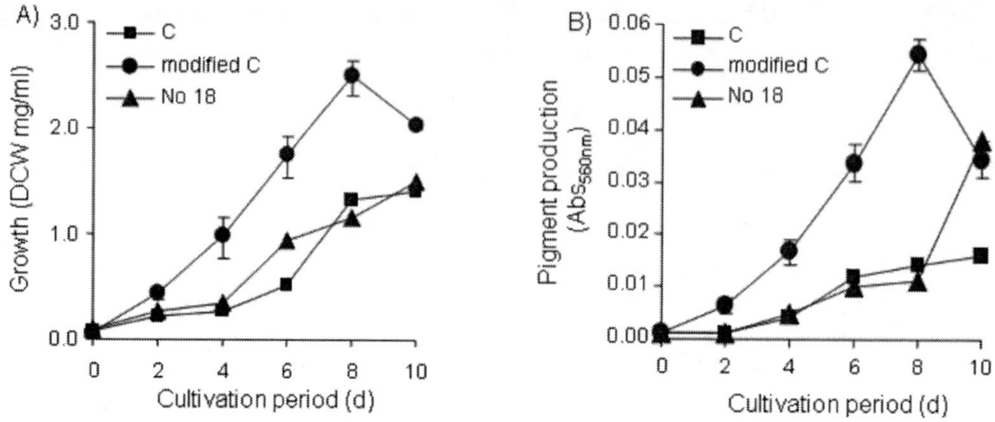

Fig. 1 Effect of different media on growth (A) and pigment production (B) of *Oscillatoria* sp. The cell was cultivated in various media in the light (100 μmol photons $m^{-2} s^{-1}$) and aeration of 1 % CO_2 in air at 25°C. DCW is dry cell weight. Values are the mean ± SDs of three independent experiments.

3. 2. Effects of light quality

Light is the primary energy source and play an important role in photosynthetic organisms including cyanobacteria. There are several reports that light quality as well as light intensity showed significant effects on growth and metabolism in cyanobacteria [8-11]. The effects of light quality on cell growth and OsPP production are shown in Figure 2. The cell growth and OsPP production were stimulated significantly by pink light irradiation.

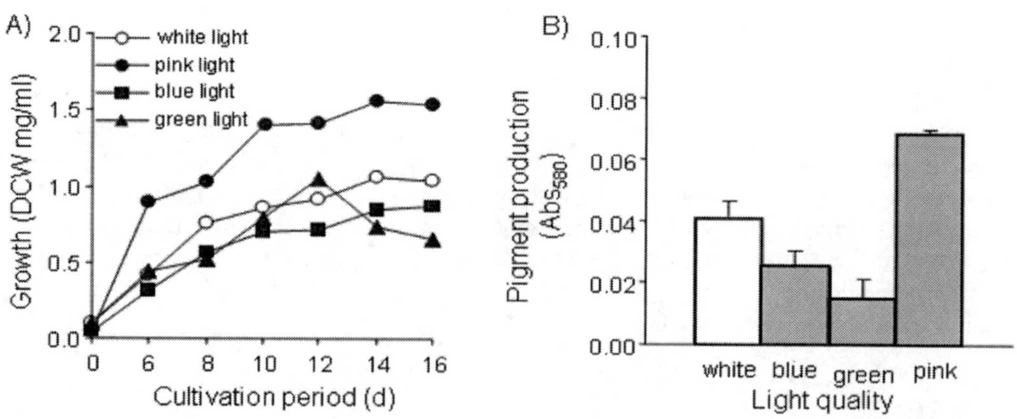

Fig. 2 Effects of light quality on growth (A) and pigment production (B) of *Oscillatoria* sp. Cells were cultivated under different wavelengths of light at 30 μmol photons $m^{-2} s^{-1}$. The production of OsPP in the medium was measured for 16-day-old *Osillatoria* sp. DCW is dry cell weight. Values are the mean ± SDs of three independent experiments.

4 Discussion

The effects of different media and light quality show significant effect on cell growth and OsPP production. Three different media C, modified C and No 18 are known as culture media for cultivation of freshwater cyanobacteria. These media were also used for cultivation of several freshwater cyanobacteria. Under modified C medium, the pink pigment in culture media was clearly observed on 6-day-old *Oscillatoria*, which is correspond to the absorbance value (Abs_{560}) of 0.02. On the other hand, the pigments from grown cells in No 18 and C media were observed on 10-day-old *Oscillatoria*, respectively.

The highest cell growth and OsPP production were obtained from grown cell in modified C medium (Fig. 1). The nitrogen concentration in modified C is higher (5 g L^{-1}) than C medium (0.1 g L^{-1}) or in No 18 (0 g L^{-1}). The correlations between nitrogen concentration and cells growth as well as on the pigment production were well documented. Nitrogen is an essential major element required for the synthesis of primary and secondary amino acids, proteins, nucleic acids, coenzymes, chlorophyll and other accessory photosynthetic pigment such as phycobiliproteins in cyanobacteria [5]. It was reported that the concentration of KNO_3 as nitrogen source was found to be an essential factor influencing the growth of *Synechocystis* sp. PCC 6701. In contrast, the growth and phycobiliproteins content in several cyanobacteria decreased during nitrogen starvation/limitation [12].

On the other hand, K_2HPO_4 as phosphorus source is also known as essential to algal growth because it was important roles in many cellular processes and it is maintaining pH of culture media by it buffering capacity [13]. Modified C medium containing phosphorus (100 mg L^{-1} as K_2HPO_4) might be appropriate concentration for growth of *Oscillatoria* strain. BG-11 medium supplemented with phosphorus (10 mg L^{-1} as K_2HPO_4) was significantly stimulated the cell growth and phycoerythrin production of *Nostochopsis lobatus* [13]. In contrast, high levels of phosphorus in medium was inhibited the growth of *Anabaena variabilis* [14]. It was reported that nitrate (KNO_3) and phosphate (K_2HPO_4) were identified as major factors of cell growth *Synechocystis* sp. PCC 6701 [12]. In this study, high cell growth and OsPP production obtained from modified C medium were agreed with these phenomena.

Light qualities as well light intensity play an important role on the cell growth and metabolite production in photosynthetic organisms, including cyanobacteria. For example, in *Rhodella reticulate,* the production of algal biomass is enhanced by about 5-6 times in response to increased light intensities from 18 to 215 $\mu E\ m^{-2}\ s^{-1}$ under white, green or red light. At low light intensity, B-phycoerythrin content is preferentially enhanced to 27 % under the influence of green light compared with red light [9]. The phycocyanin content in the cyanobacteria *Synechococcus* sp. NKBG 042902 grown under green or blue light is markedly low, while that of the cells grown under red light is high [10]. The production of marennine, a blue pigment produced by the diatom *Halsela ostrearia*, is controlled by blue-light radiation [32]. In addition, production of micosporine-like amino acids (MAAs) and scytonemin in cyanobacteria, phytoplankton and macroalgae has been observed as a response to counteract the damaging effect of UV-radiation [11]. Therefore, the individual light regimen becomes a predominant factor in affecting the productivity of algae.

It was reported that longer wavelength (540 nm) may not suitable for chlorophyll and carotene synthesis but phycocyanin and PE significantly accumulated under such wavelength. The chlorophyll synthesis was suited under white light and blue light irradiation. Longer wavelengths such as green, yellow and red lights are the main light adsorbed by photosystem II to produced highest phycocyanin and phycoerythrin content in *Spirulina fussiformis*. This is because at higher light wavelength phycobilisome is a highly efficient system for transferring energy to PS II reaction center [15]. In addition, under red or pink light radiation, most of the photosynthetic energy can be diverted into new biomass due to high carbon investment efficiency [8]. In conjunction to these phenomena, cell growth and extracellular pink pigment production of *Oscillatoria* strain were also stimulated proportionally under pink light radiation.

In conclusion, this research provides information that high cell growth and OsPP production were obtained from cultivation of *Oscillatoria* strains in modified C medium and under pink light irradiation. Modified C medium is simple in both chemical composition and preparation than the other media. Therefore, this medium has advantage for practical application on pink pigment production of *Oscillatoria* strain.

References

1. S. Singh, B.N. Kate, U.C. Banerjee. An Overview Critical Rev. Biotechnol. **25**, 73-95 (2005)
2. Karseno, K. Harada , K. Hirata. Biotechnol Lett. **31**, 999-1003 (2009)
3. C.M. Hildith, P. Balding, R. Jenkins, A.J. Smith, , L.J. Rogers. J. Appl. Phycol. **3**, 345-354 (1991)
4. A.N. Glazer. J. Appl. Phycol. **6**, 105-112 (1994)
5. S. Liotenberg, D. Campbell, R. Rippka, J. Houmard, N.T. de Marsac. Microbiol. **142**, 611-622 (1996)
6. S. J. Repka, J. Mehtonen, L. Vaitomaa, Saari, K. Sivonen. Microbiol. Ecol., **42**, 606-613 (2001)
7. K.D. Kearns, M.D. Hunter. Environ. Microbiol. **2**, 291-297 (2000).
8. N. Korbee, F.L. Figueroa, J. Aguilera. J. Photochem. Photobiol. B. **80**, 71-78 (2005).
9. S.G. Mihova, D.I. Georgiev, K.M. Minkova, A.A. Tchernov. J. Biotechnol. **48**, 251-257 (1996).
10. H.Takano, T. Arai, M. Hirano, T. Matsunaga. Appl. Microbiol. Biotechnol. **43**, 1014-1018 (1995).
11. R.P. Sinha, Klisch, M., Groniger, A., and Hader, D.P.: Ultraviolet-absorbing/screening substances in cyanobacteria, phytoplankton and macroalgae. J. Photochem. Photobiol., B, 47, 83-94 (1998).
12. Hong Seong-Joo, Lee Choul-Gyun. Biotechnol. Bioprocess Eng. **13**, 491-498 (2008).
13. U. Pandey, J. Pandey. Bioresource Technol. **99**, 4520-4523 (2008).
14. M.G. Battah, E.F. Shabana, I.A. Kobbia, H.M. Eladel. Ecotoxicol. Environ. Saf. **49**, 235-239 (2001).
15. H.K. Madhyastha, T.M. Vatsala. Biomol. Eng. **24**, 301-305 (2007).

Potential of sustainable and biological aspect of sting ray (*Dasyatis* sp.) as catch fisheries status overview in Java Sea

Arif Mahdiana[1], *Norman Arie Prayogo*[1,2]. *Isdy* Sulistyo[1]

[1]Faculty of Fisheries and Marine Science , University of Jenderal Soedirman, Purwokerto, Indonesia

[2]Center of Maritime and Bioscience, University of Jenderal Soedirman, Purwokerto, Indonesia

Abstract. Stingray fish is a demersal fish commodity in Indonesian waters which decreasing in catch. The purpose of this research is to know the potential of sustainable and some aspects of Stingray fish biology by using the Maximum Sustainable Yield (MSY) indicator as the fishing status in Java Sea. Data of catch and effort taken from production data of year 2006-2015 and primary data to know some aspect of biology is taken at TPI port of Cirebon and Tegal Specification data of fish catching tools Pari obtained from fisherman interview. The data of the catch in the analysis using Surplus Production Method (Schaefer) and observed aspects of stingray biology include: sex ratio and distribution of length of Stingray fish. The results showed the MSY value of Stingray fish was 894.275 tons / year and fMSY value of 34.716 ships / year. The sex ratio of male Stingray fish with female Stingray fish is still balanced, which is 1: 1.34 and the size of long frequency entered into the category of immature fish gonad and fish which first ripe gonad, which ranges between length 16-22 cm both male fish or female. Based on the research results can be concluded that Stingray fish catchment in Java sea still in sustainable condition.

1. Introduction

Indonesia's potential of resource is fast, expecially its potential in fisheries. The total diversity of fish strengtened by 5,8 million km^2 of seawater and 81.000 km coastal length [1,2].

Stingray fish is Indonesian catch fisheries commodity, stingray is demersal fish that included in *Chondrichthyes* class [3]. Java Sea demersal fish potential is 431.000 tons/year with and exploitation value of 56%. Stingray catchment in 2006 was 7,249. 70 tons, it was 16. 13% of total catch fisheries [4]. With stingray demand increased, stingray became main catch of fishermans and have high economical value. It's meat and fin become foods and it's skin become fashion items [5]. Because of high demand, overfishing is occuring and stingray fish stock is declining.

Potential of sustainable is a concept of fish resources management with responsibility to sustain the resources [6]. Based on Marine and Fisheries Research, Departement of Marine and Fishiries Indonesian Institution of Science 2001 [7]. Indonesian seawater fish resources MSY was 6.4 million tons / year with acceptable catch 5.1 million tons/ year (80% of MSY). Demersal fish potential of resource 1,370,090 tons/ year. Reported by Indonesin Directorate of Catch Fisheries on 2011, there was excessive catchment on 2006, causing decline in fish catchment from 925.557 kg to 841.545 kg in 2010.

The aim of this study is to give information about stingray fisheries resources management in Java Sea. The observed factors were aspect of fish, sex ratio, and fish legth structure.

2. Material and Methods

Methods used in this research are observation survey and interciew to collect primary and secondary data. The primary data is interview done to fishermen using gill net and used as supporting parameter. The secondary data is fishing effort namely total ships and stingray fish production data for 10 years from 2006 – 2015. Sample was taken using random

Corresponding Author: arifmahdiana@gmail.com

https://doi.org/10.1051/e3sconf/20184703003

sampling techinque to fishermen population that catch stingray in Java Sea. The research was done in Tegalsari Fish Auction Place, Tegal, Central Java and Kejawanan Nusantara Fisheries Port Cirebon, West Java.

Corresponding Author: *arifmahdiana@gmail.com*

Data analyzed using Schaefer production surplus model (1957) stated by Gulland (1983):

$$q = \frac{h}{f} \quad (1)$$

q = CPUE Catch Per Unit of Effort (tons/ship)
h = Catch (tons)
f = Effort (ship)
Schaefer MSY and fMSY formula:

a) $f\,\text{MSY} = -\frac{a}{2b}$ (2)

b) $MSY = \frac{a^2}{4b}$ (3)

a = intercept
b = slope
Type, sex ratio, and size of stingray fish were analyzed descriptively with table and figure.

3. Result and Discussion

3.1. Catch Per Units of Effort (CPUE)

CPUE of stingray fish from 2006 – 2015 :

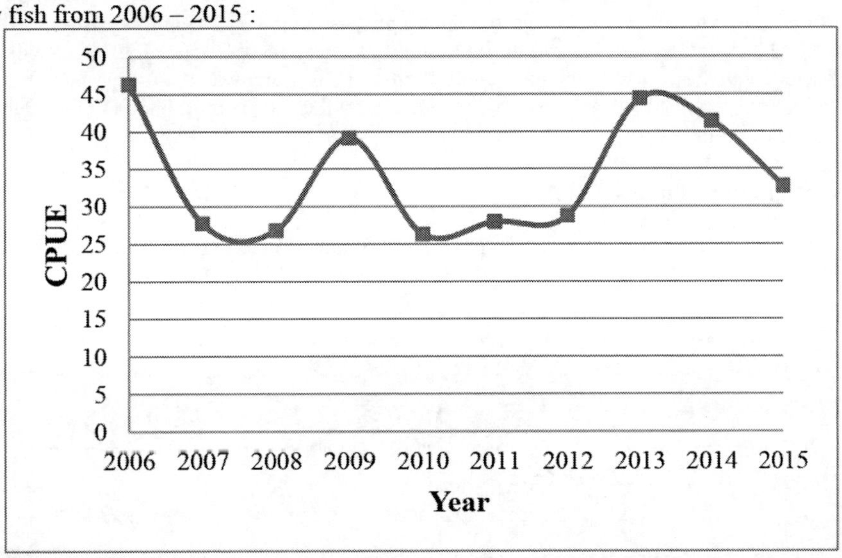

Fig. 1. Stingray Fish CPUE Graphic

The change in CPUE value (figure 2) each year was affected by changes in total ship (effort). CPUE value is inversely proportional to effort value, increase in effort would decrease CPUE value. Resources will depleted faster if effort to expoit increasing. CPUE is affected by increased effort occured to increase production. The decrease in effort will increasing CPUE.

3.2. Relation between Catch Effort and CPUE

Relation between catch and effort of stingray fish from 2006 – 2015 :

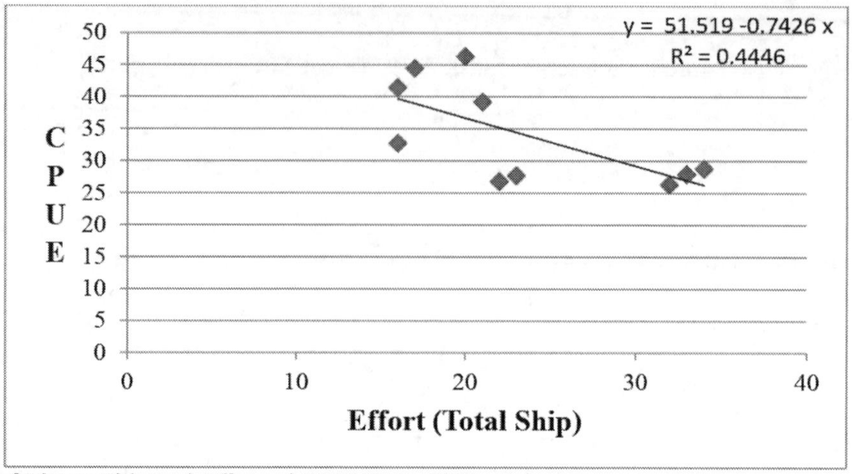

Fig. 2. Regression of Stingray Fish Catch Effort and CPUE 2006 – 2015

Regression graphic (figure 2) between catch effort and CPUE of Stingray fish showed negative relation, higher the effort value then the value of CPUE decreased. Regression analysis resulted in equation value of relation between catch effort and production Y = 51.519 -0.7426X with $R^2 = 0.4446$.

The result showed that the value of one ship in catching stingray in Kejawan Nusantara Fisheries Port was 51.519. Slope (b) value was -0.7426, negative in value can be interpreted as every increase in catch effort will decrease CPUE value by 0.7426. Increase in catch effort will decrease CPUE value [8]. R value was 0.4446 or 44% interpreted as effort had high influence over CPUE, other 56% were seasons, machine, and fishing ground. Factors that affect CPUE was far fishing ground and change of nature [9].

3.3. Maximum Sustainable Yield (MSY) Analysis

Regression was used to calculate optimum catch result (MSY) and optimum catch effort (fMSY). Schaefer model was used :

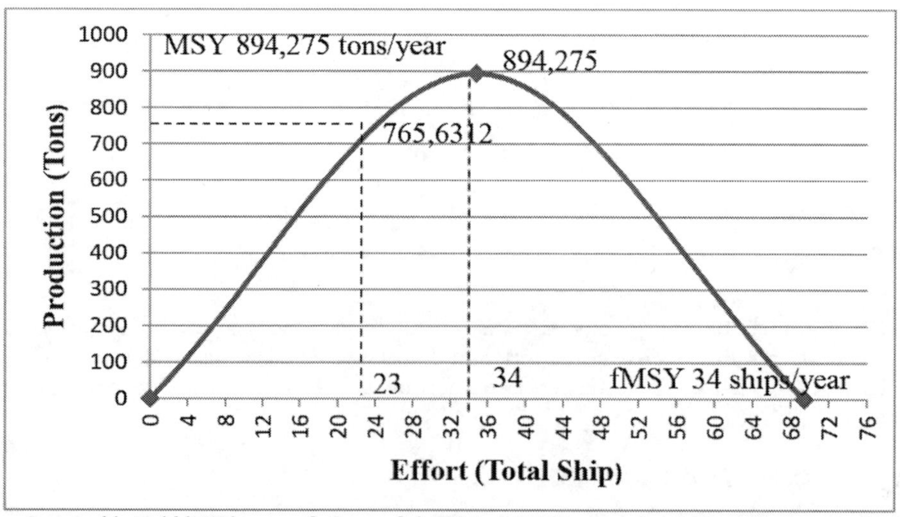

Fig. 3. Maximum Sustainable Yield (MSY) curve of stingray fish 2006 – 2015 (Source : Annual Statistic Report of Kejawanan Nusantara Fisheries Port 2015)

Based on curve (figure 3) Schaefer model, stingray fMSY value was 34 ships/ year and MSY value was 894.275 tons/year. Statistical report recording 23 ships was used. The effort of catch in 2006 – 2015 did not exceed optimum value.

Stingray fish potential of sustainable (MSY) calculated using Schaefer model was 894.275 tons/year. Based on production data and statistical report of stingray fish production value was 765.6312 tons/year. The result showed stingray fish catch was still under fishing and sustainable. Resource utilization value was 80% from MSY, if the value is lower than 80% it is called underfishing and can be further improved but not exceeding the MSY value [10].

3.4. Stingray Species

Stingray fish species catched and recorded during the research:

Table 1. Stingray species catched and recorded during research

No	Species	Local Name	Total
1.	*Dasyatis zugei*	Sharp nosed ray	163
2.	*Dasyatis kuhlii*	Oil ray	113

3.5. Stingray fish sex ratio

3.5.1. Sharp nosed ray (Dasyatis zugei) sex ratio

Recorded 77 male fish and 86 female fish catched during the research, the sex ratio of sharp nosed ray fish was 1.00 : 1.10. The result showed sharp nosed ray sex ratio was still balanced.

3.5.2. Oil ray (Dasyatis kuhlii) sex ratio

Recorded 48 male fish and 65 female fish catched during the research, the sex ratio of oil ray fish was 1.00 : 1.35. The result showed oil ray sex ratio was still balanced .

3.6. Length frequency distribution

3.6.1. Sharp nosed ray leght distribution

The length distribution of sharp nosed ray can be seen on figure 4.

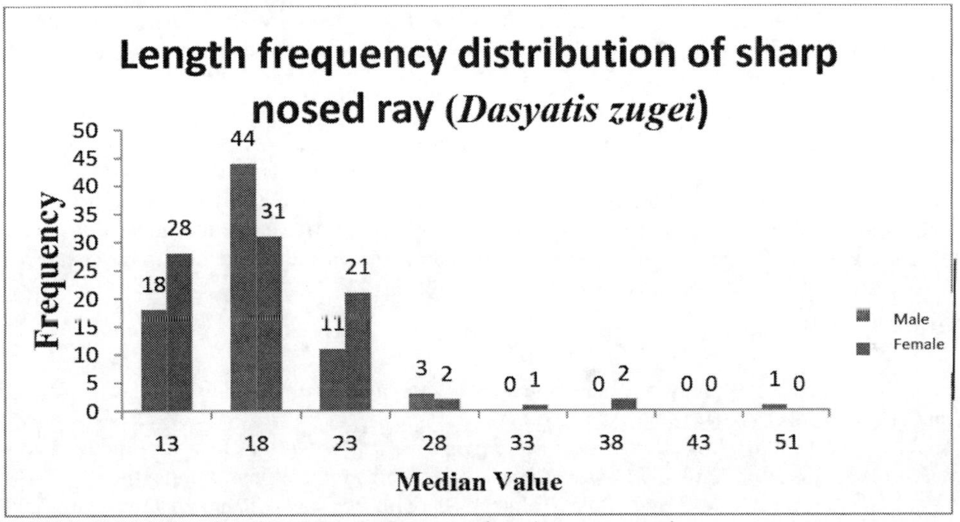

Fig.4. Graphic of legth distribution of male and female sharp nosed ray (*Dasyatis zugei*)

High distribution of sharp nosed ray length is 16 – 20 cm with the composition of 44 male and 31 female. The highest length of sharp nosed ray was 16 – 20 cm. Young (immature gonad) sharp nosed ray (*Dasyatis zugii*) length between 10 cm – 17 cm. Length of sharp nosed ray with riped gonad are 19.1 cm female and 17.8 cm. It can be concluded 16 cm – 20 cm length sharp nosed ray is fish that with immature and first riped gonad.

3.6.2. Oil Ray length distribution

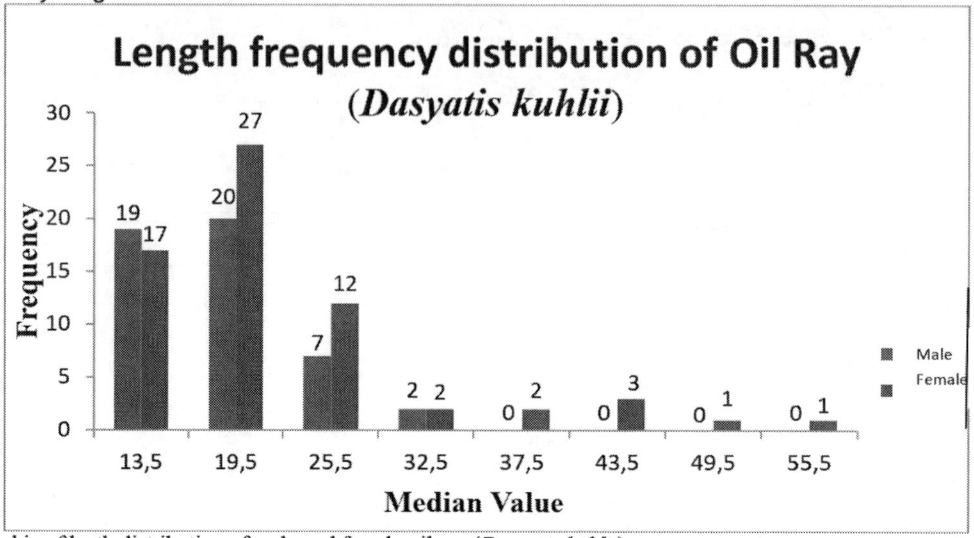

Figure 5. Graphic of legth distribution of male and female oil ray (*Dasyatis kuhlii*)

Oil ray highest leght distribution for male was 19.5 cm and for female was 17 cm -22 cm, with 20 male and 27 female fish. Oil ray population was dominated by immature gonad fish with 17 cm – 22 cm in leght. Immature gonad Oil ray fish (*Dasyatis kuhlii*) have length between 12 cm – 26 cm. The sample caught during the research dominantly oil ray fish that did not ripe.

The high number of immature gonad fish caught in the fishing activity can fisheries resources, the damage will increase if the fishes caught have low respawn rate. Catch fisheries activity that catch immature fish will damage the fish life cycle and endangering it is population.

4. Conclusion

The conclusion of this research is, from the MSY result it can be interpreted that Stingray fishing in Java Sea is still underfishing. There is two type of stingray caught during the research namely sharp nosed ray *Dasyatis zugei* and oil ray *Dasyatis kuhlii* with balanced sex ratio. The fish caught is still young (immature gonad) observed from caught fish lenght ratio. Further research and technology are needed to do fisheries activity to sustain the fish resources.

Refferences

1. Susanto, E. dan A.S Fahmi. Senyawa Fungsional dari Ikan : Aplikasinya dalam Pangan. *Jurnal Aplikasi Teknologi Pangan*. 1 (4) : 95 – 102. (2012)
2. Budiharsono, S. *Teknik Analisis Pembangunan Wilayah Pesisir dan Lautan*. Jakarta: Pradnya Paramita. (2001)
3. Nelson JS. 1994. *Fishes of The World Third Edition*. New York: John Wiley & Sons, Inc.
4. Dinas Perikanan Propinsi Jawa Barat. Buku Tahunan Statistik Perikanan Tangkap Jawa Barat 2006. Pemprov Jawa Barat, Dinas Perikanan. Bandung. (2006)
5. Widodo A.P.A. Pemanfaatan dan Pengelolaan Sumberdaya Ikan Pari di Laut Jawa. *Tesis*. Sekolah Pascasarjana Institut Pertanian Bogor. (2004)
6. Budiman. Analisis Sebaran Ikan Demersal sebagai Basis Pengelolaan Sumberdaya Pesisir di Kabupaten Kendal. *Tesis*. Sekolah Pascasarjana Universitas Diponegoro, Semarang.114 hal. (2006)
7. Purwanto. *Makalah Pengelolaan Sumberdaya Ikan*. Disajikan Pada Workshop Pengkajian Sumberdaya Ikan. (2003)
8. Cahyani, R.T., Sutrisno A., dan Bambang Y. Potensi Lestari Sumberdaya Ikan Demersal (Analisis Hasil Tangkapan Cantrang yang didaratkan di TPI Wedung Demak). *Prosding Seminar Nasional Pengelolaan Sumberdaya Alam dan Lingkungan*. 378-383 hal. (2013)
9. Prihartini, A. Analisis Tampilan Biologis Ikan Layang (*Decapterus sp*) Hasil Tangkapan *Purse Seine* yang didaratkan di PPN Pekalongan. *Tesis*. Universitas Diponegoro. Semarang. 25 hal. (2006)
10. Anugrahini, D. R. Analisis Pengaruh Penurunan Stok Ikan Terhadap Pendapatan Nelayan Kecamatan Muncar, Banyuwangi, Jawa Timur. *Tesis*. Universitas Diponegoro. Semarang. (2011)

Species richness, abundance and vertical distribution of Copepods in Banda Sea, Indonesia

Hanung Agus Mulyadi[1]* *Abdul* Basit[2], *Idha* Yulia Ikhsani[3], and *Malik* Sudin Abdul[3]

1 Research Center for Deep Sea (RCDS)-Indonesian Institute of Sciences (LIPI), Plankton Laboratory, Jl. Y. Syaranamual, Poka Ambon, Maluku, Indonesia
2 RCDS-LIPI, Physical Oceanograpy Laboratory, Jl. Y. Syaranamual, Poka Ambon, Maluku, Indonesia
3 RCDS-LIPI, Chemical Oceanography Laboratory, Jl. Y. Syaranamual, Poka Ambon, Maluku, Indonesia

Abstract. The aims of this research were to investigate the species richness and abundance of Copepods. Copepods were sampled using rosette sampler (surface, 25 m and 75 m) on November 21st – 30th, 2013, from the Banda Sea. A total of 21 taxa of Copepods belonging to three orders were identified. Calanoid was identified as the most richness order, followed by Cyclopoid and Harpacticoid. The highest value of total abundance in the surface water is 121 ind/6L, in the 25 m of depth is 221 ind/6 L, and in the 75 m of depth is 96 ind/6L. *Acrocalanus longicornis* is the most abundant taxa of Copepods, followed by *Acartia bispinosa*, and *Subeucalanus mucronatus*. The most predominant taxa of *Acartia bispinosa*, and *Subeucalanus mucronatus* showed a remarkable peak at 25 m layers.

1 Introduction

Plankton has an important role in the sustainability of fisheries resources in the Banda Sea. Some literature about oceanographic condition and plankton were conducted on Banda, such as Snellius Expedition that covered primary productivity [1]; oxygen condition [2]; and zooplankton abundance [3]. Copepods are the most important components of the zooplankton in marine ecosystems. They are the principal conduit for the flow of energy between primary producers and the higher trophic levels. Studies of zooplankton in coastal and even oceanic waters of the world have shown that copepods constitute the most important group in marine zooplankton communities [4]. The aims of this research were to investigate the species richness and abundance of copepods.

2 Material and Methods

This research was conducted from November 21st-30th 2013 used Baruna Jaya VII RV. The sampling location was in the Banda Sea from 126.9°E to 130.1°E covered 19 stations

* Corresponding author: hanu001@lipi.go.id

(Figure 1). Plankton samples were collected from six liters of seawater using rosette bottles. The plankton samples represent three different layers, surface (0 m), 25 m and 75 m. Formalin 4% was used for samples preservation. In the laboratory, water samples then filtered by using 100 mm filter. Identification processes followed several literatures [5, 6, 7].

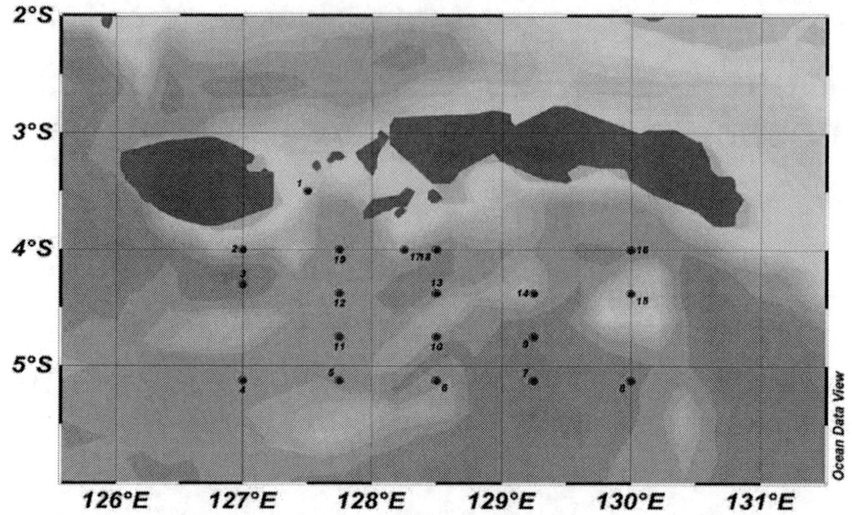

Fig 1. Sampling stations on Banda from 21st-30th November 2013.

Physical data such as temperature and salinity were measured in situ by using CTD (conductivity, temperature, depth) SBE-19 Plus. Onboard Baruna Jaya VII, the seawater samples were collected from discrete depth using rosette bottles (12 mL) that mounted on CTD carrousel. The sea water then subsampled for dissolved oxygen and nutrient content determination. Dissolved oxygen (DO) concentration was determined onboard using Winkler titration method [8]. The remain sea water then filtered through 0.45 μm filters before transferred on polyethylene bottle and stored at 4°C for nutrient analysis. The nutrient concentration was determined at Chemistry Oceanography Laboratory of Research Center for Deep Sea by using Spectrophotometer UV-Visible Shimadzu 1700. Nitrate was analyzed by standard pink azo dye method, while phosphate using standard molybdenum blue method [9]. The software Ocean Data View (ODV) was used to visualize the cross-section of copepods and hydrological condition along sampling area [10].

3 Results

3.1 Oceanographic factors

Vertical distribution of sea surface temperature (SST) and sea surface salinity (SSS) in the Banda Sea from different layer were visualized on Figure 2. SST and SSS in Banda Sea ranges from 28.2°C–30.2°C and 33.35 – 34.85 psu, respectively.

 https://doi.org/10.1051/e3sconf/20184703004

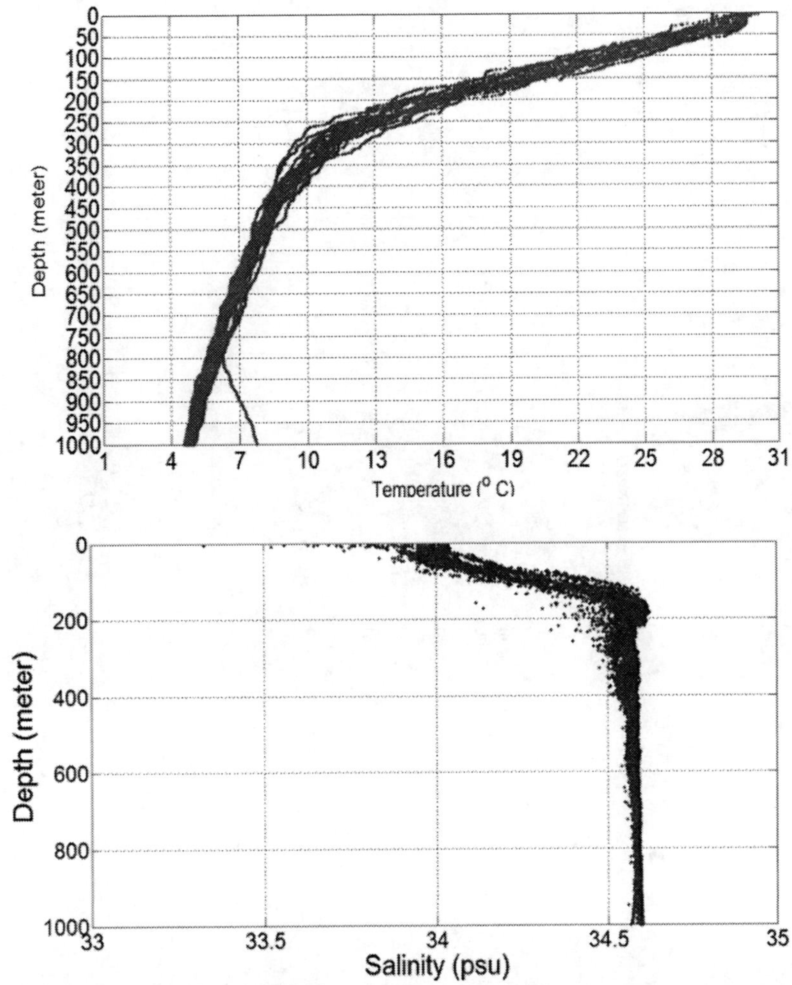

Fig 2. Vertical profile of SST (above) and salinity (below) in Banda Sea

The distribution of dissolved oxygen (DO), nitrate, and phosphate from different depth were visualized on Figure 3. The concentration of DO in the surface, 50 and 75 m depth were in a range of 5.75-7.13; 4.69-6.91 and 4.077-5.87 mgL^{-1}. The DO concentration tends to have the higher value on surface area than those on 25 or 75 m depth. For the nutrient, the concentration of nitrate and phosphate were in range of 0.0002-0.12 and 0.001-0.007 mgL^{-1} in surface area; 0.0002-0.05 and 0.001-0.008 mgL^{-1} in 25 m depth; 0.0002-0.11 and 0.001-0.1 mgL^{-1} in 75 m depth, respectively.

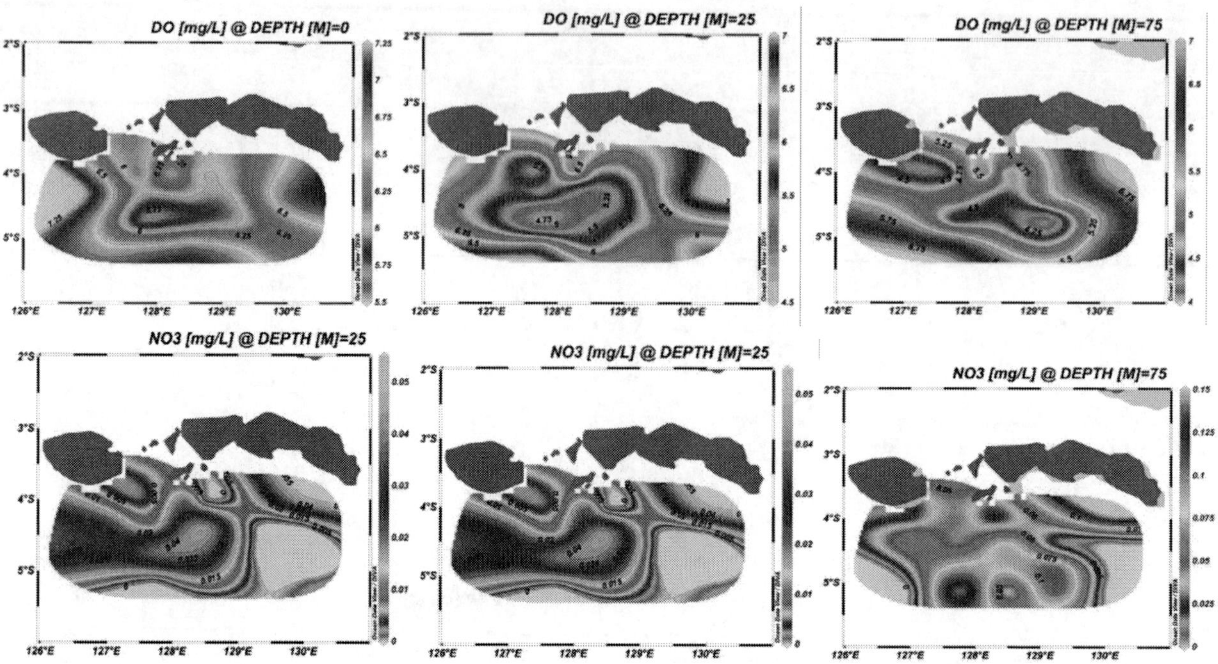
DO [mg/L] @ DEPTH [M]=0
DO [mg/L] @ DEPTH [M]=25
DO [mg/L] @ DEPTH [M]=75
NO3 [mg/L] @ DEPTH [M]=25
NO3 [mg/L] @ DEPTH [M]=25
NO3 [mg/L] @ DEPTH [M]=75
2°S
3°S
4°S
5°S
126°E
127°E
128°E
129°E
130°E

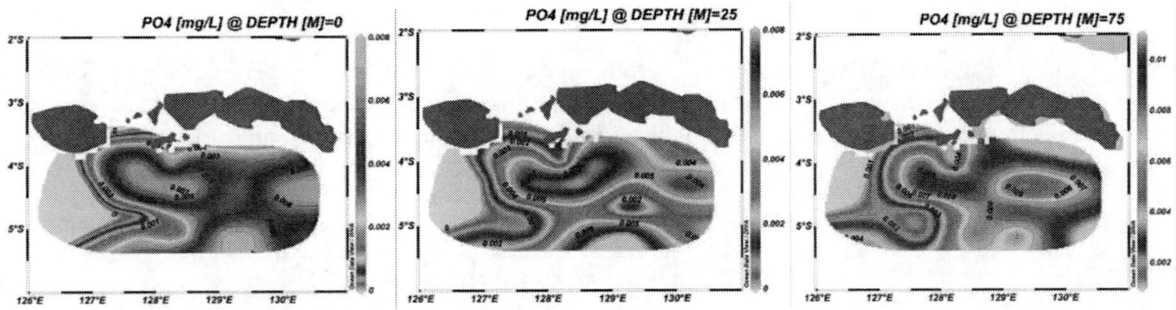

Fig 3. Oceanographic conditions in Banda Sea, 21th-30th November 2013.

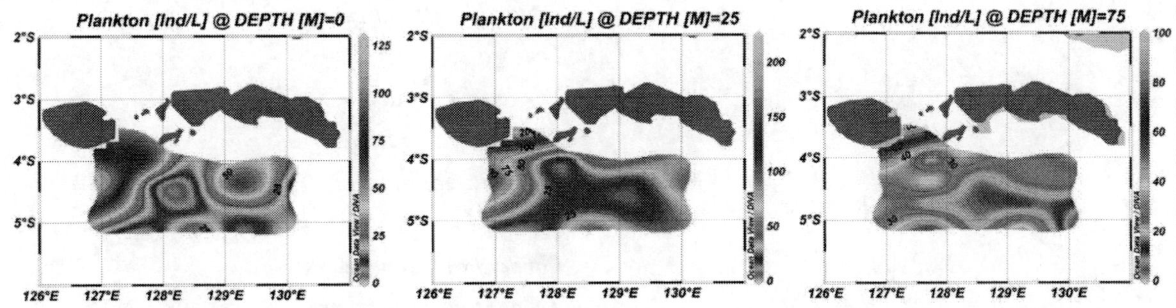

Fig 4. Abundance and vertical distribution of Copepods in Banda Sea.

3.2 Species richness

A total of 21 copepods taxa were identified, predominantly to species, over the 18 sampling occasions during the study period. These included 13 calanoid, 5 cyclopoid, and 3 harpacticoid taxa. Detail of species richness in the Banda Sea shown in Table 1.

Tabel 1. Species richness in Banda Sea, 21th-30th November 2013.

1	CALANOIDA	Candacia
2		*Centropages furcatus*
3		*Acartia bispinosa*
4		*Pontella inflatodigitata*
5		*Temora* sp
6		*Tortanus gracilis*
7		*Labidocera caudata*
8		*Euchaeta longicornis*
9		*Scolecithrix* sp
10		*Paracalanus aculeatus*
11		*Acrocalanus longicornis*
12		*Rhincalanus nasutus*
13		*Subeucalanus mucronatus*
14	CYCLOPOIDA	*Copilia quadrata*
15		*Saphirina* sp
16		*Oncaea* sp
17		*Oithona brevicornis*
18		*Corycaeus* sp
19	HARPACTICOIDA	*Macrosetella* sp
20		*Clytemnestra* sp
21		*Euterpina* sp

3.3 Abundance and vertical distribution

The abundance of copepods in the surface ranges from 7-121 ind/6L with average 34±30.34 ind/6L, in 25 meters of depth ranging from 16-221 ind/6L (44±48.43 ind/6L) and in the 75 m of depth ranging from 6-96 ind/6L (32±20.04 ind/6L) Figure 4. *Acrocalanus longicornis* is the most abundant taxa of Copepods, followed by *Acartia bispinosa*, and *Subeucalanus mucronatus* (Table 2).

Table 2. The most abundance taxa of Copepods (ind/6L) in Banda Sea.

No	Taxa		Average abundance (ind/6L)
1	CALANOIDA	*Candacia sp*	1
2		*Centropages furcatus*	2

3		*Acartia bispinosa*	7
4		*Pontella inflatodigitata*	2
5		*Temora sp*	2
6		*Tortanus gracilis*	1
7		*Labidocera caudata*	2
8		*Euchaeta longicornis*	3
9		*Scolecithrix sp*	2
10		*Paracalanus aculeatus*	2
11		*Acrocalanus longicornis*	8
12		*Rhincalanus nasutus*	2
13		*Subeucalanus mucronatus*	6
14	CYCLOPOIDA	*Copilia quadrata*	1
15		*Saphirina* sp	1
16		*Oncaea* sp	5
17		*Oithona brevicornis*	5
18		*Corycaeus* sp	3
19	HARPACTICOIDA	*Macrosetella* sp	3
20		*Clytemnestra* sp	1
21		*Euterpina* sp	3

Copepods have a high abundance in the Western area of Banda Sea rather than in Central and Eastern Banda Sea, mainly in Station 1,3 and 2 (figure 3). Station 1 has the highest average abundance (123±88,008 ind/6L), followed by Station 3 (86±46,68 ind/6L) and Station 2 (71±17, 35 ind/6L), respectively. The most predominant copepods of *Acartia bispinosa*, *Acrocalanus longicornis* and *Subeucalanus mucronatus* and showed a remarkable peak at 25 m layers (Figure 5).

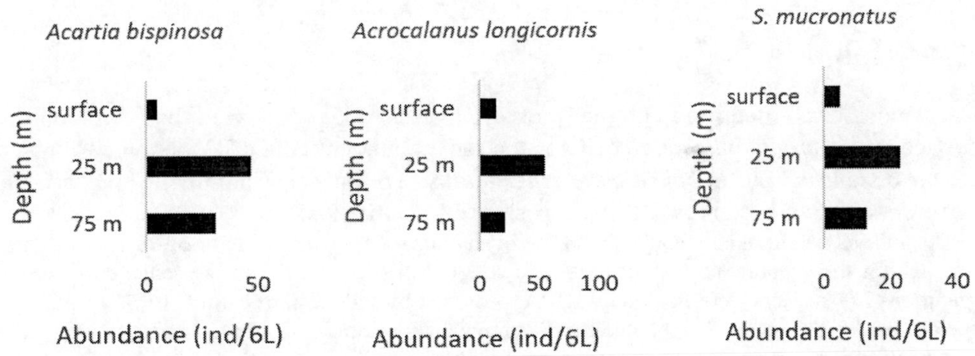

Fig 5. Abundance of predominant taxa *Acartia bispinosa*, *Acrocalanus longicornis* and *S. mucronatus* (Station 1).

Thermocline layer record on 200-500 m of depth and the halocline was occur in 100-200 m of depth. The physical conditions around the Banda Sea are influenced by monsoon [11, 12, 13]. The southeasterly winds generally expose the Banda Sea between May and October periods generating upwelling in this area. As consequence of this, the Sea Surface Temperature (SST) in the Banda Sea is relatively cold compared to the other periods. In this period, the SST varies between 25-28 °C. In contrast, the downwelling is generally observed during the northwest monsoon (December – March) when the northwesterly winds expose the Banda Sea. The relative warm SST is generally found in this period varying between 28.5-30 °C. The period between the two seasons (April and November) is called as the transition period. In this period, the winds start to change their directions. According to the result of concentration measurement, the nutrient was minimal in 25 m depth due to the most abundance of phytoplankton in this area. Nitrate was used by phytoplankton for photosynthesis processes, while phosphate was used either for photosynthesis or for tissue construction [14].

Our results show that species richness of copepods dominated by calanoid (61.91%), followed by cyclopoid (23.81%) and harpacticoid (14.29%). The species richness in the Banda Sea is similar with other copepod community structure in Piru Bay [15]; and other territorial waters [16, 17]. In the Western area of Banda Sea have the high abundance of copepods, it may be related with the nutrient upwelling. Nutrient upwelling has the hight contents of nitrate and phosphate. After that, nutrients will be utilized by phytoplankton. Further, zooplankton herbivore grazing on phytoplankton [18].

Copepods are known to vertically migrate even in the lower oxygen condition. In this study, the DO (dissolved oxygen) concentration in the Central of the Banda Sea more lower than in the Western and Eastern Banda Sea (Figure 3). It appears that the oxygen minimum concentration affected the distribution and abundance of copepods in the Banda Sea. The calanoid copepod *Acartia tonsa* identified at high abundance at a layer in well-oxygen concentration but absent at the same depths when oxygen dropped [19].

A large number of herbivorous copepods *Acartia bispinosa*, *Acrocalanus longicornis* and *Subeucalanus mucronatus* in most of station probably due to the higher abundance of phytoplankton. Grazing of herbivorous zooplankton more than 80% of phytoplankton production during summer [20]. Further, the feeding behaviour, as well as the food quantity and quality, are recognized factors that affect copepods growth [21].

4 Conclusions

Calanoid was identified as the most richness order, followed by Cyclopoid and Harpacticoid. *Acrocalanus longicornis* is the most abundant taxa of Copepods, followed by *Acartia bispinosa*, and *Subeucalanus mucronatus*. The most predominant taxa of *Acartia bispinosa*, and *Subeucalanus mucronatus* showed a remarkable peak at 25 m layer.

The authors would like to thanks Dr. Augy Syahailatua, M.Sc, the Director of Research Center for Deep Sea for the supports and opportunities to conduct the observation in the scheme of Banda Sea expedition. The surveys were funded by DIPA, provided by Indonesian Institute of Sciences in 2013. Special thanks are offered for plankton technicians, Salomy Hehakaya and La Imu for the determination analysis and the assistance in the field work, collecting samples.

References

1. W.W.W. Gieskies, G.W. Kraay, A. Nontji, D. Setiapermana, Sutomo, Neth. J. Sea. Res. **25 (4)**, 11 (1990)
2. S.B. Tijssen, M. Mulder, F.J. Wetsteyn, Neth. J. Sea Res. **25 (4)**, 15 (1990)

3. M.A. Baars, A.B. Sutomo, S.S. Oosterhuis, O.H. Arinardi, Neth. J. Sea Res. **25 (4)**, 17 (1990)
4. J.T. Turner, Zool. Stud **43 (2)**, 12 (2004).
5. Mulyadi, Treubia **32**, 167 (2002)
6. Mulyadi, *Calanoid copepods in Indonesian Waters* (Research Centre for Biology, Indonesian Institute of Sciences, (2004)
7. I.E. Yamaji, *Illustrations of the marine plankton of Japan* (1984)
8. M. Ehrhardt, K. Grasshoff, K. Kreming, T. Almgren, *Methods of seawater analysis* (Verlag Chemie Weinheim, 1983)
9. J.D.H. Strickland, T.R. Parsons. *A practical handbook of seawater analysis* (1972)
10. R. Schlitzer, Comp. and Geo. Sci. 28 (2002)
11. K. Wyrtki, *Physical oceanography of Southeast Asian Waters*, **2** (NAGA Rep, 1961)
12. A.L. Gordon, R.D. Susanto, Banda Sea surface-layer divergence, Ocean Dyn. **52 (2)**, 9 (2001)
13. R.D. Susanto, A. Ffield, A.L. Gordon, T.R. Adi, J.Geophys. Res. **117**, 16(2012)
14. R. Chester, *Marine geochemistry* (1993)
15. H.A. Mulyadi, *Prosiding seminar nasional III : hasil hasil penelitian perikanan dan kelautan FPIK UNDIP*, 6 (2013)
16. C.A. Gaonkar, V. Khrisnamurthy, A.C. Anil, Environ. Monit. Assess. **168**, 16 (2010)
17. H.M.M. Peralta, F. Md. Yusoff, Inter. J. Ecosystem. **5 (2)**, 5 (2015)
18. S. Gasparini, J. Castel, J. Plankton Res. **19**, 14 (1999)
19. M.R. Roman, A.L. Gauzens, W.K. Rhinehart, J.R. White, Limnol. Oceanogr **38**, 12 (1993)
20. G.B. Mc Manus, B.A. Costas, H.G. Dam, R.M. Lopes, S.A. Gaeta, S.M. Susini, C.H. Rosetta, Hydrobiologia **575**, 13 (2006)
21. C.A. Vargas, R.A. Martinez, R. Escribano, N.A. Lagos, J. Mar. Biological Assoc. The U.K. **90 (6)**, 13 (2010)

Diversity and abundance of giant clams in Anambas Islands, Indonesia

Syawaludin Alisyahbana Harahap[1,*], **Yogi Yanuar**[2], **and Yuwanda Ilham**[2]

[1] *Marine Department, Faculty of Fisheries and Marine Science, Padjadjaran University*
[2] *Loka NACA Pekanbaru, MMAF Republic of Indonesia*

Absract. Giant clam is one of the most common types of bivalves found in coral ecosystem. Geographically, these clams have limited distribution in the Indo-Pacific tropics, from the Red Sea to the Pacific Islands of Tuamotu. Ecologically, the clam acts as bio-filtration and natural bio-deposition. The presence of clams becomes a marker that the waters are still good and the absence of harmful pollutants. The population of this clam in Indonesian waters continues to decline. Anambas archipelago are known for its rich diversity of marine life. Since the establishment of this area became NACA as an ATP in 2014, marine tourism activities grew rapidly. Therefore it is necessary to do this research which aims to know the diversity of species and abundance of clams in the waters of Anambas Islands. Field survey was conducted in August 2015 by using Belt Transect method. The research results found 6 species with D_1 value is 0.593. The overall abundance is 0.144 indv.m^{-2}. *Tridacna crocea* has the highest abundance and relative abundance of the highest species of the other species of 60.54%. Meanwhile, the smallest relative abundance is *Tridacna gigas* which is only 2.36%.

Keywords: *Abundance, Anambas Islands, Diversity, Giant Clam*

1 Introduction

Tridacnidae or giant clam is one of the most common types of bivalves found in coral ecosystem waters. Giant clam lives in association with coral reefs by immersing themselves in the substrate (drill) or sticking to the sandy bottom of the water. These animals live in symbiosis with the dinoflagellata algae (Symbiodinium) that grow in the mantle tissue and are able to photosynthesize resulting in food and pigment [1].

Geographically, these clams have limited distribution in the Indo-Pacific tropics, from the Red Sea to the Pacific Islands of Tuamotu [2]. Around the world, this animal has two genera (Tridacna and Hippopus) consisting of nine species that are *Tridacna gigas*, *Tridacna squamosa*, *Tridacna derasa*, *Tridacna maxima*, *Tridacna crocea*, *Tridacna tevoroa*, *Tridacna rosewateri*, *Hippopus hippopus* and *Hippopus porcellanus* [3]. Seven species of giant clam apart from *T. tevoroa* and *T. rosewateri* are found in Indonesian waters [4].

Ecologically, the clams are the biota that acts as bio-filtration and natural bio-deposition. This ability is because they are a deposit feeder or suspension feeder. They are also able to filter out ammonia and soluble nitrate in seawater for the need for zooxanthellae to nitrogen for its growth process [4]. The presence of clams becomes a marker that the waters are still good and the absence of harmful pollutants.

Giant clams (Tridacnidae) have been an important traditional source of food and income for many coastal Indonesian people. Kondisi ini menyebabkan tingkat populasinya di alam semakin sedikit. This is also the basis why the giant clam is one of the marine animals that

*Corresponding author: iwalhrp@gmail.com

need to be protected throughout the world including in Indonesia. Since 1987, however, exploitation of natural stocks of giant clam has been prohibited by the Indonesian Government [5]. Government Regulation no. 7 of 1999 incorporated into seven types of clams living in Indonesia as protected animals. Nevertheless, the population of this clam in Indonesian waters continues to decline due to hunting, habitat destruction, use of fish bombs and potassium, as well as for the fulfillment of people consumption [6]. If this situation continues, it is feared that there will be a decline in population and lead to the extinction of various species of giant clam in nature. This situation has made this clam an animal that is included in appendix II list of CITES (Convention on International Trade in Endangered Species) [7].

Anambas archipelago is an archipelago located in the ecoregion of Sunda Shelf [8]. Geographically, this area is in the Natuna Sea (**Fig. 1**). This archipelago waters are known for its rich diversity of marine life. Since the establishment of this area became National Aquatic Conservation Area (NACA) as an Aquatic Tourism Park (ATP) in 2014, marine tourism activities grew rapidly.

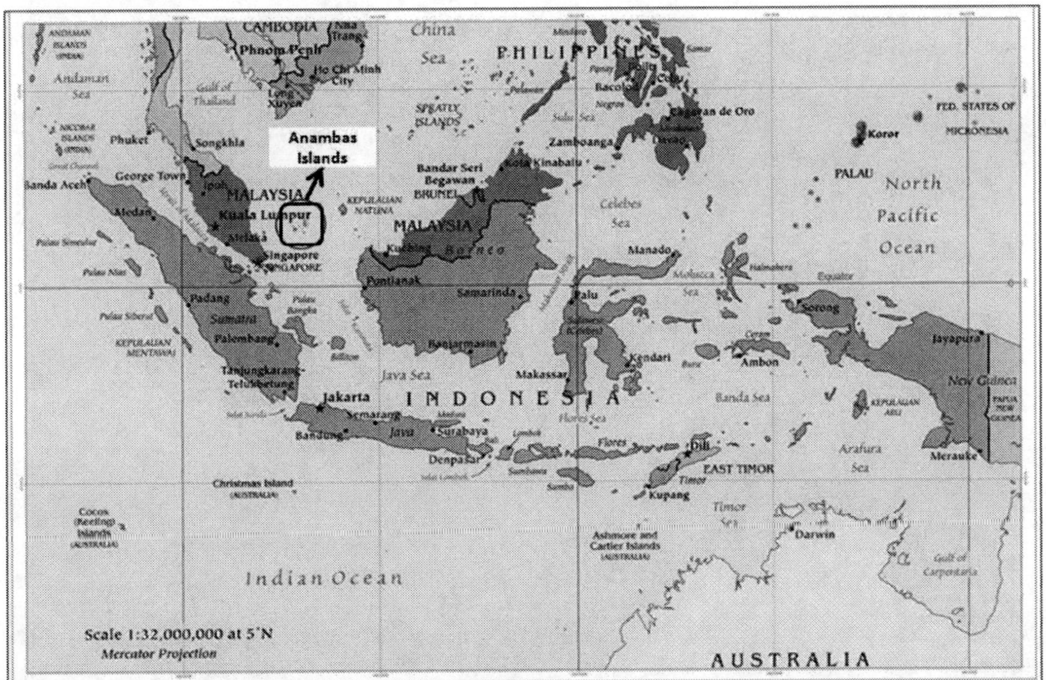

Fig. 1. Orientation map of Anambas Islands position in Indonesi

In recent years and along with these conditions, the existence of marine ecosystems feared to be degraded. This was also coupled with the use of a variety of marine life including the exploitation of tridacna for different economic commodities. The study of species diversity and abundance, or at least species richness, gives ecologists insights into the stability of communities (Walker, 1988 in [9]).

Information on giant clams collected from various locations has been described by [10], but specifically for Anambas waters do not yet exist. Therefore it is necessary to do this research which aims to know the diversity of species and abundance of clams in the waters of Anambas Islands. The results of this study are expected to be used as information for the government and other stakeholders involved in the protection, conservation and utilization of giant clams and the management of their habitats in a sustainable manner.

2 Materials and Method

2.1 Survey

Field survey was conducted in August 2015. A total of 23 sites were surveyed spread over in the waters of Anambas Islands (**Fig. 2**). The data collecting was conducted on 42 sampling points (**Table 1**).

Survey method used is a modification of Line Intercept Transect (LIT), namely Belt Transect and has been used previously by [11]. Each research station is only one transect line and pulled along the 70-meter transect line parallel to the coastline at a depth of 7-10 meters. Observations were made at every 1 meters to the right and the left of the transect line. Therefore, the area observed at each station is 140 m^2 [12].

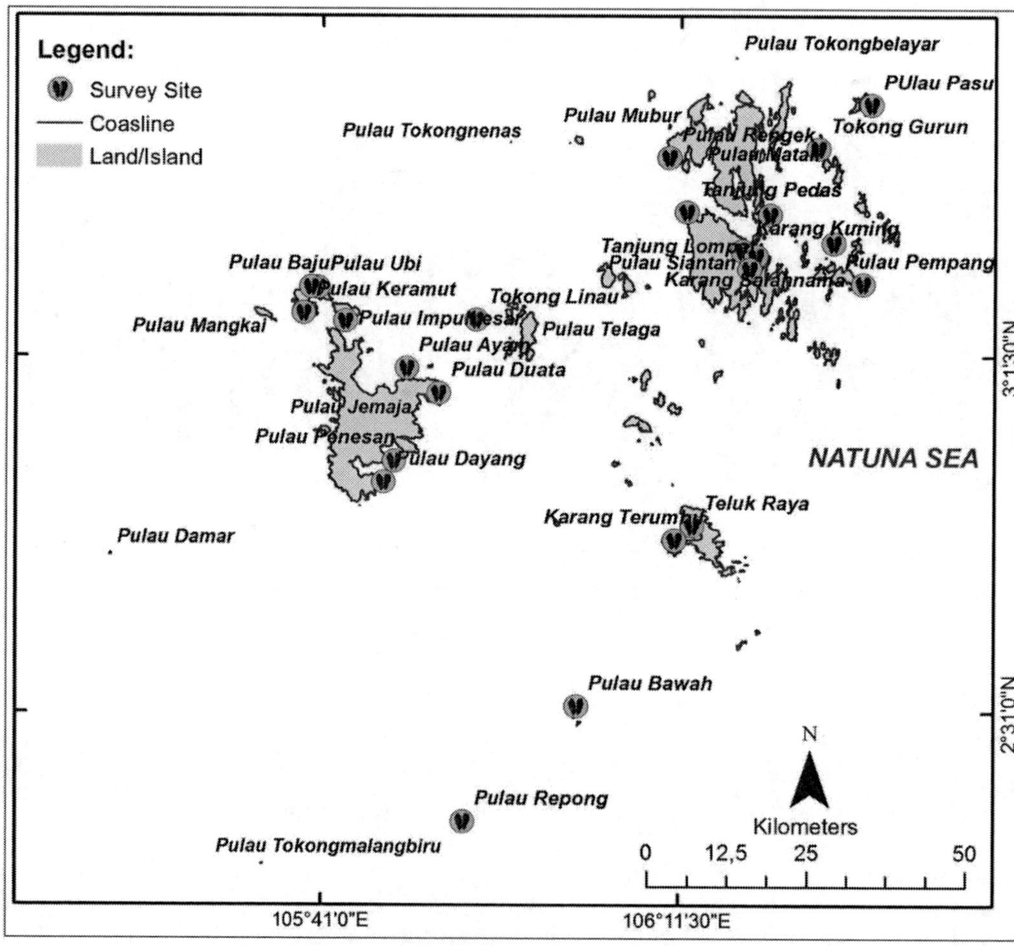

Fig. 2. Map of research location and survey sites

2.2 Diversity Index

Measuring and analyzing of diversity based on **Simpson's Index (D)**. Simpson's Diversity Index is a measure of diversity. In ecology, it is often used to quantify the biodiversity of a habitat. It takes into account the number of species present, as well as the abundance of each species. Simpson's diversity index (**D_1**) performed best when differentiating sites, but simpler indices were slightly preferable when detecting effects of land use intensity on diversity [13]. The formula to calculate species diversity according to Simpson is:

$$D_1 = 1 - \sum P_i^2 \quad (1)$$

where P_i is the proportion of individuals found in species i. For a well-sampled community, we can estimate this proportion as $P_i = n_i/N$, where ni is the number of individuals in species i and N is the total number of individuals in the community. The range is from 0 to 1, where:

- High scores (close to 1) indicate high diversity.
- Low scores (close to 0) indicate low diversity.

2.3 Abundance

There are 2 abundances analyzed, the first is the abundance and the second is the relative abundance. Calculation of giant clam abundance found in each station is calculated based on [14].

$$A = \frac{N}{area} \quad (2)$$

where A is abundance, N is total of Individu (indv.) and area is wide of sampling area (meter), meanwhile to calculate relative abundance using equation:

$$RA = \frac{n_i}{N} x\ 100\% \quad (3)$$

where RA is Relative abundance, ni is the number of individuals in species i and N is the total number of individuals in the community.

3 Results and Discussion

3.1 Species Diversity of Giant Clams

The results of research, from the 23 sites censused 849 individual of giant clams were found, comprising 6 species from 2 generas: *Tridacna gigas* (n = 20), *T. squamosa* (n = 79), *T. derasa* (n = 77), *T. maxima* (n = 125), *T. crocea* (n = 514), and *Hippopus hippopus* (n = 34). However, none of the surveyed sites were found in the six species. Maximum only found 5 species. There are only 4 sites have highest of species number such as in Karang Beretih, Teluk Raya, Pulau Bawah and Pulau Impulbesar. Meanwhile, in the sites of Karang Kuning and Karang Terumbu found only 1 species of each *Tridacna crocea* and *T. maxima*. **(Table 1).**

The number of species of clams found in this region is more than half the number of species of clams in Indonesia and even in the world. The comparison can also be seen with the waters of the Seribu Islands and Manado [2] which each has only 3 and 4 species. Even for the relatively close waters of Reef's Singapore, only 2 species are found [15]. Greater diversity of giant clam found in the central Indo-Pacific [16].

In Table 1 it can also be seen that Simpson's diversity index (D_1) at each site ranges from 0 - 0.727. The highest D_1 value are in the Karang Beretih site, whereas the lowest D_1 value is in Karang Kuning and Karang Terumbu. Overall, the giant clam diversity index based on Simpson's diversity index (D_1) in Anambas Islands waters is 0.593 (**Table 2**). The index value shows a relatively high diversity because the value is more than 0.5 on a scale of 0 to 1.

Table 1. Giant Clams Numbers and Diversity in each site of Anambas Islands Waters

Sites	Num. Of Transect	Number of Giant Clams# (Indv.)							Num. Of Species	D_1
		n1	n2	n3	n4	n5	n6	Tot.		
Pulau Pasu	2	48	0	0	0	9	0	57	2	0,266
Tokong Gurun	2	39	2	1	0	6	0	48	4	0,322
Pulau Pengerat	2	109	0	0	1	31	2	143	4	0,372
Karang Beretih	2	21	1	13	7	0	14	56	5	0,727
Karang Kuning	1	13	0	0	0	0	0	13	1	0,000
Karang Salahnama	2	7	0	0	0	2	0	9	2	0,346
Tanjung Lompat	2	5	1	0	0	2	1	9	4	0,617
Pulau Pempang	2	28	0	0	0	12	0	40	2	0,420
Teluk Raya	2	13	13	3	2	0	26	57	5	0,684
Karang Terumbu	1	0	0	0	0	1	0	1	1	0,000
Pulau Bawah	2	11	9	0	1	1	5	27	5	0,686
Pulau Repong	3	1	0	6	0	4	3	14	4	0,684
Tokong Linau	3	4	0	0	0	7	3	14	3	0,622
Pulau Impulbesar	2	112	6	19	0	30	4	171	5	0,526
Pulau Ubi	1	5	0	9	0	0	0	14	2	0,459
Pulau Baju	2	7	0	10	3	0	6	26	4	0,713
Pulau Keramut	2	37	0	1	0	12	2	52	4	0,439
Pulau Ayam	1	7	0	0	0	5	0	12	2	0,486
Pulau Dayang	2	6	0	15	0	0	8	29	3	0,614
Pulau Penesan	2	4	1	0	0	2	0	7	3	0,571
Tanjung Pedas	1	16	1	0	0	0	1	18	3	0,204
Pulau Duata	2	7	0	0	6	1	3	17	4	0,671
Pulau Rengek	1	14	0	0	0	0	1	15	2	0,124
TOTAL	**42**	**514**	**34**	**77**	**20**	**125**	**79**	**849**		

n1 = *Tridacana crocea*; n2 = *Hippopus hippopus*; n3 = *T. derasa* ; n4 = *T. gigas*; n5 = *T. Maxima;* n6 = *T. Squamosa*

Table 2. Simpson's diversity index (D_1) of giant clams in Anambas Islands Waters

No.	Species	Num. of Giant Clams (n_i)	P_i	P_i^2	D_1 $(1-\Sigma P_i^2)$
1	*T. crocea*	514	0,605	0,367	
2	*T. maxima*	125	0,147	0,022	
3	*T. squamosa*	79	0,093	0,009	
4	*T. derasa*	77	0,091	0,008	**0,593**
5	*H. hippopus*	34	0,040	0,002	
6	*T. gigas*	20	0,024	0,001	
	Σ	**N= 849**		**0,407**	

3.2 Abundance (A) and Relative Abundance (RA)

Overall, giant clams abundance from 5880 m^2 (42 transects) of total area surveyed in Anambas Islands is 0.144 indv.m^{-2} (**Table 3**). This condition still higher then other places in western Area In Indonesia like Karimunjawa Islands were ranged only from 0.03 to 0.04 indv.m^{-2} [17]. But still lower than the survey results in several other places in eastern area of Indonesia like in Cenderawasih Bay found giant clams natural populations densities were from 0.6 to 0.7 indv.m^{-2} [18].

In **Table 3** and also the graph in **Fig. 3** shows that *T. crocea* is the highest abundance of 0.087 indv.m^{-2}. *Tridacna. squamose* and *T. derasa* have the same abundance of 0,013 indv.m^{-2}. Meanwhile, the smallest abundance of 0.003 indv.m^{-2} is *T. gigas*. The data presented here indicate that giant clam abundance are low. This is very different from the results obtained in Ashmore, Mermaid and Cartier Reefs, located within the Great Barrier Reef, host populations in excess of 100 indv.m^{-2}. Another example is on Tatakoto, French Polynesia, has an abundance of 90.9 indv.m^{-2} [19, 20].

Table 3. The number of giant clams, abundance and relative abundance of each species

No.	Species	Num. of Giant Clams	Abundance (A) (indv.m^{-2})	Relatif Abundance (RA) (%)
1	*T. crocea*	514	0,087	60,542
2	*T. maxima*	125	0,021	14,723
3	*T. squamosa*	79	0,013	9,305
4	*T. derasa*	77	0,013	9,069
5	*H. hippopus*	34	0,006	4,005
6	*T. gigas*	20	0,003	2,356
	Σ	**849**	**0,144**	**100**

The results of the data analysis show that the RA of giant clam ranged between 2.36% to 60.54%. Species of *T. crocea* has the highest abundance and relative abundance of the highest species compared with the other species that is equal to 60.54%. Meanwhile, the smallest relative abundance is *T. gigas* which is only 2.36% (**Fig. 4**). *Tridacna crocea*, *T. maxima*, *T squamosa* and *T. derasa* belong to predominant species, whereas *H. hippopus* and *T. gigas* belong to subdominant species. This grouping is based on the approach presented by [21] which states that if RA> 5% belongs to predominant species, RA 2-5% belongs to subdominant species and RA <2% belongs to non-dominant species.

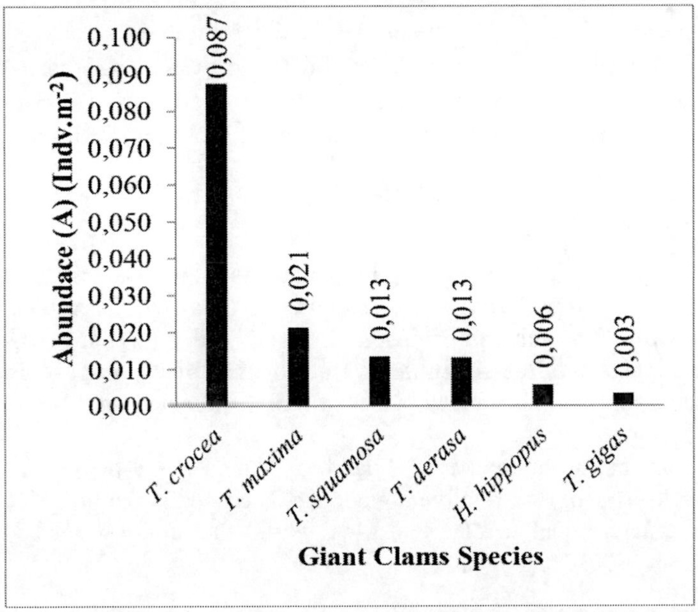

Fig. 3. Abundance of giant clam species

The composition of this abundance is similar to the observation of the population of giant clam in several places in Indonesian waters. The population is dominated by small species such as *T. crocea* and *T. squamosa*, while large species such as *T. derasa* and *T. gigas* are rarely found [22, 23]. The high abundance of giant clam species in a waters is thought to be due to habitat factors suitable for the ability of the species to adapt and survive. The substrate type is one of the key factors in this regard [23].

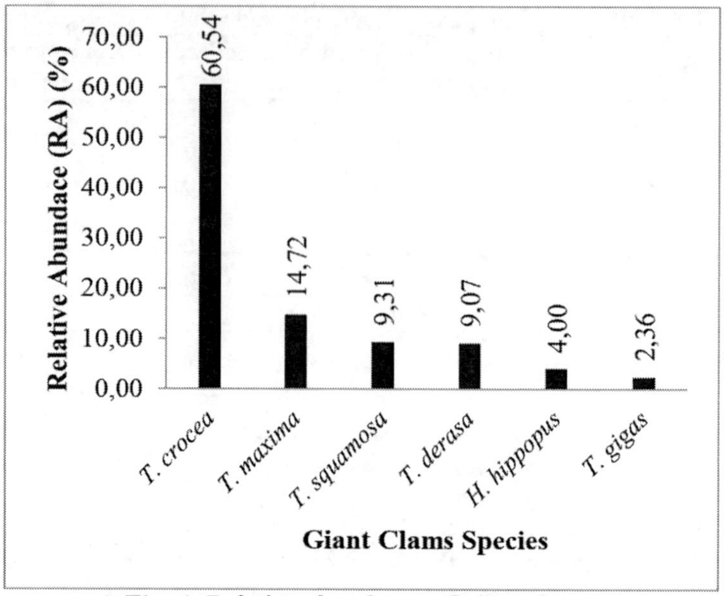

Fig. 4. Relative abundance of giant clams

All the species of giant clam identified in the Anambas Islands Waters to date vary greatly in their distribution and abundance, but all are threatened by overexploitation and habitat degradation. Even though they are listed by IUCN (*International Union for Conservation of Nature*) and international trade is CITES regulated, giant clam populations continue to decline [10].

4 Conclusions

The diversity of giant clam species found in Anambas Island waters shows 6 species from 2 generas namely *Tridacna gigas*, *T. squamosa*, *T. derasa*, *T. maxima*, *T. crocea* and *Hippopus hippopus*. The value of Simpson's diversity index (D_1) in each sites ranges from 0 - 0.727. The highest biodiversity index value located in Karang Beretih site. Overall in Anambas waters, the diversity index value of 0,593 indicates that the area has a relatively high diversity of giant clam.

Overall, giant clam abundance in Anambas Islands is 0.144 indv.m^{-2}. The abundance dominated by *T. crocea*. The relative abundance T. crocea is the highest compared with the other species that is equal to 60.54%. Meanwhile, the smallest relative abundance is *T. gigas* which is only 2.36%

Acknowledgements

Thanks are given to Loka NACA Pekanbaru which has provided material support and personnel to the researchers to carry out surveys and give permission to publish the results of data survey.

References

1. Niartiningsih A. Giant Clam, Rere Marine Life: Cultivation and Conservation Makassar: Identitas Universitas Hasanuddin (2012)
2. Yusuf C, Ambariyanto , Hartati R. Abundance of Tridacna (Family Tridacnidae) at Seribu Island And Manado Waters, Indonesia. Ilmu Kelautan **14**, 3: p. 150-154 (2009)
3. Lucas JS, Copland JW. *Giant Clams in Asia and The Pacific*. 1st ed. Canberra:

Australian Centre of International Agricultural Research (1988)

4. Mudjiono. Note Aspect of Life Giant Clam Tridacnidae Family (Mollusca, Pelecipoda). Oseana **13**, 2: p. 37-47 (1988)
5. Firdausy C, Tisdell C. *Status of, and Demand for Giant Clams in Indonesia: a Survey of Four Coastal Villages in Bali and Java*. In Tisdell C. Giant Clams in the Sustainable Development of the South Pacific. Canberra: Australian Centre for International Agricultural Research p. 101-116 (1992)
6. Setiawan H. Threats to Giant Clam Population (Tridacnidacna *sp.*) and conservation efforts in Taka Bonerate National Park. Info Teknis EBONE p. 137-147 (2013)
7. CITES. Appendices I, II and III. [Online] Available from: https://www.cites.org/eng/app/appendices.php (2017)
8. MMAF , MPAG. *Geographic Priority Sea Biodiversity For The Development of Conservatioan Area In Indonesia*. 1st ed. Huffard CL, Erdmann MV, Gunawan T, editors. Jakarta: Ministry of Marine Affairs and Fisheries-MMAF and Marine Protected Areas Governance-MPAG (2012)
9. Hamilton AJ. Species diversity or biodiversity? Journal of Environtmental Management **75**, 2005: p. 89-92 (2005)
10. Othman AS, Goh GHS, Todd PA. The distribution and status of giant clams (Family Tridacnidae) – A short review. The Raffles Bulletin of Zoology **58**, 1: p. 103-111 (2010)
11. Braley RD. Distribution and Abundance of *Tridacna gigas* and *T. derasa* in Great Barrier Reef. Micronesia **20**: p. 215–223 (1987)
12. Arbi UY, Cappenberg HAW, Sihaloho HF. *Megabenthic Monitoring*. In Suharsono , Sumadhiharga OK. *Guidance of Coral Reefs Health Monitoring*. Jakarta: COREMAP CTI LIPI p. 63 (2014)
13. Morris EK, Caruso T, Buscot F, Fischer M, Hancock C, Maier TS, et al. Choosing and using diversity indices: insights for ecological applications from the German Biodiversity Exploratories. Ecology and Evolution **4**, 18: p. 3514-3524 (2014)
14. Snedecor GW, Cochran. *Statistical Methods*. 7th ed. Ames Iowa USA: The Iowa State University Press (1980)
15. Neo ML, Todd PA. Population Density and Genetic Structure of The Giant Clams Tridacna crocea and T. squamosa on Singapore's Reefs. Aquatic Biology **14**: p. 265–275 (2012)
16. Spalding MD, Fox HE, Allen GR, Davidson N, Ferdaña ZA, Finlayson M, et al. Marine ecoregions of the world: a bioregional of coastal and shelf areas. BioScience **57**, 7: p. 573-583 (2007)
17. Hadi S. *Distribution of Giant Clams (Tridacnidae) in Burung Island, Karimunjawa*. Filed Work Practice Report. Semarang (2000)
18. Pranowo WS. *Distribution of Giant Clams (Family Tridacnidae) in Teluk Cenderawasih Marine National Park, Irian Jaya*. Undergraduate Thesis. Semarang (1998)
19. Rees M, Colquhoun J, Smith LD, Heyward AJ. *Survey of trochus, holothuria, giant clams and the coral communities at Ashmore, Cartier Reef and Mermaid Reef,*

northwestern Australia. Report to Environment Australia. Queensland (2003)

20. Gilbert A, Remoissenet G, Yan L, Andrefout S. The giant clam Tridacna maxima communities of three French Polynesia islands: Comparison of their population sizes and structures at early stages of their exploitation. ICES Journal of Marine Science **63**: p. 1573–1589 (2006)
21. Van Balen B. *Birds Counts and Bird Observation in Neighbourhood of Bogor*. The Netherlands: Wageningham University (1984)
22. Allo , Setiawan MK, Dewi H, Bisjoe IN, Nurhayati AR, Qiptiyah M. *Study of Tridacna crocea and Siganus canaliculatus ethnoecology in Takabonerate National Park*. Research Result Report. Makassar (2010)
23. Padilah M, Pratomo A, Zulfikar A. *Distribution Pattern of Giant Clam in Teluk Dalam Waters Malangrapat Village Gunung Kijang Subdistrict, Bintan Regency, Kepulauan Riau*. Undergraduate Thesis. Tanjung Pinang (2015)

 https://doi.org/10.1051/e3sconf/20184703006

SEAWEED SPECIES COMPOSITION, ABUNDANCE AND DIVERSITY IN DRINI AND KONDANG MERAK BEACH, JAVA

Taufiq Ahmad Romdoni[1], *Ajeng* Ristiani[1], *Maria* Dyah Nur Meinita[1*], *Bintang* Marhaeni[1], *Setijanto*[1]

[1]Faculty of Fisheries and Marine, Jenderal Soedirman University, Jalan dr. Soeparno GOR Soesilo Soedarman Karangwangkal, Purwokerto

Abstract. The different characteristics of subtrate might give influence on species composition, abundance and diversity of seaweed. Kondang Merak and Drini Beach are two beaches in Java Island which have different subtrate characteristic. The aim of this study is to investigate seaweed composition, abundance and diversity in Drini Beach and Kondang Merak Beach. Survey method and quadrant transect sampling technique were applied in this study. The result showed that 13 species of seaweed were found at Drini Beach and 18 species of seaweed were found in Kondang Merak Beach. The composition of the seaweed at Drini Beach and Kondang Merak Beach were dominated by Rhodophyta (69% and 56%). *Enteromorpha flexuosa* was the most abundant seaweed at Drini Beach, while at Kondang Merak Beach was dominated by *Chaetomorpha crassa*. The seaweed diversity index of Kondang Merak Beach (2,08) was higher than Drini Beach (1,27). Physical and chemical parameters were categorized as the optimal for the growth of the seaweed.

1 Introduction

Indonesia is an archipelago country located in tropical area with potential marine resources. Seaweed is the one of the coastal area's resources that has potential and can be explored. Seaweed is one of marine reources that widely spread habitat in Indonesia. Seaweed habitat distribution occur in intertidal area that has coral reef exposure. Among of the Indonesia's offshore, there are rocky Drini beach and Kondang Merak's beach are shores that has so many various species of seaweed.

Drini and Kondang Merak Beach are located in the southern of Java coastal area. Both of them are located in the different provinces, Drini is located in Special Region of Yogyakarta and Kondang Merak is located in Malang region, East Java Province. The difference of those two locations makes the enviromental characteritics becomes different. Besides the enviromental factor, human activity around those might affect the environmental characteristics. Drini beach is a tourism place that has so many visitors. Meanwhile, Kondang Merak beach has less human activity than Drini beach. It is caused by the location of Kondang Merak beach is a conservative area. Those two different

Corresponding author: *) maria.meinita@unsoed.ac.id

 https://doi.org/10.1051/e3sconf/20184703006

characteristics might influence species composition, abundance and diversity of seaweed in both of locations.

Therefore, a study on species composition, abundance and diversity of seaweed in both of the beaches is needed. The aim of this study is to determine species composition, abundance and diversity of seaweed Kondang Merak Beach and Drini Beach.

2 Materials and Methods

Sampling were conducted in Drini Beach, Gunung Kidul, Special Region of Yogyakarta (S 8°8'10.2" E 110°33'11.3") and Kondang Merak Beach, Malang Regency, East Java (S 8°23'44,5" - 8°23'54,8" E 112°30'58,6" - 112°31'19,7") (Figure 1). Biological parameters were measured consisted of seaweed composition, abundance and diversity index. Physical and chemical observed for supporting the biological data, consisted of pH, salinity, temperature, light penetration, current velocity and type of substrate.

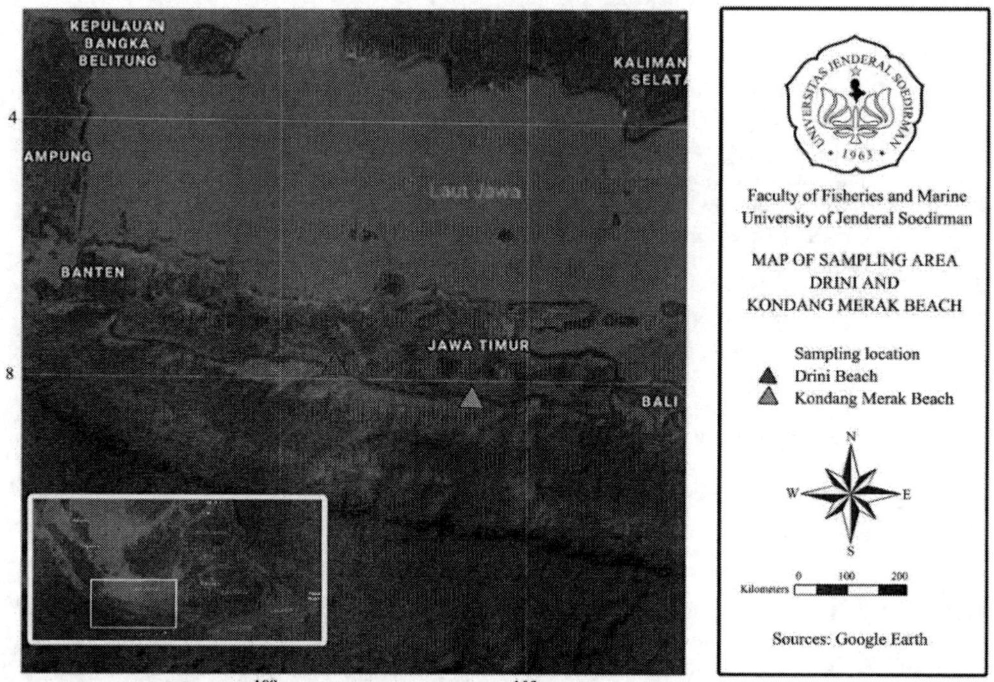

Fig 1. Map of Java Island and sampling area

Survey method with 1x1 square meter quadrant transects was applied in this study. Samples were collected during low tide period in both beaches. Each of the beach had three observation stations with triplicate. In each plot, physical, chemical and biological parameters were measured. Data were analysed using following formula [1,2] :

Abundance

$$Di = \frac{n_i}{A} \quad (1)$$

D_i = abundance species of-i (ind/m^2)
n_i = the total amount of species of-i
A = total sampling area (m^2)

Shanon Wiener diversity index:

$$H' = -\sum_{i=1}^{S} \frac{Ni}{N} \log \frac{Ni}{N} \quad (2)$$

H' = Diveristy Index
S = amount of species
Ni = number of individual of species-i
N = total number of individuals

3 Result and Discussion

Thirteen species of seaweed was identified from study areas of Drini Beach. The seaweeds were divided into 3 Division, 6 Ordo, 8 Family, 11 Genus and 13 Species. The seaweed were obtained consisted of 3 species of Phaeophyta, 1 species of Phaeophyta and 9 species of Rhodophyta. While, in Kondang Merak Beach, 18 seaweed species were found, consisted of 6 species of Chlorophyta, 2 species of Phaeophyta and 10 species of Rhodophyta. Among the 3 division of seaweed, the highest division was Rhodophyta. The Rhodophyta in Kondang Merak Beach consisted of 10 species they were *Gracilaria edulis, G. salicornia, Laurencia papilosa, Acanthophora spicifera, Acrocystis nana, Liagora* sp, *Galaxaura rugosa, Amphiroa fragilissima* dan *Gelideilla acerosa* (Table 1).

Table 1. List of identified seaweed species in Drini and Kondang Merak Beach

A. Species of seaweed obtained from Drini Beach

Class	Ordo	Family	Species
Rhodophyceae	Gracilariales	Gracilariaceae	*Gracilaria edulis*
			Gracilaria salicornia
		Cystocloniaceae	*Hypnea charoides*
	Gelidiales	Gelidiellaceae	*Gelidiella acerosa*
			Gelidium spinosum
			Cryptonema
	Ceramiales	Rhodomelaceae	*Laurencia papilosa*
			Acanthopora spicifera
			Laurencia palisada
Chlorophyceae	Cladophorales	Cladoporaceae	*Chaetomorpha crassa*
		Siphonocladaceae	*Boergesenia forbesii*
	Ulvales	Ulvaceae	*Entheromorpha flexuosa*
Phaeophyceae	Dictyotales	Dictyotaceae	*Padina tenuis*

B. Species of seaweed obtained from Kondang Merak Beach

Class	Ordo	Family	Genus	Species
Chlorophyceae	Ulvales	Ulvaceae	Ulva	*U. fasciata*
			Enteromorpha	*E. compressa*
	Bryopsidales	Caulerpaceae	Caulerpa	*C. serrulata*
		Halimedaceae	Halimeda	*H. opuntia*
	Cladophorales	Cladophoraceae	Chaetomorpha	*C. crassa*
				C. amteninna
Phaeophyceae	Dictyotales	Dictyotaceae	Padina	*Padina minor*

			Dictyota	*D. dichotoma*
Rhodophyceae	Gigartinales	Gracilariaceae	Gracilaria	*G. salicornia*
				G. edulis
		Cystocloniaceae	Hypnea	*Hypnea* sp.
	Cerameales	Rhodomelaceae	Laurencia	*L. papilosa*
			Acanthophora	*A. spicifera*
			Acrocystis	*A. nana*
	Nemaliales	Liagoraceae	Liagora	*Liagora* sp
		Galaxauraceae	Galaxaura	*G. rugosa*
	Coralinales	Coralinaceae	Amphiroa	*A. fragilissima*
	Gelidiales	Gelidiaceae	Gelidiella	*G. acerosa*

There are some differences of seaweed species which found in Drini and Kondang Merak Beach. Some species that occured in Drini were not found in Kondang Merak Beach. In the other way around, some seaweed species were found in Kondang Merak Beach but they were absent in Drini beach (Table 2).

Table 2. Comparison of seaweed recorded from Drini and Kondang Merak beach.

Species	**Drini**	**Kondang Merak**
Ulva fasciata	-	+
Enteromorpha compressa	-	+
Enteromorpha flexuosa	+	-
Caulerpa serrulata	-	+
Halimeda opuntia	-	+
Chaetomorpha crassa	+	+
Chaetomorpha anteninna	-	+
Padina minor	-	+
Padina tenuis	+	-
Dictyota dichotoma	-	+
Gracilaria salicornia	+	+
Gracilaria edulis	+	+
Hypnea	+	+
Laurencia papilosa	+	+
Laurencia palisada	+	-
Acanthophora spicifera	+	+
Acrocystis nana	-	+
Liagora sp	-	+
Galaxaura rugosa	-	+
Amphiora fragilissima	-	+

Gelidiella acerosa	+	+
Gelidium spinosum	+	-
Cryptonema	+	-
Boergesenia forbesii	+	-

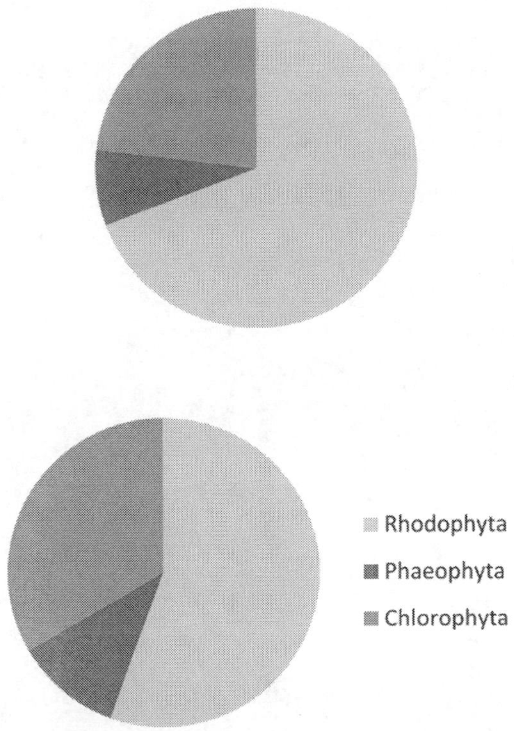

a. b.

Fig 2. Composition of Seaweed species (a) Drini Beach and (b) Kondang Merak Beach

Division of Rhodophyta was the most abundant seaweed in the both of beaches (Figure 2). Rhodophyta have wider adaptability compared with chlorophyta and phaeophyta. Rhodophyta are found in large numbers in vertical zones between the sea level to the bottom of the sea that is less suited to the Phaeophyta and Chlorophyta. Dead coral substrate or dead coral fragments are the most suitable subtrate for genus *Acanthophora, Codium, Gelidiella, Galaxaura, Jania, Amphiroa* and *Gracilaria* [3]. Present study showed that the abundance of Rhodhophyta on Drini and Kondang Merak Beach, might have relate with types of the substrate which is dominanted by rocky substrate.

 https://doi.org/10.1051/e3sconf/20184703006

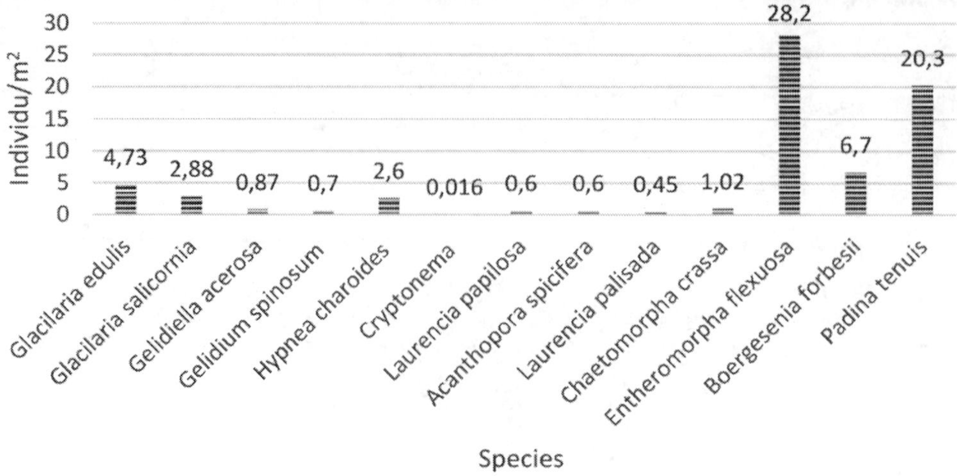

(a)

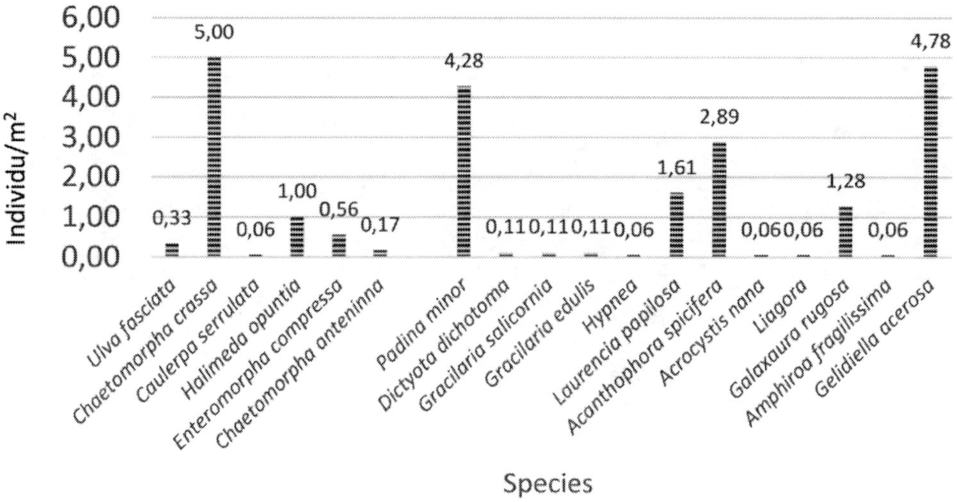

(b)

Fig 3. Abundance of seaweed (a) Drini Beach, (b) Kondang Merak Beach

Enteromorpha flexuosa was seaweed species with the highest abundance in Drini Beach. The abundance of this species might be caused by the influence of physical and chemical factors in the environment as well as the influence of the morphology of the species of seaweed. Physical and chemical environmental factors in this zone is supportive to the growth of the zone where it is passed by the huge waves. While the highest abundance of seaweed on the Kondang Merak Beach was *Chaetomorpha crassa*. This species grows on rocky substrates in the intertidal up to subtidal zone or areas that are always exposed to the air during low tide [4].

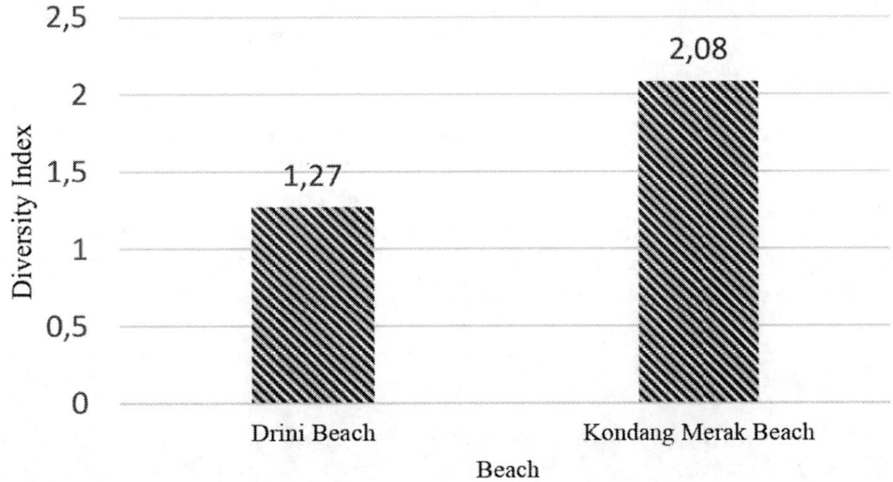

Fig 4. The diversity index were measured at Drini Beach and Kondang Merak Beach.

Diversity index of the Kondang Merak Beach is higher than Drini Beach. High and low values of the diversity index can be influenced by a number of species, the quantity of seaweed, and the balance of species distribution [5]. The level of diversity of seaweed on Drini and Kondang Merak Beach can also be caused by human activity. Drini beach as a tourism place, can have an impact on the community structure of seaweed due to human activities. Diversity of species tends to be low because of the disruption of an ecosystem [6]. Increased rates of tourists is one of the causes of the disruption of biological diversity [7].

Table 4. Comparison of physical and chemistry were measured in Drini and Kondang Merak Beach

Parameters	Drini	Kondang Merak
pH	8,0 ± 0,0	8,0 ± 0,0
Salinity (ppt)	30,0 ± 0,0	31,9 ± 0,3
Temperature (°C)	30,0 ± 0,07	28,7 ± 0,9
Light Penetration (cm)	25,0 ± 0,0	14,11 ± 4,26
Current (m/s)	0,33 ± 0,0	1,03 ± 0,11
Substrate	Rocky substrate	Rocky Substrate

The type of substrate in both Drini and Kondang Merak beach were rocky substrate. The type of substrate is an important factor that affects the existence of seaweed, this is because the substrate becomes a place for seaweed to attach themselves. The rocky substrate is stable, hence the seaweed can be strongly attached to the substrate so it is not easy to crash when high wave attack. The results showed that pH at Drini Beach and Kondang Merak still belong to a good range for seaweed growth. Seaweed can grow optimally in the pH range of 6.8 to 9.6 [7].

Kondang Merak Beach had a higher salinity than Drini Beach. Based on the salinity at Drini Beach and Kondang Merak, generally the condition of coastal waters is still good enough to support the growth of seaweed. The optimum salinity range for sea weed growth is 30-35 ppt [7, 8, 9]. Drini Beach had higher temperature than Kondang Merak Beach.

However the temperature is still within in the optimum range for seaweed grow. The high temperature might effect seaweed survival [10]. The current in Kondang Merak Beach was stronger than than Drini Beach. The movement of currents in the waters of Kondang Merak Beach might be influenced by the current pattern that occurs from the wave fragments due to the occurence of small islands in the southern part of the beach. Water movement serves to supply nutrients also help seaweed absorb nutrients, and carry out the exchange of oxygen. The result of light penetration measurement on both beaches is the result of 100% brightness measurement. Penetration of light can still penetrate to the bottom of the water. Light penetration is strongly influenced by weather conditions, measurement time, turbidity, suspended solids, and the level of vision of the researcher [11].

4 Conclusion

The different environmentalcharacteristic of beach and human activity might affect seaweed abundance, composition and diversity in Drini and Kondang Merak Beach. *Enteromorpha flexuosa* was the most abundant species in Drini Beach, while the highest abundance species in Kondang Merak Beach was *Chaetomorpha crassa*. *Rhodophyceae* was the highest class in Drini Beach and Kondang Merak Beach. Kondang Merak Beach has a seaweed diversity higher than Drini Beach.

We thank Jenderal Soedirman University for providing financial support through Riset Unggulan Terapan 2018 Research Grant.

References

1. C.J Dawes. *Marine Botany*. John Wiley and Sons, Inc., Canada, 628pp (1981)
2. V.K Dhargalkar, D. Kavlekar, *Seaweeds: A field manual. National Institute of Oceanography*, New Delhi, 36pp. (2004)
3. H. Arfah, S. Papalia, Kepadatan dan Keragaman Jenis Rumput Laut di Perairan Pesisir Teluk Weda, Propinsi Maluku Utara. *Jurnal Ilmu dan Kelautan Tropis.* **7** (2): 745-755. (2015)
4. Jr. G. C. Trono,. *Field Guide and Atlas of the Seaweed Resources of the Philippines.* Bookmark. Makati (1997)
5. F. Hadi, I. J. Zakaria, Z. Syam, Diversity of Macroalgae in Kasiak Gadang Island Nirwana Beach, Padang-West Sumatra, Indonesia. *The Journal of Tropical Life Science.* **6 (2)**: 97-100 (2016)
6. H. Sormin, G. S. Gerung, U.N.W.J. Rembet, Struktur Komunitas Rumput Laut di Pulau Mantehage, Provinsi Sulawesi Utara. *Aquatic Science & Management.* **3 (2)**: 32-37 (2015)
7. V.S.K. Chennubhotla, K.S. Rao, N. Kaliaperumal, Effect of Some Adverse Factors on the Diversity and Distribution of Marine Macro Algae Along the Indian Coasts. Seaweed Research and Utilization Association. **36 (1)**: 5-11. (2015)
8. K. Lüning, *Seaweeds: Their Environment, Biogeography and Ecohysiology.* John Willey & Sons, Inc. New York. 230 p. (1990).
9. J.W Nybakken,. *Biologi Laut Pendekatan Ekologis*. PT Gramedia Pusaka. Jakarta (1992)
10. L Sumich, *An Introduction to The Biology of Marine Life.* Wmc Brown. Dubuque. Lowa (1992)
11. H. Effendi, *Telaah Kualitas Air.* PT Kanisius. Yogyakarta. (2003)

.

Spatio-Temporal Distribution of Chlorophyll-a Using Multitemporal Landsat Image and Ground Check in Segara Anakan Lagoon

Rose Dewi [1, 3*], Muhammad Zainuri [2,4], Sutrisno Anggoro [1],
Tjahjo Winanto [3], Hadi Endrawati [5]

[1] Coastal Resources Management Doctoral Program,
Fisheries and Marine Science Department, Diponegoro University
[2] Departement of Oceanography, Fisheries and Marine Science Department, Diponegoro University
[3] Fisheries and Marine Science Department, Jenderal Soedirman University
[4] Centre of Excellent Mitigation of Natural Disaster and Coastal Rehabilitation, Diponegoro University
[5] Fisheries and Marine Science Department, Diponegoro University

Abstract. Segara Anakan lagoon (SAL) is a dynamic estuary, downstream of several rivers in Citanduy watershed, humid tropical climate and influenced monsoon, conditions that cause waters to fluctuate. There are influence of natural and anthropogenic factors that impact on the increase of aquatic organic matter, as phytoplankton indicator biota able to show its tolerance characterized by fluctuation of chlorophyll content a. The purpose of this research is to know the spatio-temporal distribution of chlorophyll- a SAL region. The chlorophyll- a measurement method uses multitemporal landsat images and field measurements. The temporal approach (time series) during 2016 refers to the monsoon wind pattern and spatially on seven different stations of ecological characteristics. Data processing using ER Mapper 7.0 software and the preparation of thematic data with Arc GIS software. Testing data accuracy using regression analysis. The results of chlorophyll-a recording of landsat images ranged from 3,213 - 3,703 (mg / m^3) and field measurements from 0.01 to 5.06 (mg / m^3). Temporally transitional seasons and on spatial approaches, stations with anthropogenic characteristics have the highest chlorophyll content. The condition is supported by the accumulation of aquatic organic materials, high rainfall and upwelling phenomenon. Test of accuracy pertained high, shown with value of r = 0,94 at equation $y = -30.07x^2 + 217.0x - 387.0$, $R^2 = 0.894$ with regression characteristic is polynomial. The high accuracy value in the equation, shows that the SAL area algorithm can be applied to obtain chlorophyll content data in the field in the previous year with the support of satellite image data in that year.

1 Introduction

Segara Anakan Lagoon (SAL) is geographically located at 7°35"- 7º46" LS and 108°45" - 109º01" BT, covering 14.221,8 ha, in district of Kampung Laut, Cilacap Regency, Central Java. SAL is located in the southern of coastal area of Java Island. The existence of Nusa Kambangan Island as a barrier, disconnects the SAL from the Indian Ocean. The lagoon is connected by two inlets (canals): Eastern Pelawangan (Eastern outlet) on the East Nusakambangan island and West Pelawangan (Western outlet). Its moist tropical climate and influence from monsoon causes the fluctuated water. The form changes happening up to now occured through various stages of environmental change. This has been caused by the SAL area does not stand alone. There are natural and anthropogenic factors, especially in the west affecting in the ecological changes of the region [1], [3], [6].

The naturally influence factor resulted in the lagoon having shallow depth (< 2m), as an estuary which receives freshwater flow in high quantity especially during rainfall, coming along with suspended sediment derived from river Citanduy as main river and largest Citanduy watershed besides river Cimeneng, river Cikonde, river Cibereum.

[1] Corresponding author: rose.83unsoed@gmail.com

They are only the parts of the lagoon leading to the Indian Ocean that has depth (10m <) [12]. Another side, the anthropogenic effect is due to the increase of comparable population to the increasing of the use of SAL land and area around Activities in Citanduy Watershed are : fishery, agriculture, industry and domestic (settlement). This gave higher pressure on the lagoons with water bodies decline and the accumulation of organic matter on the lagoon[1], [7], [12].

According to Parslow, *et al.,* [18] and Jennerjahn, et al., [12] the high-low organic materials are converted to the limiting factor of the distribution and quantity of chlorophyll-a phytoplankton, subsequently affecting its abundance. Generally, high-chlorophyll a in coastal waters as the supply of organic material from the mainland through river runoff and vice versa tends to decrease in offshore areas [8], [4]. Furthermore, Yuwono, *et al.,* (2007) and Jennerjahn, *et al.,* [13] chlorophyll -a SAL region is averaged from 1.8 to 6.1 μg / L. Chlorophyll-a is the main photosynthetic pigment in phytoplankton, as the primary biota of the recipients of the effects of demonstrated aquatic ecological changes through adaptation to the increase in organic material [14], [21]; [2].

Laboratory measurement of chlorophyll on the field (Ground check) in large waters requires the points of multiple station, relatively long time and less cost effeciency. Further remote sensing methods using satellite imagery is proposed a variety of conveniences. This technology enables to provide information quickly so it can be observed the phenomenon of waters in the range of spatial and dynamic range [9], [5]. One example of remote sensing applications that can detect distribution patterns and abundance of phytoplankton is estimated through chlorophyll-a by using landsat satellite imagery. The use of satellite of imagery will be obtained by distribution profile and chlorophyll quantity in real time, because the data information based on spectral value of object is recorded by sensor landsat [16]. Chlorophyll– a measurement ground check is to verify data from satellite image mapping. Furthermore, the accuracy test can be undergone to determine the extent to which the image can provide chlorophyll a water information. Accuracy is done by comparing the value of chlorophyll of landsat image recording with field measurement (Ground check) [20]. The purpose of this research is to know the spatio-temporal distribution of chlorophyll content of Western SAL region obtained through landsat multitemporal landscaping and ground check.

2 Materials and Methods

The measurement of chlorophyll content was carried out by using multitemporal landsat imagery and Ground check, with a year-long temporal (time series) approach (2016) based on monsoon representation. According to Nontji [15], based on Meteorological and Oceanographic data, Indonesian waters are affected by the monsoon :

- **West Season** : Desember – February
  (representing February and December).
- **Transition Season 1**: March - May
  (representing Mei).
- **East Season** : June - August
  (representing Juli)
- **Transition Season II**: September-November
  (Representing October).

The temporal approach is used as a reference time of data collection of chlorophyll-a phytoplankton in the field ground check. Furthermore, spatial approach is carried out by selecting the seven research stations that have ecologically characteristic difference of SAL area. It is aimed to evaluate the ecological influence (response) on the impact of natural factors in the station (2, 3, 4, 6) and the influence of community anthropogenic activities around the lagoon at the station (1, 5, 7). Further description of the spatial approach is shown in (Table.1).

Tabel. 1 The sampling characteristic by spacial approach of phytoplanton in Segara Anakan Lagoon, District of Cilacap.

	Characteristics	Ordinate	
1	TPI and Majingkak Port	108°48'02.4"BT	07°40'27.6"LS
2	Gate Border (Pelawangan Barat) of SAL borderlining with Hindia Ocean	108°46'56.7"BT	07°41'59.0"LS
3	Gate of Cikonde dan Cimeneng river	108°49'47.9"BT	07°40'34.6"LS
4	The Area of Mangroves	108°51'36.5"BT	07°41'44.9"LS
5	Kampung laut settlement	108°52'14.0"BT	07°42'19.5"LS
6	Conservative tourism "Mina" "Ujung Alang, district of Kampung Laut	108°52'45.4"BT	07°42'55.4"LS
7	Cultivating place by using floating net	108°48'56.0"BT	07°41'01.0"LS

2.2 The laboratorially content measurement of chlorophyll-a in Ground check

Ground check of chlorophyll a is done simultaneously as the satellite passing through the SAL region. The analysis used theory stated by Hutagalung, *et al.*, [11]. Water samples for the determination of phytoplankton chlorophyll content were started by filtration using Whatman GF / C papers with a diameter of 47 mm, placed on a filter holder (dampened with aquades, to prevent the filtered paper from attaching), the filter glass was installed. Put 500ml of water sample. Filtering assisted suction pump ± 30 cm Hg, until until the pores can not be seen. pori filter mampat. The filtrated filter was extracted by dissolving chlorophyll at 10 mL aceton by 90% and placed in a 15 mL tube, the process is carried out in a space that is impermeable to light. Followed by centrifugation process (filtrate on 15 mL tube) with 4000 rpm rotation, 30 mins to form natan (in the form of rest of filter paper carried and deposited based on test tube) and supernatant (in the form of chlorophyll solution). The supernatant is transferred into cuvet to be measured for absorbance using a spectrophotometer of the DMS-90 Uv.Vis model Variant. at 750 nm, 664 nm, 647 nm, and 630 nm wavelengths. The absorption value is used to calculate chlorophyll-a. Calculation of chlorophyll content according to Hutagalung, *et al.*, [11], calculated by the following formula:

$$\textbf{Chlorophyll - a} = \frac{\{(11.85 \times E\ 664) - (1.54 \times E\ 647) - (0.08 \times E\ 630)\} \times Ve}{Vs \times d}$$

Keterangan:
E 664 : Absorbance 664 nm - Absorbance 750 nm
E 647 : Absorbance 647 nm - Absorbance 750 nm
E 630 : Absorbance 630 nm - Absorbance 750 nm
Ve : Aceton extract volume (mL)
Vs : Volume of example of filtered (liter)

2.2 The determination of chlorophyll- a with analysis of Landscape image Landsat satellite multitemporal

The analysis is done by the following stages:

- Landsat image 8 data was obtained from the download on the site: http://earthexplorer.usgs.gov/ on the process of satellite image data download landsat masking image by using cirrus band for the freed displayed image the cloud . The process of digital satellite data processing 8, is using the manual provided by the USGS http://landsat.usgs.gov/landsat8_using_product.php.
- The correction of Geometric is then performed, with the method map to map reprojection method, is aimed to provide the actual geographic position in the image. Geoographic coordinate system NUTM, changed to SUTM for the used image is the region of Central Java.
- Radiometric Correction with Digital Number (DN) conversion to TOA Reflectance:
- Convert TOA Reflectance to a value of Chlorophyll-a.

The algorithm used to chlorophyll-a :

$$c = 0,2818\, x \left\{\frac{B5 + B6}{B4}\right\}^{3,497}$$

Explaination :
C = Consentration of chlorophyll-a (mg/ m^3)
B4 (i1) = Value of reflectansikanal 4 Landsat 8
B5 (i2) = Value of reflectansikanal 5 Landsat 8
B6 (i3) = Value of reflectansikanal 6 Landsat 8

Data processing uses ER Mapper 7.0 software, data that has been previously downloaded with tif format, then performed the merger of the band and stored in ers format (er mapper raster dataset). The data has been processed to display thematic data using Arc GIS software with output in image format (Jpeg). Based on image data in the form of digital value range and interpretation of multitemporal landsat image processing results, it can be displayed visually for further discussed descriptively [20].

2.3 The Accuracy of Cholophyl a content between the results of Landsat satellite multitemporal imagery measurements and the Ground Check (SAL) area content

After obtaining the results of measurement of chlorophyll a between the satellite landsat and field measurements (Ground check) then the mapping accuracy was conducted [16]. The mapping accuracy test is obtained from the output of regression equation between field data and landsat image data using regression equation which is adjusted to the data characteristic obtained [20]:

Liner regression : $y = a + b\,x$
Orde polynomial 2 : $y = a + bx^2 + b1\,x$
Orde polynomial 3 : $y = a + bx^3 + b1\,x^2 + b2\,x$

The close relationship between the two variables (X) and (Y) that can be seen from the correlation coefficient (r) is obtained from the root of the coefficient of determination (R^2). The value of r ranges from 0 to 1, if the result is close to +1 the relationship among the two variables is strong or there is a high correlation between the two and their respects that is close to zero, the connection is very weak [22].

3. Result and Discussion

3.1 The chlorophyll- a content during the research done in 2016 based on Landsat satellite multitemporal imagery measurements and the ground check.

The analysis of chlorophyll content parameter of Segara Anakan lagoon area in 2016 uses multitemporal landscape satellite image recording (Landsat 8) (Figure 1). Image recording while crossing the SAL area is conducted 5 times during 2016, the reference is expected to represent the picture of each season period (monsoon). Landscape image data was taken on the west season (7 February), transition season I (May 6), east season (July 25), transitional season II (October 13) and west season (December 16). The quantitative results of chlorophyll a-mark as a result of TOA reflectance conversion of Landsat imagery ranged from 3.213 - 3.703 (mg/ m^3). Especially for western season data (February), multitemporal landsat images can not support an idea of the concentration of chlorophyll content a, this is caused by the condition of the SAL area is covered by the thickness of the cloud. Allegedly in February 2016, it has a high level of rainfall. Based on the data served from Meteorology, Climatology and Geophysics Agency (BMKG) Cilacap Regency, that in February 2016 rainfall reached 437 mm, which is the highest rainfall conditions occurred in February during the last five years.

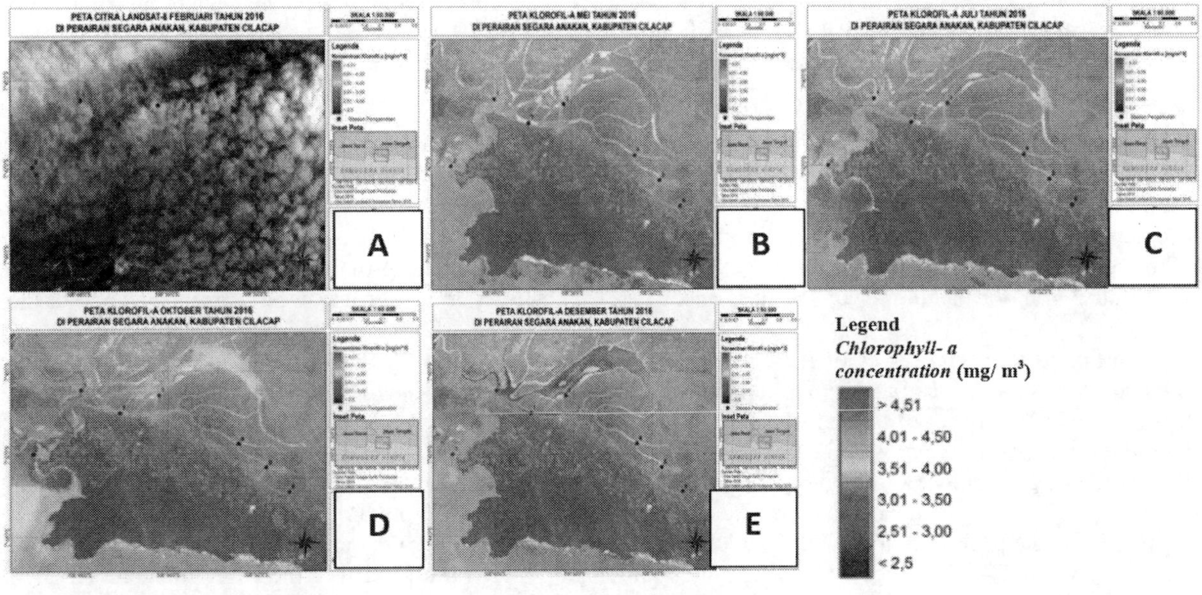

Fig. 1 Spasio-temporal chlorophyll-a content based on the result of Landsat satellite multitemporal Image Record : A) West season (February); B) Transition season I (May); C) East season (July); D) Transition season II (October); E) West season (December).

Based on the qualitative analysis, the visualization of landsat image recording and quantitative analysis of chlorophyll-a values shows that the transition season I (May) is the highest chlorophyll- a season with a range of 3.50 ± 0.09 mg / m^3. This is in line with the results of the highest temporal phytoplankton a bundance measurements occurred this season, with a quantity of (9956,57 ± 5300,23) indv / L. The distribution and abundance of phytoplankton can be estimated from its chlorophyll content through remote sensing imagery technology. Landsat imagery can provide the information of waters data according to spectral object value recorded by landsat cencor.

Nontji (2008) added that the transition season (transition) pattern direction and wind power is not stabilized, but generally, relatively, calm water conditions. Suspected phytoplankton can process photosynthesis optimally. The results of Landsatographic recording in this study strengthen the sensing results, which show that phytoplankton biomass can be described in units of chlorophyll (mg / m^3).

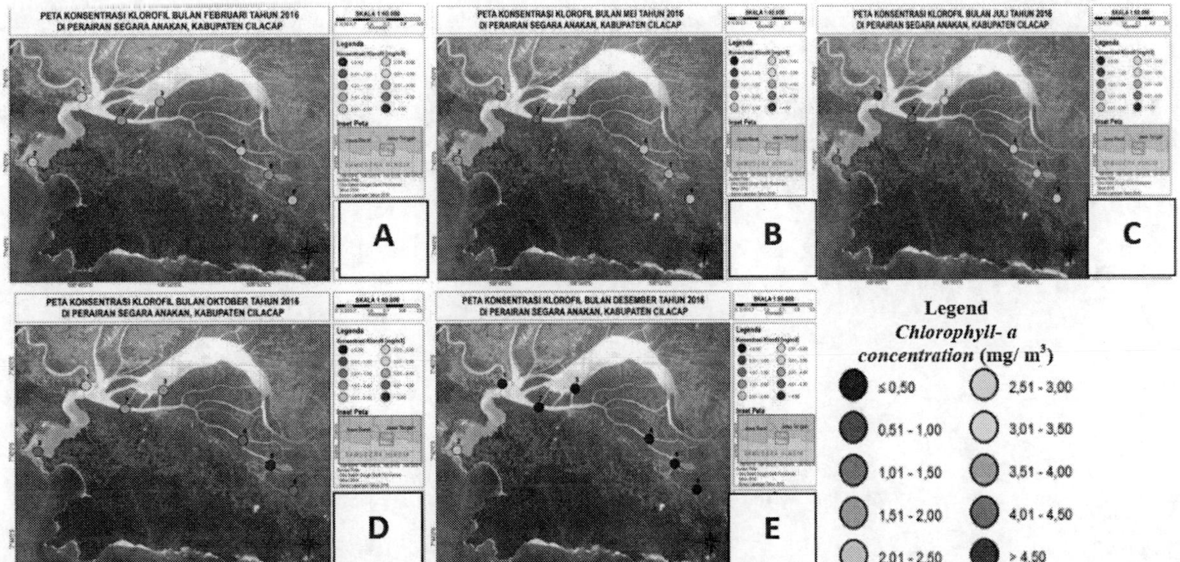

Fig. 2 Spasio-temporal chlorophyll-a content based on the result of Ground-check in Segara anakan lagoon area 2016: A) West season (February); B) Transition season I (may); C) East season (July); D) Transition season II (October); E) West season (December).

Furthermore, the highest quantity of chlorophyll-a content is followed by east season and transition II and the lowest is in west season (December). Overall it is recognised that SAL waters during 2016 are strongly influenced by high rainfall, which started in the east season (June-August) and increased again during the second transition season (September-November). Rainfall zenith occurred in October, further, It declined in November and December along with a decrease of chlorophyll-a. Despite the prevailing conditions, that in the east of the wind blows from east to west and marks the beginning of the dry season, the opposite condition in the west season that marks the rainy season.

Rainfall pattern, is a factor of fluctuation control of organic material, which was significantly influencing to fluctuation of N and P as barrier of waters affecting distribution and quantity of chlorophyll-a. [19],[14], [13]. On the other hand, the result of laboratory analysis of chlorophyll content of SAL area 2016 based on Ground check, to facilitate its spatio-temporal representation, presented in the form of thematic map (Figure 2). The overall fluctuation of chlorophyll a throughout 2016 is represented by graphs on the map (Figure 3).

The results of field measurements occur the optimality of chlorophyll-a during transitional season II, it is supported by the peak rainfall in 2016 in October reached 958 mm which is month the highest rainfall month in the last five years. The existence of ecological conditions of high rainfall, affect the high fluctuations of organic materials and quantity of chlorophyll- a. The increase of rainfall is proportional to the chlorophyll measurement data in the field and vice versa. The results of lands and field measurements ground check showed that the measurement of chlorophyll a in the east season is higher when compared to the west season. According to Nontji [16] the results of this study are in line with the results of research that occurred in the Banda Sea, where the occurrence of upwelling or lifting of water rich in organic material elements from the inner layer up to fertilize the surface layer, which occurred during the east season. While the mechanism of downwelling or the occurrence of water drowning of the surface on the deeper layers of water occurs in the west season. Furthermore, from the expedition Snellius II (1984-1985) which is a joint expedition Indonesia-Dutch. Expedition results also stated that in August there was an upwelling that caused increased water productivity, reflected by the value of phytoplankton chlorophyll ranged from 1.3 mg/ m^3. Conversely at the time of downwelling chlorophyll decreased to about 0.4 mg/ m^3. The effect of upwelling was evident throughout the southern waters of Java-Bali in August (east season) during upwelling, the phytoplankton was very fertile while in February (the western seasons) the abundance fell back.

The result of spatial analysis at research station (S) is known that the high chlorophyll a during the study occurred at (S1), (S5) and (S7), which is a station with characteristic of anthropogenic activity especially happened at transition season I, east season to season transition II. The diversity of anthropogenic activities contributes to the improvement of the organic elements of water supported by high rainfall and upwelling phenomenon. In according to the Panggabean [17] and Livingston [14] stated, rainy season occurs in the process of washing of organic material from the river mouth, mainly by the discharges of various anthropogenic activities resulting in the increase of organic material entering the coastal waters. Furthermore, phytoplankton is able to show tolerance to the condition of water instability, when an increase in organic materials. It is characterized by changes in species dominance, changes in community structure, biomass and high content of phytoplankton chlorophyll [21], [2].

Fig. 3 Thematic map of chlorophyll-a during the sampling in 2016 with the approach of spatial and temporal

3.2 The accuracy of chlorophyll- a concentration between the result of Landsat satellite multitemporal imagery measurements and ground check in SAL area.

The results of ground chlorophyll measurements are used to verify the mapping results with satellite imagery. In this study (in 2016) the chlorophyll content of the SAL region was measured by both methods, using both landsat and ground (Ground check) satellites. The measurement results show that the accuracy is high, shown by the value of r = 0.94 with the equation $y = -30.07x^2 + 217.0\ x - 387.09$, $R^2 = 0.894$ (Figure.4). Based on data of characteristic, logarithmic regression model is polynomial with data on X axis (field) and Y (Landsat satellite image). From the results show a positive pattern, that the higher value of chlorophyll of field results is proportional to the high chlorophyll in the satellite image.

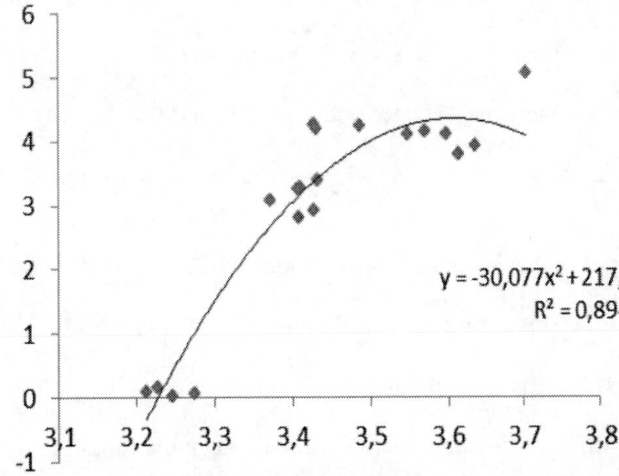

Fig. 4 Accurated result of Satellite Image of chlorophyll-a with measurements on the field in Segara anakan lagoon area

4 Conclusion

SAL region is a dynamic water region, there is influence of natural and anthropogenic factors which subsequently contribute to the accumulation of organic matter. Supported by high rainfall and upwelling phenomenon result in fluctuation of chlorophyll content a. The high value of accuracy test on the obtained equation shows that the algorithm in the SAL area can be applied to obtain chlorophyll-a content data in the previous years, for example for 15 years back 2001-2015, with the support of chlorophyll-a landsat satellite image (Y) in that year. Application of remote sensing method using chlorophyll a satellite image can then be used for the preparation of management strategy of lagoon management by Local Government, related parties and community to maintain ecological stability of lagoon.

References

1. Ardli, E.R., and M. Wolff. *Reg. Environmental Change.* **9**: 235-243 (2009)
2. Ayuningsih, M.S, I.B. Hendrart and P.W. Purnomo. *Diponegoro Journal Of Maquares. Manajement Of Aquatic Resources*. **3**, 138-147 (2014).
3. Carolita, I, E. Parwati, B. Trisakti, T. Kartika dan G. Nugroho. 2005. MAPIN XIV Annual Scientific Meeting. Effective utilization of remote sensing for enhancing the welfare of the nation. Institute Teknologi Sepuluh Nopember, Surabaya
4. Coutinho, M.T; A. C. Brito; P. Pereira; A.S. Gonçalves and M.T Moita. *Journal Estuarine, Coastal and Shelf Science*. **110**: 134-146 (2012)
5. Danoedoro, P. 2012. pp 159-366
6. Dewi, R., M. Zainuri., S.Anggoro., T.Winanto. Omni Akuatika Journal. **12**, 144-150) (2016)
7. Dsikowitzky,L, I. Nordhaus, TC. Jennerjahn, P. Khrycheva, Y.Sivatharshan, E.Yuwono, J. Schwarzbauer. *Marine Pollution Bulletin* **62**. 851–862 (2011)
8. Hatta, M. 2002. http://tumoutou.net/3_sem1_012/muh_hatta.htm. 1 April 2007.
9. Helmi, M.. Journal of Marine Science. FPIK. Diponegoro University. Semarang. (2006)
10. Herawati VE. Thesis. Diponegoro University. Semarang (2008)
11. Hutagalung HP, D. Setiapermana , SH. Riyono. P3O-LIPI (1997)
12. Jennerjahn, P. Holtermann., I. Pohlenga and B. Nasir. Research Institute University of Jenderal Soedirman. Purwokerto (2007)
13. Jennerjahn, B. Nasir, I. Pohlenga. *Reg Environ Change* **9**:259–274 (2009)
14. Livingston, R.J. Marine Science Series Boca Raton. Printed in the United State of America. Pp 327 (2001)
15. Nontji, A. LIPI Press. Jakarta. 331Pp (2008)
16. Nuriya H, Z. Hidayah, W.A Nugraha. Marine Journal. **3** (1): 60-66(2010)
17. Panggabean, L.M.G. *Oseana*. **19** (1):33-38 (1994)
18. Parslow, J.J. Hunter and A. Davidson. CSIRO Marine Research. Hobarth, Tasmania (2008)

19. Pintado, J.G; M.M. Mena, G.G. Barberá; J. Albaladejo and V.M. Castillo. *Science of the total environment*. **373**: 220-239 (2007)

20. Sidik, A., A. Agussalim, M. Rasyid Ridho. Maspari Journal. **7** (**2**): 25-32 (2015)

21. Vadrucci, M.R., E. Stanca., C. Mazzioti, N. Ungaro and A. Basset. Ecological Indicators. 34: 113-125 (2013)

22. Walpole. *Pengantar Statistika*. Jakarta: Gramedia Pustaka Utama (1995)

23. Yuwono, E; TC. Jennerjahn; I. Nordhaus; E.A. Riyanto; M.H. Sastranegara and R. Pribadi. *Asian Journal of Water Environment Pollution*. **4**, no. 1, pp: 61-70 (2007)

SUBLETHAL TOXICITY TESTS OF MERCURY (Hg) TO NILEM FISH (*Osteochilushasselti*) GILLS TISSUE DAMAGE

*Asrul*SahriSiregar[1],*Norman*AriePrayogo[1], *Emyliana* Listiowati[1], *Mahendro* Santoso[1], *Indra* Gumay Yudha[2],*Tri* WahyuniSholehah

[1]ProgramStudiManajemenSumberdayaPerairan,Fak.PerikanandanIlmuKelautan, UniversitasJenderalSoedirman, Indonesia

[2]JurusanPerikanandanIlmuKelautan,Fak.PerikanandanIlmuKelautan, UniversitasLampung, Indonesia

Abstract. *Industry development in Indonesia resulted in sewage and polluting heavy metals such as mercury (Hg). These pollutants will be harm organisms that live at aquatic environment. Fish gills could be damage due to exposure by mercury. To know the influence of mercury's toxic against aquatic organisms by using toxicity test. An aquatic organism, which is usually used as toxicity test, is nilem fish (Osteochilushasselti). The objective of the study was to determine the value of LC50-96 hours in a toxicity test of nilem fish exposed to lead acetate; to determine the changes in the fish gill and to determine the concentration of lead mercury which causes tissue damage structure in the gills of nilem fish. The design used completely randomized design (CRD) with four concentrations (0; 0.08; 0.16 and 0.24 ppm) and three replications. The objective of the study was sublethal toxicity test. Lethal toxicity concentration data were descriptive and data from sublethal toxicity test were One Way. The analysis showed that there were highest edema 10% and 17% on day 4 and 12, the highest lamella fusion 35% and 59% on day 4 and 12, the highest hyperplasia 45% and 54% on day 4 and 12, and the highest necrosis the 65% and 80% on day 4 and 12. It was concluded that the value of* LC_{50}*-96 hours in nilem fish (Osteochilushasselti) toxicity test was 0,3938 ppm; and damage of fish gill tissue changes started to occur in the lead mercury (Hg) concentration of 0,08 ppm.*

[1]Corresponding Author :asrul_sir@yahoo.com

1 Introduction

Indonesian industrial sector is growing in line with the advancement of science and technology. Increased development can reduce quality of the environment and public health, caused by the pollution. The source of pollution can be from industry, mining, agriculture, and other natural source from rock reaching river or sea, then polluting aquatic biota and it is environment.

Environmental order can be polluted or damaged by many things and most of all the cause is pollutants. One component of pollutants containing toxic substances such as heavy metals [1]. Heavy metal is a chemical element that has the potential to cause environmental pollution problems. One of them is mercury which is widely used for various industries such as gold mining. Alfian[2] stated that mercury is widely used in agriculture, industry and medical purpose. Mercury is used in agricultural sector for fungal and pest extermination such as fungicide, herbicide, and pesticide. Mercury is used in industrial sector for anti-fungi in paints and paper industry, and plastic synthetic. Mercury usage in medical sector for syphilis treatment, calomel (HgCl) used for wound cleanser but not used anymore because it is known to be toxic, while in cosmetic commonly used for whitening cream [3].

Mercury and it is derivatives are the most harmful metal pollutant for aquatic biota and aquatic environments. One of the aquatic biota affected by mercury is fish. Fish can accumulate mercury in the form of methyl mercury, either through food or directly from the water. Based on Stokes and Wren [4] research, shown that 68–80% of mercury exist on environment were accumulated in fish body. Continuous accumulation of mercury metals will causing death on fish. This will endanger life of aquatic organism especially fish and damage the food chain in aquatic environment. The impact caused by the accumulation of mercury in fish shown on Dalimunthe[5] research, in BatangGadis River Mandaling Natal Regency, mercury (Hg) accumulates in the body of mercury (Hg) accumulates in the body of catten fish (Mystacoleucusmarginatus) which ranges from 0.00128 - 0,01097 ppm. It shows that mercury in river water has passed the threshold value of 0.001 mg / l based on Indonesian Republic Government Regulation No. 82 of 2001 on Water Quality Management and Water Pollution Control.

Research on mercury in mercury chloride compound itself has been done previously using carp (Cyprinuscarpio) with size 10-13 cm but still limited to liver damage. Based on research Destiany [6], that the concentration of 0.02 ppm mercury chloride has caused damage to liver cells, while the LC50-96 hour value of 0.24 ppm. However information on the lethal and subletal effects of nilem fish has limited(20,21). Therefore, toxicity test of lethal and mercury subletal in freshwater fish that is nilem fish (Osteochilushasselti) need to be studied.

This study will examine the value of sub-lethal toxicity (LC50-96 hours) of mercury metals in nilem fish; How big is the influence of concentration and length of mercury (Hg) exposure to the damage of gill tissue of nilem fish (Osteochilushasselti) at sub lethal condition (19). The benefit of this research is to provide information about histology of nilem fish gill (Osteochilushasselti) due to exposure to heavy metals mercury. In addition, this information can be used to determine the range of LC50-96 h values in the toxicity test of mercury-exposed nilem fish (Hg) so as to determine the standard of mercury quality in the waters.

2Research Methods

The research was focused on sub lethal toxicity test. Experimental method used with Completely Randomized Design (CRD). This study used nilem fish with length of 10-13 cm. Mercury was used by diluting HgCl2 at determined concentration in 1 liter of water, after mercury dissolved it was mixed in to test medium of 30L water, each treatment repeated 3 times.

In lethal toxicity test, the main parameter is percentage of fish mortality with series of mercury treatment. In sub lethal toxicity test, the main parameter observed is fish gill microanatomy with series of mercury treatment based on LC50-96 hour result. Observed supporting parameter was temperature, pH, and dissolved oxygen.

2.1. Media Preparation

The research was undertaken at laboratory, 18 aquariums with a capacity of 50L were prepared, washed and dried expose those to the sunlight, then filled with 30L of a mixture and dilution of mercury and water with different Hg concentration per treatment an then aerators installed. Nilem fish acclimated at 5 days, one day before used and during treatment periods fish were fastened. The size of nilem fish used is 10-13cm and for evenness purpose the larger fish should not more than 1½ small fish [7].

2.2. Sub Lethal Test

Sub lethal test used 12 aquariums with capacity of 50L. Each was filled with 30L of test medium and 15 fish. Sub lethal observation was done by taking fish gill in 2 period of time (4th and 12th day) one of each treatment to observe fish's gill histology.
Microscopic observation ofnilem fish (*Osteochilushasselti*) gill histology done to gill's filament, compound microscope used at 40x10 magnification scale. The extend damage to the gill tissue of nilem(*Osteochilushasselti*) was analyzed based on gill structure changes in lamella edema in which the lamella epithelial cell release from the underlying tissue. Hyperplasia is the occurrence of abnormal increase of the total cells in lamella and fusion of lamella which is the loss of secondary lamella structure and filament damage.

2.3. Physic and Chemical Parameter Measurement

Measured parameter were temperature, pH, and dissolved oxygen. Water physical and chemical quality was measured from designed treatment media.

2.4. Research Time and Place

The research was done from January until April 2015 at Faculty of Biology JendralSoedirman University, Microanatomy Laboratory Faculty of Veterinary, GadjaMada University, Yogyakarta, and Aquatic Research Management Laboratory Faculty of Fishery and Marine Culture JendralSoedirmanUniversityl, Purwokerto.

2.5. Data Analysis

Data obtained from the observation of tissue damage nilem fish (Osteochilushasselti) in the form of qualitative data. Percentage of gill tissue damage rate was analyzed descriptively and quantitatively based on the change of gill structure edema, hyperplasia, lamia fuse and necrosis. Gill tissue damage percentage calculation was done by observing gill lamella, at microscope with 400x magnifier. The result from damaged gill tissue analyzed with One Way ANOVA to determine concentration used damaging tissue or not.

3Results and Discussion

3.1. Test Medium Water Quality

Water quality measured were physical and chemical parameters such as temperature, pH and dissolved oxygen. Temperature and pH measurement were done at the start, middle, and end of the research, dissolved oxygen measurement was done at start and end of the research. The purpose of measurement was to monitor water quality for nilem fish during the research. The result of parameter measurement can be seen at Table 1.

Table 1. Parameter measurement result

Parameter	Value	Quality Standard
Temperature (^{o}C)	25-29	20-30
pH	7,40-8,56	7-8,5
DO (mg/L)	4,8-7,8	5,4-6,2

Based on the result, temperature, pH, and dissolved oxygen during the research were still in tolerated value for fish to live.

3.2. Sub-lethal Toxicity Test

The sub-lethal test was done after lethal toxicity test (LC_{50-96}). The concentration for sub-lethal test were 0%, 20%,40% and 60%. The result from LC_{50-96} indicates that higher the concentration of mercury causing more severe gill damage.

According to Palar [1], the process of mercury entering gill is mercury with other metals ion will form ion that solublein fats. The metal ions that can soluble in fats can penetrate gill cell membrane and enter the gill. Connel[8] state that toxic element intake on the aquatic organism can happen by 3 methods: (1) Respiration, (2) Contact with toxic element, and (3) Feed, digested water particle. Dissolved toxic and organic compound causing irritation on gills and gills lamella, this disturbed the respiration process.

The level of damaged gill tissue <30% means light damage. The level of damaged gill tissue 30%-70% means moderate damage. The level of damaged gill tissue >70% means severe damage.

The damage of nilem gill tissue exposed to mercury for 4 and 12 days with different concentration show on figure 3. The lightest damage is edema. Edema is the first phase of gill degeneration showed by swelling. The edema damage presentation is shown on figure 1.

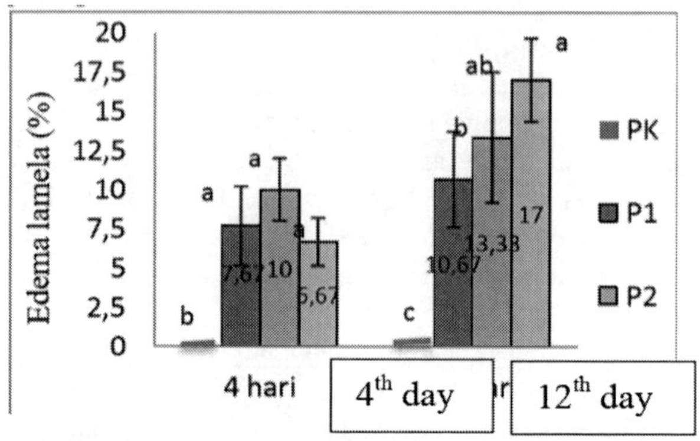

Fig 1. The edema damage presentation in day 4 and day 12

The damage value caused by edema show significant difference (P<0,05) on treatment concentration 0 mg/L to 0,08 mg/L, 0,16 mg/L and 0,24 mg/L in day 4, but in day 12 only on treatment 0 mg/L to 0,08 mg/L and 0,24 mg/L that show significant difference, treatment concentration 0,16 mg/L did not show significant difference (P>0,05). The significant difference (P>0,05)means that mercury treatment concentration is affecting edema presentation on nilem fish gill. Robert [9]states that edema on lamella caused by contact with high concentration chemicals that mainly metals. Edema will inhibit gas diffusion efficiency and causing death on organism. Gas diffusion is inhibited because of narrowing secondary gill lamella.

Mercury concentration of 0,16 ppm on day 4th observation showed high edema. Likewise, when the observation on the 12th day of 0.24 ppm treatment showed high edema. This is because on the 4th day observation the mercury concentration of 0.24 ppm has not affected the edema of the gills of nilem fish, but the longer the gill is exposed to heavy metals mercury then there is an increase of edema in the gill tissue, as happened during the observation of the 12th day of treatment 0.24 ppm of mercury has an effect on edema. This is supported by Dalimunthe [5] study which states that edema with 30-70% of field extent also occurs in Cencen fish (Mystacoleucusmarginatus) taken from Sungai BatangGadis Kalimantan, with a content of Hg in water of 0.5 ppm.

The other damages that occur are hyperplasia. Hyperplasia occurs in the treatment of 0.08 ppm, 0.16 ppm and 0.24 ppm. Hyperplasia is excessive tissue formation due to the increase in the number of cells [9]. The percentage of hyperplasia in the gill tissue of nilem fish was observed on the 4th and 12th days showed in figure 2.

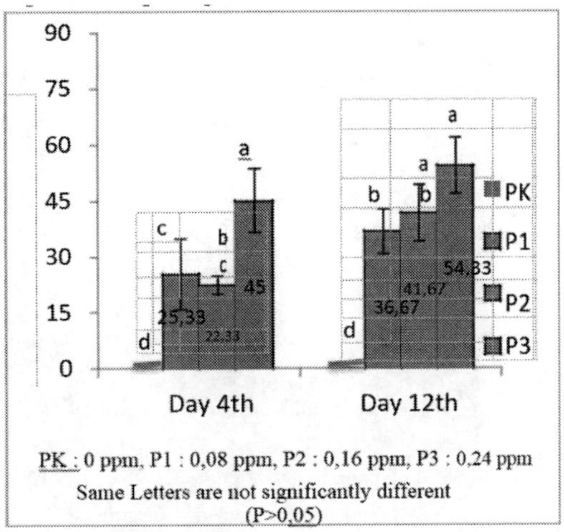

Fig 2. Percentage of hyperplasia on day 4th and 12th.

The 4th day the average of hyperplasia at the treatment of 0.08 ppm and 0.24 ppm showed significant difference (P <0.05) to the control, whereas in the treatment of 0.16 ppm there was no significant difference (P> 0 , 05) to the 0.08 ppm treatment. The 12th day, the 0.08 ppm and 0.24 ppm treatments showed significantly different values (P <0.05) to the control, while the 0.16 ppm treatment showed no significant difference (P> 0.05) treatment 0.24 ppm. Significantly different concentration (P <0.05) could be interpreted that the percentage value of hyperplasia in the gills of nilem fish was influenced by the treatment of mercury concentration. Concentrations were not significantly different (P> 0.05) can be interpreted that the treatment of concentration does not affect the percentage value of hyperplasia in the gills of nilem fish.

The observation of the 4th and 12th days the gill at the highest treatment (0.24 ppm) showed the greatest hyperplasia. This is because irritated nilem fish gill due to the large content of heavy metals mercury and characterized by an increase in the number of mucus cells in the base lamella. Concentration does not affect the percentage value of hyperplasia in gill nilem fish.

The addition of the number of secondary lamella epithelium causes the secondary lamella to become enlarged and coincident, the result between secondary lamella to stick together and unite. This makes the gill lamella appear larger than normal and the gill is no longer clearly visible difference between the primary and secondary lamella [10]. This is in accordance with the research of Yuniar et al., [11], that on the observation of the 10th day of Hg treatment of 0.15 ppm, the (Oreochromisniloticus) gill experienced the greatest hyperplasia. Hyperplasia makes the gill lamella appear larger than normal and the gill is no longer clearly visible between the primary lamellae and the secondary lamellae.

The 4th-day observation at the lowest treatment (0.08 ppm) had slight hyperplasia. This is because at a concentration of 0.08 ppm in a time of 4 days exposure has not caused significant hyperplasia symptoms, in addition, 0.08 ppm treatment did not cause hyperplasia damage because it has become fusion lamella. This is supported by a statement from Robert [9] that hyperplasia occurs at a lower irritation level and is usually accompanied by an increase in the number of mucus cells in the lamella base and results in lamella fusion.

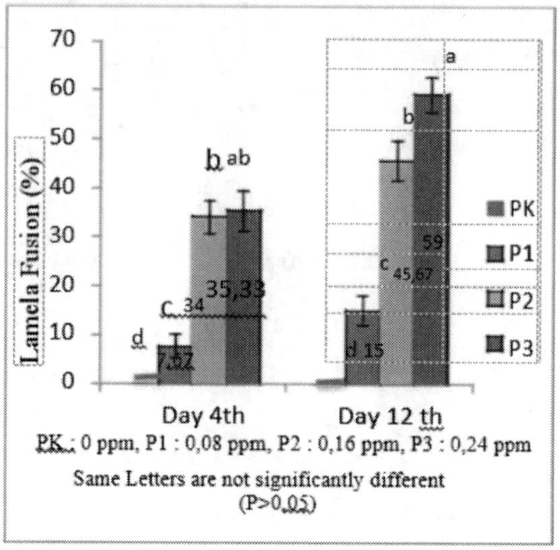

Fig 3. Percentage of Lamella Fusion on day 4th and 12th.

The next damage is the fusion of lamella found in the three treatments concentration, 0.08 ppm, 0.16 ppm and 0.24 ppm. The highest lamella fusion on day 4th was found in the treatment of 0.24 ppm, which ranged from 35.33%, then the lowest average was shown in the control treatment which ranged from 0%. While on the 12th day, the highest lamella fusion was found in the treatment of 0.24 ppm, which is about 59% and the lowest average was shown at 0 ppm. This shows the treatment of 0.24 ppm to moderate damage that is about 30-70% of the field width of view.

Lamella fusion in nilem fish gill due to mercury metal tends to increase at treatment 0.08 ppm, 0.16 ppm and 0.24 ppm. The highest lamella fusion was found in the treatment of 0.24 ppm, ie on the 4th and 12th day observations with values of 35% and 59%. On day 4, lamella fusion of 0.08 ppm and 0.16 ppm mercury treatment showed significant differences in control treatment ($P < 0.05$), whereas in the treatment of 0.24 ppm did not show any significant difference ($p > 0.05$) to the treatment of 0.16 ppm. On day 12, lamella fusion of mercury treatment 0.08 ppm, 0.16 ppm and 0.24 ppm showed significant differences in control treatment ($P < 0.05$). A significantly different concentration ($P < 0.05$) can be interpreted that the concentration treatment does not affect the percentage value of lamella fusion in the gills of nilem fish. Concentrations that were not significantly different ($P > 0.05$) could be interpreted that the concentration treatment did not affect the percentage value of lamella fusion in the nilem fish gill.

The 4th and 12th day observations at the highest mercury treatment (0.24 ppm) experienced the greatest lamella fusion. This is caused by the amount of edema at a concentration of 0.24 ppm which can trigger the occurrence of fusion of lamella that attaches two parts of secondary lamella. This is on accordance to the Dalimunthe[5] study that looked at cencen fish (Mystacoleucusmarginatus) gill histology that occurred in a primary lamella fusion of 30% of the viewing area of Hg concentration of 0.04 ppm.

Lamela fusion is caused by gill organs in contact with mercury ions, so the gills release mucus for protection against mercury toxicity, but the resulting mucus closes the lamella surface of the gill so that the exchange of O2 and CO2 is inhibited. According to Sukarni et al., [12] statement that excess mucus is one of the mucous gland's responses to protect gills from mercury, but if the resulting mucus is excessive it will be negative for the fish so that oxygen uptake from the water will become obstructed.

The observations of the 4th and 12th days of the nilem fish gill at the lowest treatment concentration (0.08 ppm) experienced slight lamella fusion. This is because at these concentrations, nilem fish not releasing much mucus out of mucus cells that can lead to increased lamella fusion. This is supported by a statement from Mulyani et al. [13] that lamela fusion can occur because the mucus cells in the lamella are increased in number resulting in the fusion between the secondary lamella.

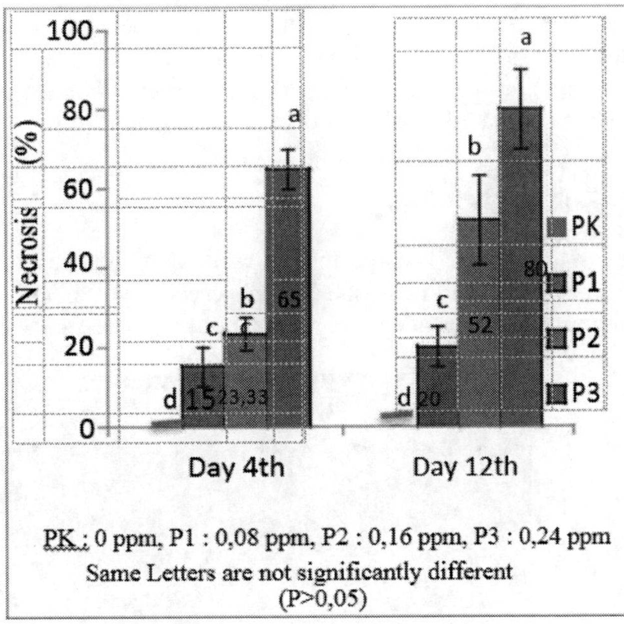

Fig 4. Percentage Necrosis on day 4th and 12th.

Other damage to the gills of nilem fish is necrosis. Necrosis is cell death that occurs due to hyperplasia and excessive secondary lamela fusion, so that the gill tissue is not intact anymore or in other words necrosis occurs accompanied by the death of a biota. According to Plumb [14] necrosis is characterized by the presence of death of cells or tissues accompanying cell degeneration in each necrosis of the gill tissue of the observed nilem on the 4th and 12th days can be seen in Figure 4.

The necrosis percentage value on day 4 showed significant difference ($P < 0.05$) in 0.08 ppm and 0.24 ppm for 0 ppm (control) treatment concentration, while 0.16 ppm treatment did not show the significant difference ($P > 0,05$) to 0,24 ppm treatment. A significantly different concentration ($P < 0.05$) can be interpreted that the concentration treatment does not affect the percentage value of necrosis in nilem fish gill. Concentrations that were not significantly different ($P > 0.05$) could be interpreted that the treatment of concentration did not affect the value of percentage of necrosis in gill nilem fish.

Necrosis on the nilem fish gill tends to increase at 0.08 ppm, 0.16 ppm and 0.24 ppm. The observation in day 4th and day 12th the gill at the highest treatment concentration (0.24 ppm) experienced the greatest necrosis.

This is because the concentration of mercury in the water is too high and there is continuous absorption of mercury ions into the gill tissue. Exposure to heavy metals with high doses can cause damage to gill cells, because the gill tissues are the first that have contact with heavy metals in the fish body [10]. This is supported by the Nirmala [15] study, that with 0.3 ppm mercury dose during 72 hours exposure on gold fish (Cyprinuscarpio) seen a lot of damage to the lamella suffering from necrosis (> 70%). The necrosis seen in these observations proves that fish contaminated by heavy metals Hg inhibit the oxygen for cell growth and development due to the destruction of the fish's circulatory system where the blood that carries the nutrient throughout the body is disrupted, resulting in deaths in the gill tissue.

The condition gill on the 12th day at mercury treatment concentration of 0.24 ppm experienced a severe level of damage compared to 0.08 ppm and 0.16 ppm. The shape of the gill lamella was destroyed and deformed severely. Most gill epithelial are detached from the organs. Before deformation occurs, the epithelium is edematous. This is in accordance to the study of Yuniar et al. [11] that mercury toxicity of 0.15 ppm in Nile fish (Oreochromisniloticus) at day 20th exposure, gill conditions experienced very severe necrosis levels when compared with day 4th and day10th exposure.

3.3 Nilem Fish Gill Destruction

The degree of histologic symptoms of gill damage observed included edema, hyperplasia, lamella fusion and necrosis. The observed level of gill tissue damage from 3 treatments (0.08 ppm, 0.16 ppm and 0.24 ppm) can be seen in Figures 5, 6 and 7.

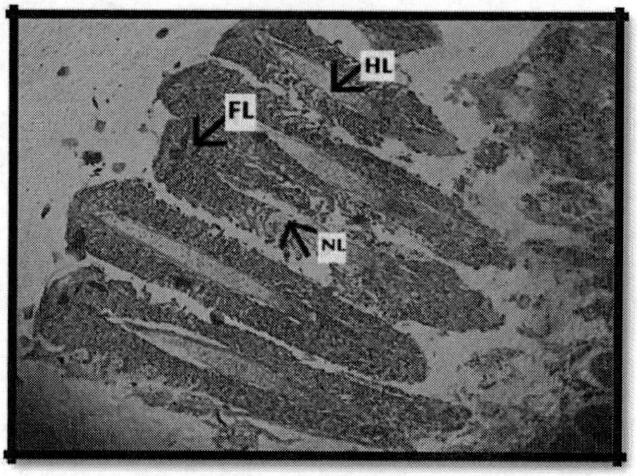

Fig 5. The histologic features of nilem fish gill (Osteochilushasselti) Hg concentration 0.08 ppm with 400x magnification. FL: Secondary lamella fusion, HL: Primary lamella hyperplasia, NL: Secondary lamella necrosis.

The effect of mercury with concentrations of 0.08 and 0.16 ppm (Figure 5) resulted in damage to the structure of gill tissue in the form of lamella fusion. Lamella fusion is characterized by the presence of fusion (attachment) between the secondary lamella due to secondary lamella epithelial hyperplasia. At this concentration there is an increase in damage from the concentration of 0.08 ppm, this is in accordance with the statement of Lu [16] which states that the higher the concentration of a toxic material in the water then the damage will occur more and more. Excessive hyperplasia and edema at a concentration of 0.08 ppm may cause lamella fusion. Lamella fusion is the attachment of two parts of the secondary lamella. In addition, lamella fusion is also caused by the excessive mucus in the gills that will close the secondary lamella. This excess mucus is one of the mucous gland's responses to protect the gills from mercury, but if the resulting mucus is excessive it will be negative so that oxygen uptake from water will be inhibited [17].

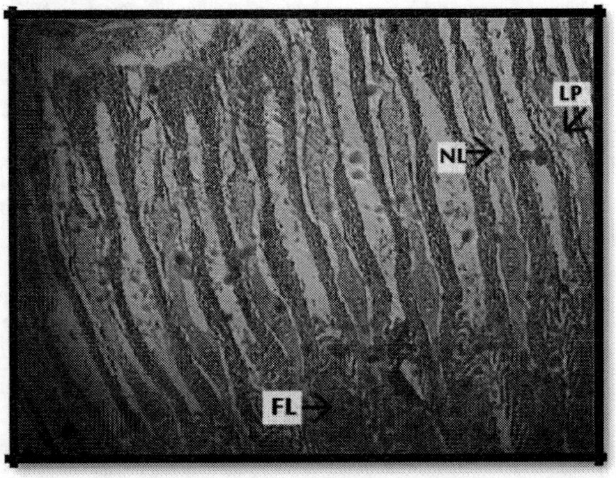

Fig 6. Histologic features of nilem fish gill (Osteochilushasselti) Hg concentration 0.16 ppm with 400x magnification. LP: Bent Primary Lamella, FL: Secondary lamella fusion, NL: Primary lamella necrosis.

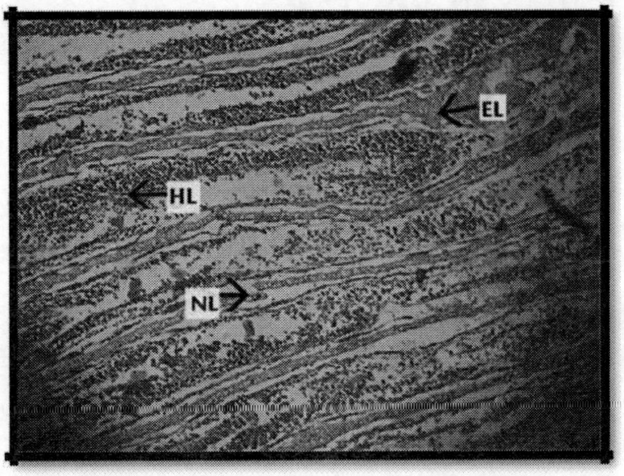

Fig 7. Histologic features of gill fish nilem (Osteochilushasselti) Hg concentration 0.24 ppm with 400 xmagnification. NL: Primary lamella necrosis, HL: Primary laryngeal hyperplasia, EL: Primary lamella edema.

The description of the structure of the gill tissue at concentrations of 0.16 ppm and 0.24 ppm (Fig. 6 and Fig. 7) contained edema and is still in moderate criteria [18]. Other damages seen at concentrations of 0.24 ppm are hyperplasia. Hyperplasia is excessive tissue formation due to the increase in the number of cells. Hyperplasia occurs with the addition of the number of secondary lamella epithelium so that the secondary lamella is enlarged and coincided, consequently between secondary lamella stick together and fused. This makes the gill lamella appear larger than normal and the gill is no longer clearly visible difference between primary and secondary lamellae. The presence of hyperplasia will reduce the surface area of contact between the gill tissue with oxygen, resulting in the gas and ion exchange process during respiration is inhibited [15].

4 Conclusions

Based on the results and discussion it can be concluded that the sub-lethal toxicity effect on the condition of nilem fish gill tissue (Osteochilushasselti), there is edema 10%, 45% hyperplasia, 35% lamella fusion and necrosis of 65% on day 4. While on the 12th day, there was edema of 17%, 54% hyperplasia, 59% lamella fusion and necrosis by 80%. The longer the mercury exposure time and the higher the mercury concentration exposed to the nilem fish, indicating an increase in gill tissue damage. Based on the results of this study suggested further research on gill tissue damage due to the influence of mercury in living organisms with longer exposed time or with varying mercury concentrations. The use of other heavy metals of the same concentration in this study may also be applied, so that it can be used as a comparison.

Reference

1 Palar, H. *PencemarandanToksikologiLogamBerat*. (Jakarta, RinekaCipta, 2008)

2 Alfian, Z. *PidatoPengukuhanJabatan Guru BesarTetapdalambidangIlmu Kimia AnalitikpadaFakultasMatematikadanIlmuPengetahuanAlam*.(Universitas Sumatra Utara, Medan,2006)

3 Widowati, W. *EfekToksikLogam*. (Andi, Yogyakarta, 2008)

4 Stokes, P. M., dan Wren, C. *SCOPEJournal*.**16,** 255-277 (1987)

5 Dalimunthe, L.S. *Skripsi*.(FakultasPertanian,Universitas Sumatera Utara, 2010)

6 Destiany, M. Skripsi. (UniversitasNegeriSemarang, Semarang,2007)

7 APHA (American Public Health Association).*Standard Method for TheExamination of Water and Waste Water.21 th ed.*(American Publik Health AssociatoinInc, New York,2005)

8 Connel,D.W.*BioakumulasiSenyawaXenobiotik*. (Jakarta, UI Press,1995)

9 Robert, R.J. *Fish Pathology*. USA: W. B.Saunders.Finney, D.J. *ProbitAnalisis;A StatisticalTreatment Of The Sigmoid Response Curve*, (Cambridge at the University Press, 2009)

10 Rennika, Aunurohim, Nurlita, A. *JurnalSainsdanSeniPomits*. **2**, 132-137 (2013)

11 P. Sukardi,, Hana, H., Prayogo, N.A., Harisam, T., Winanto, T. ActaScientiarum - Animal Sciences 40,e38335. (2018)

12 Sukarni, M., danNursyam.JurnalExpo Life Sci.**2**, 6–12 (2012)

13 Siregar, A.S., Prayogo, N.A. European Zoological Journal 84(1), pp. 436-443 (2017)

14 Plumb, J.A. **6**,118-121 (1994)

15 Nirmala, K. JurnalAkuakultur Indonesia. **11**, 38-48 (2007)

16 Lu,C.F.*ToksikologiDasar*. (Jakarta,Universitas Indonesia, 2008)

17 Lagler, K.F. Bardach, J. E, Miller, R.R, Passino, D.R.M. *Ichtiology. Second Edition*. (New York, John Wiley and Sons,1997)

18 andjung, S.D. *Paper presented in theInternational Seminar on Human Ecology, Tourism, and Sustainable Development* (Denpasar, 1982

19 N.A Prayogo, g.e. wijayanti, murwantoko, m. Kawaichi and p. Astuti.8(6) : 591-597. (2012)
20 N.A Prayogo, G.E. Wijayanti, I. Sulistyo, P. Sukardi. Biodiversitas.**17**, 523-530 (2016)
21 N.A. Prayogo, A. Siregar, P. Sukardi. Turkish journal fishery and aquatic science. **16**, 1003-1009 (2016)

Classification of shallow water seabed profile based on Landsat 8 imagery and in-situ data. Case study in Gili Matra Island Lombok, Indonesia

Ratih Ayustina[1], *Zahra* Aulia[1], *Haji* Mustakin[1], *Fiyesti* Alam[1], *Amron* Amron[*1], *Doddy* Yuwono[2], *Triguardi* Ahmad[1], *Aldo* Prayogo[1], and *Fitra* Sari[1]

[1)]Department of Marine Science, Faculty of Fisheries and Marine Science Jenderal Soedirman University, Purwokerto Indonesia
[2)]Center for Thematics Mapping and Integration, Geospatial Information Agency, Cibinong Indonesia

Abstract. Shallow water seabed profile has considerable potential resources so the availability of information which very important for coastal resources. The use of remote sensing techniques is considered to provide coastal information effective and efficient. This research aimed to determine the shallow water seabed profile based on Landsat 8 Imagery and its accuracy related to the in situ data. Methods of this research are satellite mage pre-processing, image classification, field survey, image classification, and accuracy assesment . Therefore, 6 classification of shallow water seabed profile, there are rubble (R), seagrass mixed sand (MIX -SG/SD), coral reefs mixed rubble (MIX-C/RB), rubble mixed dead coral (MIX-RB/DC), sand mixed rubble (MIX-SD/RB), and sand mixed seagrass (MIX-SD/SG), respectevely. The result of this classification has an accuracy value 80%.

1 Introduction

The shallow water seabed profiles potentially coastal resources, so it must be managed well in order to be optimally and sustainably utilized. Potential coastal areas such as fishery assets, marine tourism, energy, maritime industry and marine development activities[1]. One example of its management can be done using remote sensing for later mapping [2]. According Diaz *et al.* [3] by being able to map the basic objects of shallow water would be useful to assess the state of resources in the area.

One way to get information on coastal resources is to use remote sensing. The use of remote sensing is done because the conventional method is considered less efficient in providing a profile of shallow seabed profile in addition to require a long time is also a relatively expensive cost considering that the region is generally located in remote areas and access difficult. Therefore, the use of remote sensing satellite imagery as an alternative is the most ideal way to answer those needs and is considered effective and efficient for

*[1] Corresponding Author: twoone_brond@scientist.com

obtaining spatial information, whose results are useful as basic information in the preparation and preparation of coastal area management plans [4]. Through this technology, information of an object can be obtained using satellite sensors and then analyzed and presented in spatial and thematic form in accordance with the purpose of the activity [5].

The Landsat-8 Satellite Imagery is considered a potential satellite as a medium for monitoring and obtaining information on shallow underwater objects [6], which has short recording times about 16 days [7] and having a blue spectral band to identify shallow sea floor objects [8]. The advantages of this satellite image also have medium resolution on the multispectral band of 30 for the study of aquatic objects [9]. The use of Landsat-8 Satellite Imagery in this study is considered appropriate to obtain information on shallow water sea bed profiles in Gili Matra. This is in accordance with the direction of this study which aims to determine the shallow water sea bed profile of Gili Matra (Gili Meno, Gili Air, and Gili Trawangan), Lombok NTB and its accuracy based on Landsat-8 satellite images and in situ data.

2 Materials and Methods

2.1. Study Site

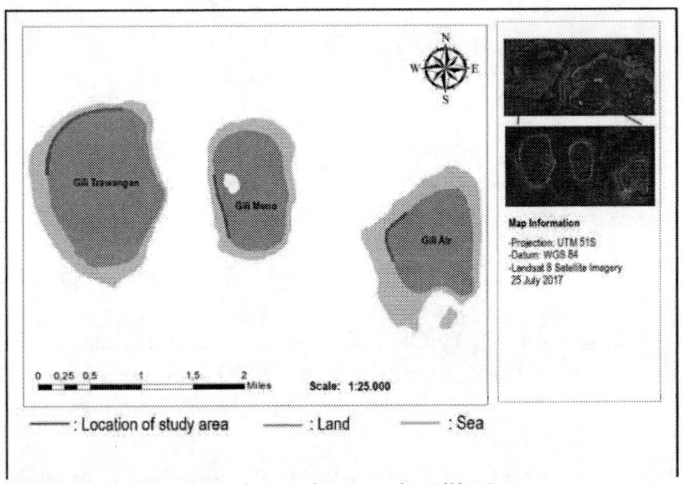

Fig 1. Index map showing location of mapping area in Gili Matra

The study area in this study is located on Gili Matra Islands covering Gili Meno, Gili Air and Gili Trawangan (**Figure 1**). In accordance with the database owned by Indonesian Marine and Fisheries Minsistry, Taman Wisata Perairan Gili Meno, Air dan Trawangan with an area of 2,954 hectares, which covers the land area of Gili Air ± 175 ha with the circumference of the island ± 5 km, Gili Meno ± 150 ha with around the island ± 4 km and Gili Trawangan ± 340 ha with a circumference of the island ± 7.5 km and the rest is marine waters. Geographically TWP Gili Matra is located at 8° 20'00"- 8° 23'00" S and 116°00'00" - 116° 08'00" E. Site selection is based on the diversity and variability of the shallow water seabed profile of pre-survey image processing results assessed as representative for mapping by Landsat-8 OLI record of July 25, 2017. Gili Matra waters have high marine resources potential including their biotas because the waters of Lombok have warm water temperatures and their tidal characteristics are influenced by the cyclone winds in the Indian Ocean. Thus, water conditions in the western seasons of high air humidity, good irradiation, low salinity, and sediment composition are composed of medium to rough sand, coral fragments and other biota.

2.2. Satellite Image Pre Processing

Satellite image pre-processing consists of radiometric correction, reflectance correction, water column correction, masking and cropping. Radiometric correction is performed to improve image quality due to atmospheric disturbances, such as haze scattering or other object scattering [10]. Radiometric correction includes radiance correction and reflection correction (Convertion From Radiance at Sensor to ToA Reflectance). The correction of radians done is::

$$L\lambda' = LQcal + AL \qquad (1)$$

where :
$L\lambda'$: radian spectral on sensor (W/m^2 str μm)
$Qcal$: pixel value (DN)
M_L : rescalling constant
(RADIANCE_MULTI_BAND_x,band used)
A_L : adders constant
(RADIANCE_ADD_BAND_x, band used)

Meanwhile, for reflectance correction on using the following formula::

$$\rho\lambda' = \rho Qcal + A\rho/Cos\,(\theta s) \qquad (2)$$

where:
$\rho\lambda'$:the result of processing without the retrieval correction, $\rho\lambda'$ does not contain correction for the sun angle
$Qcal$:pixel value (DN)
$M\rho$:rescalling constant
(RADIANCE_MULTI_BAND_x, band used)
$A\rho$: konstanta penambah
(RADCIANCE_ADD_BAND_x, band used)
$Cos(\theta s)$: sun zenith angle (90 - angle of sun elevation), where (Cos (Rad (90-angle of sun elevation)).

Image sharpening to see shallow water seabed profile using an approach of band transformation, the Depth Invariant Index (DII) algorithm [34]. The algorithm uses band 2 and band 3. The basic use of band 2 and band 3 is that both bands have good penetration into the water column [35]. The algorithm used following the BIG (2014) are:

$$Y=(ln\ Band1-Band1\ Deep) - ki/kj*(lnBand2- Band2\ Deep) \qquad (3)$$

where:

Y	: result of lyzenga algorithm
(ln(Band1)	: natural blue band logarithm
(ki/kj)	: variance and covariance values on blue and green bands
(ln(Band2)	: natural green band logarithm.
Band Deep	: pixel value of the band-i in the deep sea

The process of masking and cropping aims to limit the area of image analysis. In the image analysis for shallow water, it is necessary to separate the land and water, so that the analysis process is only done on the water zone. Masking and cropping can be done using either the pixel value threshold on a given object to be digitally analyzed or through a digitization process [10]. Then do the sampling area at least 30 dots. Point retrieval is done by using the same colored substrate point that allegedly represents the shallow water seabed profile [11].

2.3. Field Survey

The field survey was conducted on 07-11 August 2017. Field data was collected by determining the observation point that was considered to have represented each object based on the variation of the seabed profile condition of the waters of the processed satellite images true color [12]. Field data were collected using Rapid Reef Assessment (RRA) techniques to obtain shallow water sea bed profile [13]. Intake of the shallow water seabed profile is done by photographing the object in the field every 15 meters once continuously within the RRA area, this is due to the results obtained because Landsat 8 image has a spatial resolution of 30 meters [6] so that sampling every 15 to anticipate the pixel errors . Furthermore, recording the number of photos and GPS together with the recording of objects observed in the field [14]. Then the GPS is placed in the housing and tied up with the data taker (cultivated GPS remains on the surface) to record tracking during the data retrieval of baseline objects [15].

2.4. Image Classification

The classification taken is a supervised classification, because according to [16] high resolution image is suitable to use guided classification, this process includes the analyst doing the training on the image pixel area that has been sampled the object when the field survey. Sampling is expected to be evenly distributed so that the computer device is able to recognize the pixel as a particular object. Then use the maximum likelihood classifier to display image classification using field data obtained from photographs in the field [12]. Classes built for image classification refer to Ball *et al.* [17] as **Table 1**

Table 1. Table of Classification

Class	Characteristics
Coral + Rubble (MIX-C/RB)	Dominant coral, but has rubble
Sand + Seagrass (MIX-SD/SG)	Dominant sand but has seagrass
Rubble (RB)	100% Rubble
Seagrass + Sand (MIX-SG/SD)	Dominant seagrass but has sand
Sand+ Rubble (MIX-SD/RB)	Dominant sand but has rubble
Rubble + Dead Coral (MIC-RB/DC)	Dominant rubble but has dead coral

 https://doi.org/10.1051/e3sconf/20184704002

2.5. Accuracy Assesment

Data obtained from the field survey are for accuracy assesment, re-classification and re-analysis to determine the baseline profile of the waters. The results of the accuracy assesment affect the level of user confidence in the data type and the method of analysis [18]. The accuracy assesment can be done using the table of confusion matrix as shown in **Table 2**. The total samples obtained in the field, some (50%) are used in digital multispectral classification process and others (50%) are used for accuracy assesment's classification [19].

Table 2. Table confussion matrix based on[33]:

1	2	k	n_{1+}	
1	n_{11}	n_{12}	n_{1k}	n_{1+}
2	n_{21}	n_{22}	n_{2k}	n_{2+}
K	n_{k1}	n_{k2}	n_{kk}	n_{k+}
	n_{+1}	n_{+2}	n_{+k}	N

3. Result and Discussion

3.1. Shallow water seabed profile

Sampling data in the field was obtained 193 points and there were 156 points correctly classified into satellite image class. The sample class is built based on the objects obtained in the field. Based on samples taken in the field, Rubble and MIX-SG / SD classes have the Highest number of classified and most-sampled samples, while the MIX-SD/SG class has the lowest number of classified and accurate samples (**Table 3**)

Table 3. Detailed samples of shallow water seabed profile based on field data.

Class	Characteristics	Classified Sample	ccurated Sample
Coral + *Rubble* (MIX-C/RB)	Dominant coral, but has rubble	37	33
Sand + Seagrass (MIX-SD/SG)	Dominant sand but has seagrass	12	4
Rubble (RB)	100% Rubble	50	48
Seagrass + Sand (MIX-SG/SD)	Dominant seagrass but has sand	50	47
Sand+ Rubble (MIX-SD/RB)	Dominant sand but has rubble	13	6
Rubble + Dead Coral (MIC-RB/DC)	Dominant rubble but has dead coral	31	18
Total		193	156

This is the appearance of the shallow water seabed profield in the field recorded by underwater cameras which has been classified into 6 classes, ie:

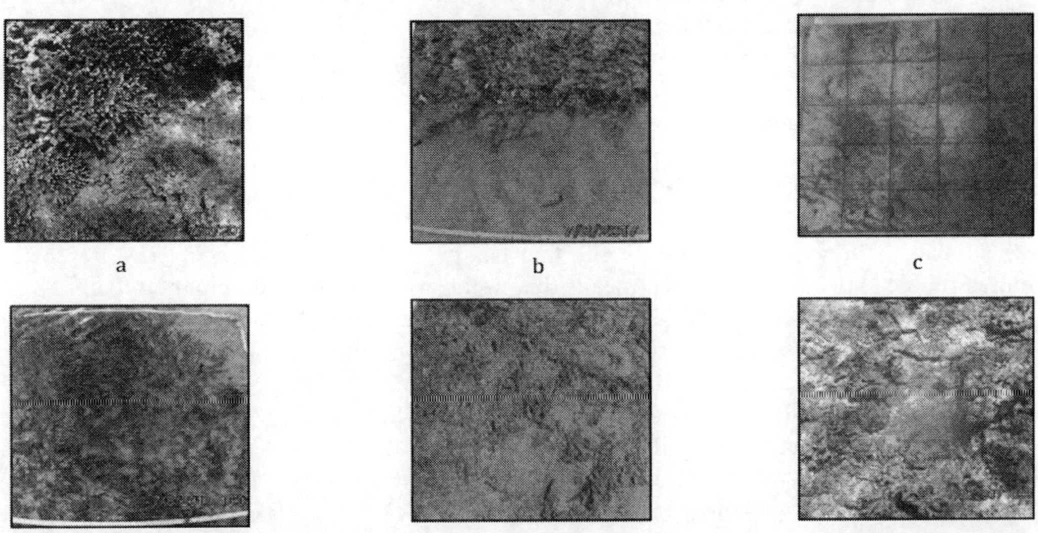

Fig 2. An overview of the common object variations appears in the field and is used as a reference for constructing the base substrate cover class in guided classification (a) MIX-C/RB (b) MIX-SD/SG (c) RB (d) MIX-SG/SD (e) MIX-SD/RB (f) MIX-RB/DC

Based on observation and data collection in the field, the dominant object is found rubble and the mixture between seagrass and base substrate (sand). The amount of rubble found in Gili Matra is caused by environmental conditions that are experiencing global warming and human activities are less conservative such as placing anchors at random when the ship is docked, and tourists who are less used to doing snorkeling so that many seabed objects are damaged by the activity on while swimming. This is consistent with the opinion of Setiawan *et al* [20], during 2012 to 2017 in Gili Matra there are many dead corals found due to extreme high or low water temperatures, and this is associated with global warming where corals include animals with low temperature tolerance. While seagrass and substrate are easy to find in Gili Matra because its distribution is very wide along the coast of Gili Matra. Seagrasses in the coastal waters of Lombok island can be found in intertidal areas almost along the coast including on small islands such as Gili Air, Gili Meno, and Gili Trawangan

The differences in data points in the field and correctly classified points can be due to misclassification due to the difference in the delicate spectral values, as well as the object classes in the field that cohere its existence [21]. The distance of sampling in the adjacent field is also the cause of the result of an accurate object on the map. So that, in some samples of objects with a narrow distribution in the field the results are less than optimal [18] , consequently the number of points that are accurate and classified there are differences.

The most classified and accurate class of samples can be due to the relatively uniform homogeneity and spatial distribution in the field. Highly classified and accurate grades are reinforced by field conditions stated by Setiawan *et al.* [20] he stated that Gili Matra dominantly found rubble and gratitude statement (2015) that seagrasses tend to be in the intertidal region of Gili Matra. In the image aspect, Landsat-8 Satellite Imagery is suitable for mapping seagrass and basic substrate on a wide scale [22,23,24]

The used of the class naming in this study is the result of interpretation in the field, because there has not been a standard naming class to a more detailed level in benthic object mapping [25]. This is in line with what the Geospatial Information Agency (BIG) discloses only 4 general classes for mapping the shallow water seabed profile, ie coral reefs, seagrass beds, macroalgae and substrate (BIG, 2014). Regardless of the naming context, however, the use of Landsat-8 Satellite Imaagery in this research is one of the efforts of optimizing remote sensing media in monitoring and providing coastal information in accordance with the research done by Hafizt *et al.* [18] benthic habitat in shallow waters by using Landsat-8 Satellite Image.

3.2. Classification of Shallow Water Seabed Profile in Gili Matra

The shallow water seabed profile that has been classified into 6 classes are corals mixed rubble (MIX-C / RB) which means that in the area corals are dominant but algae still exist, sand mixed seagrass (MIX-SD / SG) which means that in the area sand are dominant but seagrass still exist, rubble class (RB) which means that in that area dominated by rubble, seagrass mixed sand (MIX-SG / SD) which means that in the area seagrass are dominant but sand still exist, sand mixed rubble (MIX-SD / RB) which means that in the sand are dominant but rubble still exist, rubble mixed dead coral (MIX-RB / DC) which means that in the area rubble are dominant but dead coral still exist, (**Figure 3**).

Based on the classification, the distribution of MIX-C / RB classes tends to be on the outer edge, because the coral reefs in Gili Matra include the fringing reef group that is near the coral reef and parallel to the coastline. Some examples of the types of edge reefs present in Indonesia are in the Mentawai, Pangandaran and Parangtritis areas on the southern coast of Java Island, Lombok and Sumbawa, East Nusa Tenggara and north and west of Papua [1]. While MIX-RB / DC and rubble classes tend to be found in flat areas caused by waves and strong currents carrying rubble and dead corals toward the shore. This is in accordance with the statement of McKenzie and Yoshida [26] that coral reef ecosystems in Gili Matra are often encountered dead coral this is caused by anthropogenic activities such as cruise lines where the decline of the anchor can damage the coral reef ecosystem. Furthermore the class, MIX-SG / SD tends to be close to the mainland where the seagrass habitat is adjacent to the shore. Locations of seagrass beds in accordance with the statement of Nurdin *et al.* [27]where the presence of seagrass cover and sand are in shallow waters with a depth ranging from 0-1,5 meters. Shallow and clear areas where seagrasses breed. As flowering plants that acclimatise to drown in the sea, the seagrasses flourish in the waters with the foundation of mud, sand, gravel and dead coral fractures with the seabed still penetrating sunlight.

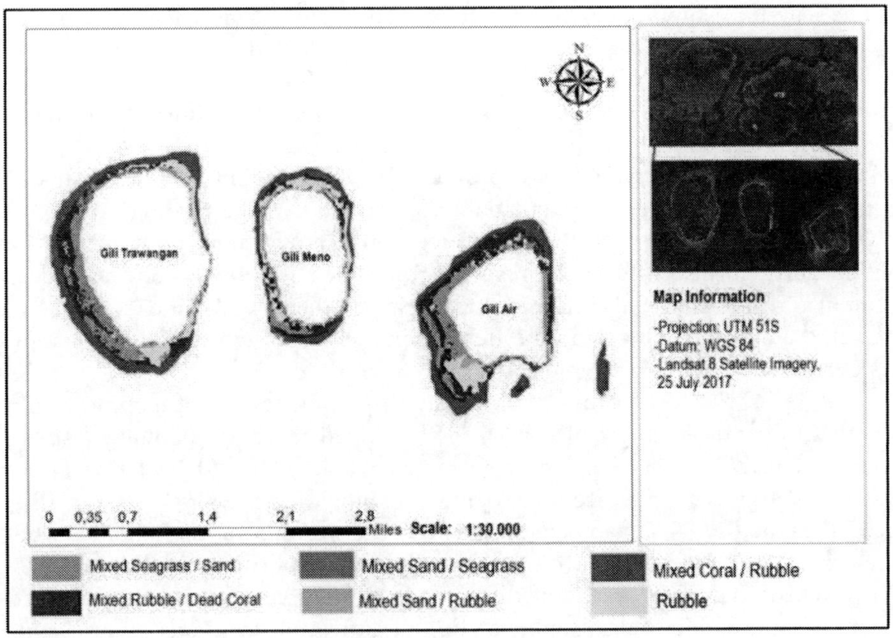

Fig. 3. Map of classification shallow water seabed profile in Gili Matra

3.3. Accuracy Assesment of Shallow Water Seabed Profile

The overall accuracy of the baseline shallow end surface of shallow seabed profiles from Landsat-8 Satellite Imagery is 80%. The lowest accuracy user is the MIX-SD / SG class and the highest is Rubble. Meanwhile, the lowest Accuracy Producer is MIX-SD / RB and the highest is MIX-C / RB (**Table** 4).

Table 4. Accuracy Assesment Based On Landsat 8 Satellite Imagery

	mix-C/RB	mix-SD/SG	RB	mix-SG/SD	mix-SD/RB	mix-RB/DC	UA
mix-C/RB	33	0	2	0	0	2	**89,18 %**
mix-SD/SG	0	4	2	4	2	0	**33%**
RB	2	0	48	0	0	0	**96%**
mix-SG/SD	0	1	2	47	0	0	**94%**
mix-SD/RB	0	1	1	5	6	0	**46,15 %**
mix-RB/DC	4	1	4	4	0	18	**58%**
PA	**84,61 %**	**57,14 %**	**81,35 %**	**78,3 %**	**25%**	**90%**	
OA				**80%**			

The overall accuracy obtained in this study is 80%, it is already high [28], referring to the statement of Green *et al.* [4]that the accuracy can be used is an overall accuracy test of more than 60%, and also reinforced by the opinion of Ambodo *et al.* [28] that the accuracy test of 60% can be said to be good. The results of good accuracy in this study can be caused by several possibilities, one of them is about the the recording time from satellite imagery which processed, does not vary too much with the field survey, where the Landsat 8 Satelite Imagery recording in July 2017 and field survey conducted in August 2017. This is reinforced by the opinion Harvini [29] which states that the results of image recording with the time of research conducted affect the test accuracy, because of the possibility that the field conditions have changed.

The low level of user accuracy and producer accuracy could be due to difficulty of position matching between imagery and field data such as image position shift with GPS causing deplacement of observed object location [30]. In addition, the number of sample variations in the field also affects the results of accuracy [4]. Image errors in describing objects can also result in low accuracy values because objects that are described by image do not match the observed objects in the field, then the proportion of too few samples in a sample is compared with other samples [18].

According to [4] that the accuracy of the shallow marine habitat mapping that can be used is with an overall accuracy of> 60%. [18] mapped the basic habitat of shallow sea waters using Landsat-8 imagery to produce accuracy of 47.57% with 7 classes of cover. In other research, produced 76% accurate benthic habitat maps with 4 classes [8], while Marlina [18] resulted in 84% map accuracy with 5 classes of coral reef habitat, while Phin *et al.* [31] yielded 57% accuracy with a relatively high classification class of 20 classes. The emergence of mapping accuracy differences from several studies using Landsat-8 Satellite imagery was due to differences in classification methods, number of field observation points, number of benthic habitat classes and study sites..

As stated by Hafizt *et al.* [18] that if by using Landsat-8 Satellite Imagery capable of achieving high accuracy, then in observing the dynamics of changing coastal conditions of Indonesia can utilize Landsat-8 Satellite Imagery that can record data on a regular basis. In addition, Landsat-8 Satellite Imagery that is open access is also suitable for mapping coastal conditions of tropical regions such as Indonesia with the object of the sea floor tends to be encountered are coral reefs and sea grass beds [32].

4. Conclusion

The use of Landsat-8 Satellite Imagery to map the shallow bottom profile profile in Gili Matra resulted in 6 classes of shallow bottom substrate covering the rubble mixed rock (MIX-K / RB), mixed seagrass sand (MIX-SD / PL), rubble (RB), sand mixture (MIX-PL / SD), mixed rubble (MIX-SD / RB), dead coral rubble (MIX-RB / DC) and Landsat-8 Satellite Imagery are said to be good enough to map shallow sea floor objects in coastal areas.

The authors would like to thank the Head of Gili Matra Village who has been pleased to provide accommodation to us while collecting data in Gili Matra, Mr. Amron and Mr. Doddy who helped to review and review the manuscript of this journal, to the expedition team HIMAKEL III Jenderal Soedirman University for support and trust to us, as well as all parties who have helped in the data collection and writing of this journal.

References

1 Dahuri R. *Naskah Orasi Ilmiah Pengukuhan Guru Besar Tetap Ilmu Pengelolaan Sumberdaya Pesisir dan Lautan*. (Institut Pertanian Bogor, Bogor, 2003)

2 Lillesand, dan Kiefer. *Penginderaan Jauh dan Interpretasi Citra*. (Gadjah Mada University Press, Yogyakarta, 1999)

3 J.R. Diaz, M. Solan, M.R. Valente. J. Environ Manag. **73**, 165-181 (2004)

4 E. P. Green, P.J. Mumby, A.J. Edwards, and C.D. Clark. *Remote Sensing Handbook for Tropical Coastal Management*. (UNESCO, Paris, 2000)

5 A.F. Syah. Jurnal Kelautan**. 03,** 18-28 (2010)

6 USGS. *Landsat 8 (L8) Data Users Handbook*. (Department of the Interior U.S, U.S. Geological Survey, 2015)

7 G. Sitanggang. *Berita Dirgantara*. **11,** 47-58 (2010)

8 H.S. Murti, and P. Wicaksono. *Majalah Globe.* **16,** 133-120 (2014)

9 J. Jensen. *Remote Sensing of The Environment: an Earth Resource Perspective, 2nd edition*. (Pearson Prentice Hall, United States of America, 2007)

10 B. Prayuda. *Panduan Teknis Pemetaan Habitat Dasar Perairan Laut Dangkal.* (Critc Coremap II LIPI, Jakarta, 2014)

11 T. Hariyanto, and A. Lingga. *GEOID*. **01,** 171–175 (2016)

12 P. Wicaksono. *12th Biennial Conference of Pan Oceanic Remote Sensing Conference* (PORSEC 2014, Bali Indonesia, 2014)

13 Dahlan. *J. of Fish Dev.* **1,** 61-82 (2014)

14 A.B.M.A. Suriadi, Y. Siswantoro, S. Hartini, G.B. Saputro, D.M. Yuwono, D. Suhendra. *Pedoman Survei dan Pemetaan Terumbu Karang Pusat Survei Sumberdaya Alam Laut*. (Badan Koordinasi Survei dan Pemetaan Nasional, Bogor, 2005)

15 C.M. Roelfsma, S.R. Phinn. *A manual for Conducting Georefenced Photo Transect Surveys to Assess the Benthos of Coral Reef and Seagrass Habitats*. (University of Queensland, Queensland: Centre, 2009)

16 P. Danoedoro. *Pengantar Penginderaan Jauh Digital*. (Penerbit ANDI, Yogyakarta, 2012)

17 D. Ball, S. Blake, A. Plummer. *Review of Marine Habitat Classification System*. (Parks Victoria, Melbourne, 2006)

18 M. Hafizt, Y.M. Iswaril, and B. Prayudah. Oseanologi dan Limnologi di Indonesia. **2,**1-13 (2017)

19 P. Wicaksono, A.P. Aryaguna, and H. Akhyar. Pemetaan. *Seminar Nasional Pengelolaan Pesisir & Daerah Aliran Sungai ke-1*. (Universitas Gadjah Mada, Yogyakarta, 2015)

20 Setiawan. J. Fish. Mar. Sci. **1,** 39-54 (2017)

21 F. Mafazi. *Pemetaan Habitat Bentik Menggunakan Citra Quickbird di Sebagian Pulau Kemujan, Kepulauan Karimunjawa* (UGM, Yogyakarta, 2014)

22 R. Shofiyati. *Informatika Pertanian.* **19**, 109-124 (2010)

23 I.H. Supriyadi. Oseanologi dan Limnologi di Indonesia. **36,** 147-164 (2010)

24 I.M. Sakaruddin. *Komposisi Jenis, Kerapatan, Persen Penutupan, dan Luas Penutupan Lamun di Perairan Pulau Panjang Tahun 1990-2010.* Skripsi. (Fakultas Perikanan dan Ilmu Kelautan. Institut Pertanian Bogor, 2011)

25 M. Hafizt. *Inventarisasi Habitat Bentik Menggunakan Teknologi Pengindraan Jauh untuk Mendukung Program Pengelolaan Pesisir Terpadu di Pulau Kemujan Kepulauan Karimunjawa.* (Universitas Gadjah Mada, Yogyakarta, 2015)

26 L.J. McKenzie, and R.L. Yoshida. T*he Nature Concervacy, Coral Triangle Center, Sanur.* (Seagrass-Watch HQ Cairns. Bali, 2009)

27 N. Nurdin, T. Komatsu, C., Rani, G. Arafat, S.A.M. Akbar. *Proceeding of SPIE Asia Pasific Remote Sensing,* (Kyoto, Japan, 2012)

28 P. Ambodo, and R.H. Jatmiko. IPI Jurnal Bumi Indonesia. **1,** 9-18 (2012)

29 W. Harvini. J.Bhumi. **3,** 98-110 (2017)

30 V.P. Siregar. J. Ilm.Tekno.Kel.Trop. **2**, 19-30 (2010)

31 V.D. Tran, S. Phinn, C. Roelfsma. Asian J. Geoinformatics. **12,** 2 (2012)

32 N. Radiarta. Jurnal RIS Akuakultur. **2**, 271-280 (2007)

33 G.R. Congalton. Int. J. of Wildland Fire. **10**, 321-328 (2001)

34 D.R. Lyzenga. *Applied Optics*. **17**, 379–383 (1978)

35 R.E. Mount. *Photogrammetric Enginering and Remote Sensing*. **71,** 1407–1415 (2006)

Mumby, P.J., Green, E.P., Clarck, C.D., Edwards, A.J. 1999. Digital Analysis of Multispectral Airbone Imagery of Coral Reefs. *Coral Reefs.***17**:59-69.

Nybakken, J.W. 1988. *Biologi Laut: Suatu Pendekatan Ekologis*. [Terjemahan dari: Marine Biology: An ecologycal approach, 3rd ed]. Eidman HM, Koesoebiono, Bengen DG, Hutomo M & Sukarjo S (penerjemah). PT Gramedia. Jakarta. 459 hlm.

Syukur. 2015. Distribusi, Keragaman Jenis Lamun (*Seagrass)* dan Status Konservasinya di Pulau Lombok. *Jurnal Biologi Tropis.* **15(2)**: 171-182.

Wicaksono, P., Hafizt, M. 2013. Mapping Seagrass from Space: Adressing the Complexity of Seagrass LAI Mapping. *Europan Journal of Remote Sensing*. **46** : 18-39.

Associated Of Coral Reef With Reef Fish In Northern And Southern Gili Air Island And Gili Trawangan Island Lombok

Trisna Wahyu A P[1], *Helmi* Mubarak[1], *Damar* Lazuardy Rolian[1], *Hanson* Geraldi Pardede[1], *Prabowo*[1], *Dhananta* Rizka Safitri[1], *Kharisma* Zunika Putri[1], *Muhammad* Iqbal Kenedi[1], *Ainia* Gita Tiara Shanti[1], *Satria* Nur Afrizal[1], *Niken* Larasayu Santoso[1], *Karina* Maharani[1], *Ria* Widyaningrum[1], *Azzam* Helmi Muflih[1], *Rifki* Nur Faizan[1], *Adini* Safitri[1], and *Tjahjo* Winanto[1]

[1)] Marine Science Department, Fisheries and Marine Science Faculty, Jenderal Soedirman University, Indonesia

Abstract. Damage to coral reef ecosystems is a major problem on the islands of Gili Air and Gili Trawangan. This will have an impact on the presence of reef fish in the area. This study aims to look how much relation caused by associated between live coral cover and reef fish in Gili Air and Gili Trawangan island.. This research method using Line Intersept Transect (LIT), Underwater Visual Census and simple linear regression analysis to know the relation. The percentage of coral cover on Gili Air Island and Gili Trawangan Island on reef flats (1-5 m) is 11.75% and 11.67% respectively, on the reef slopes (6-10 m) the percentage is 50.4% and 48.9%. In addition we observed the existence of 11 families reef fish. The abundance of reef fish on Gili Air and Gili Trawangan islands on reefs flat with an average of 0,406 ind / m² and 0,137 ind / m², on the reef slope on average - respectively 0,434 ind / m² and 0,274 ind / m². The determinant value in Southern part of both island indicates a value close to +1 and in Northern part of both island indicates a value close to -1.

1 INTRODUCTION

Gili Matra is an archipelago made up of three islands, namely Gili Meno, Gili Air and Gili Trawangan located in Lombok, Indonesia. Gili Matra have a great attention of tourists, both domestic and foreign because of coral reefs and mangrove vegetation. as a frequently visited place by tourists, the balance of the ecosystem is an asset of Gili Matra. Status of the Park waters also make travel demand in the region Gili Matra TWP increases. Since declared a conservation area in 1993, tourism activity has grown rapidly, and on the other hand causes the degradation of ecosystems [31]. Coral reefs in the area of Gili Matra is a center of attraction for tourists in the coastal and marine activities, such as snorkeling and diving. Interest in dive increasing since 1995. However, the damage began to occur due to tourist activities that are less cautious.

Coral reefs are a group of individual corals. Light intensity and water quality was good requirements for living coral reefs. Changes in oceanographic conditions can cause the coral susceptible to disease. Corals can be infected to the disease caused by three factors that is the pathogen, genetics, and environmental. Rapid changes in environmental conditions can exacerbate infections because the pathogens are able to reproduce faster [12]. Coral reefs become the habitats of other organism such as sea grass, algae, crustaceans, sponges, mollusks, fish, corals and other biota [9]. The existence of coral reefs that serve as habitat for reef fish has associated with the number of reef fish [26].

According Hallacher [13], reef fish is a group of life fish associated with coral reef ecosystem. Found as many as 113 reef fish families and the most is Perciformes order. The top ten major families of reef fish are Gobiidae, Labridae, Pomacentridae, Apogonidae, Bleniidae, Serranidae, Murraenidae, Syngnathidae, Chaetodontidae, and Lutjanidae.

When viewed in terms of ecology and utilization, reef fish can be grouped into three namely, indicator fish, mayor fish, target fish. Target fish is a group of fish that can be consumed and usually hunted fishermen, indicators fish is a group of fish whose lives have strong associations to the corals and Mayor fish is small fish, 5 to 25 cm generally [1].

Damage to coral reef ecosystems is a major problem in the tourist park waters of Gili Air and Trawangan. Recorded at 55.39% condition of conservation area in Gili Indah Lombok is a bad condition [6]. In 2003, amounted to 52.94% of corals were damaged [4]. Results of research conducted Ahyadi and Jufri [2] showed that until the end of 2008 reached 75% of coral damage in the area of Lombok Sea Tourist Park. Damage to coral reefs caused the presence of groups reef fish is threatened, and in the case of fishing activities. This study aims to look at how much relation caused by associated between live coral cover and reef fish in Gili Air and Gili Trawangan island.

2 METHODOLOGY

This research method is a survey that is research by collecting data directly to the variables that the object of research. This method is also called the research assessment (exploratory), the research is still open and still looking for the elements, characteristics or traits.

2.1 Time and Place

The study was conducted on a 1 to 15 August 2017 with a location in the coral reef island of Gili Air and Gili Trawangan, Lombok west. Observations were made on two stations consisting of north and south. Determination of observation stations in the waters of Gili Matra based areas leeward and windward where the area leeward is more sheltered southern part by the waves so that the current in the area is quieter, while the area windward is an area that is exposed by the waves so that it has a faster current The sampling was conducted at two different depths are 1-5 meters (reef flat) and 6-10 meters (reef slope).

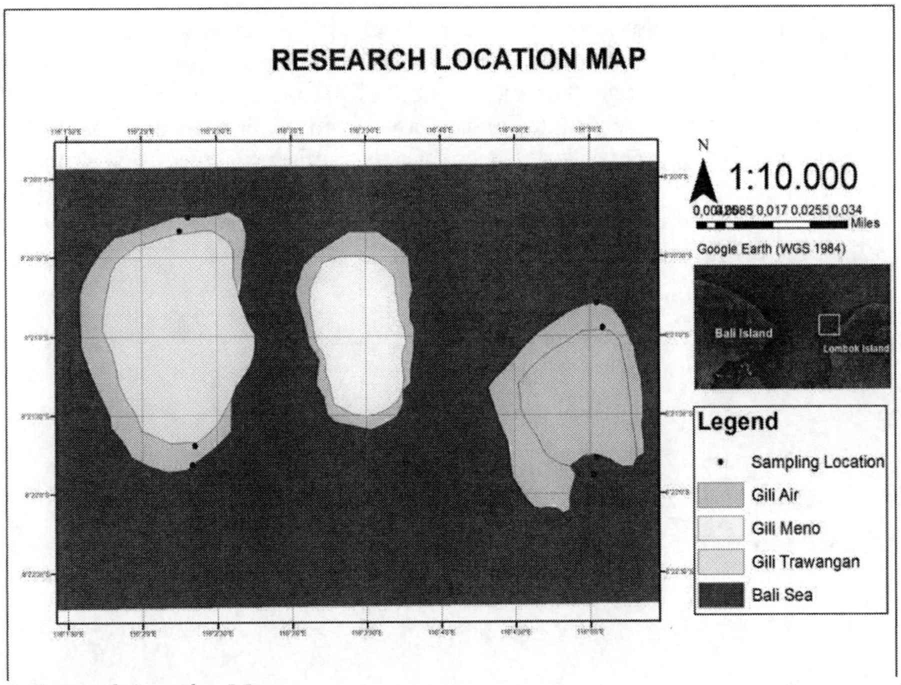

Fig 1. Research Location Map

2.2 Data Collection

2.2.1 *Coral Cover*

Determining coral cover, using Line Intercept Transect (LIT) method to see the growth of live reef [11]. Corals observed at a depth of 1 - 5 meters using snorkling and a depth of 6-10 meters using SCUBA. Data retrieval begins by placing a roll meter horizontally and drawn parallel to the coastline and drawn along 50 m, then record the shapes of coral growth are only exposed transects on the datasheet.

2.2.2 *Reef Fish*

Reef fish Sampling was done by using Underwater Visual Census following the methods of English *et al.*[11], that is using belt transects on Line Intercept Transect. The recording was made on a 50 m transect line while conducting a census of reef fish with the observation distance of 2.5 m to 2.5 m to the right and left, so that the observed area measuring 350m (70m x 5m). implementation of the UVC in the sea, fish species and number of fish recorded in the data sheet watertight. Fish species have been recorded, then identified using pictorial guide of Setiawan [29] and Allen *et al.*[3].

2.2.3 *Water Quality*

Measurements were performed insitu. Sampling of water quality consist of temperature, salinity, currnet velocity, pH, brightness. Each measurment, using a different tool. The measurment of temperature is using thermometer tool by dipping the thermometer into the water and wait a few minutes until the mercury stops at a certain temperature. The maesurment of salinity is using refraktometer by putting water on the glass and read the scale on refraktometer. The measurment of current velocity is using strap and bottle. The strap and bottle are released and allowed to follow the movement of the mass flow of sea water and write the time and range. The measurment of pH using Universal pH by putting litmus paper into water for a few minutes and remove from water then reas the scale. The last measurment that is brightness using secci disk by putting the secchi disk into water until no longer can be seen, at that time write how deep that secchi disk. Then, pull the secchi disk off the water until the secci disk has been seen and write how deep that secchi disk. After that, use the formula to see the brightness.

2.3 Data Analysis

2.3.1 *Percentage Coral Cover*

Calculation coral cover percentage using the formula English *et al.*,[11], namely:

$$Ni = \frac{li}{L} x100\% \quad (1)$$

Description:
Ni = Percentage of coral cover (%)
li = Total length lifeform living coral
L = Total length of transects

Status reef coral is divided into 4 categories including 0% - 24.9% categorized as poor, 25% - 49.9% categorized as moderate, 50% - 74.9% categorized as good, and 75% - 100% categorized as very good [11].

2.3.2 *Reef Fish Abundance*

The reef fish abundance is expressed by a number of individuals per area using Odum's formula [26], namely:

$$Xi = \frac{ni}{A} \quad (2)$$

Description:
Xi = abundance of individuals (ind / m2)
ni = number of fish species to -i
A = area transect (m2)

2.3.3 *Analysis the Relation Between Percentage Coral Cover With Abundance Of Reef Fish*

Analysis used to see the effect of the percentage of coral cover and abundance of reef fish that is simple linear regression analysis, where the independent variable (X) is the percentage of coral cover and the dependent variable (Y) is the abundance reef fish. Sarwono [29] explains, the value of coefficient (R^2) is used to examine the relationship between variables. The coefficient value close to -1 has a weak influence between the two variables or not at all, while a value close to +1 coefficient has a positive influence.

3 RESULTS AND DISCUSSION

3.1 The Percentage Of Coral Cover

Results obtained from the research that has been done using method Line Intercept Transect the percentage of coral cover located on the island of Gili Air and Trawangan can be seen in Figure 2.

https://doi.org/10.1051/e3sconf/20184704003

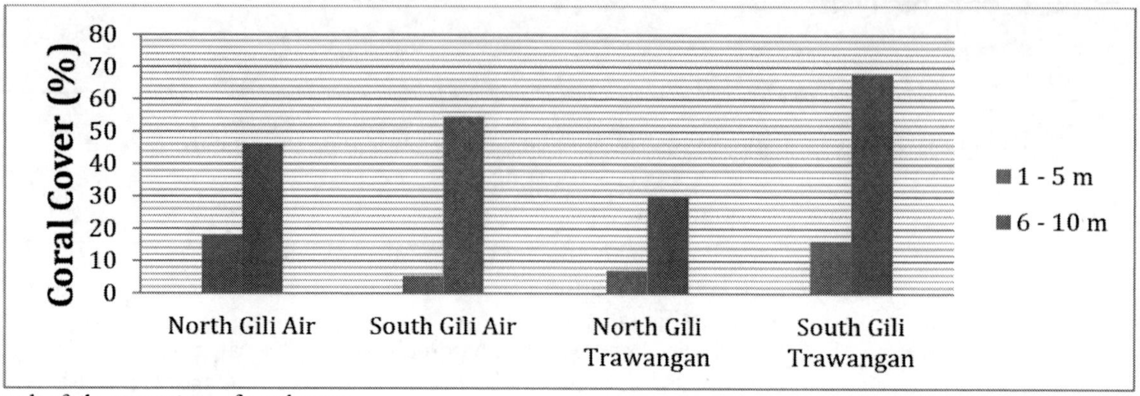

Fig 2. Graph of the percentage of coral cover

The result show, on a reef flat area (1-5 m), the highest percentage was found in the northern part of Gili Air (18.1%), while the lowest percentage was found in the southern part of Gili Air (5,4%). The highest percentage on reef slope area (6-10 m) obtained in the southern part of Gili Trawangan (67.8%), while the lowest percentage was found in the northern part of Gili Trawangan (30%).

Based on data from the graph as a whole, the southern part of the island has the higher percentage than the northern part of the islands, If the terms of the depth, at a depth of 1-5 m has lower percentage or can be categorized as bad condition compared to a depth of 6 - 10 m which can be categorized as good condition. The low percentage of coral cover at a depth of 1-5 m impact from the number of anthropogenic activities so much coral damaged by trampled by tourists who lacked knowledge about the importance of the reef environment. the percentage of live coral cover is affected by natural interactions found in the waters surrounding ecosystems and human activity. Gili Air Island and Gili Trawangan Islands Tourism Region is a place where many visitors ecotourism activities that directly or indirectly affect coral reef ecosystems. Various studies showed to us that, coral decline may be facilitated or accompanied by a detrimental increase in algal turf biomass [7, 10, 17, 23, 34]. Cernohorsky *et al.*, [8] said that small herbivorous fishes can play an important role in preventing the accumulation of algal biomass on coral reefs.

3.2 The Abundance Of Reef Fish

The results calculation of reef fish abundance in this research has been done by using Underwater Visual Census can be seen in Figure 3.

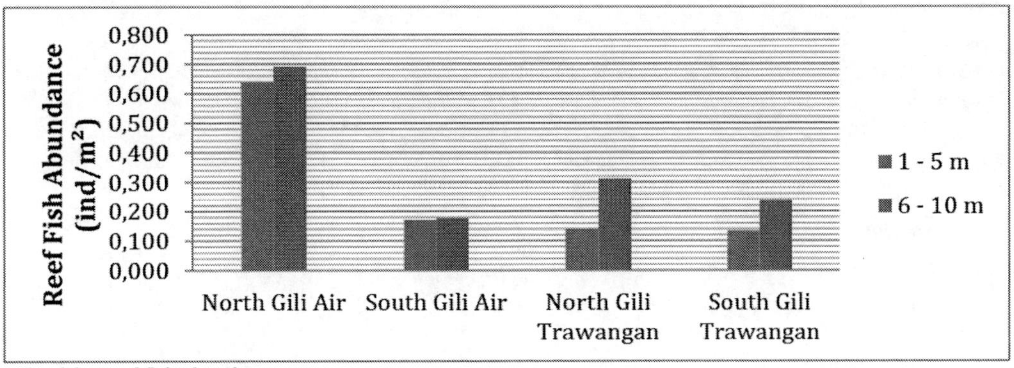

Fig 3. Graph of the reef fish abundance

The results show, on reef slope area (6-10 m), the highest fish abundance was found in the northern part of Gili water (0.691 ind / m²), while the lowest abundance of fish found in the Southern part of Gili Air (0.177 ind / m²). The highest fish abundance on reef flat area (1-5 m) was found in the northern part of Gili Air (0.640 ind / m²), while the lowest percentage was found in the southern part of Gili Trawangan (0.134 ind / m²).

According to the criteria [35], the category of an abundance of reef fish including "slightly" if the transect density less than 0,2 ind / m2; including "a lot" if the density is between 0.2 to

0.4 ind / m2; and including the "abundant" if the density over than 0,4 ind / m2. Based on these criteria, the Northern part of Gili Air was categorized abundant while 3 other side was categorized slightly

Our study showed evidence that Northern part of Gili Air can be effective for conservation of commercially important coral reef fishes, even in the face of high tourist and fishing activity pressure. For example, higher densities of the three most abundant commercially important coral reef fish families or subfamilies (Acanthuridae, Pomacentridae, and Chaetodontidae) were recorded in Northern part of Gili Air Island. The overall, bad condition of reef fish abundance is in 3 location that is southern part of both islan an northen part of Gili Trawangan Island. Some research has showed that reef fish behavior can be affected by the presence of perceived threats, including fishing activities [18]. Fishes, for example, may stay inside or close to corals most of the time if the fish senses existence of people activities.

The existence of reef fish greatly influenced by factors both physical and chemical oceanography. Water conditions were not good to encourage fish to migrate to another, causing the abundance that is in place can be a little. The existence of coral reefs make the ecosystem in the area is stable, so there is no imbalance between extinction and speciation. Based on the overall chart data at two depths, the northern part of the island of Gili Air and Gili Trawangan has a higher abundance of fish from the southern part of the islands. This is due, in the northern area of the islands have coral cover is low so that the abundance is high but the diversityis a bit, and and vice versa in the southern part of both islands have high coral cover so the abundance tends to be small but high diversity.

Table 1. Reef fish Species found at Gili Trawangan Island and Gili Air Island

Species	Gili Trawangan				Gili Air			
	1 - 5 m (Depth)		6 - 10 m (Depth)		1 - 5 m (Depth)		6 - 10 m (Depth)	
	North ern	South ern	North ern	South ern	North ern	South ern	Northern	Southern
Abudefduf bangalensis							11	
Abudefduf septemfasciatus	5							
Abudefduf sexcafiatus	2				35	10		
Abudefduf vaigiensis					20			
Acanthurus auranticavus	2							
Acanthurus blochii					11		60	
Acanthurus japonicus			2					
Acanthurus leucopareius								4
Acanthurus leucosternon			1					
Acanthurus lineatus	5				11		11	
Acanthurus triostegus	3						7	
Acanthurus tristis			10			1		
Acanthurus blochii						15		
Acanthurus leucocheilus			5				3	1
Achanthurus nigroris			4					
Amblyglyphidodon aureus				3				
Amphiprion sandaracinos				2	3			
Balistapus undulatus							5	
Chaetodon collare					4			
Chaetodon selene							1	
Chaetodon speculum				4				
Chaetodon triangulum							3	
Chaetodon Tricinctus							1	
Chaetodon vagabundus	5				9			
Cheilinus trilobatus							3	

Cheilodipterus artus							1	
Cheilodipterus isostigmus					5		6	
Chlorurus bleekeri							15	
Chlorurus microrhinos			8					
Chlorurus troschelii					11			3
Chromis atripectoralis					9			
Chromis caudalis			10			13		13
Chromis dimidiata								1
Chromis iomelas	2							
Chromis margaritifer	15				10			
Chromis morgantilei							1	
Chromis xanthura							3	
Chrysiptera Kuiteri							2	
Chrysiptera taupou		18	20		35	18	47	30
Cryptosentrus sp						1		
Ctenochaetus cyanocheilus			4		3			
Ctenochaetus striatus	3							
Fistularia commersoni	1						3	
Halichoeres hotulanus							1	
Hemiglyphidodon plagiometopon		1		1	10			
Heniochus diphreutes					5		1	
Heniochus various								2
Labroides bicolor							35	
Lutjanus kasmira						2		
Melichthys indicus			7					
Neoglyphidodon crossi							1	
Neoglyphidodon oxyodon					26			
Neoniphon aurolineatus				50				
Neopomacentrus violascens			1					
Ostracion meleagris	1							
Parapercis hexophtalma			2					
Pomacentrus auriventris		25	23					
Pomacentrus cuneattus	1							

Pomacentrus moluccensis				10	5		
Pomacentrus similis	1						
Premnas blaculeatus			2				
Pygoplites diacanthus				5			
Scarus niger						1	
Scarus rivulatus							6
Scoloposis billeniatus							1
Siganus virgatus						1	
Siganus vulpinus		1		2	12		
Zanclus cornutus		2	10	6		19	1
Zebrasoma rostratum	3						

A total of 69 fish species were identified within this study (Table 1). Family of fish that are found on the island of Gili Air, there are Pomacentridae, Chaetodontidae, and Scaridae, but some families was founded but not many.. There are Lutjanidae, Acanthuridae, Gobiidae, Zanclidae, Nemipteridae, Fistulariidae, Apogonidae, and Balistidae. Gili Trawangan island also has many family of fish are found such as Acanthuridae and Pomacentridae, but some families was founded but not many. There are Zanclidae, Ostraciidae, Fistulariidae, Apogonidae, Scaridae, and Balistidae. almost all of which can be found is herbivore fish. In addition, many of the species, especially those with comparatively overall low densities, were found in both location. The distribution of reef fish are highly restricted with oceanography condition and living coral [21, 19, 24].

Other factors such as destructive fishing methods is a negative factor for the resilience process coral. On the contrary, the positive factors that are used as protection and conservation are community policies and initiatives [25, 31]. As reported by Suana and Ahyadi, [32], the use of ecosystems in Gili Matra is as a tourist attraction of the sea and fisheries activities that affect each other on coral reef ecosystems. In recent decades, diving and fishing activities have been increasing since 1993. The complexity in resource management in Gili Matra has led to a variety of environmental problems. Various studies have also demonstrated that Acanthuridae predominated in the shallow reef habitats (< 5 m depth) [20], In this case, maybe the Northern part of Gili Air provide a good algal turf food source and has a better oceanography condition. Most relatively undisturbed reefs support a high biomass of herbivorous fish, but a low biomass of their algal turf food source: it is the high productivity of algal turfs rather than their standing crop that maintains a high standing stock of grazers [14 – 16, 22, 28).

3.3 Percentage Coral Cover Relationship With Reef Fish Abundance

The Calculation results using simple linear regression equation be obtained in the northern part of both island, namely: $y = 0{,}4133x + 0.0694$ with determinant value (R^2) 0,425, while in the southern part of both island, namely: $y = 5{,}919x - 0.06091$ with a value determinant (R^2) 0.5909. The difference value of the determinant indicated differences oceanographic characteristics that may affect the ecosystem in the region. Determinant value indicates a value close to +1 in both parts of the island which means there is a strong relationship between the percentage of coral cover with an abundance of reef fish. Instead of determinant value indicates a value close to -1 which means there is a weak relationship between the percentage of coral cover with an abundance of reef fish.

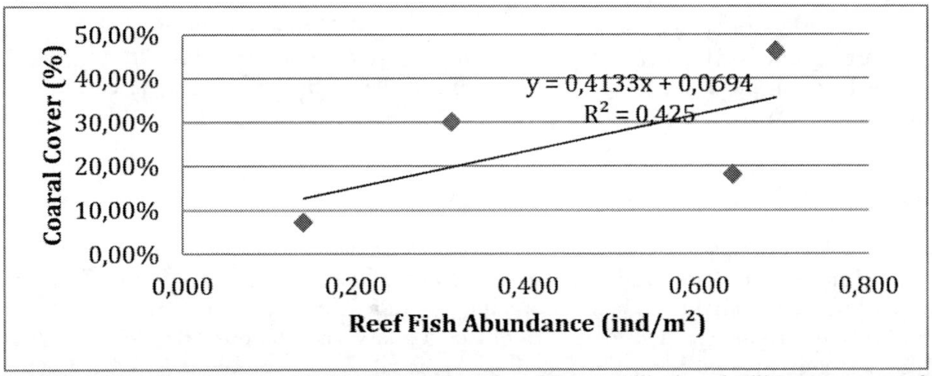

Fig 4. Graph the relationship of coral cover with abundance of coral reef fish on Gili Air and Gili Trawangan northern part

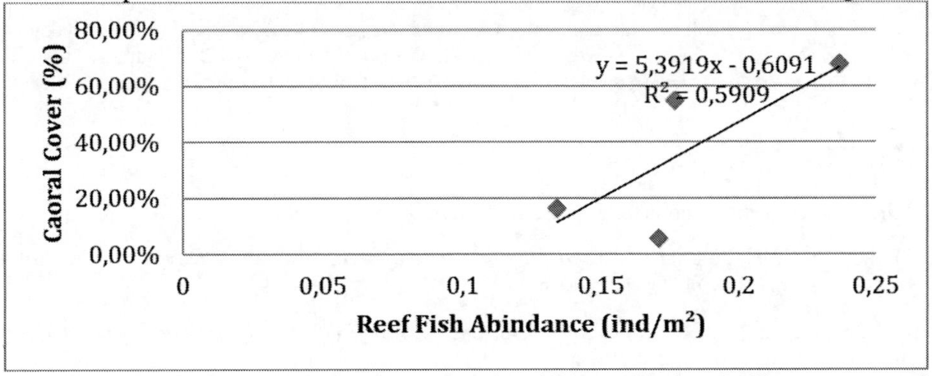

Fig 5. Graph the relationship of coral cover with abundance of coral reef fish on Gili Air and Gili Trawangan southern part

Based on the research that has been done, the chart above shows, the greater the percentage of coral cover is found then the there is great abundance of coral reef areas. The magnitude of the percentage influenced by several factors such as natural interactions found in the waters surrounding ecosystem, the human activity and the physical parameters of the water chemistry that research shows the results or the value that is optimal for the growth and development of coral reefs.

3.4 Water Quality

The Result of measurment from the field, water current velocity was greatest in the leeward site (South) on both island, probably due to strong tidal currents. But there was different current velocity in the windward site (North). This study was done coinciding with the eastern monsoon. The consequently is in Asia the air pressure is low and in Australia the air pressure is high. it makes the wind blow from Australia to Asia. the wind that blows brings a lot of effects for what it passes like currents, waves and tides. Differences in wind direction also affect the difference in current velocity in coastal areas. Water flow and turbulence determine the rate of nutrient supply to algae and are major drivers of productivity [35].

Table 2. Water Quality Data

No.	Location	Temperature (° C)	Salinity (ppt)	pH	Brightness (meters)	Current velocity (m/s)
1	North Gili Air	25	32	7	5	0,19
2	South Gili Air	25.8	32	8	5	0,23
3	North Gili Trawangan	26.4	33	7	5	0,15
4	South Gili Trawangan	27	32	8	5	0,2

Based on the measurement water quality that, on the whole is still in a tolerant condition of marine ecosystem. Good environmental conditions can make the growth of corals and biota association grow well. Value of the sea water quality standard for marine life which is set by the Environment Decree No. 51 year 2004 that is 33-34 ppt, and according Dahuri *et al.* [9], generally at a salinity of 30-35 ppt salinity is good for coral reefs to grow. According to Anwar *et al.* [4] explains that a good temperature range for fish is 25 ° C - 32 °C. The pH values obtained are included in the ideal category for the life of the organism as an ideal category generally has a pH between 7 to 8.5.

4 CONCLUSION

Based on the results of this study concluded that the percentage of coral cover at the reef flat was categorized as poor and at reef slope was categorized as good. The North side of Gili Air was categorized abundant, however 3 other side was categorized slightly. There are a strong relationship between the percentage of coral cover with an abundance of reef fish in the southern part of both island and weak relationship in the norhtern part. It means the higher value percentage of coral cover can found, the higher abundance of reef fish in that areas.

The authors would like to thank to National Water Conservation Area Centers (BKKPN) Archipelago of Gili Matra for the facility as well and Marine Science Department from Soedirman University. The author also thankful to Scientific Communication In Fisheries And Marine Sciences (SCIFIMAS) 2018 that give the opportunities to publish the manuscript.

REFERENCES

1. M. Adrim, M Hutomo. Netherlands J. Sea Res. **23**, 85-93. (1989)
2. H. Ahyadi, A. Jufri, *Report of Research and Development. Regional Development Planning Agency, West Nusa Tenggara, Mataram.* 121p. (2008)
3. G. R. Allen, Steene, P. Human, and N. DeLoach. *Reef Fish Identification. Tropical Pacific. Florida.* (New World Publications, Inc. 484p, 2007)
4. J. Anwar, A. J. Whitten, S. J. Damanik, N. Hisyam. Ecosystem Ecology of Sumatra.Yogyakarta: Gadjah Mada University Press. 653 p. (1984)
5. I. Bakhtiar. Jur Bio Tro. **5,** 1-9 (2004)
6. I. Bakhtiar, T. Karnan, A. Hidayat, Arianto, E. Bursan, and Susiono. *Activity reports Development and Conservation Business Improvement Area in Persons and Natural Resources Conservation Konservasi.Unit, West Nusa Tenggara, Mataram*. 20p. (2000)
7. C.L. Birrell, L.J. McCook, B.L. Willis. Mar. Pollut. Bull. **51**, 408–414. (2005)
8. N.H. Cernohorsky, T.R. McClanahan, I. Babu, M. Horsák. Ind. Ocean. Coral Reefs. **34**,1023 – 1035 (2015)
9. R. Dahuri, J Rais., S. P. Ginting, M.J. Sitepu. *Management of Coastal and Marine Resources Integrated. PT. Pradnya Paramita*. 305p. (2008).
10. G. Diaz-Pulido, L.J. McCook. Coral Reefs. **23**, 225–233 (2004)
11. S. English, C. Wilkinson, V. Baker. *Survey Manual For Tropical Marine Resources.* (Townsville: Autralian Institute of Marine Science, 390p, 1997)
12. Estradivari. *The Reefs Jakarta Report: Long-term observations.* (Indonesia, Jakarta, p: 134-135. 2007)
13. L. E. Hallacher, The Ecology of Coral Reef Fishes.http://www.kmec.uhh.hawaii.edu/Quest Info/Coral20Reef 20 Fishes20-20May 202003. pdf. Access 27 September 2017. (2003)
14. B.G. Hatcher. Trends Ecol. Evol . **3**, 106–111 (1988)
15. B.G. Hatcher. Trends Ecol. Evol. **5**, 149–155 (1990)
16. B.G. Hatcher, A.W.D. Larkum. J. Exp. Mar. Biol. Ecol. **69**, 61–84 (1983)
17. T.P. Hughes, M.J. Rodrigues, D.R. Bellwood, D. Ceccarelli, O. Hoegh-Guldberg, L. McCook, N., Pratchett, M.S. Moltschaniwskyj, R.S. Steneck, B. Willis. Curr. Biol. **17**, 360–365. (2007)
18. F.A. Januchowski-Hartley, N.A. Graham, D.A. Feary, T. Morove, J.E. Cinner. PLoS One **6**, e22761 (2011)

19. M.N. Lavides, N.V. Polunin, S.M. Stead, D.G. Tabaranza, M.T. Comeros, J.R. Dongallo. Environ. Conserv. **36**, 235–244 (2009)

20. M. S. Lewis. J. Exp. Mar. Biol. Ecol,. **81**, 215-228 (1985)

21. S. Mamauag, P. Aliño, R. Gonzales, M. Deocadez. Philipp. Agric. Sci. **92**, 370–387 (2009)

22. L.J. McCook. Coral Reefs. **18**, 357–367 (1999)

23. J.W. McManus, J.F. Polsenberg. Prog. Oceanogr. **60**, 263–279 (2004)

24. C.L. Nañola Jr, P.M. Aliño, K.E. Carpenter. Environ. Biol. Fishes **90**, 405–420 (2011)

25. D. Obura, and G. Grimsditch. *Resilience Assessment of Coral Reefs Rapid assessment protocol for coral reefs, focusing on coral bleaching and thermal stress* (p. 71). (IUCNResilienceScience Group Working Paper Series – No 5, Gland, Switzerland. 2009)

26. E. P. Odum. *Fundamentals of Ecology*. (W.B. Sounders Company, Ltd. Philadelphia, 1971)

27. A. B. Rondonuwu, J. L. Tombokan, and U. N. W. J. Rembet. Jur Ilm Pla. **1**, 87 – 92 (2013)

28. G.R. Russ. Coral Reef. **22,** 63–6. (2003)

29. J. Sarwono. *Quantitative and Qualitative Research Methods.* (Yogyakarta, Science Graha 286 p, 2006)

30. F. Setiawan. *Identification Field Guide Reef Fish and Marine Invertebrates*. (Manado 350 p, 2010)

31. B. Steneck. *How to kill a coral reef: Lessons from the Caribbean*. http://www.reefresilience. org/ Toolkit Coral/C3a1_Herbivory.html. Access 29 April 2018. (2012).

32. I. W. Suana, H. Ahyadi. Jour of Coas Dev. **16**, 94-101 (2012)

33. R. Sukarno, N. Aziz, Darsono, K. Moosa, M. Hutomo, Martosewojo., and Romimohtarto. *Sumber Daya, Permasalahan dan Pengelolaannya.* (Proyek Penelitian Potensi Sumber Daya Alam Indonesia, 109 p, 2003)

34. M.J.A. Vermeij, I. van Moorselaar, S. Engelhard, C. Hörnlein, S.M. Vonk, P.M. Visser, R.K.F. PLoS One 5, 1547–1551 (2010)

35. S.L. Williams, R.C. Carpenter. J. Exp. Mar. Biol. Ecol. **226**, 293–316 (1998)

E3S Web of Conferences **47**, 04004 (2018) https://doi.org/10.1051/e3sconf/20184704004
SCiFiMaS 2018

THE EFFECT IN THE WASTEWATER TREATMENT AT SOYBEAN CURD OF CONTACT TIME MODIFICATION OF ARTIFICIAL WETLAND USING *SSF* BY USING *Schoenoplectus corymbosus* TO IMPROVE WATER QUALITY

Arif Mahdiana1, *Asrul* Sahri Siregar[1], *Sonny* Januar C, *Norman* Arie Prayogo[1],

[1]Program Studi Manajemen Sumberdaya Perairan, Fakultas Perikanan dan Ilmu Kelautan, Universitas Jenderal Soedirman

Abstract. Soybean curd industrial is rapidly growing industry. Characteristic of effluent from soybean curd industrial may contains a lot of organic materials and suspended solids that give negative impact to water. One of the soybean curd industry pollution prevention is a modification of artificial wetlands with sub surface flow system using recirculation by *S. corymbosus* plant to reduce the value of BOD, TSS and nitrate in soybean curd wastewater. Early of research procedure is prepare wetland construction consist of 8 part box container with a capacity of 75 L, then filled in the basic medium gravel 5 cm, clay 10 cm, and then planted a *S. corymbosus* by 30 pieces/construction, then drained soybean curd waste that has been diluted through the recirculation pipe for 16 days in the land of green house plant physiology. The result of optimal decline BOD occured on day 12th ie 99,7, TSS decline value optimal ie 89,3% on day 12th, and declined of optimal nitrate on day 12th ie 89,7%. The decline value of BOD, TSS and nitrate until day 16th has been standart quality unless TSS according to Perda Jateng no.5 tahun 2012 and PP no.82 tahun 2001.

[1]Corresponding Author: setiawanlara@gmail.com

https://doi.org/10.1051/e3sconf/20184704004

1 INTRODUCTION

Soybean curd industry is one of the home industry that rapidly growing in every city or country [1]. Banyumas Regency, Central Java is having 524 soybean curd industry [2].

Soybean curd waste characteristic contains organic materials and suspension solid such as BOD ranged between 3630-3650 mg/L in value, TSS 8935-13824 mg/L and 4 – 5 value of pH [3].

The high potential of pollution due soybean curd wastewater, it is necessary to control the pollution. One of the biological methods is using artificial wetlands that are quite effective in reducing BOD, COD, TSS, NO3, sulfide, fat, and chromium levels [4].

Modification of artificial wetland with sub surface flow system using recirculation has some advantages that using plants to process wastewater (a) physical process sedimentation and filtration removal mechanism (b) physical and chemical process adsorption and phosphor and heavy metals precipitation removal mechanism (c) biochemical process degrading organic materials, nitrification, denitrification, anaerobic decomposition and absorption by water plants removal mechanism. The advantage of recirculation pattern are it can reduce TSS value to ± 90%, and Total Nitrogen value to 50% in domestic waste treatment [5].

Based on previous wetland research using Scirpus grossus SSF system with detention time of 72 days showed the reduction of TSS value to 100% in household waste treatment [6]. Munazah and Soewondo [3] research stated that *Scirpus grossus* can reduce BOD value to 85,07% with 5 days contact time. The previous research result is became basis of researcher to use *Scirpus* with different species *Schoenoplectus corymbosus* (*Scirpus corymbosus*) in soybean curd wastewater treatment.

2 RESEARCH METHODS

2.1 Research Methods and Design

The research design used was completely randomized design (CRD) with 5 treatment: 0 day, 4 day, 8 day 12 day and 16 day contact with soybean curd wastewater and 4 replicates each treatment.

2.2 Research Parameters

The determinants parameter in this research was the decreasing value of Biological Oxygen Demand (BOD5), Total Suspended Solid (TSS) and nitrate (NO_3). The supporting parameters were temperature, pH, and dissolved oxygen.

2.3 Research Procedure

2.3.1 Media Preparation

Research was started by preparing 8 container box. Container filled with 5 cm of medium gravel, 5 cm of sand, and 10 cm of clay then planted with 30 *S. corymbosus* / container, flowed by diluted soybean curd wastewater through recirculation pipe. Modification Artificial Wetland by SSF system.

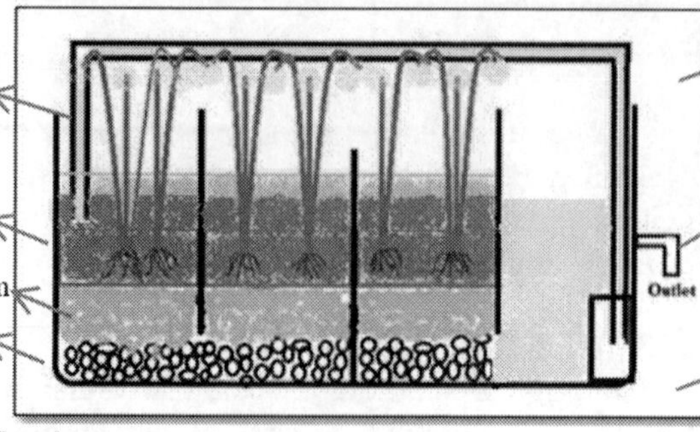

Fig 1. Wetland Modification

2.3.1 Research Implementation

Research implementation was divided into 4 phase:

1) Wastewater Pre-Treatment
   Soybean curd wastewater taken from Kaliputih soybean industry, the wastewater was filtered, precipitated, and neutralized before it was diluted.
2) Plants Acclimatization
   Chosen *S.corymbosus* was that growth in group. Plant was acclimated by planted it in the wetland construction to make adaptation in wetland environment.
3) Preliminary Test
   Soybean curd wastewater was diluted with concentration of 50% and 75%. Construction that already flowed with soybean curd wastewater was put on greenhouse for 1 week.
4) Treatment
   Wetland constructions that had filled with media and plant were set. Constructions was set with a slope of ± 1%, determinants and supporting parameters were measured for 16 days.

Table 1. Determinant Parameters Measurement

No	Parameters	Source
1.	BOD	[16]
2.	TSS	[16]
3.	Nitrate	[17]

2.4 Research Time and Place

The research was done in June at Plants Physiology Green House, Faculty of Biology, Jendral Soedirman University and DO and BOD5 was done at Water Quality Management Laboratory, Major of Fishery and Marine , Faculty of Science and Technique, Jendral Soedirman University. Parameters measurement such as temperature and pH was done in situ , and TSS and Nitrate measurement was done ex situ in Semarang Wahana Laboratory.

2.5. Data Analysis

Data on reduction of BOD, TSS, and Nitrate was obtained from soybean curd wastewater treatment using artificial wetland modification SSF system recirculation with *S. corymbosus* were analyzed by:

1) Comparative descriptive to determine artificial wetland modification SSF system recirculation with *S. corymbosus* capability to reduce the value of BOD, TSS, and Nitrate in soybean curd wastewater.
2) F test and effectiveness level test, F test was used to determine the significance of every treatment given, with 1 – 5 % error rate. If ANOVA had showed significant difference or highly significant difference, the analysis can be progressed into Least Significance Different (LSD) test. Effectiveness of wastewater contact time with *S. corymbosus* to reduce the value of BOD, TSS, and NO_3 measured by Effectiveness formula [15].

$$E\ (\%) = \frac{Co - Ce}{Ce}\ x\ 100\% \qquad (1)$$

Keterangan :
E = Effectiveness value (%)
C_0 = Initial concentration (mg/L)
C_e = Later concentration (mg/L)

3) Descriptive comparative analysis was used to compare BOD, TSS, and nitrate reduced value with standard environment quality according to Perda Jateng no.5 th 2012 dan PP no.82 th 2001 (Table 2) to measure artificial wetland type SSF with *S. corymbosus* successfulness.

Table 2. Soybean wastewater standard quality

No	Parameters	Maximum Value
1.	Temperatur	38°C *
2.	BOD_5	150 mg/L *
3.	TSS	100 mg/L *
4.	pH	6,0-9,0 *
5.	Nitrat	20 mg/L **
6.	Debit maksimum	20 m^3/ton kedelai *

Exp. * Perda Jateng no.5 th 2012
** PP no.82 th 2001

3 RESULT AND DISSCUSION

3.1 System Capability in Reducing BOD, TSS, and Nitrate Value.

Wastewater treatment using artificial wetland affected by environmental condition such as, temperature, pH, and dissolved oxygen because they are fundamental factor of living being. Hammer [7] stated CTW can effectively purify wastewater with same process as natural wetland consisted of plants, soil, and microbial community, but within controlled environment condition. Environment condition is optimal for wastewater treatment because it support used microorganism and aquatic plant growth, so they can reduce soybean curd wastewater BOD, TSS, and Nitrate value.

Wastewater treatment mechanism had two main focuses, first was recirculation process that happened in construction and media that its used (soil, sand, and gravel). Munazah and Soewondo [3] stated that elimination mechanism in constructed wetland were filtration and sedimentation by media, biological elimination by microorganism at soil and root. Plants accelerate the process by giving good environment for microorganism growth at ryzosphere and can transfer oxygen to root and soil so there was exchange in aerobic process (anoxic and anaerobic zone) that can make different microbes growth and eliminate pathogen [3].

3.2 BOD Value Reduction

BOD is measurement of required oxygen by aerobic microorganism to decomposing organic materials in water for five days incubated without light. Based on research result, constructed wetland capable on reducing soybean curd wastewater BOD value **(Fig 2).**

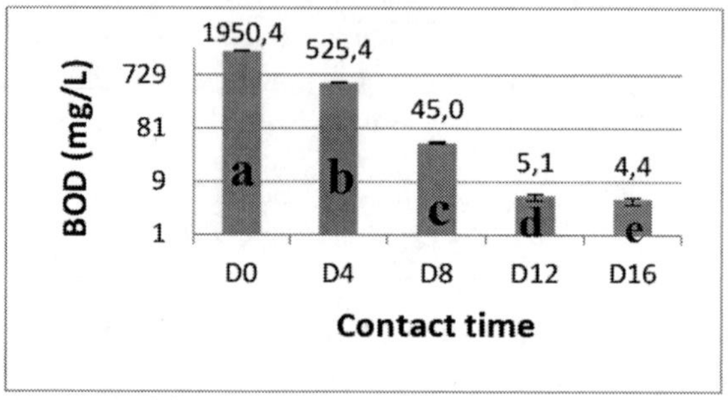

Fig 2. BOD Effectiveness Graphic

Based on picture 2, from D0 to D4 BOD value was reduced 1425 mg/l (59%) with effectiveness value of 73,1%. Oxygen value was 0,2 so it was anaerobic process. In this condition denitrification process occur that formed nitrate converted into nitrogen gas ($NO_3 \rightarrow N_2$) [8]. The nitrogen will be released and released on sediment [9].

BOD value from D0 to D8 was reduced 1905,4 mg/l (71%) with effectiveness value of 97,7%. Longer contact time show there still reducing process from initial treatment because microorganism need time to consume organic materials, increased DO value from 5,0 mg/L to 8,0 mg/L. Increased DO value made microorganism can consume more organic materials up to 97,7%. Contact time from D4 to D8 made 480,4 mg/L and effectiveness 24% difference in value. This happened in 4 days because decomposition process by microorganism began to stabilized, and existing microorganism began to make contact with the waste. Munazah dan Soewondo [3] stated, biological process happened because of contact between waste and microorganism can reach 91,53% in 5 days per 4000 mg/L soybean curd waste load of influent.

BOD value from D0 to D12 was reduced 1945,3 mg/l (85,4%) with effectiveness value of 99,7%. The BOD value was declined optimally because microorganism whose growth had reached stationary phase had degraded 99,7% BOD value in soybean curd wastewater. It is allegedly in addition to the decomposition of organic materials that occur in the wetland construction system because of microorganisms in the media and microorganisms attached to the stem of the plant *S.corymbosus*. Stability of oxygen value of 8,0 mg/L was one of the factor that organic materials by oxidation process by microorganism was happened in the soybean curd wastewater. Contact time from D8 to D12 made 39,9 mg/L and effectiveness 40% difference in value. Wardhana [10] stated, aerobic microorganism that need oxygen in decomposing organic materials can effectively decomposing when its oxygen and nutrient needs were fulfilled.

BOD value from D0 to D16 was reduced 1946 mg/l (97%) with effectiveness value of 98,8%. It is suspected that microorganism regeneration growth activity that occur in the system is less supportive because the availability of nutrients needed for degrading organic materials were limited although oxygen availability still supportive. Contact time from D12 to D16 made 0,7 mg/L make effectiveness 0,1% difference in value. Population decline level was happened because lack of nutrient made decline in bacterial population occur, in this time dead cells is higher than living cells [11].

Based on F test, shown there was significant difference in D0, D4, D8, D12, and D16 treatment in reducing soybean curd wastewater BOD value. $P < 0{,}005$ followed by LSD test. Based on LSD test the H0 control treatment had a significant difference on the H4 contact time treatment, it is suspected because of the activity of microorganisms, the increased value of dissolved oxygen and pH. The difference in treatment of D0 with D8 had significant response, due to the addition of longer contact time and increased in dissolved oxygen. The difference of control treatment (D0) with D12 had significant result, it is caused by microorganism performance reaching optimum point that is at stationary phase. This shows the value reached 99.7% and a decrease of 0.1% from H12 to H16. The longer the contact time that occurs then the process of degradation of organic pollutant parameters can take longer so that the performance of the rector will be better and the effluent concentration also decreased [3].

3.3 TSS Value Reduction

TSS are suspended solids that can be distinguished from their size, in example colloid suspended particles and suspended particles. The value of TSS contained in the soybean curd wastewater can be derived by the effectiveness of the system in experimental artificial wetland research with S. corymbosus plant seen in **Fig 3.**

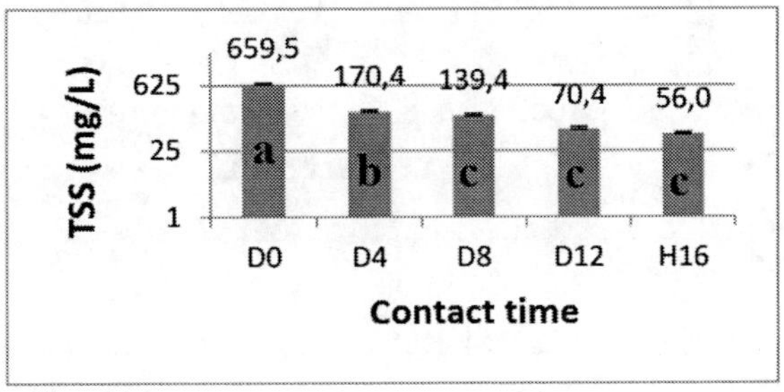

Fig 3. TSS Effectiveness Graphic

Based on picture 3, from D0 to D4 TSS value was reduced 489,2 mg/L (40%) with effectiveness value of 74,2%. It is suspected that wetland media (soil, sand, and gravel) and plants as attached places for microorganism that consume available organic materials. Slow water movement which made filtration process still unstable.

TSS value from D0 to D8 was reduced 520,1 mg/l (61%) with effectiveness value of 78,9%. Due 8 days wetland system condition, media, and plant began to stabilize, filtration and sedimentation process began to occur in waste. In addition, available organic materials will settle at the bottom and will get used by microorganism and inorganic materials will be released because of recirculation process and gravity. Contact time from D4 to D8 made 30,9 mg/L and effectiveness 4% difference in value. Sedimentation process occurs inside the reactor because of gravitation, while filtration process occur when particle was filtered mechanically as they pass through soil and root [12].

TSS value from D0 to D12 was reduced 589,2 mg/l (62,4%) with effectiveness value of 89,3%. The TSS value was declined optimally because 89,3% TSS value in soybean curd wastewater had been degraded. This is occur because of longer contact happened between wastewater with media and plant, so with the result that filtration and sedimentation process optimally done. Contact time from D8 to D12 made 69 mg/L and effectiveness 11% difference in value. It is efficiency was increased along with detention time [13].

TSS value from D0 to D16 was reduced 603,6 mg/L (60,9%) with effectiveness value of 91,5%. This shown that optimal reduction was happened in contact days 12, so difference in value between D12 and D16 was less significant. Contact time from D12 to D16 only 14,4 mg/L and effectiveness 2% difference in value. Prolonged detention time affect wastewater TSS reduction [12].

Based on F test, shown there was significant difference in D0, D4, D8, D12, and D16 treatment in reducing soybean curd wastewater TSS value. P <0,005 followed by LSD test.

Based on LSD test the H0 control treatment had a significant difference on the H4 contact time treatment due to the existence of media as a place for microorganisms to attach and water movement is still slow. Difference in treatment between control with D8, D12, and D16 gave same respond equal to the decrease in TSS value, indicated by the same letter. This is due to the longer treatment contact time will increase the efficiency of elimination process. The longer the contact time made greater sedimentation and filtration process [12].

3.4 Nitrate Value Reduction

Artificial wetland with *S. corymbosus* system is shown capable on reduce nitrate value in soybean curd wastewater **(Fig 4)**

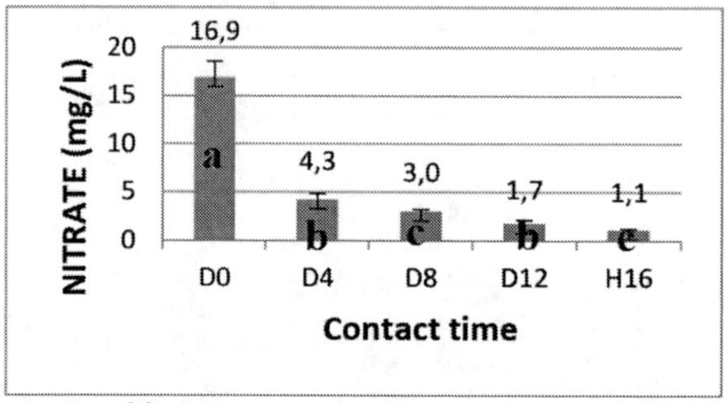

Fig 4. Nitrate Effectiveness Graphic

Based on picture 4, from D0 to D4 BOD value was reduced 12,6 mg/l (59%) with effectiveness value of 74,8%. Oxygen value was 0,2 so it was anaerobic process. In this condition denitrification process occur that formed nitrate converted into nitrogen gas ($NO_3 \rightarrow N_2$) [8]. The nitrogen will be released and released on sediment [9].

Nitrate value from D0 to D8 was reduced 13,9 mg/L (52,6%) with effectiveness value of 82,1%. This decreased in value was due the activity of bacteria with oxidized wetland system due stabilized oxygen supply which is increased to 8 mg/L, so that Nitrosomonas bacteria that oxidize ammonia to nitrite and Nitrobacter that oxidize nitrite to nitrate began to work on aerobic condition. The activity of the S.corymbosus plant is getting stronger by utilizing the nitrate produced from the nitrification process. Contact time from D4 to D8 made 1,3 mg/L make effectiveness 8% difference in value. This aerobic condition triggers nitrification process, ammonium nitrogen converted to nitrate ($NH_4 + \longrightarrow NO_3$) [8].

Nitrate value from D0 to D12 was reduced 15,2 mg/L (66,2%) with effectiveness value of 89,3%. The nitrate value was declined optimally because good media and stable supply of oxygen. The absorption of nitrogen by *S.corymbosus* stem, allows microorganisms growth around the stem. Contact time from D8 to D12 made 1,31 mg/L and effectiveness 7% difference in value. Nitrogen will be taken by plants in mineral form and became plant biomass [9] and plants use nitrate and ammonium as nutrition that can be stored in inorganic nitrogen form.

Nitrate value from D0 to D16 was reduced 15,8 mg/L (%) with effectiveness value of 93,5%. It is suspected to occur because available nutrient was used in 12 days of contact and optimally used. Contact time from D12 to D16 made 0,6 mg/L make effectiveness 4% difference in value.

Based on F test using minitab version 16 software, shown there was significant difference in D0, D4, D8, D12, and D16 treatment in reducing soybean curd wastewater nitrate value. P <0,5 followed by LSD test. Based on LSD test shown D0 to D4 and D12 contact time treatments provide the same response, indicated by the same letter, as well as on the treatment of contact times D8 and D16. The control treatment D0 gave a real response to the treatment of contact time D4 and D12 due to denitrification and nitrification process. Control D0 treatment shown significant response to D8 and D16. That is because the condition of wetland system and oxygen supply had been stabilized

3.5 CTW Processed Quality Standard

Based on research result of soybean curd wastewater treatment using sub surface flow system with *S.corymbosus* can reduce BOD value at 16 days contact time with average value of 58,8 mg/L, the value met the quality standard because the value was < 150 mg/L [18]. TSS reduce in value until 16 days of contact time with average value if 267,9 mg/L, the value did not meet the quality standard because the value was > 100 mg/L [18]. Nitrate reduce in value until 16 days of contact time with average value if 5,7 mg/L, the value met the quality standard because the value was < 20 mg/L (PP no.82 th 2001).

4 CONCLUSIONS

Modified artificial wetland using sub surface flow system can reduce BOD, TSS, and nitrate value in soybean cur wastewater. Most optimal contact time for reducing BOD, TSS, and nitrate value by was at the 12th day each 99,7%: 89,3%; and 89,7%. BOD and nitrate reduced value met with quality standard except for TSS. Based on research result, impact of contact time between modified artificial wetland using SSF system with *S. corymbosus* in soybean curd wastewater treatment do not use fish as experimental animal, so the need for further research to determine the effect of modified artificial wetland used.

REFERENCES

1 Jasmiati., Sofia, A. Journal of Enviromental Science. **2** (2010)
2 Dinas Perindustrian Perdagangan dan Koperasi Kabupaten Banyumas. 2014. Diakses tanggal 23 Oktober 2014.
3 Munazah, A. R. dan Soewondo, P. **4**, 1-8 (2008)
4 Ningsih, D. Martianto, T., Harjanto, S. *Pemanfaatan Air Olahan Wetland di Sitimulyo dalam Pembuatan Pikel Kulit Domba*. **28**, (2012)
5 Prigent, S., Paing, J., Andres, Y., dan Chazarenc, F. Effects of a saturated layer and recirculation on nitrogen treatment performances of a single stage Vertical Flow Constructed Wetland (VFCW). Water Science & Technology. **68**,7 (2013)
6 Al-Badawi, I.A., Abdullah, S.R.S., Anuar, N., Suja, F., Idris, M.. *Water Science & Technology*. **68.**10.
7 Hammer. Constructed Wetland for Wastewater Treatment: Municipal, Industrial and Agricultural. (Chelsea : Lewis Publisher, 1989)
8 Hastuti, E., dan Agustien, R.R. **8**, (2013)
9 Wetland International. 2003. *The use of constructed wetlands for wastewater treatment*. Malaysia Office. 24 pp.
10 Wardhana, W.A. Dampak Pencemaran Lingkungan. (Penerbit Andi Yogyakarta, 2004)
11 Brock, T.D. and Madigan, M.T. Biology of Microorganism. (Prentice Hal International Inc, Englewos Clifs, New Jersey, 1991)

12 Masturah, A., Darmayanti, L., Lilis Y. *Pengolahan Air Limbah Domestik Menggunakan Tanaman Alisma plantago dalam Sistem Lahan basah Buatan aliran Bawah Permukaan*. (Pekanbaru, Riau, 2014)
13 Abdulgani, H., Izzati, M., Sudarno. *Semnas Pengelolaan Sumberdaya Alam dan Lingkungan* (2013)
15 Kasam., Yulianto, A., Sukma, T. Logika. **2,** 3-16 (2005)
16 APHA, AWWA, AWEF. *Standard Methods for The Examination of Water and Wastewater* . (Washington. 2005
17 SNI 06-2480. *Cara Uji nitrat (NO_3-N) dengan Spektofotometer UV-Visible*. (Badan Standarisasi Nasional, 1991)
18 Perda Jateng No.5 Tahun Peraturan Daerah Jawa Tengah No.5 tahun 2012 Tentang Baku Mutu Limbah Cair Tahu. (2012)

Acute effects of *crude oil*for three common mangrove seedling in segara anakan nature reserve (sanr) cilacap, Indonesia

Taufan Harisam[1,4*)], *Sugiyono*[2)], *Rudhi* Pribadi [3)], *Asrul* Sari Siregar[1)], *Sri* Marnani[1], *Dewi* Wisudyanti Budi Hastuti[1], *Norman* Arie Prayogo[1,4)]

[1]Fisheries and Marine Science Faculty, Jenderal Soedirman University, Indonesia
[2]Biology Faculty, Jenderal Soedirman University, Indonesia
[3]Fisheries and Marine Science Faculty, Diponegoro University, Indonesia
[4]Centre for Maritime Bioscience Studies, Institute for Research and Community Service, Jenderal Soedirman University, Indonesia

Abstract.The growth of three mangrove species seedlings,*Bruguiera gymnorrhiza*, *Rhizophora apiculata*and *Ceriops decandra* in sediments contaminated by *crude oil*. Thisresearchwas designed to evaluate the acute effects of *crude oil* on the growth and survival performance of mangrove seedling in 4-week field experiment. Three treatments(sub plot) were compared with the control (no oil added) usingi.e. 100 ml, 120 ml and 140 ml *crude oil* applied once, in a split-plot design with nine replicates. The result showed the extent of survival has maximum for *B. gymnorrhiza* while it was minimum for *C. decandra*. The growth of mangrove seedlings was decrease with higher concentrations of *crude oil*. Stem height of *B. gymnorrhiza* seedling higher than *R. apiculata* and *C. decandra*.We observed, severe damage in the epidermis and the cortex at the root of all three types of mangrove seedlings when crude oil applied at 140 ml.

1 Introduction

The Segara Anakan Nature Reserve (SANR) is a mangrove fringed shallow coastal lagoon, and it is the largest still existing mangrovearea in Java, Indonesia [29]. The SANR is located along the southern coast of Java, Cilacap district, Indonesia, surrounded by an area of mangrove swamps, tributaries, and intertidal lands. Such an ecosystem serves a unique and a productive marine nursery and abundant aquatic ecosystem because the influence of the Indian Ocean may enter lagoon troughs in the western and eastern passages (Fig.1). Consisting of a mangrove fringed lagoon, SANR and its surrounding environment in the Cilacap coastal area are considered a unique ecological feature in Java, Indonesia. As an ecosystem, mangroves are important for overall ecological function as well as the economiesof populations living along the coast, where fisheries are a vital source of food and income [25].

Oil pollution poses a serious health risk to coastal ecosystems throughout the world. Many studies have investigated the effects of oil pollution on coastal marshes[15], [21] and mangroves [1], [7]. Crude oil is composed of up to 17,000 organic compounds, of which about 75% are hydrocarbons [11]. The light and medium compounds in oil remain at the surface and volatilize or degrade over time, while those with high molecular weight are deposited on sediments and retain toxic properties for years [22]. Despite their hydrophobic nature, polycyclic aromatic hydrocarbons (PAHs) can bio-concentrate and are toxic and mutagenic to organisms [7], [17].

Mangroves, which inhabit low wave energy, sheltered locations in intertidal zones of tropical and subtropical regions, are highly vulnerable to oil spills [17]. Oil can penetrate into soft sediments and coat aerial roots [11], [21], leading to oxygen deficiency, suffocation and growth irregularities in the shortterm and mortality in the long-term [31], [3]. Oil can persist in the mangrove environment under typical reduced conditions. Several studies have documented the effect of different types of crude oil on mangrove species under laboratory [27], [31] and field conditions [25][18].

Research on oil contamination have frequently concentrated on the mass mortality of plants after large-scale crude oil spills [11], [9]. Some studies focused on the effects of oil contamination on seedling establishment and growth under sub-lethal concentrations [21]. Responses of mangroves to oil treatment include reduction in foliage production and total plant biomass (Mille et al., 2006; Ke et al., 2011), reduction in the rate of

Corresponding author : taufan.ltd@gmail.com

photosynthesis and respiration [22], [18], increased mutation [3], development of anomalous growth forms [23], damage to root cell membranes [27], [13], impairment of transpiration [27], and increased mortality (Tam et al., 2005). In SANR area, oil pollution is one common pollution type that would cause a devastating impact on mangrove wetlands [24].SANR are vulnerable to contamination *crude oil* as an international marine are traversed carrier tankers and *crude oil* refineries from domestic crude oil industry.There is no available information by nowadays in SANR regarding to theeffects of *crude oil*exposure to mangrove system. Thus, our research aims to observe the possible effect of crude oil contaminationon survival rate, growth andanatomical structure of mangrove roots of *Rhizophora apiculata,B. gymnorhiza* and *C. decandra*.Sampling and preliminary treatment were carried out in September and October of 2017 in the SANR, Indonesia, and the analysis was conducted in the Fisheries and Marine Science Laboratory of Jenderal Soedirman University, Indonesia.

2. Materials and methods

2.1. Description of study area

The study conducted at Segara Anakan Nature Reserve (SANR) Cilacap, Central Java and lies within longitude 108°46'-109°03' and latitute 07°34'-07°47' (Figure 1). The climate of the area is basically that of equatorial tropical rainfall occurring almost through the year except the mouths of December, January and February, which are not completely free from rainfall in some years. The annual rainfall of the area is about 3,444mm (11). Annual mean air temperature is 31.3 °C the highest monthly mean temperature was 29.7 °C (in August), and the lowest monthly mean temperature is 27.5 °C (in January). The surface seawater temperature values range between 25.9 °C and 30.6 °C, and the salinity of the seawater ranges between 8‰ and 20‰. The hydrology of the lagoon is governed by seasonally varying river runoff mainly of the Citanduy River in the west and tidal exchange with the Indian Ocean through two channels in the western (Plawangan) and eastern parts of the lagoon. The mixed and predominantly semidiurnal tide ranges between 0.4 m during neap tide and 1.9 m during spring tide [29].The Citanduy is the fifth largest river of Java in terms of discharge which is estimated to 227 m^3 s^{-1} (dry season 171 m^3 s^{-1}, rainy season 283 m^3 s^{-1}; or 195 m^3 s^{-1} on an annual average [29].

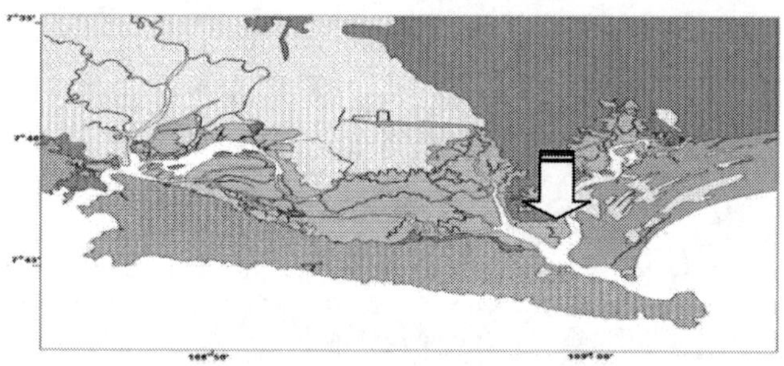

Fig 1. Location of sampling sites in the Segara Anakan Nature Reserve.

2.2. Nursery preparation andSeedling establishment

The sampling sites were established in a relatively undisturbedtidallyinundatedmangrovewetlandCilacap. Surface soil form the study area was collected (0 - 20 cm depth) during tidal recession. The sediment surface samples (5 kg) were weighed and potted in polyethylene bags (40 x 60 cm), leaving 10 cm at the upper end for irrigation of water. Each bag was labeled.

Propagules of *B. Gymnorrhiza, R. apiculata,* and *C. decandra*were planted in 17 cm diameter × 15 cm height, 24 cm × 21 cm and 11 cm x 12 cm in plastic pots. All pots contained amixture of sand, potting soil and compost (1:2:1 volume based). Mangrove seedlings in good condition were carefully uprooted using hand trowel and transplanted into thepottedbags,ensuringthatthetherewasnorootdamage.Seedlings were grown for 2 months to eliminate effects of transplant shock. The seedlings were arranged in 9rowsofparalleltriplicatesat1mintervalsforeachtreatment (0 ml, 100 ml, 120 ml and 140 ml) Plate1.

2.3 Growth analysis

The stem height and leaf sizeof each mangrove seedlingswere monitored weekly for 4 weeks.The stem heights of *R. apiculata,B. gymnorrhiza*, and *C. decandra*seedlings were measured from the top of the seedlings where the stem emerged to the bottom of the mostdistal opened pairs of leaves, and the leaf area were measured individually using vernier calipers[4], [5].

24 Root structure

Anatomical damage resulting from oil pollution of the roots of three species were determined on plants of all Treatments on 4 week. Typical transections of live mediate and fine roots were observed with an Microscope (Boeco, Germany).

2.5 Treatments

Treatment was by applying the *crude oil* that commenced at the end of 60 day acclimation period. The crude oil constitutes of n-alkane-containing oil such as saturates (56%), aromatics (31%), polars (11%), and asphaltenes (2%), it also has 35.3° API gravity and contains 0.1% sulphur content. The accute treatment, consisted of a one-time application of 120 mL crude oil add on the surface of the mud [4], [5]. Stem height at the first inter-node and leaf area were monitored weekly for 4 weeks. Any yellowing of leaves and seedling survivalwere recorded.

2.6 Data analyses

Mean and standard deviation (S.D.) values of triplicates were calculated. Before subject to parametric statistical analyses, the data were tested for the assumptions of normality of distribution and homogeneity of variance. All tested data satisfied the assumptions and no data transformation was necessary. A parametric two-way analysis of variance (ANOVA) with repeated measures was used to compare the effects of oil dose and exposure time on plant growth, including the increment of stem height and change of leaf area. The effects of crude oil dose on mangrove plants were evaluated by one-way multivariate analysis of variance (MANOVA). If the statistical results are significant at $p \leq 0.05$, Tukey'smultiple comparisons as post-hoc tests were applied to locate the differences. All statistical analyses were performed using the software, Statistical Package for Social Sciences (SPSS) 11.0 for Windows, SPSS Inc. IL, USA

3. Results

3.1. Seedling Survival

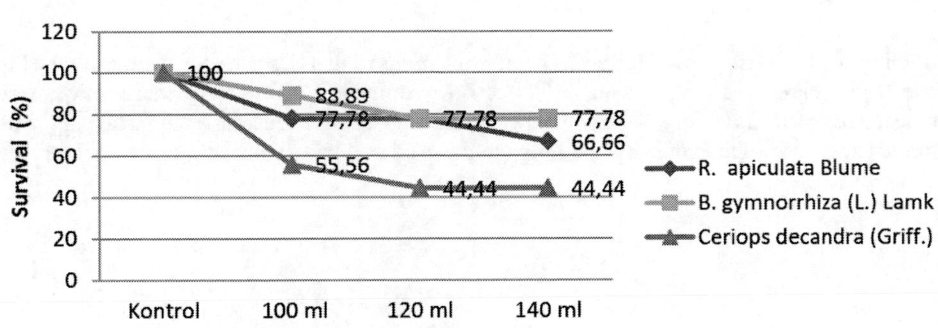

Fig. 2. Effects of oiling on seedling survival

Among the three mangrove species, *B. gymnorrhiza* was the most tolerantspeciestospentlubricatingcrude oil,with100%, 88.89%, 77.78% and 77.78%survival oftheseedlingsinthetreatmentswith 0 ml, 100 ml, 120 ml and 120 ml, respectively *R. apiculata* ranked No. 2 in

 https://doi.org/10.1051/e3sconf/20184704005

oil tolerability, with 66.66 % seedlings survived at 140 ml oil. *C. decandra* the most vulnerable species to oil contaminationwith 100%, 55.56 %, 44.44% and 44.44 % survival of the seedling for all treatment. Most of the resident ground dwelling macrofauna were killed by oil.

3.2. Growth response

3.2.1. Stem growth

Stem growth of three seedling mangrove was reduced by acute effect of crude oil (Fig. 3). For the always in field group, control plants displayed relatively linear growth over the study period. Growth rates of one-time oil treatments were less than that of the control.

The growth of all mangrove species, even exposed to the lowest oil dose (100 ml),was reduced compared to that in the control.Crude oil dose effects on stem growth (expressed as stem height increment) of the three mangrove species were significant ($F=0.657$, p=0.005). *R. apiculata*, in which significant reduction in the degree of stem height increment was observed ($F=4.344$, p=0.005).

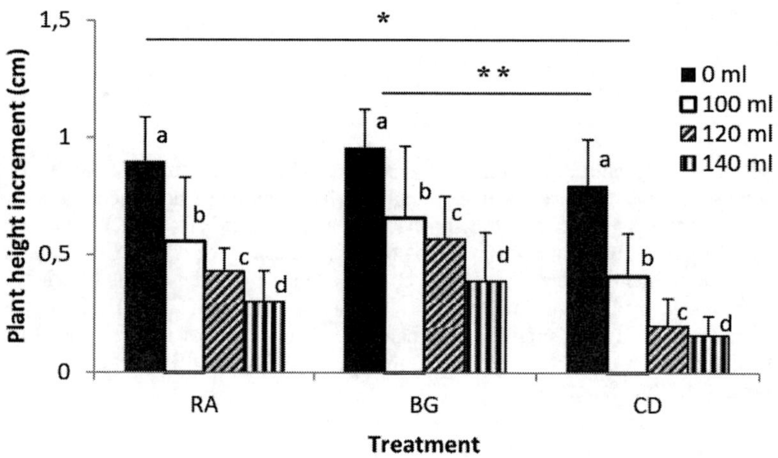

Fig. 3. Effects of oiling on seedling height increment (final seedling height – initial) (*: P<0.05, **: 0.001). RA: *Rhizophora apiculata,* BG: *Bruguiera gymnorrhiza*, CD: *Ceriops decandra.*

3.1.2 The leaf size

Leaf size of three seedlings mangrove followed a pattern similar to that describe for stem growth (Fig. 4). For all weeks of the study, there was a significant effect of oil treatment, with tipically more leaves size by control plants. The leaf size of *R. apiculata, B. Gymnorrhiza* and *C. Decandra*decreased significantly with increasing oil doses.Acute exposure of the mangrove seedlings to the crude oil had significant effect on leaf size.

 https://doi.org/10.1051/e3sconf/20184704005

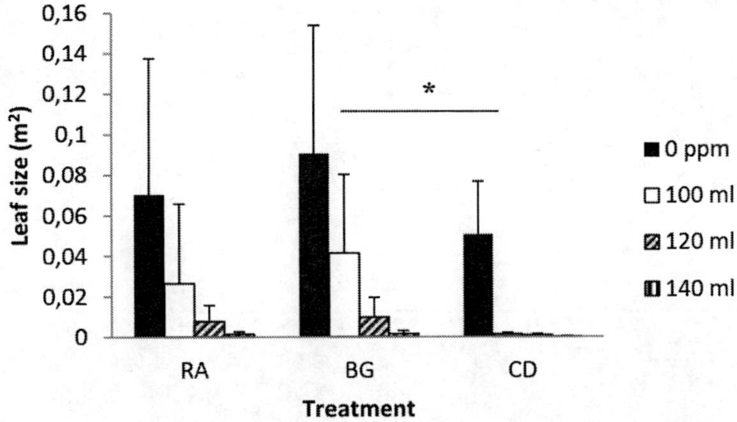

Fig. 4. Effects of oiling on plant leaf size (*:P<0.05). Bars with differentletters are significantly different at P ≤ 0.05 using Tukey-Kramer multiple comparisonstest.

3.1.3Root structure

Damage by base-oiling was found to the roots of *B.gymnorrhiza*, *R. apiculata*and *C. decandra* (Fig.5). The worst damage occurred at a concentration of 140 ml, damage occurred in the epidermis and the cortex on the three types of mangrove seedling roots. The roots of oiledseedlings were black in colour in contrast to the brownish andwhitish root systems in unoiled seedlings of *B.gymnorrhiza*, *R. apiculata*and *C. decandra*, respectively, and a strong oily smell wasnoted. From microscope observation oftypical root transections, oiling crude oil resulted in damageto conducting tissues (epidermis and cortex) in fine roots but not in mediate roots.

 https://doi.org/10.1051/e3sconf/20184704005

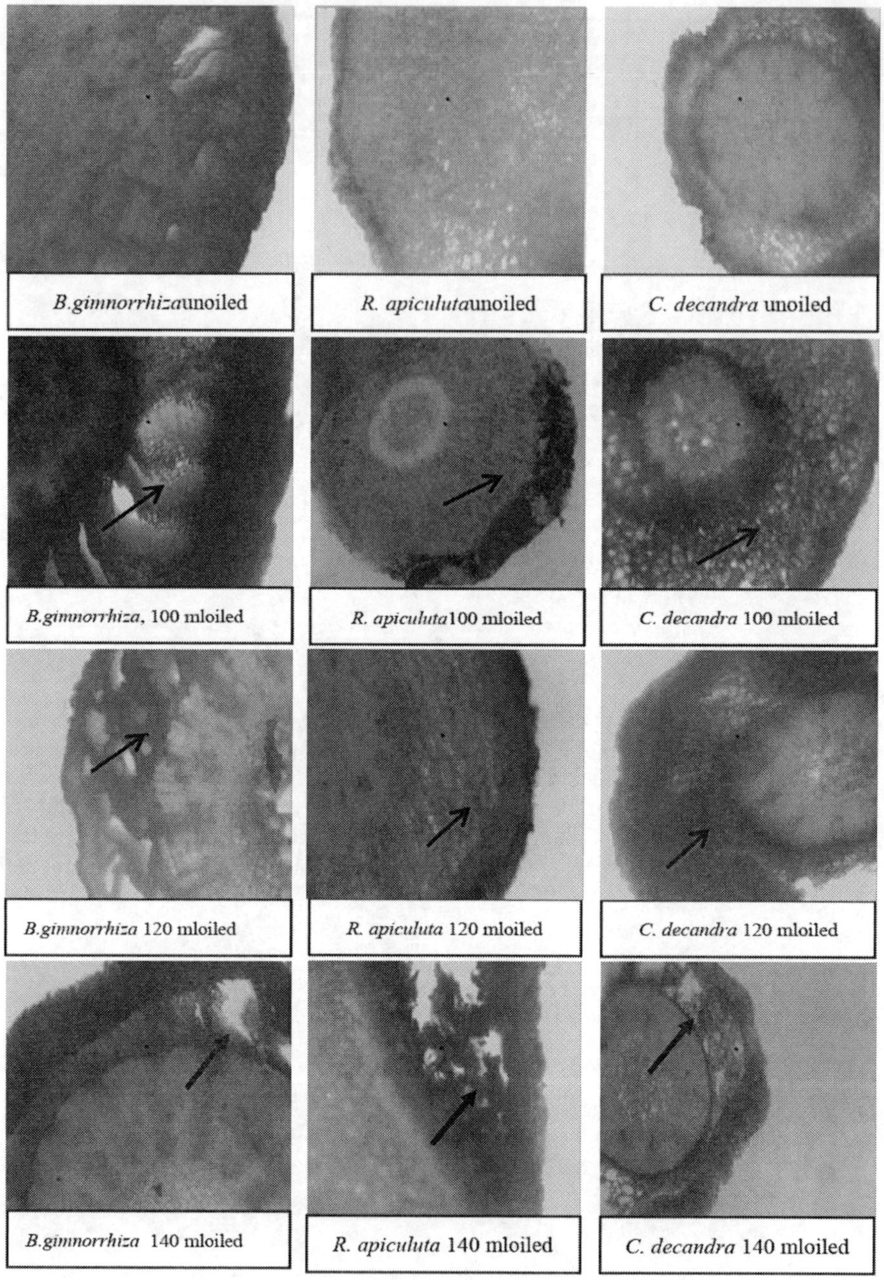

Fig. 5. Effects of oiling seedling on typical transections of fine roots (Scale: 1:40).

4. Discussion

Oil dose is one of the most important variables that determine the degree of damage to seedlings mangroves. A one-time addition of 120 mL oil, an acute oiling with sediment saturation, reduced the survival and growth of red mangrove seedlings more significantly than those received weekly chronic oiling (weekly addition of 15 mL oil for 8 weeks) even though the accumulative amount of oil given to plants in the weekly treatment was equivalent to the quantity given in the one-time treatment [21].

This research has clearly demonstrated that seedling mangroves are extremely sensitive andvulnerabletooilcontamination,as reported alsoinseveralothersstudies[7], [11], [25]. Oileffectsonseedling mangrovesare complex becauseofthelong-livednatureofthetreesandtheir tendency toexhibitshorttermeffects,aswellassub-lethal responses, whichmaypersistfordecades.Shorttermadverse effects suchassenescence, leafwilting,anddefoliation,reported in previousstudies[7], [4], were also observedinoilcontaminatedtrees of *R. apiculataB. gymnorrhiza*and *C. decandra*in thisresearch.

Our field research on the effects of crude oil exposed to different treatments (acute) indicated considerable variation in seedling reaction ranging from survival, growth response and root histology. In this field research we have shown that a form of crude oil, can adversely affect seedlings mangrovesurvival, growth, and root histology. It is clear from our experiment that *C. Decandra* are more sensitive to lubricating oil than are seedlings of *B.gymnorrhiza*and*R.apiculata*. This research has demonstrated that trees and seedlings of *B. gymnorrhiza*, but not of *R. Apiculata* and *C. Decandra*, have the ability to initiateadventitious roots following oiling of stem sand the root system.In *A. marina* and *Aegiceras corniculatum*, lubricatingcrude oilwasshowntodamage the xylemvesselsoffineroot [4]. The primary effect of toxic hydrocarbons works probably via disruption of membranes in the conducting tissue [12], [30].

Oiling crude oil resulted in damageto conducting tissues (epidermis and cortex) in fine roots but not in mediate roots.The greatest damageat a concentration of 140 ml. Destruction of conducting tissues in fine roots will undoubtedly have resulted in the reduction of transpiration, a frequently reported effect of oils on mangroves [11], 27].

In addition to physiological and morphological effects on mangrove seedling, oil also alters the chemical and physical characteristics of the sediments [10], including decreases in dissolved oxygen concentrations, pH, salinity, redox and potentials of the interstitial water [23]. The highly reducing soil redox potentials increase oxygen stress in roots [6], [19], reduce root respiration, water relations and disrupt ion and subsequently growth [12].

5. Conclusions

B. gymnorrhiza was the mosttolerantspeciesto crude oil in accute effect softthereductionin andmortality,followedby *R. Apiculata* were themostvulnerablespeciestooiltoxicity and *C. Decandra* are more sensitive to lubricating crude oil in accute effect.

The work described in this paper was supported by research grants from BLU Jenderal Soedirman University, Indonesia.

References

1. Andersen, L.E., Melville, F., Jolley, D., 2008. An assessment of an oil spill in Gladstone, Australia – impacts on intertidal areas at one month post-spill. Mar. Pollut. Bull. 57, 607–615.

2. Bjorlykke, K., 2011. Petroleum Geoscience: From Sedimentary Environments to Rock Physics. Springer, New York (517 pp.).*Bruguiera gymnorrhiza* to oil contamination. Flora 205, 357–362.

3. Cavalcante, R. M., Sousa, F. W., Nascimento, R. F., Silveira, E. R., Freire, G. S. S., 2009. The impact of urbanization on tropical mangroves (Fortaleza, Brazil): evidence from PAH distribution in sediments. J. Environ. Manag. 91, 328–335.
4. Chindah. A.C.. S.A. Braide. J. Amakiri and J. Onokurhefe. 2007. Effect of Crude Oil on The Development of Mangrove (*Rhizophora mangle* L.) Seedling from Niger Delta. Nigeria.*J. Revista UDO Agricola.*7. 181-194.
5. Chindah. A.C.. S.A. Braide. J. Amakiri and J. Onokurhefe. 2008.

6. Dicks, B.,1986.Oil and the black mangrove Avicennia marina in the Northern Red Sea. Mar. Pollut.Bull.7,500–503.

7. Duke, N. C., Ellison, J., Burns, K. A., 1998. Surveys of Oil Spill Incidents Affecting Mangrove Habitat in Australia: A Preliminary Assessment of Incidents, Impacts on Mangroves, and Recovery of Deforested Areas. Australian Petroleum Production and Exploration Association (APPEA), Canberra, 646–654.

8. Duke, N.C., Pinzon, Z.S., Prada, M.C., 1998. Large-scale damage to mangrove forests following two large oil spills in Panama. Biotropica 29, 2–14.

9. Duke, N.C., Ramsay, M.A., Swannell, R.P.J., Shipton, W.A., Hill, R.T., 2000. Effect of bioremediation on the microbial community in oiled mangrove sedimentsMar. Pollut. Bull. 41, 413–419.

10. Farias, C.O., Hamacher, C., Wagener, A.D.L.R., Scofield, A.D.L., 2008. Origin and degradation of hydrocarbons in mangrove sediments (Rio de Janeiro, Brazil) contaminated by an oil spill. Org. Geochem. 39, 289–307.

11. Getter, C. D., Ballou, T. G., Koons, C. B., 1985. Effects of dispersed oil on mangroves: synthesis of a seven year study. Mar. Pollut. Bull. 16, 318–324.

12. Gilfillan, E.S., Page, D.S., Bass, A.E., Foster, J.C., Fickett, P.M., Ellis, W.G., Rusk, S., Brown, A.C., 1989. The use of Na–K ratios in leaf tissues to determine effects of petroleum on salt exclusion in marine halophytes. Mar. Pollut. Bull. 20, 272–276.

13. Kang, F., Chen, D., Gao, Y., Zhang, Y., 2010. Distribution of polycyclic aromatic hydrocarbons in subcellular root tissues of ryegrass (*Lolium multiflorum* Lam.). Plant Biol. 10, 210.

14. Ke, L., Zhang, C., Wong, Y. S., Tam, N. F. Y., 2011. Dose and accumulative effects of spent

15. Lin, Q., Mendelssohn, I.A., Suidan, M.T., Lee, K.,Venosa, A.D. 2002. The dose response relationship between No.2 fuel oil and the growth of the saltmarsh grass, Spartina alterniflora. Mar. Pollut. Bull. 44,897–902.

16. Meudec, A., Dussauze, J., Deslandes, E., Poupart, N., 2006. 44,8–92.

17. Mille, G., Guiliano, M., Asia, L.,Malleret, L., Jalaluddin, N., 2006. Sources of hydrocarbons in sediments of the Bay of Fort de France (Martinique). Chemosphere 64, 1062–1073.

18. Naidoo, G., Naidoo, Y., Achar, P., 2010. Responses of the mangroves *Avicennia marina* andPAHs into internal shoot tissues of a halophytic plant artificially exposed to petroleum polluted-sediments. Chemosphere 65, 474–81.

19. Pezeshki, S.R., DeLaune, R.D., Jugsujinda, A., 2001. The effects of crude oil and the effectiveness of cleaner application following oiling on US Gulf of Mexico coastal marsh plants. Environ. Pollut. 112, 483–489.

20. Pezeshki, S.R., Hester, M.W., Lin, Q., Nyman, J.A., 2000. The effects of oil spill and clean up on dominant US Gulf coast marsh macrophytes: a review. Environ. Pollut. 8, 129–139.

21. Proffitt, C. E., Devlin, D. J., Lindsey, M., 1995. Effects of oil on mangrove seedlings grown under different environmental conditions. Mar. Pollut. Bull. 30, 788–793.

22. Reed, M., Johansen, O., Brandvik, P.J., Daling, P., Lewis, A., Fiocco, R., Mackay, D., Prentiki, R., 1999. Oil spill modeling towards the close of the 20th century: overview of the State of the Art. Spill Sci. Technol. Bull. 5, 3–16.

23. Suprayogi, B., Murray, F., 1999. A field experiment of the physical and chemical effects of two oils on mangroves. Environ. Exp. Bot. 42, 221–229.

24. Syakti, AD., Nuning V.H., Endang H., Anne P., Pierre D., 2013. Source apportionment of sedimentary hydrocarbons in the Segara Anakan Nature Reserve, Indonesia. Marine Pollution Bulletin 74 (2013) 141–148.

25. Tam, N. F. Y.,Wong, T. W. Y.,Wong, Y. S., 2005. A case study on fuel oil contamination in a mangrove swamp in Hong Kong. Mar. Pollut. Bull. 51, 1092–1100.

26. Watts, A. W., Ballestero, T. P., Gardner, K. H., 2006. Uptake of polycyclic aromatic hydrocarbons (PAHs) in saltmarsh plants Spartina alterniflora grown in contaminated sediments. Chemosphere 62, 1253–1260.

27. Youssef T, Ghanem A, 2002. Salt secretion and stomatal behaviour in Avicennia marina seedlings fumigated with the volatile fraction of light Arabian crude oil[J]. Environ Poll, 116: 215–223.

28. Youssef, T., and Ghanem, A., 2002b. Physiological responses and accumulation of polynuclear aromatic hydrocarbons in *Avicennia marina* seedlings treated with the water soluble fraction of light Arabian crude oil. Environmentalist 22, 149–159.

29. Yuwono, E; TC. Jennerjahn; I. Nordhaus; E.A. Riyanto; M.H. Sastranegara and R. Pribadi. Asian Journal of Water Environment Pollution. 4, no. 1, pp: 61-70 (2007)

30. Zhang, C. G., Leung, K. K.,Wong, Y. S., Tam, N. F. Y., 2007. Germination, growth and physiological responses of mangrove plant (*Bruguiera gymnorrhiza*) to lubricating oil pollution. Environ. Exp. Bot. 60, 127–136.

Occurrence of beach debris in Tunda Island, Banten, Indonesia

Adinda Maharani[1,3], *Noir* P Purba[2*] and *Ibnu* Faizal[2]

[1]Komitmen Research Group, Padjajaran University, Indonesia
[2]Marine Research Laboratory (MEAL), Padjajaran University, Indonesia
[3]HIMAIKA-FPIK, Padjajaran University, Indonesia
FPIK-UNPAD, Jl. Ir. Sukarno km. 21 West Java-INDONESIA
[*]Corresponding author, e-mail: noir.purba@unpad.ac.id

Abstract Marine Debris is one of the spotlight issues of the world. The existence of marine debris can disrupt the ecosystem. Marine Debris carried by the current to coastal areas and accumulated along the coast. The study area of this research was on the coastline of TundaIsland, which located between Ban-ten and Lampung Provinces. These two provinces are part of two major islands of Indonesia, Java and Sumatra Island which have a high density of population. Household needs produce every day directly proportional to the population. This study is to record what types of garbage contained along the coast. There are 8 stations used in this study using International Coastal Cleanup (ICC) form with an area of 100 m2 collection at each station. From 800m2 area coverage, 1,234 items collected with 47 kg weight of debris and the density 0.072 kg/m2 with abundance 0.75 items/m2. This result suggests that household needs give the big proportion of debris such as Food wrapper, cigarettes butts, and foam pieces. In addition, the study indicates that there is a strong relation between human activities and spreading of debris.

1 INTRODUCTION

Beaches located on smaall island in Indonesia are areas with different characteristics, requiring suistanable and integrated environmental management[1,2]. The biodiversity found in small islands is very diverse, but small islands are vulnerable to damage such as natural disasters or pollution. Limitations of garbage disposal areas, resulting in the processing of infrastructure on the small island is still lacking [3].

Marine litter is found in all oceans of the world, from the poles to the equator, and from continental coastlines to small islands.The geographical position will affect the presence of marine waste [4]. When many people move closer to the coast, will result in increased marine waste continuously [5].

The accumulation of plastic waste is a problem across the world's oceans, coastlines and marine ecosystems, because they have ecological and economic impacts as they can disrupt ecosystem stability and problems for marine animals and birds [6,7,8] and reduce the value - environmental aesthetic values.

Marine debris is one issue that is in the spotlight in the world today because it causes a bad impact on the sea environment, social-economic and public health [9,10]. Marine Debris is a solid material produced or processed directly or indirectly, intentionally or unintentionally, disposed of or abandoned in the marine environment [11]. Marine debris is generally the result of anthropogenic (human) activity. The negative impact of marine waste is greatly felt by the neglected small islands [12,13].

The study of Marine Debris in Indonesia is important to develop Indonesia is the second largest contributor of waste in the world with a total waste of 3.2 million tons [4]. The increase of marine waste in the territorial ocean of Indonesia is very fast, to make a small island in the middle of the sea. [14,15,16]

The marine debris is sourced from the sea and the land produces a gradual buildup in the marine and coastal environments, which destroy the coast, float on the surface, float in the water column, and even found on the deep seafloor. Over the past five or six decades, seawater pollution and seawater contamination in the world, wider coastal waters and open oceans by plastics and other synthetic, non-biodegradable materials has become an ever-increasing phenomenon [17,18].Some fractions of increased amounts of post-powder plastic waste must have escaped recycling and waste streams and made their way into the global ocean [19].

2 METHODS

2.1 Site Description

Tunda Island is a small island is located in Java Sea, in the northern of Banten Gulf. Administratively, the island is included in the district of Serang, Banten.Tunda Island is a vast island of about 300 acres. In 2007, the population reached 3,000 people Defer Island.Geographyc Located TundaIsland in 5°48'43" LS dan 106°16'47"[20].Tunda Island actually already a very attractive tourism objects. The beauty of the beach and sea further yet unexplored.TundaIsland has three marine ecosystems. That are coral, seagrass and mangrove. Though sailing activities with a view of the sea and dolphins, fishing, snorkeling, diving, see the sunset and sunrise is an attractive tourist potential. Only a few investors investing or conducting business in the tourism industry in Tunda Island . So that local governments should continue to explore the potential of this exciting unspoiled[21].

2.2 Data Collection

There are 8 points around Tunda. The location is part of the island that is not submerged in sea water. Location was taken around the coastal area of Tunda Island. Each location represents the ecosystem and coastal characteristics of Tunda Island.

 https://doi.org/10.1051/e3sconf/20184704006

Fig1. Tunda Island Location and Stations Located

The method used in the collection of waste data in mangroves using the form of International Coastal Cleanup is downloaded from (http://www.oceanconservancy.org/our-work/international-coastal-cleanup/dataform.pdf) for garbage collection. There are 4 contents in the form: garbage type, weight, location and width of garbage area. The width of the width area using a 100 meters long line of transects. Tools and materials used are garbage bags, GPS, roll meters. There are 6 categories of junk written in the form ie; A. most likely to find the goods, b. fishing gear, c. packaging materials, d. personal hygiene, e. other waste, f. Small trash less than 2.5 cm).

3 RESULTS AND DISCUSSION

Waste collection is picked up at 6 point stations in a mangrove forest. There are 6 categories of trash found based on the ICC form iecategoria) most likely to find items, b) fishing gear, c) packaging material, d) personal hygiene, e) other trash, f) tiny trash less than 2.5 cm. However, there is one category of garbage found on the island of UntungJawa and not included in the form so that the addition of categories, namely g) clothes. Trash types most likely to find items are the most commonly found waste, as this type of garbage includes packs of ma-right packaging which are the basic needs of the general public for daily consumption. So this garbage will continue to be produced every day. The second category is fishing gear. Fishing gear is a tool waste fishing equipment used by the fishermen. Packaging Material is a type of waste packaging materials. The next category is personal hygiene; this garbage is a personal hygiene item. Tiny trash less than 2.5 cm is garbage that size less than 2.5 cm and in the form of small pieces of large garbage. The last category that is an additional unusual category category is Clothes, this garbage in the form of clothing or personal items (clothing, sandals, etc.).

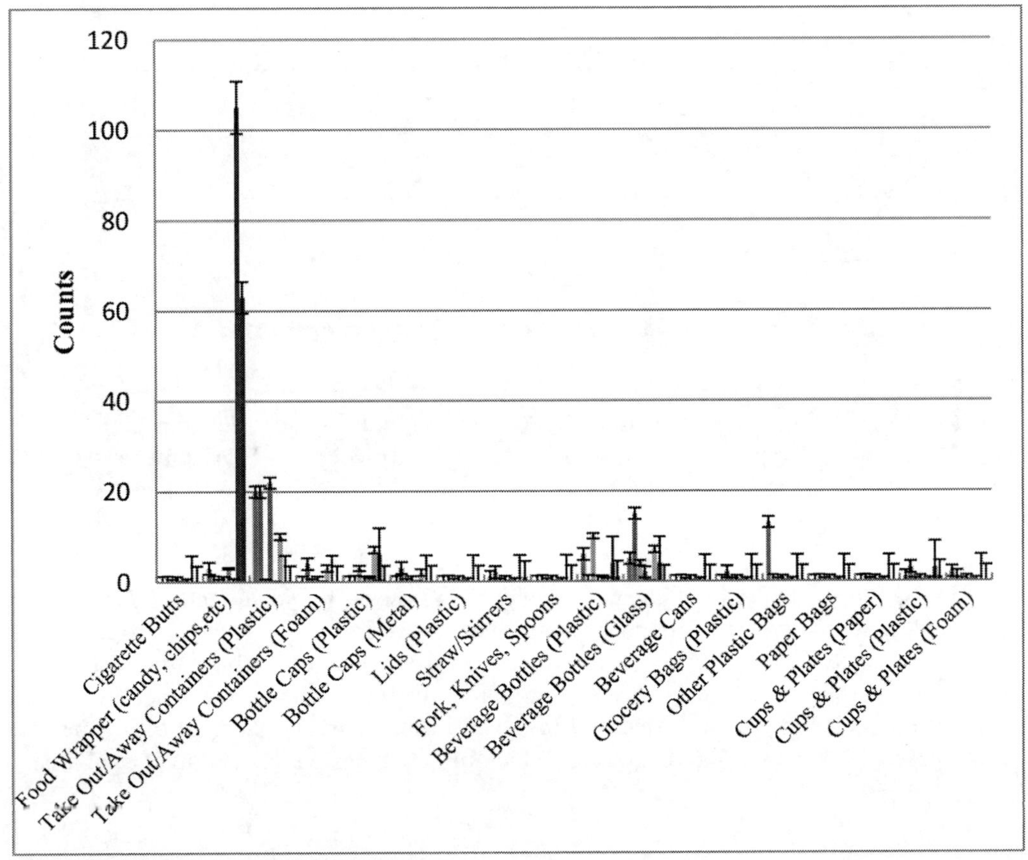

Fig 2. Most Likely To Find Items Chart

Figure 2 is a Most Likely To Find Items chart in 8 stations. This category of garbage includes waste that is commonly found in coastal areas. Household waste is most often consumed by people. that this type of garbage scattered at each point station. Of the 18 types of waste contained in this category, the most common type of waste is food wrapper with the total amount collected from all stations is 174 pieces. This garbage is mostly found in station 7, from 174 pieces 105 pieces are in station 7. Based on Figure 2 most frequently found waste is type A2 is Food wrapper. This garbage is often generated by the community or tourists while playing on the beach. The stalls incoastal area sell food with wrap so that the amount of waste is to be the most widely found.

https://doi.org/10.1051/e3sconf/20184704006

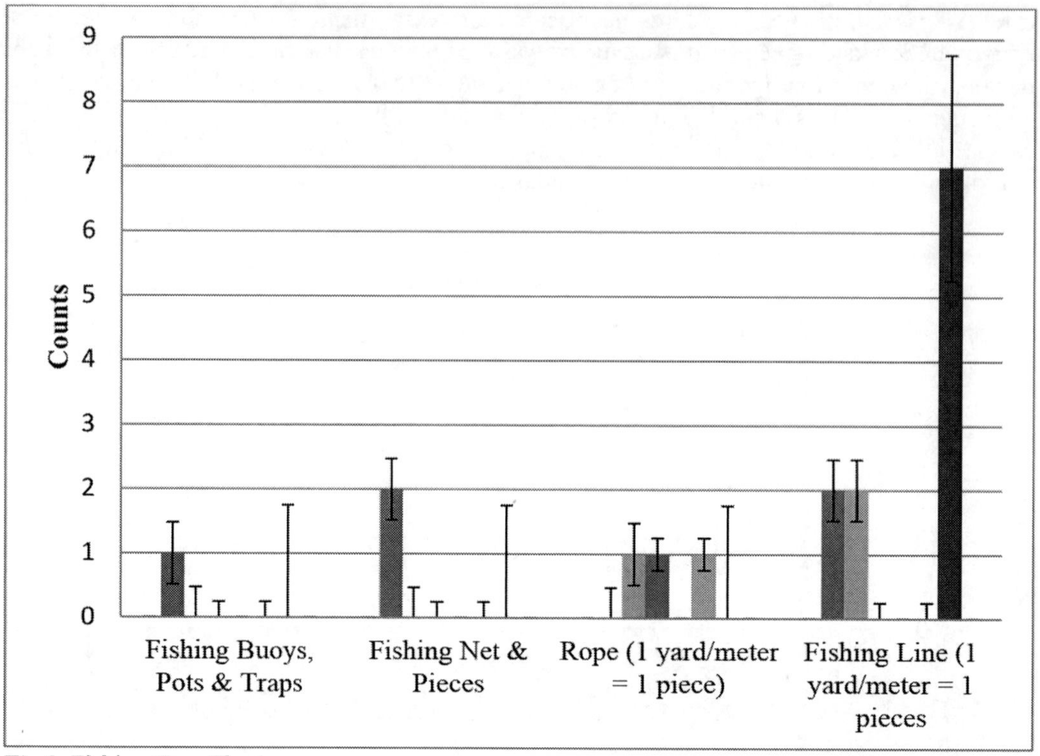

Fig 3. Fishing Gear Chart

Figure 3. shows the Fishing Gear category. This type is found in station 7 with total waste 7 pieces with a type of garbage Fishing line (1 yard/meter = 1 piece) or rope net + 1 meter. Most found trash is Fishing Line, there are 12 pieces of garbage found from all stations. This garbage is produced from fishing equipment of fishermen dumped in the sea, or caught during fishing, then drifted and involved in the coast. Because of the shape of a mesh net and rope, make this garbage easily stuck in the mangrove or sand in coastal area.

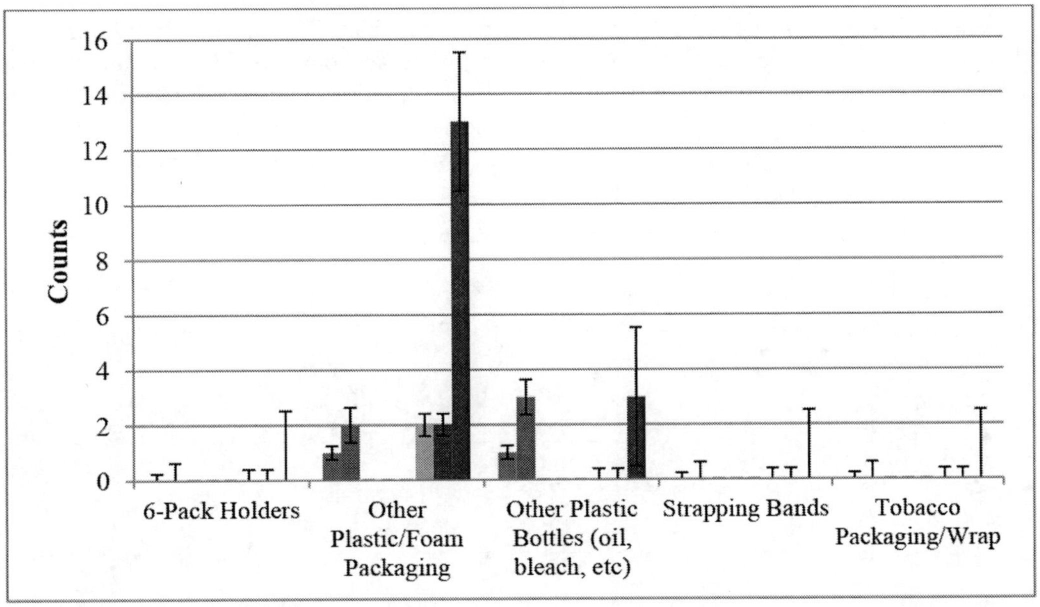

Fig 4. Packaging Material Chart

Figure 4. shows the Packaging Material Chartcategory chart. The most common type of waste from this category is Other plastic/foam packaging with the total number found in all stations is 20 pieces. Packaging Material category at most available at the station 8. There are some small stalls owned by residents, so it can be analyzed Foam Packing is from small stalls selling packaged food packets.

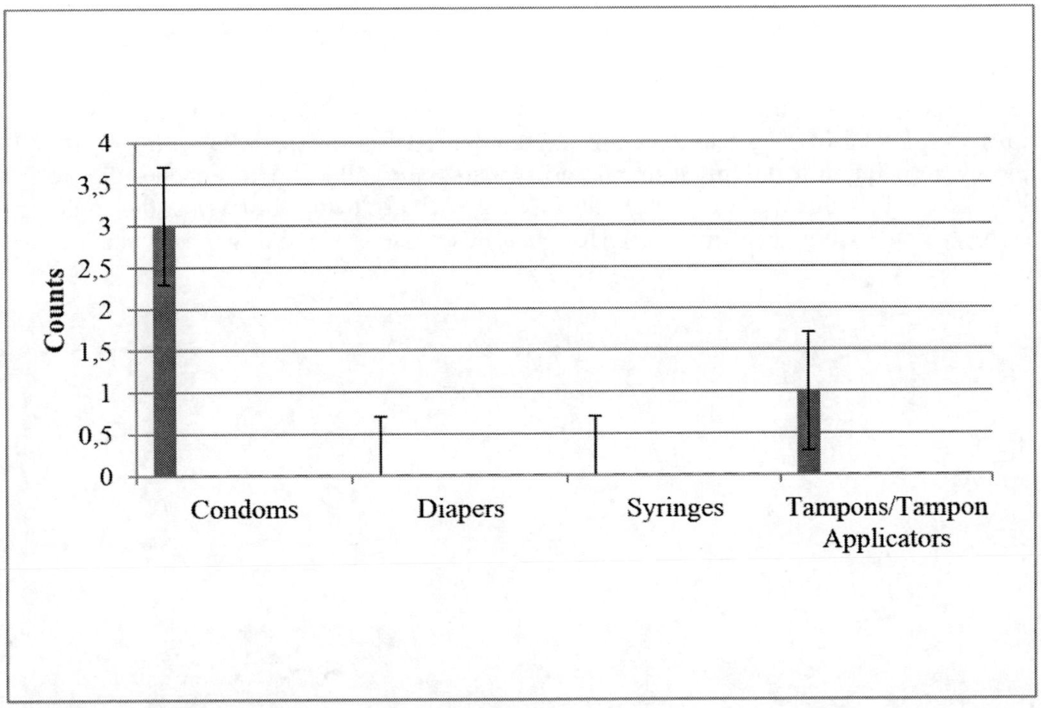

Fig 5. Personal Hygine Chart

Figure 5. shows the personal hygiene category trash chart. The most widely found from this category is Condoms with the number of 3 pieces and only found in station 1. Total waste of this category found in all stations only 4 pieces. Type of waste category of Personal Hygiene is indeed rarely found in the coast because the species is not commonly used by surrounding communities. In addition, this type of garbage is slightly dangerous and cannot be thrown away carelessly like needles and Tampon syringes.

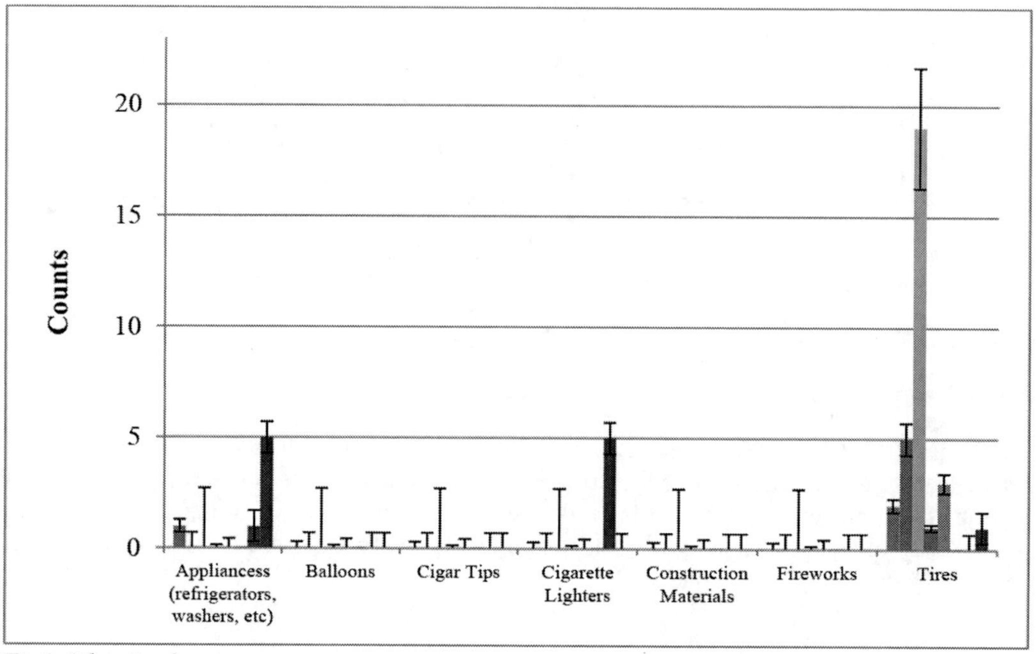

Fig6. Other Trash

Figure 6. shows the other trash category chart. The most common types of waste are tires with a total of 31 pieces. Tires most commonly found in station 3 with a total of 19 pieces. In addition to the often found garbage tires are balloons and matches. There is a little privilege in this type of garbage, the rare type of trash is type of trash or the equipment (refrigerator, washing machine, etc.). However, this waste is found in station 1 and 8.

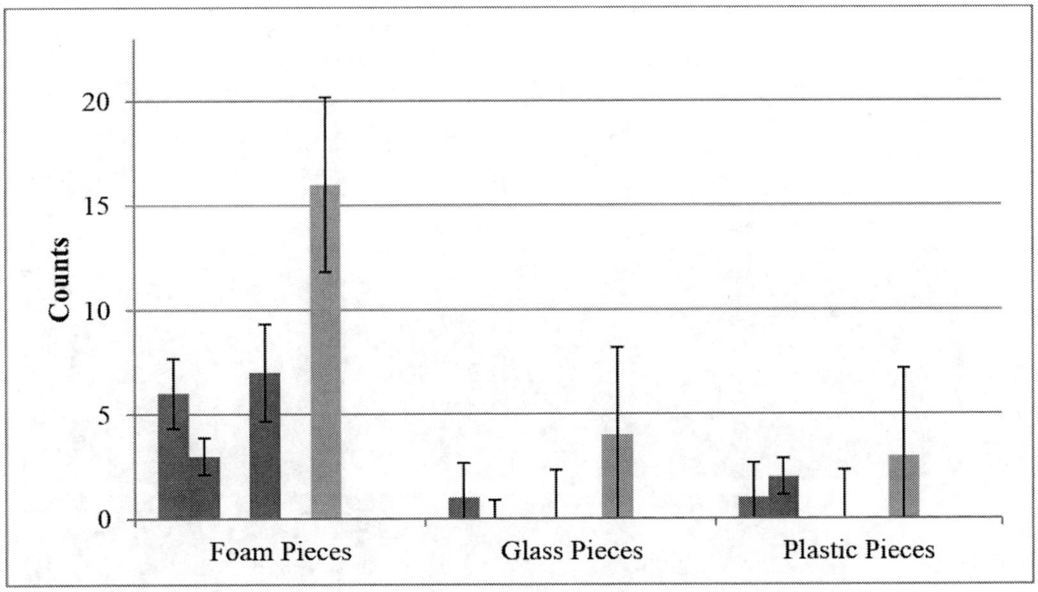

Fig 7. Tiny Trash

Figure 7. shows the tiny trash category chart. Total trash found throughout the station is 42 pieces. The most popular tiny trash category is foam pieces with 32 pieces. Station 6 is a station with the Foam Pieces at most with the amount of garbage found as many as 19 pieces.

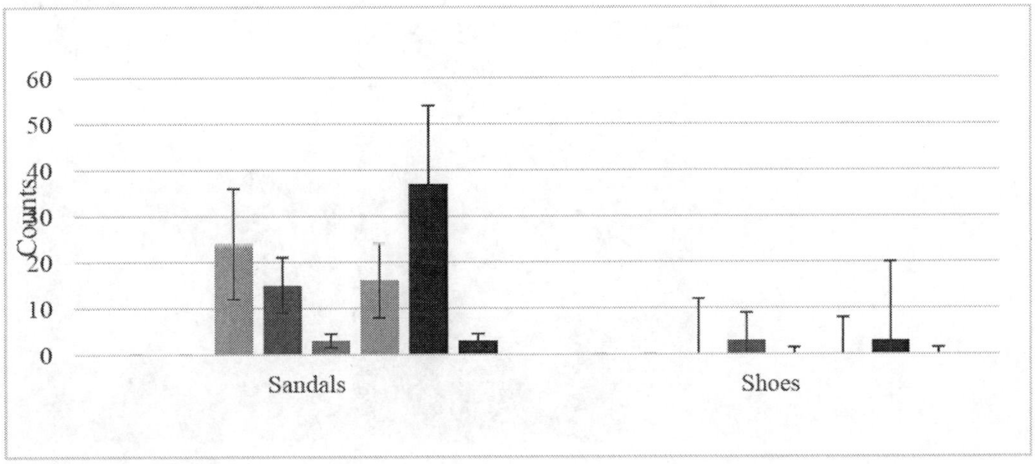

Fig 8. Personal Trash

Figure 8. shows the personal trash graph. The most common type of garbage from this category is sandals with the amount of garbage found in all stations is 98 pieces. Personal Trash category found in station 7 with the number of garbage found 37 pieces. Total trash found in all stations is 104 pieces. This category is an additional category because the garbage is found in most stations but there is no special category in the ICC form. This waste is categorized as Human Trash.

Figure 9 is a visible condition of trash on the coast of the island. Pulau Tunda is one of the islands located in the northern part of Java Sea. The island is adjacent to the Province of Lampung located on the island of Sumatra and Banten Province located in Java. This causes the water period in the waters to be affected by two periods of water, the period of water carried from the waters of Sumatra and the waters of Java.

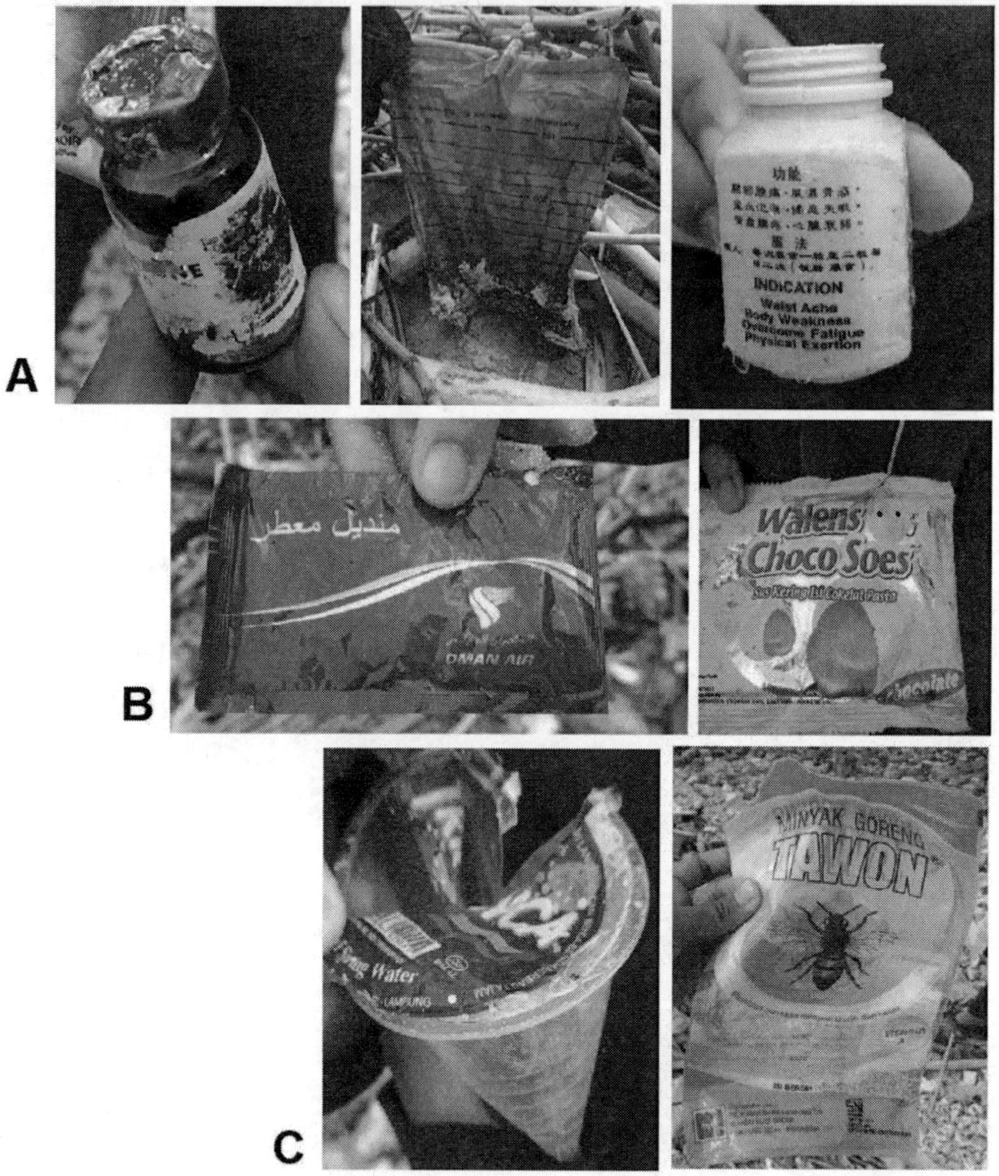

Fig 9. Trash Found in Tunda Island

Marine or marine wastes are known as Marine Debris is mostly about 80 % coming from terrestrial sources. Broadly speaking, marine debris is divided into two types based on the source of the land and from the sea. Trash originating from the mainland and then carried away by rivers, ocean currents, and storms, and waste disposal in the oceans, whether intentional or unintentional is the source of marine debris [22].

According to Greenpeace in 2006, the amount of marine waste that reaches an area greatly relies on ocean currents. Other influencing factors are the distance to the source, debris type (eg resistance), a form of waste (density and size), physical-chemical parameters, coastal geomorphology, wind, rain and storm, and tides. There are two factors that influence the distribution of marine debris in the sea, natural and human factors.

Much of the marine debris in the marine environment, particularly in high-populated coastal areas, enter through vast quantities of ship activities, fishing activities, industrial discharges, and river water flow [23,24]. When in the ocean, floating marine debris can be concentrated due to natural processes occurring along the lines of convergence between discrete water masses, at the center of a gyre, or onshore and submerged rock areas where the area is the vital habitat of marine species. For example, water flow along eddies or down welling areas can accumulate floating trash so as to provide food for seabirds and turtles [24,25].

Local-scale phenomena that need to be taken into account in the marine debris movement are the currents (tidal currents and residual currents), wind, rainfall, and natural disasters. Coastal is an area affected by phenomena at sea and landed. So the problem of garbage in the sea and on land will affect the coastal areas.

Tunda Island is a small island located in the Java Sea, north of Banten Bay and separated from the waters of the Thousand Islands, causing the island to be relatively open. The island is located on the western tip of Java Island and adjacent to Lampung Province.

The waters of Banten Bay lie on the main axis of the monsoon wind. In the western period (December - February) the wind in this region is blowing from the west/northwest direction. As for the east wind season (June - August), the wind blows from the east/southeast. This position has consequences for the movement of horizontal water masses (currents) in the region. In the west season the current will move to the east, and in the east season, the current will move west. It is the flow that affects the movement of waste in the sea.

In the east season, the current will move from east to west. So the current will bring a lot of garbage located in the eastern part of Tunda Island, as part of the thousand islands. The waste distribution of the thousand islands is also influenced by the island of Java precisely Jakarta. So it could be the origin of waste that docked in Tunda comes from the island of Java. Besides, Tunda Island is adjacent to Banten Province. So the garbage from Banten province took part in supplying the garbage in Pulau Tunda.

First, the trash comes from the consumption of the residents of Tunda itself, which is thrown away and then drifted by wind or rain to the coast. Both of this garbage come from Lampung Province which carried by ocean currents and anchored in Pulau Tunda. The current carrying the waste is estimated due to the west monsoon. This western wind moves from the west (Sumatra Island) to the East (Java Island). Friction between the wind and the surface of the ocean will cause a current. The direction of the current will follow the wind. So the current will move from west to east. The western part of the island of Tunda is Lampung Province. Therefore, the waste can come from Lampung.

In addition, there is some evidence of waste found not from the island of Tunda. The axis can be seen from Figures 9.1 and 9.2. Figure 9.1 is a medical waste that is infusion and liquid medicine. The waste is usually produced by the hospital. While in Pulau Tunda there is no hospital. Hospitals are listed in big cities such as Banten or Lampung. So it can be ascertained the origin of this garbage is from Java or Sumatra, which is wasted by ocean currents. While picture 11.2 is garbage with foreign language writing. The first picture reads Arabic with Oman Air. The second image of garbage with writing the Chinese language. Both trash is carried by the ocean currents. The current influencing the movement of the waste is very diverse because Indonesian waters are influenced by many world and local currents starting from geostrophic currents, South Khatulistiwa Flow, Java Beach Flow, Aruh Indonesia and Monsoon Flow. So that the movement of sea waste can be waved off until finally will be anchored on the coast. Many factors that affect the things mentioned, especially the topography and geography of the beach.

4. CONCLUSIONS

It can be concluded that the distribution of garbage in Tunda Island is influenced by some local currents ie currents influenced by east and west monsoon angina. The distribution of waste in this island is influenced by the production of garbage coming from two big provinces namely Lampung and Banten. The two provinces are located on the islands of Sumatra and Java. The most commonly found waste in all stations is food wrappers.

We thanks to Demand Driven Research COREMAP CTI 2018 and PusatPenelitianOseanografi LIPI (021 64713850), its funders and observers for field data collection. We thank to aculty Fisheries and Marine Science UniversitasPadjadjaran, Marine Research Laboratory (MEAL) FPIK UNPAD, Bandung Diving School (SDB) for being safe, supportive and understanding. We also thank to Marine Science UniversitasPadjadjaran Student who has been collected data trough Integrated Research and Community Services Program in 2018.

REFERENCES

1 D.Ferrol-Schulte, P.Gorris, W. Baitoningsih, D. S.Adhuri, & S. C.Ferse.Mar Pol, **52**, 163-171(2015).

2 N. Rukminasari, Yusran. Jurnal Panrita Abdi.**1**, (2016)

3 C. Sur, J. M. Abbott, R. Ambo-Rappe, N.Asriani, S. O. Hameed, B. M.Jellison, and S. L. Williams. Frontiers in Marine Science.**5**, 2 (2018).

4 J. R. Jambeck. Science.**347**, 768-771(2015)

5 Factsheet. Impact of marine litter.(2012)

6 C. M.Boerger, G. L.Lattin, S. L. Moore, & C. J. Moore. Mar.Poll. Bull. **60** , 2275 – 2278 (2010)

7 V. Hidalgo-Ruz,& M.Thiel. Mar. Environ.Res **87-88**, 12-18 (2013)

8 S. B. Sheavly, *Sixth Meeting of the UN Open-ended Informal Consultative Processes on Oceans & the Law of the Sea. Marine debris – an overview of a critical issue for our oceans.* June 6-10, (2005)

9 A. D. Nash. Mar. Poll. Bull. **24**, 150–156 (1992)

10 A.McIlgorm, H. F. Campbell, and M. J. Rul.*A report to the Asia-Pacific Economic Cooperation Marine Resource Conservation Working Group by the National Marine Science*, (Centre University of Coffs Harbour, 2008).

11 CSIRO. Marine Debris: Sources, Distribution And Fate Of Plastic And Other Refuse – And Its Impact On Ocean And Coastal Wildlife. Www.Csiro.Au. (2014)

12 B.Glaeser, and M. Glaser .Res.Hum. Ecol. **17**, 135–147 (2010)

13 A. R.Farhan, and S. Lim. Ocean Coast. Manag. **54**, 391–400 (2011).

14 S. M. Evans, M. Dawson, J. Day, C. L. J.Frid, M. E. Gill, L. A.Pattisinat.Mar. Poll. Bull. **30**, 109–115 (1995)

15 P.Uneputty, S. M. Evans. Mar. Environ. Res. **44**, 233–242 (1997)

16 N. G. Willoughby, H.Sangkoyo, B. O.Lakaseru. Mar. Poll. Bull. 34, 469–478 (1997)

17 R. C.Thompson, C. J.Moore, F. S.vom Saal and S. H. Swan. Phil. Trans. R. Soc. B **364**,2153-2166(2009)

18 M. R.Gregory Environmental implications of plastic debris in marine settings—entanglement, ingestion, smothering, hangers-on, hitch-hiking and alien invasions. Phil. Trans. R. Soc. B **364**, 2013–2025(2009)

19 E. R.Zettler, J. M.Tracy, and, A.Linda. Environ. Sci. Technol.**47**, 7137-7146(2013)

20 N. P.Zamani.Ilmu Dan TeknologiKelautanTropis 7, 273 – 286 (2015)

21 Lintangkawuryan. AEBMR. **52,**(2018)

22 United Nation Environment Program (UNEP). *Negative Socio-Cultural Impact For Tourism.* Www.Uneptie.Org/Pc/Tourism/Soc-Drabacks.Htm. (2001)

23 T.R.Dixon, and T.J. Dixon. Mar. Pollut. Bull. **12,** 289-295 (1981a)**12**, 53-56 (1981b)

24 D. W.Laist.Mar. Pollut. Bull. **18**, 319-326 (1987)

25 A.Car.Rips, FADS, and Little Loggerheads. Bioscience **36**, 92-100. 1987.. Mar. Pollut. Bull.**18**, 352-356 (1986)

PHOSPHORUS LOADING FROM FISH FARMING ACTIVITIES TO WADASLINTANG RESERVOIR WATERS

Agatha Sih Piranti[1]*, *Diana* RUS Rahayu[1], *Gentur* Waluyo [2]

[1] Department of Environmental Biology, Faculty of Biology, Unsoed, Karangwangkal Purwokerto 53122, Central Java, Indonesia
[2] Departments of Geology, Faculty of Engineering Unsoed, Blater Purbalingga, 53371, Central Java, Indonesia

Abstract. Wadaslintang Reservoir located in Wonosobo Regency of Central Java provides to fish farming area using floating net cage. There are 2 feeding system applied that are semi intensively system managed by local community and super intensively managed by PT. Aquafarm. Both different feeding systems resulted in the amount of pollutant load coming into the reservoirs in the form of total P derived from the remaining feed and feces of fish entering into the reservoir. This can trigger the eutrophication condition of the reservoir waters. The purposes of this study were 1) to calculate the capacity of Wadaslintang reservoir to phosphorus load based on its morphometry and hydrology, 2) to evaluate the capacity of wadaslintang reservoir to Phosphorus load from floating net cage activities in wadaslintang reservoir, and 3) to design the program on the reducing load of Total Phosphorus from floating net activities in wadaslintang reservoir. The research was conducted by survey method in the Wadaslintang Reservoir during January 2016 in both locations of floating net cage fish farming. Parameters measured were P content in fish and P content of feed used in both cultivation systems, and P concentration of reservoir water. The amount of pollutant load coming into the reservoir and its impact on the P concentration of water was calculated using the formula. The results showed that Wadaslintang Reservoir to phosphorus load capacity was 106,56 ton P/year. The Phosphorus load from floating net cage was 216,2 ton P/year. The P load from PT. Aquafarm have exceeded the capacity as much as 70,74 ton P/year. In general the existence of fish farming system using floating net cage in Wadaslintang Reservoir has contributed to increase the concentration of P equal to 43,9% or in other words as much as 50% of P load that goes into reservoir of wadaslintang. Programs designed to reduce the phosphorus load could be done by reducing the number of floating net and decreasing all the activities in the catchment area up to 50% to fulfill phosphorus standard of 100 mg/m^3.

Corresponding author: agatha.piranti@gmail.com

1 INTRODUCTION

Wadaslintang Reservoir is located in Wadaslintang sub-district, Wonosobo Regency of Central Java Province, Indonesia. As a wadaslintang dam ecosystem functions as a reservoir of water from its catchment area, so the impact of human activities on DTA such as settlement, livestock, agriculture and industry and mining affect the water quality of the reservoir. Activities in reservoir water such, as fish cultivation using KJA then the waste is also a source of pollutant water reservoir. Therefore, the pollutant load comes from various sources, both external (coming from the catchment area) and from the internal reservoir must be calculated so that the incoming waste does not exceed its capacity. The pollution load from various sources will increase along with the increase of population and other activities in the catchment area

The main function of the Wadaslintang Reservoir is to fulfill the irrigation water requirement for 33.279 ha of paddy field, which are located in Kebumen and Purworejo districts. Drinking water and water for fish cultivation area using floating net cages. As a function as fish farming area using floating net cage, there are 2 (two) cultivation method which is used based on ownership and feeding method applied in Wadaslintang Reservoir which is semi intensively done by Community around reservoir called *Forum Komunikasi Tani Nelayan* (FKTN) *Sejahtera Mandiri* and super intensively by PT. Aquafarm. Semi-intensive fish farming requires a moderate level of inputs and fish production is increased by the use of fertiliser and/or supplementary feeding. Intensive fish farming involves a high level of inputs and stocking the ponds with as many fish as possible. The fish are fed supplementary feed, while natural food production plays a minor role [1]. Super intensive feeding system conducted by PT Aquafarm done with feeding in adlibitum for 15 minutes until the fish full and no appetite again.

Fish cultivation activities using KJA intensively will bring the consequences that many feeds are not utilized by pet fish and will fall to the bottom of the waters. Lukman & Hidayat [2] stated that the remaining feed in the form of fish feces that fell into the waters about 50% of the feed given. Abery *et al* [3] stated that wastewater feed is wasted to waters estimated at about 30-40%. Sukadi [4]. The feed given to fish does fish consume only 70% and the remaining 30% will be released to the lake water bodies as pollutants or waste. Therefore, Garno [5] states that fish feed is the highest contributor of organic materials in lakes/reservoirs (80%) in producing environmental impacts of water quality degradation caused by eutrophication conditions.

Decrease in water quality due to eutrophication will reduce water function and disrupt the ecosystem in it. Human activity is the most influential factor on the improvement of organic matter. The organic material will decompose and increase the phosphorus and nitrogen elements in the water. Fishery cultivation activity using karamba is the most human activity done in reservoir wadaslintang so it is suspected to cause the entry of organic materials, especially the phosphorus element large enough to the body of water. The high concentration of total phosphorus (total-P) indicates that the organic material entering the water is also high. Phosphorus in the waters of the lake becomes a determinant of growth for phytoplankton or Alga. Phytoplankton or algae can act as one ecological parameter that can describe the condition of a waters and is also an important biotic component in the metabolism of water bodies, since it is the primary link in the food chain of aquatic ecosystems. Changes in the number and types of plankton populations in the waters can illustrate the state of the structure of aquatic communities. Abundance of phytoplankton that is too large and dominated by certain species alone can be used as an indicator that the eutrophication waters.

Control of eutrophication can be done by limiting the phosphorus entering the reservoir [6]. The limitation of phosphorus can be done by reducing the amount of karamba if its impact has exceeded the carrying capacity of the lake / reservoir as well as the management of good and proper fish feeding [7]. The management of aquaculture using good and proper karamba will increase the productivity of the fishery itself. Therefore, the activity or production of aquaculture in the waters of the lake should be determined based on the load capacity of the phosphorus contamination. The capacity of lake phosphorus pollution load is the ability of lake water to receive input load of phosphorus contamination without causing lake water to be polluted based on environmental characteristic and condition around it ie lake morphology and hydrology covering area, volume, depth and water discharge; quality and trophic status; water quality standard requirements and allocation of pollution load from various sources and types of water. In order for the existence of the reservoir and its function can be sustainable then the water resources need to be maintained so that its quality can meet the required quality standard. Therefore, the purpose of this study is 1) to calculate the capacity of Wadaslintang Reservoir to phosphorus load based on its morphometry and hydrology, 2) to evaluate the capacity of Wadaslintang Reservoir to Phosphorus load from floating net cage activities in Wadaslintang Reservoir, and 3) to design the program on the reducing load of Total Phosphorus from floating net activities in Wadaslintang Reservoir.

2 MATERIALS AND METHODS

The research was conducted in Wadaslintang Reservoir at the area of floating net cage activities owned by the community called Forum Komunikasi Tani Nelayan (FKTN) Sejahtera Mandiri and the other owned by PT. Aquafarm (Figure 1).

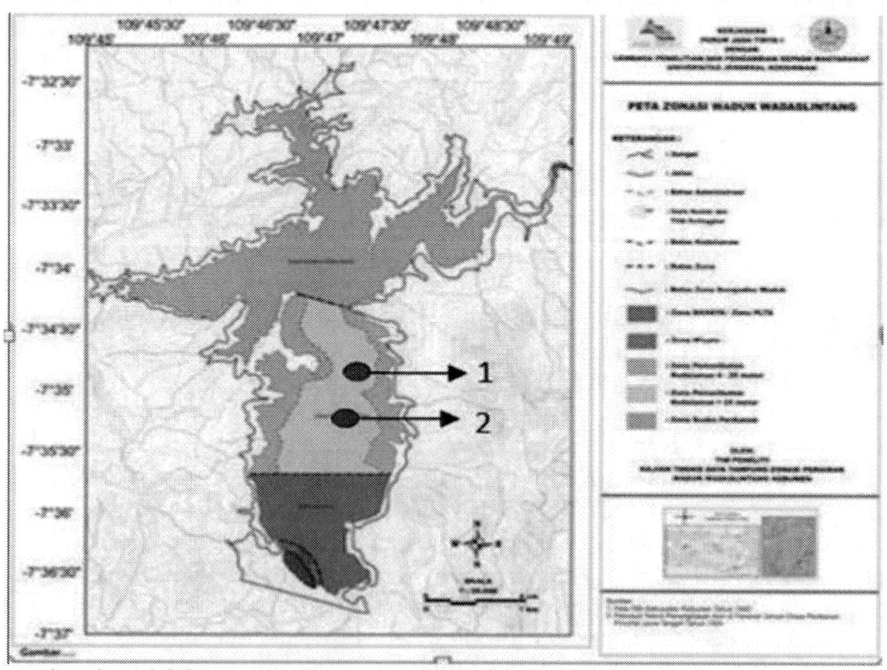

Source : Piranti *et al*, [8]

Fig 1. Sampling Location in the area of Floating Net Cage in Wadaslintang Reservoir

Sampling was conducted in Januari 2016 by random sampling. Sampling of aquatic fish, sediments and fish taken in the floating net cage area of community (FKTN Sejahtera mandiri) and PT. Aquafarm with 3 replicates each places (floating net cage unit). The parameters measured were Total P concentration in water, sediment and fish. The aquaculture data was obtained by taking direct interviews with the head of FKTN Mandiri Sejahtera (Mr. Nurdin) and PT Aquafarm. Cultivation obtained related to cultivation activities including all data especially regarding to the type and amount of feed given to fish and fish production data from each floating net cage system. Data of FCR (Feed Convertion Ratio) was obtained by calculating the ratio between the amount of feed given with the amount of fish produced (tons of feed/tons of fish). Morphometric of reservoir data including total area, volume, depth, outflow, flushing rate, and retention time. Water quality standards and trophic status of lakes/reservoirs are also needed in determining reservoir capacity.

The capacity of pollution loads of lake water and/or reservoirs depends on the characteristics and conditions of the surrounding environment, including morphology and hydrology of lakes and/or reservoirs, water quality and water quality requirements or standards as intended, and allocation of water pollution load from external and internal sources. The calculation of morphology-hydrology of the reservoirs, the allocation of pollutants entering the reservoir, the water pollution load capacity of the lake / reservoir, the capacity of pollution load of aquaculture cultivation on the lake/reservoir, Phosphor waste as the total P parameter of the remaining feed and fish metabolism waste, total production of KJA cultivation fish and the amount of feed according to capacity was found using the formula based [9] as follows.

1. Morphology dan Hydrology of Reservoir

$$\check{Z} = 100 \times V \tag{1}$$

Ž : mean depth (m)
V : water volume (juta m3)
A : area (Ha)

$$\rho = Qo / V \tag{2}$$

ρ : flushing rate (per year)
Q : outflow debit (million m^3 / tahun)

2. Load Allocation of Phosphorus (P)
Utilization of lakes only for cultivation of fisheries and agriculture or other activities that are not sensitive to level of P-total :

$$\Delta [P]d = [P]f - [P]i \tag{3}$$

The Utilization of Multipurpose Lakes Based on Water Quality Standard

$\Delta [P]d = [P]STD - [P]i - [P]DAS$ (4)

Δ [P]d : alokasi beban P-total budidaya ikan (mg P/m3)

[P]f : maximum P-total requirement in water (mg P / m^3)
[P]STD : maximum P-total requirement based on Water Quality Standard (mg P/m^3)
[P]CA : P-total load allocation from Catchment area other than fish farming (mg P/m^3)
[P]i : P-total of lake water monitoring (mg/m^3)

3. The Capacity of pollution load from aquaculture
$L\ fish = \Delta [P]\ \check{Z}\ \rho / (1 - R\ fish)$ (5)
$R\ fish = x + [(1-x)R]$ (6)

R = 1 / (1 + 0,747 ρ 0,507) (7)

$$La\ fish = L\ fish \times A \quad (8)$$

L fish : capacity of P-total from fish waste per unit area (g/m^2.year)
La fish : capacity of P-total fish waste in waters (g/year)
R : P total bounding in sediment
R fish : proportion of P-total dissolved from sediment into water
X : proporsi total P-total bound to sediment permanently (45-55%)

4. Feed and P-Total waste from aquaculture

$$P_{LP} = FCR \times Pfeed - Pfish \quad (9)$$

P_{LP} : P-total load from fish waste to reservoir (kg P/ton ikan)
FCR : Feed Conversion Ratio (ton feed/ton fish)
Pfeed : P-total contet in feed (kg P/ton feed)
Pikan : P-total content in fish (kg P/ton fish)

3 RESULTS AND DISCUSSION

The morphometric and hydrological character of the reservoir will determine the processes by which physical, chemical, and biological processes occur in these waters. The physical and chemical characteristic and Capacity of phosphorus load of the Wadaslintang Reservoir is presented in Table 1.

Table 1. Reservoir Capacity of Phosphorus Load based on Morphometri and Hydrology

Subjects	Parameters	Symbol	Unit	Value (+185 m)
Reservoir morphometry	Area	A	Ha	1330
	Volume	V	juta m^3	415
	Mean depth	Ž	m	31,20
Reservoir hydrology	Total outflow	Q_o	Million m^3/year	473,04
	Flushing rate	ρ	year^{-1}	1,140
	Water Residence Time	Tw	year	0,877
P standard and P coefficient bounding	Pstd	Pstd	mg/m^3	100
	coefficient of constants	R		0,558
Reservoir Capacity	Reservoir capacity of P-total from fish waste per unit area	Li	gP/m^2/year	8,012
	Reservoir Capacity of P-total based on Morphology and hydrology	La	ton P/year	106,563

The average depth of the reservoir is 31.20 meters. Water depth effect on abiotic transport processes related to mass-balances (eg, sedimentation, resuspension, diffusion, burial, mixing and outflow) and also the changes of abiotic factors among lakes influence key foodweb processes concerning production, predation and biomasses of functional groups of organisms (phytoplankton, zooplankton, fish, etc.) [11].

The flushing rate of the Wadaslintang Reservoir was 1.140 million m^3/year or the water retention time was 0.9 years. This duration of water residence depends on the morphometry and hydrology and associated with the characteristic of inflow and outflow of the reservoir.

This characteristic will affect by the function of reservoirs which further result both phosphorus accumulation in sediments and losses within lakes then bound to sedimentary rocks [12]. This losses are most commonly conceptualized as a function of the area of hydraulic loading rate or mean lake depth and a putative particle settling velocity [13]. However, this phosphorus can be released into the water column when the pH is in an acidic situation [14].

The form in which phosphorus is supplied to the lake (i.e. inorganic or organic, dissolved or particulate) probably has a particularly important impact on phosphorus retention. If phosphorus is supplied predominantly as soluble reactive P or as a constituent of animal or human wastes, it can be readily incorporated into the biologic cycle of lakes [15]. Conversely, if it is supplied predominantly in particulate form as P sorbed onto or incorporated into inorganic particles, it is much less bioavailable to phytoplankton or bacteria. Furthermore, particulate, mineral-bound phosphorus will also settle out of the water column very rapidly and be lost to the sediments. Lakes with long hydraulic retention times trap substantially less phosphorus. Conversely, lakes with short hydraulic retention times trap substantially more TP but the large prediction error is probably due to inadequate lake TP input budgets and the myriad of factors that are known to influence TP losses to lake sediments [13]. However, the phosphorus bound in the sediment will be released into a dissolved state when the condition without oxygen or internal processes of eutrophication [16].

In accordance with the requirements of water quality of lakes and reservoirs stated on the government regulation No. 28/2009 concerning with the Capacity of Water Pollution Load using parameters of P-total in eutrophic trophic status, showed that the water quality of reservoir should meet the Total P standard of 100 ug/l. This condition will be met if the reservoir could fulfill the P load of 106.66 P/year (Table 1).

Wadaslintang Reservoir is used as a field of fish farming using floating net cages. Based on the method of feeding, it is divided into 2 systems namely semi intensive and super intensive system. Semi-intensive system is a fish cultivation system that feeds do not rely on artificial feed, which is done by fishermen group joined in Forum Komunikasi Tani Nelayan (FKTN) Sejahtera Mandiri. Super intensive feeding system conducted by PT Aquafarm. This feeding system is just relied in artificial feed. Types of feed used is artificial food (pellet) with 25-30% protein content. Feeding is done in adlibitum for 15 minutes until fish is not eat anymore. This method give possibility that the huge amount of pellet residue that was not consumed by the fish accumulates in the bottom of the water and further contribute to pollution of phosphorus load in the reservoir.

Phosphorus loading is the amount of phosphorus entering the waters caused by fish culture waste using floating net cage. Calculation of the amount of phosphorus loading into the waters can be used to estimate the capacity of the reservoir to meet the requirements of trophic status specified in accordance with its designation. The load capacity of phosphorus contamination shows the maximum concentration of phosphorus element which can be accommodated by Wadaslintang Reservoir based on morphology and hydrology. The load capacity of the phosphorus contents of the Wadaslintang Reservoir was calculated based on Minister of Environment Regulation No. 28 of 2009 about the capacity of pollution load. Table 2 shows the amount of waste from fishery activities that goes to the reservoirs of floating net cage activities managed by community and PT. Aquafarm in Wadaslintang Reservoir.

Table 2. The load of Phosphorus Load from Aquaculture waste in Wadaslintang Reservoir

Kadar P	**mg P /100 gr fish biomasa**	**g P/ton fish biomasa**	**kg P/ton fish biomasa**	**FCR**	P load **(Plp) Kg P/ton ikan**
Fish Phosphorus content from communities floating net cage	139,2	1392	1,392	1,5	**39,140**
Fish Phosphorus content from PT. Aquafarm floating net cage	156,43	1564,3	1,564	1,88	**49,236**
Fish Feed Phosphorus content	2702,12	27021,2	27,021		

Based on fish production and fish feed consumption calculation, it could be explained that Phosphorus waste from floating net cage of community released to reservoir was up to 39.14 Kg/ton fish. Fish production in floating net cage of community was up to 1.000 tons of fish/year, so the total waste load (Total P) into the reservoir is 39.1 tons/year. For PT. Aquafarm floating net cage the amount of waste to reservoir was up to 49.236 kg/ton fish. Currently fish production was of 3.600 tons fish/year so that the total load of P into the reservoir was up to 177.3 tons/year.

The calculation of capacity of P load from aquaculture activities to Wadaslintang reservoir is presented in Table 3.

Table 3. Wadaslintang Reservoir Capacity to Phosphorus Load and reducing program of Phosphorus

Subjects	**Parameters**	**Symbols**	**Units**	**Floating net**		**Total**
				Community	**Aquafarm**	
Feed dan P load from floating net cage activity to reservoir water	Total feed consumtion	LP	ton feed/year	1.502,9	6.770,7	8.273,6
	Food Convertion Ratio	FCR	Ton feed/ton fish	1,5	1,88	
	Feed P content	P feed	Kg P/ton feed			27,02
	Fish P content	P ikan	Kg P/ton fish	1.392	1.564	
	P bound to sediment	R		0,556	0,556	
	P- Total load to reservoir water	P_{LP}	Kg P/ton ikan	39,14	49,236	
	Proporsi of Phosphorus resuspension from sediment	Rikan		0,778	0,778	
	Concentration of P in water caused by aquaculture	dP	mg/m^3	**18,4**	**83,2**	**102,6**
	Monitoring of P in water	dP obsv	mg/m^3			**231,5**

Load and reservoir capacity of P-total from aquaculture waste	Reservoir Capacity of P-Total coming from aquaculture waste per unit area	L ikan,	g P/m^2/thn	2,95	13,33	
	Reservoir capacity of P coming from fish waste	La ikan	gP/tahun	3.923,5	17.728,9	21.652,4
	P-Total Load from aquaculture waste to reservoir	La ikan	ton P/year	39,2	177,3	**216,5**
Natural capacity, exceeded P load, and P reducing program	Capacity of reservoir based on morfology and hydrology	La reservoir	ton P/year	106,563	106,563	.06,563
	Exceesive P-Total		ton P/year	- 67.36	+ 70.74	+109.9
	Phosphorus reducing Program		%		39,89	50,76

The waste load from the community floating net cage was 39.1 tons/year. It had give impact to Total P content in the reservoir to become 18.4 mg/m^3. For the waste load coming from PT Aquafarm was 177.3 ton/year, raises Total P levels in the water reservoir of 83.2 mg/m3. The monitoring result of Total P in reservoir water during this study was 231,5mg/m^3, whereas the concentration of P in caused by aquaculture impact was 102,6mg/m^3. It meant that there was increasing concentration of Phosphorus up to 128,9mg/m^3 might come from human activities in the catchment area beside aquaculture. It could be said that the existence of fish farming system using floating net cage in wadaslintang reservoir has contributed total P reservoir equal to 43,9% from total concentration of P in water.

The impact of aquaculture in the Wadaslintang Reservoir has caused the entry of phosphorus into reservoir of 216.5 kg P/ton fish (Table 3). Based on the calculation that the total load of P-total to the reservoir is 39.2 ton P/year for floating net cage of community and 177.3 ton P/year for floating net cage of PT. Aquafarm. While based on morphometry and hydrology, the reservoir capacity should be 106.56 ton P/year. It means that the presence of floating net cage in the Wadaslintang Reservoir has caused the total concentration of P to exceed its capacity. Based on the amount of feed given by floating net cage of FKTN *Sejahtera Mandiri* has not exceeded the reservoir capacity, but the floating net cage of Aquafarm of P load from the feed waste has exceeded the capacity of 70,74 ton P/year or equal to 39,89% from the requirement. So when summed the incoming P burden from the aquaculture activity of 216.5 ton/year and has exceeded its carrying capacity of 109.9 ton/ year or 50% of the total P load into the Wadaslintang Reservoir.

The magnitude load from floating net cage from PT. Aquafarm due to feeding system of adlibitum. Actually, it should be given to the larval phase until the size of the seed instead of enlargement phase. If it is given to the larval phase and seeds of larval, their stomach capacity are still very limited so that by feeding them crisply and always available in unlimited quantities, so that they can eat at anytime. Excessive feeding in phase after larvae or seeds will have an impact that decreases water quality and may then be detrimental to aquatic systems in aquaculture activities. Excessive fish feed and not consumed by the fish causes the accumulation of feed in the bottom of the water. According to Abery *et al* [3] the cage culture operations in Saguling, Cirata and Jatiluhur reservoirs resulted in substantial nutrient loading, due to that the feed discharged into the waters reached 30% - 40%. Degradation of the accumulation of fish feed and fish feces and fish urine that accumulate in the bottom waters under certain conditions will release toxic materials such as ammonia, H_2S that can affect the health of fish cultivation.

The amount of P-total Load to High total phosphorus concentration (total-P) indicates that the organic material entering the water is also high. The limitation of phosphorus can be done by reducing the number of karamba that has passed the threshold and the management of fish feeding is good and appropriate [17]. The management of aquaculture using good and proper karamba will increase the productivity of the fishery itself. Therefore, the activity or production of aquaculture in the waters of the lake should be determined based on the load capacity of the phosphorus contamination. The capacity of lake phosphorus pollution load is the ability of lake water to receive input load of phosphorus contamination without causing lake water to be polluted based on the characteristics and environmental conditions around the lake morphology and hydrology covering area, volume, depth and water discharge; quality and trophic status; water quality standard requirements and allocation of pollution load from various sources and types of water [9]. The load capacity of phosphorus pollution of Rawapening Lake should be determined as control of phosphorus pollution in waters for fishery activity using karamba.

In general, the challenges associated with the floating net cage system include the increase of nutrient content in waters from inedible feed residues, fish excretion and feces, as well as the possible impacts on water quality, environment and ecosystem health conditions [3]. Based on research conducted by [10] fish farming production in floating net cage (417,9 ton/year) is still within the permissible limit of watershed reserve capacity (443,9 ton/year), so that KJA cultivation production in Waduk Wadaslintang still possible raised. However, at the time of this study, the production has reached 4600 tons/year, which means that it has increased 10 times and the carrying capacity of the waters has exceeded 50% of the total production allowed.

4 CONCLUSIONS

Wadaslintang Reservoir to phosphorus load capacity based on its morphometry and hydrology was about 106,56 ton P/year. The Phosphorus load from floating net cage was 216,2 ton P/year (39,2 ton/year from FKTN sejahtera mandiri and 177,3 ton P/year from floating net cage of PT. Aquafarm). The P load from PT. Aquafarm have exceeded the capacity as much as 70,74 ton P/year. In general the existence of fish farming system using floating net cage in Wadaslintang Reservoir has contributed to increase the concentration of P water equal to 43,9% or in other words as much as 50% of P load that goes into reservoir of wadaslintang. Programs designed to reduce the phosphorus load in the wadaslintang could be done by reducing the number of floating net and decreasing all the activities in the catchment area up to 50% to fulfill phosphorus standard of 100 mg/m^3.

Acknowledgement Perum Jasa Tirta I to funding this study. Unsoed for supported the implementation of this study. I would like to express my special thanks of gratitude to Mr. Badruddin Machbub for doing Phosphorus Loading calculation as well as our research team (Gentur Waluyo and

Diana RUSR) who gave me the golden opportunity to do this wonderful project on the topic (Technical study on the Capacity and Zonation On Wadaslintang Reservoir Waters), which also helped me in doing a lot of Research and I came to know about so many new things, I am really thankful to them. Secondly I would also like to thank the committee of International Conference of Kripik SciFimas Faculty of Fishery and Marine science of Unsoed who have give me opportunity to involve in, and this manuscript could be critically reviewed to become eligible being published to a reputable journal.

REFERENCES

1 Carballo, Eira; Assiah van Eer, Ton van Schie, Aldin Hilbrands. (2008). *Small-scale freshwater fish farming.* Agromisa Foundation and CTA, Wageningen

2 Lukman dan Hidayat. 2002. Pembebanan dan Distribusi Organik di Waduk Cirata. *Jurnal Teknologi Lingkungan. P3TL-BPPT.* Vol. 3 (2): 129 – 135.

3 Abery, N.W; Sukadi, F; Budhiman, A.A.; Kartamihardja, E.S.; Koeshendrajana, S.; De Silva, S.S. 2005. Fisheries and cage culture of three reservoir in west Java, Indonesia : A case study of ambitious development and resulting interactions. Fisheries Management and Ecology 12 (5) : 315 - 330

4 Sukadi, M. F. (2016). Ketahanan Dalam Air Dan Pelepasan Nitrogen & Fosfor Ke Air Media Dari Berbagai Pakan Ikan Air Tawar. *Jurnal Riset Akuakultur*, 5(1), 1-12.

5 Garno, Y.S. 2002. Beban pencemaran limbah perikanan budidaya dan eutrofikasi di perairan waduk pada DAS Citarum. *J. Tek. Ling. P3TL-BBPT* 3 : 112-120.

6 Samudra, S. R. and Soeproboowati, Tri Retnaningsih, Izzati, M. (2013) 'Daya tampung beban pencemaran fosfor untuk budidaya perikanan danau rawapening', *Prosiding Penyelamatan Ekosistem danau Rawapening*, (June).

7 Erlania, Rusmaedi, Prasetio AB, H. J. (2010) 'Dampak Manajemen Pakan Dari Kegiatan Budidaya Ikan Nila (Oreochromis niloticus) Di Keramba Jaring Apung Terhadap Kualitas', *prosiding forum Inovasi Teknologi Akuakultur*, (2002), pp. 621–631.

8 Piranti, A.S; Waluyo, G.; Rahayu, D.R.U.S.; Carmudi; Christiani. (2016). Kajian Teknis Daya tampung dan Zonasi Perairan Waduk wadaslintang. Laporan Penelitian Kerjasama Perum Jasa Tirta I - LPPM Unsoed Purwokerto.

9 Machbub, B. (2010) '3_Model_perhitungan_daya_tampung_Badruddin.pdf'. *Jurnal Sumber Daya Air Vol* 6 No 2, pp. 129–144.

10 Widyastuti, E., Piranti, A. S. and Rahayu, D.R.U.S (2009) 'Monitoring Status Daya Dukung (Monitoring of Carrying Capacity Status of Wadaslintang Reservoir on Cage Net)', *J. manusia dan Lingkungan*, 16(3), pp. 133–140.

11 Hakanson, L. (2015) 'The Importance of Lake Morphometry for the Structure and Function of Lakes', *Internat. Rev. Hydrobiol*, 90 (4)(August 2005), pp. 433–461. doi: 10.1002/*iroh*.200410775.

12 Teodorowicz, M. (2013) 'Surface water quality and intensive fish culture', *Arch. Pol.Fish*, 21, pp. 65–111. doi: 10.2478/aopf-2013-0007.

13 Brett, M. T. and Benjamin, M. M. (2008) 'A review and reassessment of lake phosphorus retention and the nutrient loading concept', *Freshwater Biology*, 53, pp. 194–211. doi: 10.1111/j.1365-2427.2007.01862.x.

14 Horne A.J. and Golmand, C.R.,. 1992. Limnology. McGraw Hill Company. New York.

15 Reynolds, C., & Davies, P., (2001).Sources and Bioavailability of Phosphorus Fractions in Freshwater: A British Perspective. *Biological Reviews*, 76 (1), 27 - 64

16 Smolders, A. J. P. *et al.* (2006) 'Internal eutrophication: How it works and what to do about it—a review', *Chemistry and Ecology*, 22(2), pp. 93–111. doi: 10.1080/02757540600579730.

17 Karakassis, I., Pitta, P. and Krom, M. D. (2005) 'Contribution of fish farming to the nutrient loading of the *Mediterranean* *', 69(2), pp. 313–321.

PRESSURES AND STATUS OF SEAGRASS ECOSYSTEM IN THE COASTAL AREAS OF NORTH BINTAN, INDONESIA

Ita Karlina[1]*, Fery Kurniawan[2] and Fadhliyah Idris[1]

[1]Marine Science Departement,Faculty of Marine Science and Fisheries, Raja Ali Haji Maritime University, Jl. PoliteknikSenggarang 29111 Tanjungpinang,
[2]Center for Coastal and Marine Resources Studies, Bogor Agricultural University, Indonesia

Abstract. This study investigates the status and anthropogenic pressures on seagrass ecosystem. Urbanization and tourism in the coastal areas of Bintan, Indonesia were increasing in the recent years. They have become new pressures to intertidal ecosystem and habitats, particularly seagrass ecosystem. Seagrass meadows are the important ecosystem and habitats in Bintan region. They provide a wide range of ecosystem services, including for small-scale fisheries and have being Dugongs (Dugong dugon) habitats for food. Currently, the sea grass status is less healthy or damaged with 11 species. The anthropogenic disturbance processes have affected the spatial distribution, percent coverage, biodiversity, and community structure of sea grass. Moreover, several species are hard to find. Directly, sea grass meadows are impacted by introduced coastal development (i.e. settlement area, tourism accommodation, port, etc.), land-based pollution, reclamation, boating traffic, dredging activities and tourism activities. Sea grass conservation area and marine spatial planning based on the land- and seascape connectivity is important for conservation efforts and sustainable management of sea grass.

1 Introduction

Understanding of the relations between sea grass ecosystem as ecological systems and social system is developing. Sea grass ecosystem provide significant number of function and benefit for human and nature (multiple ecosystem services) as a coupled social-ecological system (SES), including recreational fishing [19], tourism[10], fisheries [2], [20], food security [3], [8], carbon storage and climate change mitigation [15], habitat function [16], conservation function, especially for dugongs (Dugong dugon) as endangered species [8], etc.

In Indonesia, sea grass beds are under widespread pressure to decline with the main threats i.e. coastal development, land reclamation, deforestation, seaweed farming, fishing, garbage dumping, oil pollution, run-off, marine culture and public awareness[18] [13], whereas in some areas, the coastal communities and traditional fisherman are actual dependent on the existence of sea grass ecosystems [18], [19], [2], [17], [7]. The destruction of the sea grass ecosystem will change the SES of coastal areas, ecological health, habitat function and human well-being[2], [19], [20], [3], [16], [8], [6]. Moreover, some other pressures that threaten the sea grass are tourism, dredging, boating activities and sedimentation [10], [9], [12]. The aim of the study is to investigate the status and anthropogenic pressures on sea grass ecosystem.

2 Materials and methods

2.1 Study Area

Urbanization and tourism in the coastal areas of Bintan, Indonesia were increasing in the recent years. They have become new pressures to intertidal ecosystem and habitats, particularly sea grass ecosystem. Sea grass meadows are the important ecosystem and habitats in Bintan region. They provide a wide range of ecosystem services, including for small-scale fisheries and have being Dugongs (Dugong dugon) habitats for food. Efforts to conserve seagrass beds once existed, but are no longer running due to dependence on project funds.

*Corresponding author, e-mail: itakarlina@umrah.ac.id

 https://doi.org/10.1051/e3sconf/20184704008

2.2 Pressure factors identification

The identification of pressure factors in Bintan is done through in-depth interviews with experts. Experts were selected from academic institutions, government agencies, non-government organizations and local communities experienced at least 3 years in sea grass studies, area managers and a minimum of 10 years for local communities. In addition, the desk study is also conducted from research papers on sea grass conducted at the study site. The questions to the expert focus on the pressures on the sea grass and the recommendations needed.

2.3 Sea grass sampling

This study was focused at sea grass beds ecosystem in North Bintan area. There are four observation stations were set based on the level of complexity in SES and the different condition of sea grass ecosystem (Fig. 1). The process of data collection was completed systematically by using line transect method by placing quadrant 1 x 1 m [11].

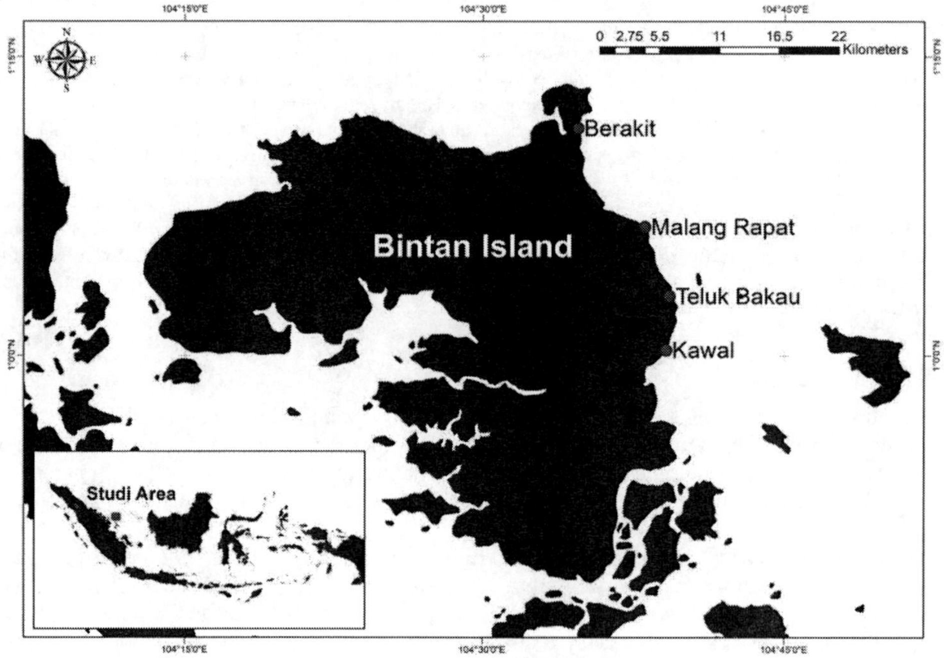

Fig 1. Study area.

2.4 Environmental Quality

To determine the sea grass status caused by anthropogenic, several parameters of environmental quality were analyzed, including temperature, pH, salinity, nitrate, phosphate and substrate. The substrate sample was taken using a paralon pipe to a depth of 20 cm. Measurements and sampling of environmental quality parameters are taken at the sea grass condition observation stations. The temperature, pH and salinity are measured in-situ using portable equipment, while the analysis of nitrate, phosphate and the fraction of the sediment were done in the laboratory using the standard protocol of American Public Health Association – APHA [1].

2.5 Data analysis

The qualitative data were analyzed descriptively, whereas descriptive statistics and one-way ANOVA were used to compare the sea grass density between the four sites. Statistics analyses were achieved with PASW Statistics 18 (Predictive Analytics Software). To obtain a quantitative measure of the sea grass structure of the community was conducted the analysis of Shannon's diversity index (H') [5], Simpson' evenness index (E) and dominance index (C) [14].

3 Results and discussion

3.1 Pressure

Some of the activities that are pressure factors on the sea grass ecosystem are as follows:

1. Tourism: beach and berkarang (taking shells and other biota in the sea grass ecosystem)

Beach tourism is quite popular in North Bintan. In some tourism actors' perspective, the existence of sea grasses is considered disruption the beauty of the beach and coastal tourism activities, so that many sea grasses are discarded. In addition, the tourist activities that directly interact with the sea grass to make the seagrass leaves cut. The other pressure of the tourism is the activities of berkarang, which is the activity of taking shells and marine biota in the seagrass ecosystem during low tide.

2. Oil pollution

The pollution of black oil or sludge oil occurs annually in the northern seasons (December-March). The waste is able to contaminate the intertidal area, including the sea grass ecosystem. Many sea grasses and association biota are covered with oil. Oil pollution is a major threat to the sea grass ecosystem in Bintan [18].

3. Illegal port

Illegal ports are traditional ports built by communities without government permission or are not included in coastal development planning. The existence of ports can increase the activity of ship traffic and waste generated from the seepage of fuel, paint and washing the ship, thus making the sea grass ecosystem disrupted and damaged.

4. Reclamation

Reclamation activities take place along Kawal beach, and in some beach areas of Bakau, Trikora and Berakit. Reclamation generally aims to support coastal development, both for resorts and hotels to provision tourism, as well as settlements. The reclamation will convert the sea grass area.

5. Marine debris

Marine debris usually comes in the north season, either plastic waste (plastic bottle, plastic bag, plastic pact, etc.), organic (grass, wood, etc.), other household waste. The garbage is buried on the beach and in the sea grass.

6. Boating activities

Boating activities can directly affect seagrass beds, either because of ship traffic or off anchors. The anchor fishing activities of the fishing and the tourist boats still have not been arranged, so it can take off any place, including in the sea grass meadow.

7. Fisheries

Small-scale fishermen in North Bintan do a lot of fishing activities around the sea grass ecosystem, either using nets or hand-line fishing, including the activities of berkarang (collecting shellfish). In fact, the current fishing around the sea grass has also become a recreational activity [19]. This activity is also regarded as a pressure on the ecosystem because it is not matched by good regulation and conservation efforts.

3.2 Sea grass ecosystem status

In Bintan region, the sea grass status in Bintan is currently less healthy or damaged [13]or with a percentage of cover ranging from 30 to 59.9% based on Decree of the Minister of Environment No. 200/2004 on Criteria of Damage Standard and Guidelines on Sea grass Status Determination [18] also report that status of sea grass in Bintan is some loss and/or degradation.

The status can also be seen from the sea grass distribution and density. Based on the observation result obtained 10 types of sea grass only, i.e. Enhallusacoroides (Ea), Thallasiahempricii (Th), Cymodocearotundata (Cr), C. serrulata (Cs), Syringodiumisoetifolium (Si), Hallodulleunivervis (Hu), H. pinovolia (Hp), Thalassodendroncilliatum (Tc), Hallophilaovalis (Ho) and H. spinulosa (Hs). In fact, previous observations mentioned that the type of sea grass in Bintan reached 11 species. Furthermore, not all species are found in every observation location; only in Berakit found 10 species of sea grass, whereas in Malang Rapat there are 9 species, 8 species in TelukBakau and 4 species in Kawal (Table 1). It shows that Kawal, TelukBakau and Malang Rapat have a moderate diversity index (1.45, 2.75 and 2.95), and only Berakit has a high diversity index value (3.76) (Table 2).

Table 1. Seagrass distribution in the coastal area of north Bintan.

No.	Species	Kawal	TelukBakau	Malang Rapat	Berakit
1	*Enhallusacoroides*(Ea)	+	+	+	+
2	*Thallasiahempricii*(Th)	+	+	+	+
3	*Cymodocearotundata*(Cr)	+	+	+	+
4	*C. serrulata*(Cs)	+	+	+	+
5	*Syringodiumisoetifolium*(Si)	-	+	+	+
6	*Hallodulleunivervis*(Hu)	-	+	+	+
7	*H. pinovolia*(Hp)	-	-	+	+
8	*Thalassodendroncilliatum*(Tc)	-	+	+	+
9	*Hallophilaovalis*(Ho)	-	+	+	+
10	*H. spinulosa*(Hs)	-	-	-	+
Total		4	8	9	10

+: present; -: absent.

Table 2. Seagrass density in the coastal area of north Bintan.

Locations	Seagrass density (range per m^2)										H'	E	C
	Ea	*Th*	*Cr*	*Cs*	*Si*	*Hu*	*Tc*	*Ho*	*Hp*	*Hs*			
Kawal	58-80	24-30	0-20	0-15	0	0	0	0	0	0	1.45	0.27	0.52
TelukBakau	20-25	24-30	30-40	43-55	30-36	0-55	2-8	9-11	0	0	2.75	0.85	0.17
Malang Rapat	10-30	14-21	24-40	16-29	0-24	0-8	0-7	0-26	0-21	0	2.94	0.87	0.17
Berakit	40-55	0-38	15-21	19-26	14-29	9-27	2-8	0-28	16-28	0-4	3.76	1.17	0.14
Mean	40	23	24	25	17	12	3	9	8	1	-	-	-
Std. Deviation	23.22	11.58	13.25	17.25	15.11	19.54	3.66	11.82	11.67	1.41	-	-	-
P-value	**0.04**	0.85	0.17	**0.04**	0.12	0.57	0.58	0.73	0.73	0.48	-	-	-

Ea:*Enhallusacoroides*; Th: *Thallasiahempricii*; Cr: *Cymodocearotundata*; Cs: *C. serrulata*; Si: *Syringodiumisoetifolium*; Hu: *Hallodulleunivervis*; Hp: *H. pinovolia*; Tc: *Thalassodendroncilliatum*; Ho: *Hallophilaovalis*; and Hs: *H. spinulosa.*Significant p-values are in bold ($p < 0.05$).

Additionally, the highest average density value is only 40 ind. m2 -1 with the type of EA, although the other types average only from 1 to 23 ind. m2 -1 (Table 2). The value indicates the condition of sea grass ecosystem is very rare and unhealthy because decreasing sea grass density negatively influences associated fauna [16], [18]. The analysis statistics demonstrated that the location showed a significant difference ($p < 0.05$) on the seagrass density of species of Ea and Cs, however other species are not significant difference (Table 2). That is, the density and distribution of the species are uneven across all observation locations. The evenness index values show the low category in Kawal area (0.27) and high category in TelukBakau, Malang and Barakit area (0.85, 0.87, 1.17). This condition also greatly affects the composition of the species. Generally, the dominating species is EA, followed by Cr, Cs and Th (Fig. 2). The highest dominance occurred in Kawal area (0.52), while in TelukBakau, Malang Rapat and Berakit are relatively low (Table 2). The differences caused by several factors referred substrate, coastal depth, waters condition and coastal activities or the factors of pressure to seagrass meadow.

The environmental conditions at North Bintan are less appropriate for seagrass ecosystems. Generally, seagrasses grew on substrate of sand and muddy sand category. Based on a few parameter of environmental quality, the parameter of temperature (31-32 °C), pH (6-6.9) and salinity (29.30-32.33 ‰) are out of range of marine water quality standard, therefore nitrate and phosphate parameters are exceeded the standard ranged from 5.88 to 9.00 mg l-1 and 6.00 to 7.00 mg l-1, respectively (Table 3). Pollution sources affecting water quality are estimated to come from settlements and tourism accommodation, including hotels and resorts. BPS Bintan Regency (2017) recorded that residents in the areas of Berakit, Kawal, TelukBakau and Malang Rapat reached 11.452 people, while tourism accommodation has reached 42 units. Consequently, the result shows that anthropogenic factors are the main supplier of the pollution on environmental.

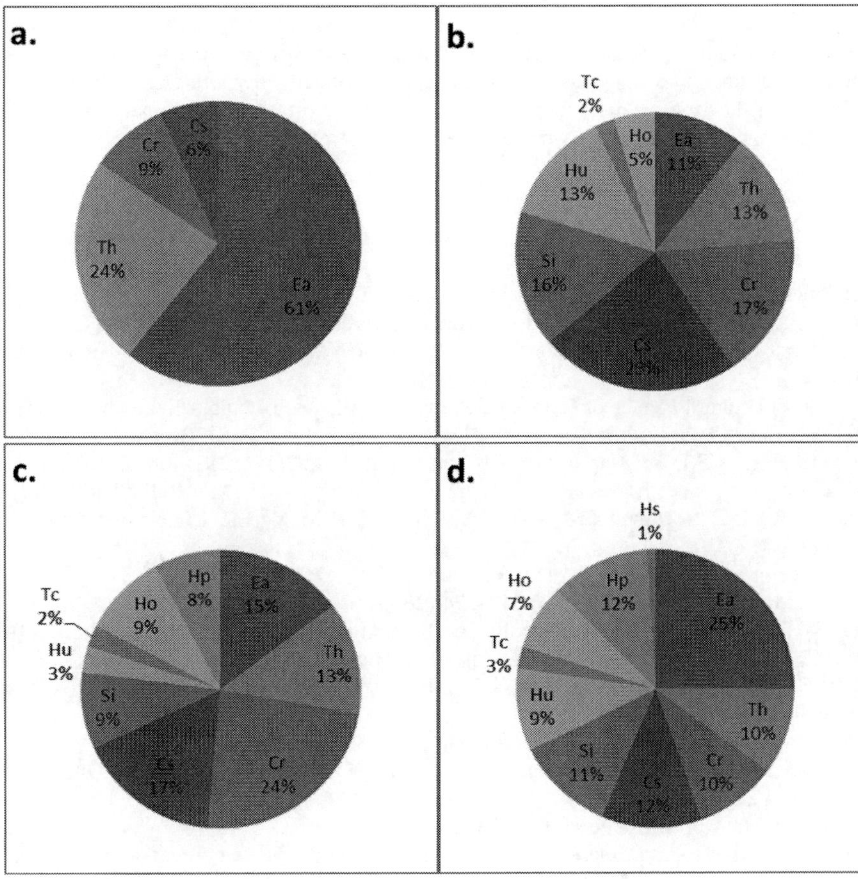

Fig 2.The composition of seagrass species at each station. a. Kawal; b. TelukBakau; c. Malang Rapat; and d.Berakit.

Table 3. Range of environmental condition at seagrass ecosystem in the coastal area of north Bintan.

Environmental quality parameters	Range	Quality standard[a]
Temperature (°C)	31–32	28-30
pH	6.0–6.9	7-8.5
Salinity (‰)	29.30–32.33	33-34
Nitrate (mg/l)	5.88–9.00	0.008
Phosphate (mg/l)	6.00–7.00	0.015
Substrate	Sand–muddy sand	-

[a]: Based on Kepmen LH No. 51/2004, specifically for seagrass (marine biota category). All parameters of environmental quality are out of range and exceeded the standard of marine water quality.

4 Conclusions

In Bintan, seagrass meadows provide multiple ecosystem services, namely provisioning services, regulating services, and cultural services. Averagely, the economic value of seagrass for recreational fishing is estimated at around IDR 351,179.56 ha-1 year-1 [19]. Seagrass ecosystem areas have supported the fisherman income from fish catch, with total revenue IDR 202,124 day-1 or 12 kg day-1 in Malang Rapat Village and IDR 193,151 day-1 or 13.8 kg day-1 in Berakit village [2]. However, there are many pressures that threaten the existence of seagrass ecosystem, i.e. tourism, oil pollution, illegal port, reclamation, marine debris, boating activities and fisheries. For that, conservation efforts are a crucial need to protect the function of seagrass ecosystem services. Specifically, sea grass conservation area and marine spatial planning based on the land- and seascape connectivity is important for sustainable management of seagrass, including blue carbon strategies. Adaptive co-management is well-defined to build the processes of decision making and public awareness.

References

1. [APHA] American Public Health Association. American Public Health Association (APHA) (2012)
2. Arkham MN, Adrianto L, Wardiatno Y. JurnalIlmudanTeknologiKelautanTropis**7(2)** (2016)
3. Baker S, Paddock J, Smith AM, Unsworth RKF, Cullen-Unsworth LC, Hertler H. Ecosystem Services **11**:12-21.(2015)
4. [BPS KabupatenBintan]. Center of Indonesia-Statistic–Statistics of Bintan Regency. 2017Brower JE, Zar JH. Field and Laboratory Methods for General Ecology. Dubuque, Iowa (US): Brown Company Publihsers (1998)
5. Campagne CS, Salles JM, Boissery P, Deter J..Mar Pollut Bull **97**(1-2):391-400 (2015)
6. Cullen-Unsworth LC, Jones BL, Seary R, Newman R, Unsworth RKF. Mar Pollut Bull(2017)
7. Cullen-Unsworth LC, Nordlund LM, Paddock J, Baker S, McKenzie LJ, Unsworth RK.. Mar Pollut Bull **83(2)**:387-97 (2014)
8. Cuttriss AK, Prince JB, Castley JG. Aquatic Botany **108**:41-47(2013)
9. Daby D. Environmental Pollution **125(3)**:313-324 (2003)
10. English S, Wilkinson C, Baker V. Townsville (AU): Australian Institute of Marine Science(1997)
11. Freeman AS, Short FT, Isnain I, Razak FA, Coles RG. Biological Conservation **141(12)**:2993-3005(2008)
12. Hernawan UE, Sjafrie NDM, Supriyadi IH, Suyarso, Iswari MY, Anggraini K, Rahmat. Jakarta (ID): PuslitOseanografi – LIPI (2017)
13. Krebs CJ. New York (US): Harper Collins Publishers (1989)
14. Mazarrasa I, Samper-Villarreal J, Serrano O, Lavery PS, Lovelock CE, Marba N, Duarte CM, Cortes J. Mar Pollut Bull (2018)
15. McCloskey RM, Unsworth RK. PeerJ 3:e1053 (2015)
16. Sjafrie NDM, Adrianto L, Damar A, Boer M. Environment, Development and Sustainability **20(2)**:865-881.(2017)
17. Unsworth RKF, Ambo-Rappe R, Jones BL, La Nafie YA, Irawan A, Hernawan UE, Moore AM, Cullen-Unsworth LC. Sci Total Environ **634**:279-286 (2018)
18. Wahyudin Y, Kusumastanto T, Adrianto L, Wardiatno Y. Ecological Economics **148**:22-35 (2018)
19. Wawo M. IOP Conference Series: Earth and Environmental Science **89**:012023 (2017)

Responses of Herbivorous Fishes on Coral Reef Cover in Outer Island Indonesia (Study Case: Natuna Island)

Risandi Dwirama Putra[1], *Ani* Suryanti[2,1] , *Dedi* Kurniawan, [3], *Arif* Pratomo[1], *Henky* Irawan[3], *Tengku* Said Raja'I[4], *Rika* Kurniawan[1] , *Ginanjar* Pratama[5] , and Jumsurizal[5]

[1] Marine Science Departement, Faculty of Marine Science and Fisheries, Raja Ali Haji Maritime University, Jl. Politeknik, Senggarang 29111, Tanjungpinang, Indonesia
[2] Sosio-Economic Departement, Faculty of Marine Science and Fisheries, Raja Ali Haji Maritime University, Jl. Politeknik, Senggarang 29111, Tanjungpinang, Indonesia
[3] Aquatic Resources Management Departement, Faculty of Marine Science and Fisheries, Raja Ali Haji Maritime University, Jl. Politeknik, Senggarang 29111, Tanjungpinang, Indonesia
4 Aquaculture Departement, Faculty of Marine Science and Fisheries, Raja Ali Haji Maritime University, Jl. Politeknik, Senggarang 29111, Tanjungpinang, Indonesia
5 Fisheries Products Technology Departement, Faculty of Marine Science and Fisheries, Raja Ali Haji Maritime University, Jl. Politeknik, Senggarang 29111, Tanjungpinang, Indonesia

Abstract. Coral reefs are one of the most highly productive marine ecosystems, with the largest transfer of energy attributed to the trophic interaction between herbivores and algae. Rapid demographic growth, leading to transmigration to small islands such as Natuna Island where located on outer Island Indonesia.The aim of the present study is to test for significant associations between herbivore fish species traits and Habitat complexity was derived from coral reef cover. Method to record each observed herbivore fish species with UVC (Underwater Visual Census) using SCUBA diving equipment along modification line transects. A total of 39 different fish species belonging to 3 families' herbivore fish were identified. Total biomass of herbivorous fish correlated with Dead Coral Algae (DCA). Biplot of the first two axes for the nonmetric multidimensional a scaling (NMDS) analysis for family and grazer showed the dominance of herbivore fish grazer.

1 Introduction

Coral reefs are one of the richest and diverse and ecologically important marine ecosystems [1,2,3,4] and their feature has extraordinary biodiversity and a complex structure of interconnections between organisms and their environment [5,6] Among reef habitats, the richest and diverse coral reef ecosystem is found in the tropics in Indo-West Pacific region with the highest level of biodiversity particular around the islands of the Indonesia, Philippines, and Papua New Guinea [7,8,9] Unfortunately, the richest and

[1] Corresponding author: yanti.ajb@umrah.ac.id

diverse this marine ecosystem, coral reef are also among the most threated ecosystem [10] with uncertain future [6]. The degradation of coral reef habitats of marine ecosystem has become a worldwide trend [1,11,12,13], many coral reefs have lost their typical community [14] and increasingly degraded due to anthropogenic threats and stressors on global scale [15,16,17].

A root cause of coral reef degradation due to anthropogenic threat is demographic growth worldwide [10]. Demographic growth cause increasing rapid growth in coastal areas. coastal areas become the main consideration in the placement of residence status from demographic worldwide due to the natural highly diverse coral reef ecosystem service such as coastal protection, food for coastal communities and income from tourism [18], and this consideration becomes worse without any strict regulation of the sustainability of coral reef ecosystems that through disturbance to coastal zones associated with coral environments [10].

The degradation coral reef ecosystem due to rapid demographic growth also occurs in Indonesia and the worse effect occurs in Small Island. Rapid demographic growth, leading to transmigration to small islands such as Natuna Island where located on outer Island Indonesia. Monitoring and controlling outer island area is limited, this condition can cause increasing threatening and degrade on the coral reef ecosystem. Threats to Indonesia's coral reef resources can be divided into two main types: acute threats; and chronic stresses [19] and most numerous incident threatening in coral reef ecosystem in outer island was acute threats by destructive fishing including the use of bombs, cyanide poison [20], coral mining and sand mining [21]. Acute threats from destructive fishing practice and sand mining cause dramatic damage and declining water clarity and quality have been associated with increased presence of macro alga growth [2,8] and in the long term, this effect will also cause phase shift on coral reef ecosystem. Shifts are characterized by the passage from a stage of coral dominance to alternative stages of coral depletion and increased cover of algae [16]. Increasing in algae cover is harmful coral growth because the relationship between non-endosymbiotic algae and corals is reciprocally negative [16] There are several varieties of increasing the shift from live coral to algal dominance as a major concern in assessments of outer island coral reef health; overfishing, pollution and coastal community that give impact pollution and nutrient enrichment [10] and all these varieties are suspected In small Islands Indonesia such as Natuna Island.

Evidence over the last 40 years has revealed a widespread loss of coral cover in tropical coastal waters of the Indo-West Pacific. In some areas, entire coral reefs have resulted in significant modification of the structure and functioning of an exploited ecosystem [22,23]. Coral reef has been overgrown and killed by fast-growing species of macro algae [24] and algal cover jumped from being slightly higher than corals [14], this condition linear correlation with destructive fishing practice and overfishing can cause degrade herbivore fish abundance very fast [25] and resulting in decrease in total biomass [26,17]. Abundance and biomass each functional species of the herbivorous reef fish group was also used as an indicator of the potential for coral reef recovery [20]. Herbivorous fish play a critical role and critical functional groups (CFGs) for maintaining coral reefs in the recovery of coral assemblages by limiting and control growth of algae communities [15] that are competing with reef-building corals such as for space and light [16,27,28,29]. Generally, coral reefs are a good example of understanding how herbivores control algae biomass and growth [15] which prevents algae from taking over [5].

Many studies in tropical areas such as in Indonesia water show how grazers from herbivore fish play the main role limiting and controlling the abundance of epilithic algal (EA) [4,30] and how overfishing can lead to trophic cascade effects driving ecosystem shifts [31]. The EA is one of the most nutritious food resources on coral reefs with high productivity rates that support a diverse and abundant assemblage of herbivore fish grazers

[32]. All goods and services that coral reef ecosystems provide rely on the status of their herbivore fish habitats [15] and spatial feeding patterns, herbivore fish can move between reefs and are less likely to feed at the same place [14].

Based on their foraging behavior, herbivorous reef fish can be differentiated into four main groups; which are scrapers/small excavators, large excavators/borders, grazers/detritivores, and browsers. Each category plays different and complementary roles in coral resilience in terms of how they feed benthic algae, what algae they consume, and their impact on the underlying substratum (i.e. hard corals). Moreover, the availability of stable substrates such as rock, dead coral, and coralline algae also plays an important role in providing a surface for coral planula to settle and grow [27,33]. Algae are involved in the bio-construction of coral reefs and can, therefore, be considered as bio-indicator of health for coral reefs as with hard corals [10]. In this research, we use dead coral with algae (DCA) cover as an indicator of coral reef health depending herbivore fish biomass as an indicator as well as biodiversity in outer island regional scale. The aim of the present study is to test for significant associations between herbivore fish species traits and Habitat complexity was derived from coral reef cover.

2 Methodology

2.1. Study area

Sampling for the present study occurred from Augustus - September 2016 distributed on reef location of the north (N) to the southwest (SW) located in one of outer island Indonesia, Natuna Island (3°56'31.74" N, 108°12'27.18" E). A total of eleven survey sites reef were studied in 2016. In general, there are five sites in South Natuna (SW: 4 sites, SE: 1 site) and six sites in North Natuna (NE: 3 sites, N: 1 site, NW: 2 sites).

Tabel 1. Location and direction of the study sites.

Site ID	Longitude	Latitude	Location	Direction from central Natuna Island
CR.001	108°0'11.52" E	3°47'23.28" N	Sedanau Island	Southwest (SW)
CR.002	108°5'18.24" E	3°43'49.08" N	Sedanau.Timur	Southwest (SW)
CR.003	108°6'22.68" E	3°40'22.48" N	Pulau Tiga	Southwest (SW)
CR.004	108°4'45.84" E	3°34'43.68" N	Sabang Mawang	Southwest (SW)
CR.005	108°25'59.52" E	3°52'18.48" N	Cemaga	Southeast (SE)
CR.006	108°21'25.85" E	4°0'10.08" N	Sepempang	Northeast (NE)
CR.007	108°20'12.12" E	4°2'28.68" N	Tanjung	Northeast (NE)
CR.008	108°18'25.38" E	4°3'35.54" N	Kelanga	Northeast (NE)
CR.009	108°13'47.28" E	4°13'28.92" N	Pengadah	North (N)
CR.010	108°9'13.68" E	4°9'55.44"	Kelarik Utara	Northwest (NW)

		N		
CR.011	108°5'02.04" E	4°7'03.00" N	Kelarik	Northwest (NW)

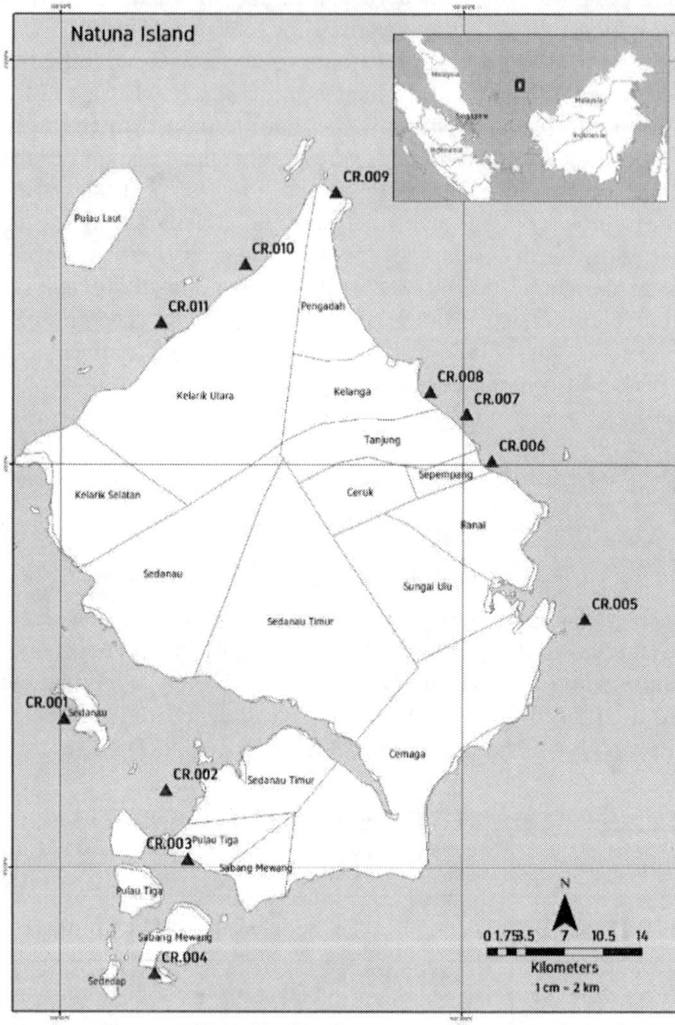

Fig. 1. Map of the study area and sampling sites.

2.2. Characterization of habitat complexity

Habitat complexity was derived from benthic cover based on the following substrate: hard coral (HC), dead coral (DC) and dead coral with algae (DCA). A principal focus of percent cover from 3 standard benthic categories HC, DC and DCA of its functional roles as a key builder of reef structure and a key provider of complex, rigid habitat for much another reef biodiversity especially for herbivore fish diversity [25,34]. Cover by substrate type was evaluated from 11 sites underwater photographic transects (UPT) [35,36,37] of 50 m length. The digital photograph was taken parallel on quadrat transect modification by [38] with dimension 58 x 44 cm at a height of 1.20 m, with approximately 1 m of distance between each photo. The UPT using Nikon D90 underwater digital camera mounted, for a total of 50 photos per site and 550 total photos for this study. The analysis of each photo was carried out using the software CPCe 4.1 (Coral Point Count with excel extension) [39] to estimate the percent cover of dominant benthic categories at each reef sites (live coral, dead coral with algae, and recently dead coral) [37].

2.3. Reef fish community structure

To record fish species diversity, abundance, and their size length to estimate biomass, performed diurnal. Diurnal increase in temperature, as a result of heating, and dissolved oxygen, as a result of photosynthesis on the reef, seem most likely amongst environmental variables in diurnal condition correlate with the feeding activity patterns and herbivore fish abundance [39]. Method to record each observed herbivore fish species with UVC (Underwater Visual Census) using SCUBA diving equipment along modification line transects based on [41,42,43] with 70 m long at 5 – 10 m depth. The observer within 5 m on either side of transect was identified herbivore fishes. The UVC method is a nondestructive, capture-independent method, superior to collection techniques diversity and abundance of reef fishes although underwater visual censuses require little post-processing, whereas collections take many hours to complete [44]. This study was used three families herbivore fish that are most easily and rapidly identifiable by underwater visual census methods (Siganidae, Scaridae, and Acanthuridae) [10,45,41]. Herbivore fish species were characterized based on their diet information and feeding behavior, obtain from extensive published data [27] This study focuses on four functional groups of herbivorous reef fishes that each play different and complementary roles in coral reef resilience: scrapers/small excavators, large excavators/bio-erodes, grazers/detritivores, and browsers. For the herbivore fish community: diversity (number of species), abundance (number of individuals) and biomass per transect, was calculated using the length-weight relationship by [10,20].

$$W = a\,LT^{b} \tag{1}$$

Where W is the weight (g), LT the fork length of herbivore fish (cm) and a and b are coefficients specific to each species from FishBase data [27]. The total biomass for each station corresponds to the total weight of all fish per unit area (g/m^2)

2.4. Statistical Analysis

2.4.1. Herbivore fish Indices

In the present study, the data herbivore fish were analyzed followed by [46] using Shannon-Wiener Index (*H'*), Margalef richness index (d), Pielou's evenness (*J'*) index and Simpson's dominance index (D).

2.4.2. Cluster analysis

Cluster analysis was used to discriminate between herbivore fish species and the similarity between sites. The zero adjusted Bray-Curtis Dissimilarity index was chosen for the analysis, as it the most appropriate distance matrix for non-binary ecology abundance data [47]. The transformed abundance matrix was then converted to a Bray-Curtis coefficient similarity matrix and a resultant dendrogram was generated with of similar quadrat groupings was based on sites dominant location (North Natuna and South Natuna) location and used to characterize zones that were predicted (a priori) to correlate with herbivores fish abundance. Bray-Curtis coefficient of similarity analysis which is more robust than techniques using maximum or minimum distances between sites. Cluster dendrogram provides two types of *p*-value: AU (Approximately Unbiased) and BP (Bootstrap Probability) for assessing the uncertainty in hierarchical cluster analysis for location habitat.

2.4.3. Meta Data Analysis.

Meta-analysis **show** effect sizes and confidence intervals from herbivore fish species community, as well as grazer type overall, mean effects and confidence intervals from its abundant. Forest plots from this studies used the meta-analytic equivalent of error bar or confidence interval plots, and are useful for showing mean effect size and variation among herbivore fish from grazer type [48]. Meta-analysis forest plots are drawn with the dependent qualitative from abundant herbivore fish species (Y) axis horizontal, while the vertical axis is no quantitative from herbivore fish grazer (Excavator, Grazer, and Scraper) [49].

2.4.4 Nonmetric multidimensional scaling (nMDS) Analysis

Visuals spatial and temporal patterns in the herbivore fish family community (Acanthuriade, Siganidae, and Scaridae) and grazer type (Excavator, Grazer, and Scraper) datasheet were used Non-metric multidimensional scaling (nMDS) solutions [50]. nMDS was performed for each site separately, as well as on spatially and temporally averaged datasets in herbivore fish communities and grazer types. Both herbivore fish and grazer type datasets were converted to relative diversity (i.e. row standardized) prior to analyses. Two dimensions were chosen in the nMDS and in all ordinations, stress values were less than 0.5. To identify herbivore family that was principally responsible for determining the grazer type dissimilarities. Herbivore fish family scores were calculated as weighted averages of the grazer scores. The weighted averages were 'expanded' so that their biased weighted variance was equal to the biased weighted variance of the corresponding grazer type [50].

2.4.4 Corresponding Analysis (CA)

Corresponding analysis has been applied to indicate the main herbivore fish family variable (total biomass and density) factor in coral reef habitat benthic (Hard Coral, Dead Coral, Dead Coral Algae) of the study area considering the variable load size and the relationship between variable and the sampling point partition based on modification from [51]. Corresponding analysis is a robust multivariate statistical method of study of relationships between variables and relationship between. This analysis is a method to analyze many complicated variables, on this studies herbivore fish abundance in every family based on grazer type and coral reef habitat benthic. This variable can be compressed into two principal component variables. Subsequently, the relationships between herbivore fish abundance and coral reef habitat benthic can be easily analyzed and explained [51,52].

2.4.6 Regression Analysis (RA)

Linear regression was used to investigate the effect on the relation between herbivore fish total biomass on sampling sites and coral reef benthic habitat after analyzed from the corresponding analysis (CA). Regression analysis has a function to determine the effect of herbivore fish on total biomass recorded at each site and the main component type of coral reef habitat benthic [53].

3 Result and Discussion

We found 21 species from Scaridae family, 6 species from Acanthuridae, and 9 species from Siganidae family on 11 sites in Natuna Island. Daisy parrotfish (*Chlororus sordidus*) from Scaridae had the highest mean density in the Natuna Island coral reef ecosystem (17.45 ± 8.93). Other species with significantly higher density including Java Parrotfish (*Scarus hypselopterus*) and Barrhead Spinefoot (*Siganus virgatus*) have significant mean density difference (12.27 ± 7.20) and (11.78 ± 7.12) respectively. The highest biomass herbivore fish in Natuna Island are Daisy parrotfish (*Chlororus sordidus*) (with highest mean density) (2.159 ± 1.535). Orange-blotch Parrotfish (*Chlororus bowersi*) have different IUCN status in coral reef ecosystem habitat with category NT (Near Threatened), another herbivore fish species categorized as LC (Least-Concerned). The lowest biomass herbivore fish in Natuna Island are Steephead Parrotfish (*Chlorurus michrohinos*) and Scribbled Rabbitfish (*Siganus spinus*) with biomass (0.002 ± 0.007) and (0.003 ± 0.008) respectively.

Table 2. List of herbivore fish recorded associated coral reef with density and biomass from Natuna Island

Family	Species	Grazer type	Common Name	IUCN Status	Mean Density (Ind/area*)	Biomass (kg/area*)
Scaridae	*Cetoscarus bicolor* (Ruppel, 1829)	Excavator	Bicolour parrotfish	LC	2.67 ± 0.58	0.065 ± 0.116
	Chlorurus capistratoides (Bleeker, 1847)	Scaper/small excavator	Pink-margined parrotfish	LC	2.67 ± 2.08	0.089 ± 0.154
	Chlorurus spilurus (Valenciennes, 1840)	Scaper/small excavator	Bullethead Parrotfish	LC	2.20 ± 1.30	0.189 ± 0.309
	Chlororus bleekeri (de Beaufort, 1940)	Scaper/small excavator	Bleeker's Parrotfish	LC	7.20 ± 5.12	1.003 ± 0.889
	Chlororus bowersi (Snyder, 1909)	Scaper/small excavator	Orange-blotch Parrotfish	NT	3.00 ± 2.88	0.130 ± 0.159
	Chlororus sordidus (Forsskål, 1775)	Scaper/small excavator	Daisy Parrotfish	LC	17.45 ± 8.93	2.159 ± 1.535
	Chlororus michrohinos (Bleeker, 1854)	Excavator	Steephead Parrotfish	LC	2.67 ± 0.58	0.002 ± 0.007
	Scarus schlegeli (Bleeker, 1861)	Scaper/small excavator	Yellowband parrotfish	LC	1.75 ± 0.96	0.107 ± 0.208
	Scarus dimidiatus (Bleeker, 1859)	Scaper/small excavator	Yellowbarred Parrotfish	LC	2.55 ± 3.08	0.118 ± 0.148
	Scarus forsteni (Bleeker, 1861)	Scaper/small excavator	Big belly Parrotfish	LC	2.00 ± 0.00	0.055 ± 0.127
	Scarus frenatus (Lacepède, 1802)	Scaper/small excavator	Bridled Parrotfish	LC	1.75 ± 0.96	0.230 ± 0.416
	Scarus ghobban (Forsskål, 1775)	Scaper/small excavator	Blue-barred parrotfish	LC	3.17 ± 1.17	0.501 ± 0.722
	Scarus globiceps (Valenciennes, 1840)	Scaper/small excavator	Globehead parrotfish	LC	9.50 ± 10.61	0.088 ± 0.261
	Scarus hypeselopterus (Bleeker, 1853)	Scaper/small excavator	Java Parrotfish	LC	12.27 ± 7.20	0.370 ± 0.248
	Scarus niger (Forsskål, 1775)	Scaper/small excavator	Dusky Parrotfish	LC	8.71 ± 4.50	0.650 ± 0.905
	Scarus oviceps (Valenciennes, 1840)	Scaper/small excavator	Dark Capped Parrotfish	LC	4.80 ± 3.63	0.225 ± 0.381
	Scarus xanthopleura (Bleeker, 1853)	Scaper/small excavator	Red Parrotfish	LC	0.09 ± 0.30	0.012 ± 0.038
	Scarus prasiognatus (Valenciennes, 1840)	Scaper/small excavator	Blue-faced Parrotfish	LC	2.50 ± 2.12	0.041 ± 0.092
	Scarus quoyi (Valenciennes, 1840)	Scaper/small excavator	Green-blotched Parrotfish	LC	6.88 ± 3.72	0.459 ± 0.497

	Scarus rivulatus (Valenciennes, 1840)	Scaper/small excavator	Surf Parrotfish	LC	9.38 ± 5.26	0.515 ± 0.509
	Scarus spinus (Kner, 1868)	Scaper/small excavator	Greensnout Parrotfish	LC	2.80 ± 2.05	0.025 ± 0.046
Acanthuridae	*Acanthurus lineatus* (Linnaeus, 1758)	Grazer/Detritivore	Lined Surgeonfish	LC	3.67 ± 1.53	0.027 ± 0.055
	Achanturus triostegus (Linnaeus, 1758)	Grazer/Detritivore	Convict Surgeonfish	LC	2.00 ± 1.41	0.029 ± 0.095
	Achanturus nigrofuscus (Forsskål, 1775)	Grazer/Detritivore	Blackspot Surgeonfish	LC	7.00 ± 1.73	0.152 ± 0.360
	Ctenochaetus striatus (Quoy & Gaimard, 1825)	Grazer/Detritivore	Striped Bristletooth	LC	2.50 ± 0.71	0.116 ± 0.386
	Naso lituratus (Forster, 1801)	Browsers	Orange Spine Surgeonfish	LC	4.00 ± 2.00	0.190 ± 0.379
	Zebrasoma scopus (Cuvier, 1829)	Grazer/Detritivore	Brushtail Tang	LC	1.00 ± 0.00	0.046 ± 0.154
Siganidae	*Siganus corallinus* (Valenciennes, 1835)	Grazer/Detritivore	Blue-spotted Spinefoot	LC	3.67 ± 3.27	0.326 ± 0.513
	Siganus argenteus (Quoy & Gaimard, 1825)	Grazer/Detritivore	Forktail Rabbitfish	LC	1.50 ± 0.71	0.105 ± 0.249
	Siganus doliatus (Guérin-Méneville, 1829)	Grazer/Detritivore	Barred Spinefoot	LC	1.00 ± 0.00	0.027 ± 0.089
	Siganus guttatus (Bloch, 1787)	Grazer/Detritivore	Golden Rabbitfish	LC	3.50 ± 0.71	0.083 ± 0.185
	Siganus punctassiumus (Fowler & Bean, 1929)	Grazer/Detritivore	Fine-spotted Rabbitfish	LC	1.50 ± 0.71	0.045 ± 0.149
	Siganus puellus (Schlegel, 1852)	Grazer/Detritivore	Masked Rabbitfish	LC	1.50 ± 0.58	0.154 ± 0.418
	Siganus spinus (Linnaeus, 1758)	Grazer/Detritivore	Scribbled Rabbitfish	LC	1.50 ± 0.71	0.003 ± 0.008
	Siganus Virgatus (Valenciennes, 1835)	Grazer/Detritivore	Barhead spinefoot	LC	11.78 ± 7.12	0.812 ± 0.735
	Siganus Vulpinus (Schlegel & Müller, 1845)	Grazer/Detritivore	Foxface Rabbitfish	LC	4.71 ± 2.87	0.278 ± 0.292

Note: *Area of study is 350m^2 from UVC method with 70 m long line transect within 5 m on either side of the transect. Mean density and Biomass from 11 (eleven) survey sites reef. IUCN status from http://www.iucnredlist.org based on Red List Category & Criteria: (LC-Least Concern, NT-Near Threatened). Grazer type of herbivore fish (Scaridae, Siganidae, and Acanthuridae) from data (Green and Bellwood 2009; Froese and Pauly, 2012). Format the table modification based on (Jayaprabha, Purusothaman, and Srinivasan 2018)

Table 3. Sites variation of diversity indices calculated for herbivore fish in Natuna Island (2016)

Sites	Herbivore fish indices				Diversity ($n/350m^2$)	Abundance ($Ind/350m^2$)	m ± Sd
	H'	*J'*	*D*	*d'*			
CR1	2.499	0.882	0.093	7.916	17	105	2.838 ± 4.776
CR2	2.380	0.840	0.112	7.557	17	131	3.541 ± 6.619
CR3	2.486	0.877	0.106	7.793	17	113	3.054 ± 5.553
CR4	1.971	0.856	0.148	4.392	10	112	3.027 ± 6.685
CR5	1.916	0.872	0.162	4.413	9	65	1.757 ±4.159
CR6	2.118	0.920	0.117	5.444	10	45	1.216 ± 2.485
CR7	2.525	0.891	0.092	7.682	17	121	3.270 ± 5.440
CR8	2.711	0.938	0.062	9.343	18	66	1.784 ±2.440
CR9	2.432	0.858	0.113	7.966	17	102	2.757 ±5.236
CR10	1.987	0.828	0.164	5.092	11	92	2.486 ±5.867
CR11	2.523	0.910	0.080	8.305	16	64	1.730 ±2.765

Note: *H'* – Shannon Wiener index; *J'* – Evenness; *D* - Simpson's dominance index; *d'* – Margalef index

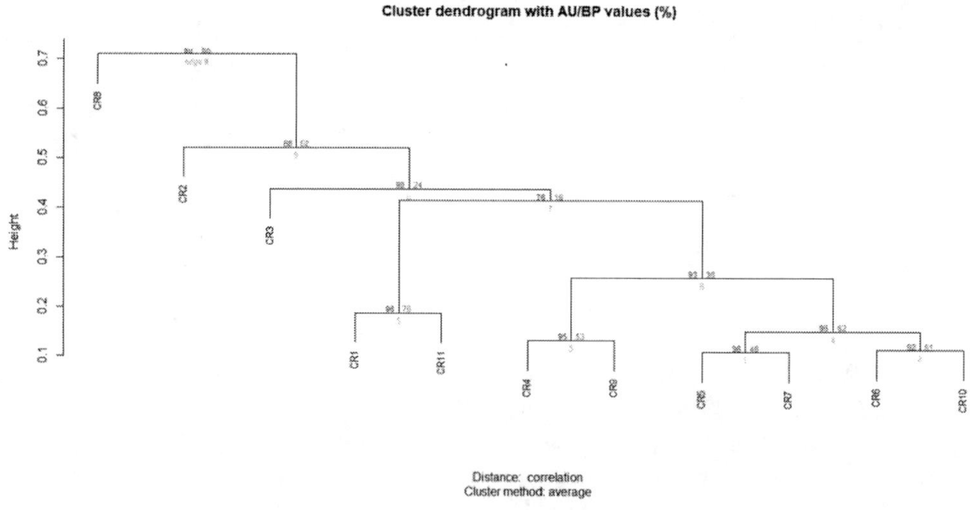

Fig 2. Dendrogram showing similarity in herbivore fish composition of coral reef sites recorded in AU/BP

Dendrogram analysis indicated that the herbivore fish species mainly responsible for the dissimilarity in abundance between the sampling sites (Fig. 2). Fishing has major effects both directly and indirectly on the environment, diversity, and productivity of communities. The average similarity of dendrogram in clustering sites sample showed three group clustering similarity (group 1, CR6-CR10; CR5-CR7, group 2, CR4-CR9, group 3 CR1-CR11). Similarity on herbivore fish abundance and diversity each site with Approximately Unbiased and Bootstrap Probability (AU/BP) to group 1 showed CR6-CR10 and CR5-CR7 have AU/BP (92/51) % and (96/46) % respectively. Group 2 similarity on clustering sampling CR5-CR7 showed AU/BP value (95/53) % and group 3 to CR1-CR11 with AU/BP value (96/70) %

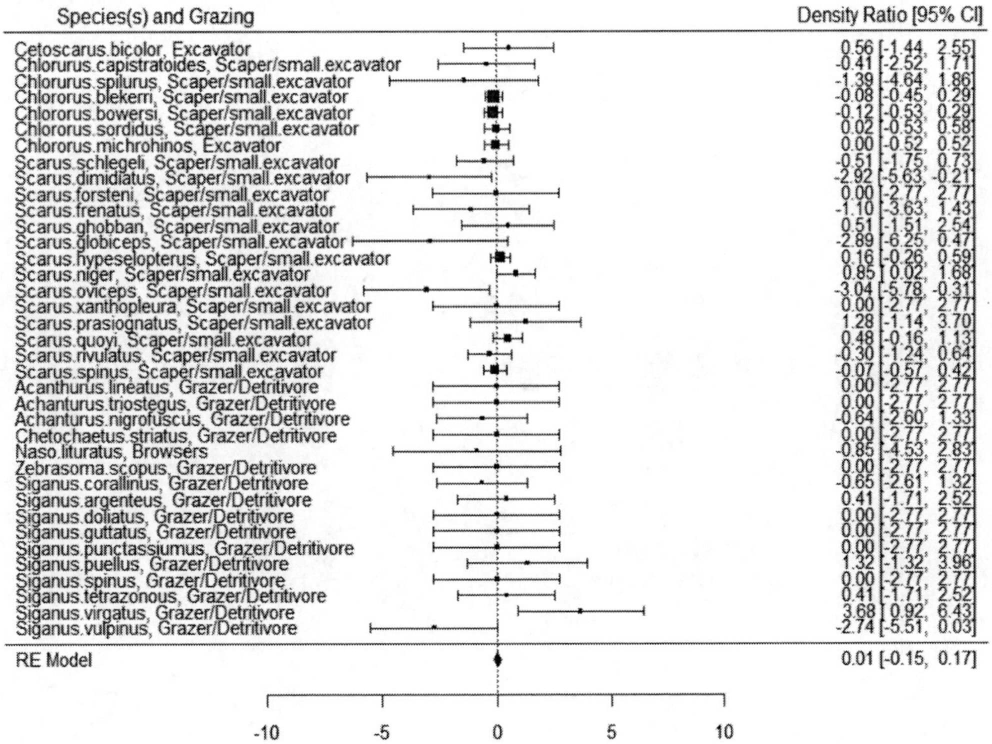

Fig. 3. Forest plots used the meta-analytic equivalent of density herbivore fishes based on the type of grazing on a coral reef in 11 sites

Thirty-nine herbivore evaluations were included (11 coral reef sites and 3 grazer types). Overall form meta-data analysis there was no statistically significant grazer type's bias for included herbivore evaluations (Fig.3). Only two species from Siganidae Family - *Siganus virgatus* (mean = 3.68, 95% CI [0.92, 6.43]) and Scaridae family – *Scarus niger* (mean = 0.85, 95% CI [0.02, 1.68]) reported differences in two species fish population for herbivore fish grazer type that were *Siganus vulpinus* (mean = -2.74, 95% CI [-5.51, 0.03]) and *Scarus oviceps* (mean = -3.04, 95% CI [-5.78, -0.31]) .The grand mean (Fig 1.), mean = 0.01, 95% CI [-0.15,0.17] and mean effect size for herbivore fish grazer. *Chlorurus bleekeri, Chlorurus bowersi, Scarus hypselopterus* have higher abundant from herbivore fish population in 11 sites with meta-analysis evaluation (mean = -0.08, 95% CI [-0.45, -0.29]), (mean = -0.12, 95% CI [-0.53, -0.29]), and (mean = -0.16, 95% CI [-0.26, -0.59]) respectively

 https://doi.org/10.1051/e3sconf/20184704009

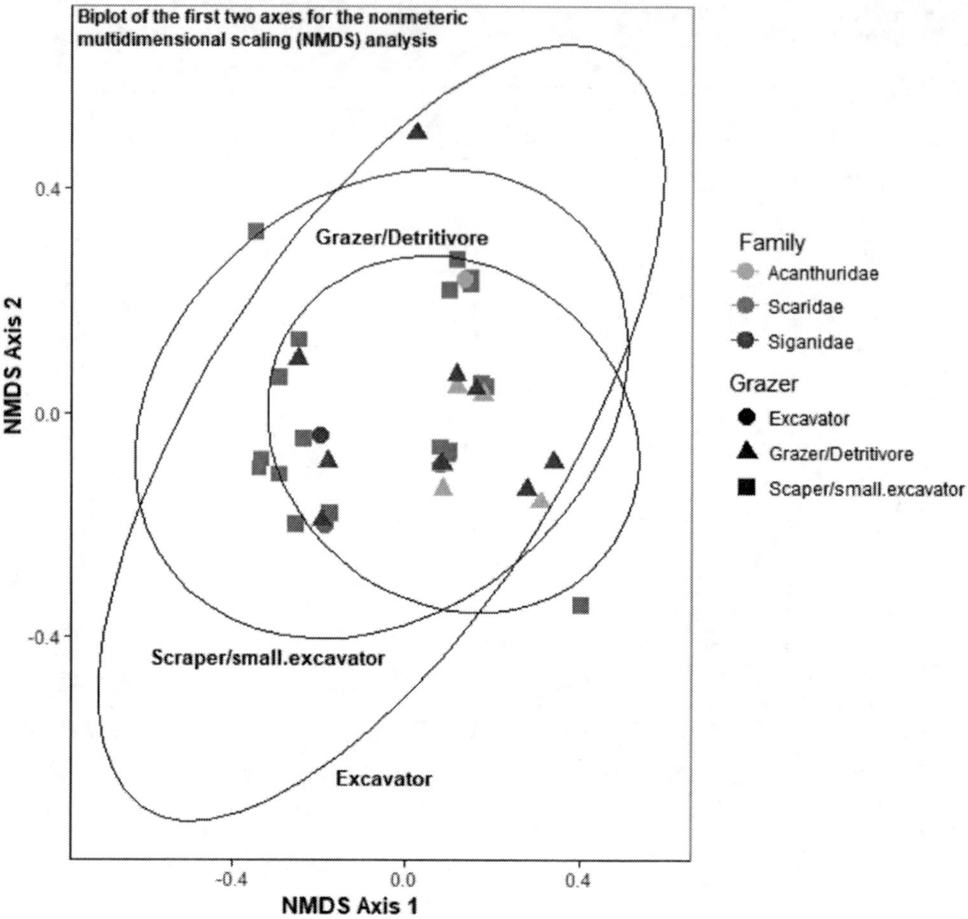

Fig. 4. Biplot of the first two axes for the nonmetric multidimensional a scaling (NMDS) analysis for Herbivore fish grazer

nMDs ordination of herbivore fish community showed there is no clear separation of grazer types. Ellipse representing the standard deviation of grazer types from herbivore fish family. There was significant overlap in every grazer type. This overlap deviation showed similarity component of variance due to sites location and habitat between herbivore fish communities could be attributed to the similarity of paired grazer types. nMDS ordination of herbivore fish community and grazer types revealed a gradient in the structure of theses herbivore fish on coral ecosystem across his habitat consume alga on a coral reef (Fig. 4). Generally, Grazer/Detrivore was more similar to Scraper/small excavator types of herbivore fish abundance on coral reef habitat. However, some herbivore fish family within a particular grazer type were just as dissimilar to each other as they were to reefs site in another coral reef ecosystem habitat type such *Excavator*. These herbivore family similar related to grazer type so that herbivore fish diversity within each coral reef site sampling were generally more similar to each other grazer type although one species form Scaridae Family *Chlorurus micrhohinos* has different grazer type. This pattern was consistent for ordinations of herbivore fish its grazer type.

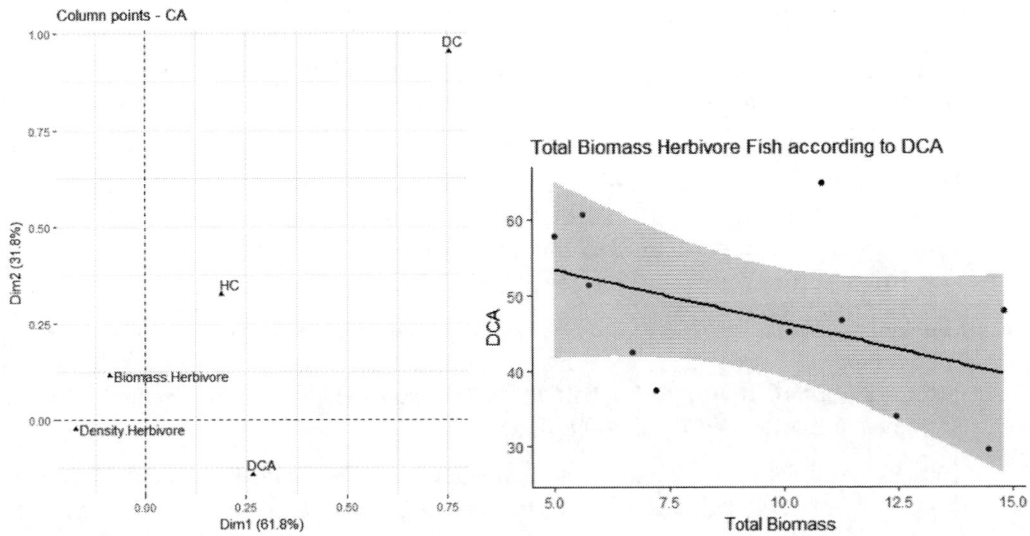

Residuals:				
Min	1Q	Median	3Q	Max
-12.8421	-8.7734	-0.8998	6.3413	19.7209
Coefficients:				
	Estimate Std.	Error	t value	Pr(>|t|)
(Intercept)	60.4902	9.2323	6.552	0.000105 ***
Total.biomass	-1.413	0.9168	-1.541	0.157637

Residual standard error: 10.43 on 9 degrees of freedom
Multiple R-squared: 0.2088, Adjusted R-squared: 0.1209
F-statistic: 2.376 on 1 and 9 DF, p-value: 0.1576

Fig. **5.** Corresponding Analysis Herbivore Fish variable data (Biomass and Density) according to Coral Benthic Substrate type (Hard Coral, Dead Coral and Dead Coral with Algae), Regression analysis (RA) Total biomass herbivore fish according to Dead Coral Algae (DCA)

Corresponding Analysis herbivore fish variable data (Total biomass and density) according to the coral benthic substrate (DCA, DC, and HC) showed establishes repeatable relationships between Dead Coral Algae (DCA) and Biomass herbivore (Fig 5.a) as well as between density herbivore fish and DC. This both relationships showed opposite correlation. Dead Coral Algae on coral reef benthic decrease with increasing total biomass on herbivore fish family (Fig 5b). From regression analysis showed multiple R-squared from this analysis 0.2088 with Adjusted R-squared 0.1209.

References

1 Draisma, G.A. Stefano, F.W.P Reine, M.C. Sekar, Herandarudewi, and B.W. Hoeksema. Macroalgal Diversity along an Inshore-Offshore Environmental Gradient in the Jakarta Bay" Thousand Islands Reef Complex, Indonesia." [https://doi.org/10.1016/j.ecss.2017.11.010]. Estuarine, Coastal and Shelf Science.

200, 258–69 (2018)

2 Kench, S. Paul, S.D. Owen. *Coral Reef Systems and the Complexity of Hazards. Coastal and Marine Hazards, Risks, and Disasters*. [https://doi.org/10.1016/B978-0-12-396483-0.00015-7]. Elsevier Inc. (2014)

3 S. Schmidt-Roach, P. Lundgren, K..J. Miller, G. Gerlach, A.M.E. Noreen, N. Andreakis. "Assessing Hidden Species Diversity in the Coral Pocillopora Damicornis from Eastern Australia.. [https://doi.org/10.1007/s00338-012-0959-z]. Coral Reefs **321**, 161–72 (2013)

4 Marshell, Alyssa, P.J. Mumby. The Role of Surgeonfish (Acanthuridae) in Maintaining Algal Turf Biomass on Coral Reefs. [https://doi.org/10.1016/j.jembe.2015.09.002]. Journal of Experimental Marine Biology and Ecology.**473**, 152–60 (2015)

5 Kubicek, Andreas, and Hauke Reuter. Mechanics of Multiple Feedbacks in Benthic Coral Reef Communities. [https://doi.org/10.1016/j.ecolmodel.2016.02.018]. Ecological Modelling **329**, 29–40 (2016)

6 R. Sebastián, Carlos, T.R. McClanahan. Description and Validation of Production Processes in the Coral Reef Ecosystem Model CAFFEE (Coral-Algae-Fish-Fisheries Ecosystem Energetics) with a Fisheries Closure and Climatic Disturbance. [https://doi.org/10.1016/j.ecolmodel.2013.05.012]. Ecological Modelling **263,** 326–48 (2013)

7 Roberts, Callum M., Colin J. McClean, John E.N. Veron, Julie P. Hawkins, Gerald R. Allen, Don E. McAllister, Cristina G. Mittermeier. Marine Biodiversity Hotspots and Conservation Priorities for Tropical Reefs. [https://doi.org/10.1126/science.1067728] Science. **295**, 1280–84 (2002)

8 Z. Dubinsky. Coral Reef Ecosystems. Edited by Robert A. Meyers. New York: *Springer*. (1990)

9 E.L. Peñaflor, W.J. Skirving, A.E. Strong, S.F. Heron, L.T. David. Sea-Surface Temperature and Thermal Stress in the Coral Triangle over the Past Two Decades. [https://doi.org/10.1007/s00338-009-0522-8]. Coral Reefs. **28**, 841–50 (2009)

10 P.Chabanet, L. Bigot, J.B. Nicet, P. Durville, L. Massé, T. Mulochau, C. Russo, E. Tessier, D. Obura. Coral Reef Monitoring in the Iles Eparses, Mozambique Channel (2011-2013). [https://doi.org/10.1016/j.actao.2015.10.010]. Acta Oecologica. **72**, 62–71 (2016)

11 C.B Edwards, A.M Friedlander, A.G Green, M.J Hardt, E. Sala, H.P Sweatman, I.D Williams, B. Zgliczynski, S.A. Sandin, J.E. Smith. Global Assessment of the Status of Coral Reef Herbivorous Fishes : Evidence for Fishing Effects Global Assessment of the Status of Coral Reef Herbivorous Fishes : Evidence for Fishing Effects. Proc. R. Soc. B. 7–11. (2014)

12 Bruno, F. John, E.R. Selig. Regional Decline of Coral Cover in the Indo-Pacific: Timing, Extent, and Subregional Comparisons. [https://doi.org/10.1371/journal.pone.0000711.15]. PLoS ONE. **2** (2007)

13 M. Nystrom, C. Folke, F. Moberg. Coral Reef Disturbance and Resilence in a Human-Dominated Environment. [https://doi.org/doi:10.1016/S0169-5347(00)01948-0]. Trends in Ecology & Evolution. **15**, 413–17 (2000)

14 Victoria-Salazar, Isael, A. Héctor Hernández-Arana, J.A. Meave, Miguel A. Ruiz-Zárate, A. Vega-Zepeda, J.P. Carricart-Ganivet, H. López-Adame. Did the Community Structure of a Coral Reef Patch Affected by a Ship Grounding Recover after 15 Years? Merging Historical and Recent Data Sets.. [https://doi.org/10.1016/j.ocecoaman.2017.04.009]. Ocean and Coastal Management **144**, 59–70 (2017)

15 G.Fernández, C.D. Paulo, E.A. Serrão, A.H. Engelen. Limited Differences in Fish and Benthic Communities and Possible Cascading Effects inside and Outside a Protected Marine Area in Sagres (SW Portugal).. [https://doi.org/10.1016/j.marenvres.2015.12.003]. Marine Environmental Research **114**, 12–23 (2016)

16 Ramos, C.A. Colaço, R.K.P. de Kikuchi, F.D. Amaral, J.E. Fauth. A Test of Herbivory-Mediated Coral-Algae Interaction on a Brazilian Reef during a Bleaching Event. [https://doi.org/10.1016/j.jembe.2014.03.005]. Journal of Experimental Marine Biology and Ecology**. 456**, 1–7 (2014)

17 S.R Floeter. B.S. Halpern, C.E.L. Ferreira. Effects of Fishing and Protection on Brazilian Reef Fishes.. [https://doi.org/10.1016/j.biocon.2005.10.005]. Biological Conservation. **128**, 391–402 (2006)

18 Powell, Abigail, D. Pelletier, T. Jones, D. Mallet. The Impacts of Short-Term Temporal Factors on the Magnitude and Direction of Marine Protected Area Effects Detected in Reef Fish Monitoring. [https://doi.org/10.1016/j.gecco.2016.09.006]. Global Ecology and Conservation **8**, 263–76 (2016)

19 Edinger, N. Evan, J. Jompa, G.V. Limmon, W. Widjatmoko, M.J. Risk. Reef Degradation and Coral Biodiversity in Indonesia: Effects of Land-Based Pollution, Destructive Fishing Practices and Changes over Time. [https://doi.org/10.1016/S0025-326X(98)00047-2]. Marine Pollution Bulletin. **36**, 617–30 (1998)

20 Putra, M.I. Herwata, S. Afatta, J. Wilson, A. Muljadi, I. Yusidarta. Coral Reef Resilience in 17 Islands Marine Recreation Park, Riung – An Assessment of Functional Groups of Herbivorous Fish and Benthic Substrate.. [https://doi.org/10.1016/j.proenv.2015.01.035]. Procedia Environmental Sciences **23** 230–39 (2015)

21 Cleary, Daniel F.R. Linking Fish Species Traits to Environmental Conditions in the Jakarta Bay-Pulau Seribu Coral Reef System. [https://doi.org/10.1016/j.marpolbul.2017.06.054]. Marine Pollution Bulletin. **122**, 259–62 (2017)

22 Dulvy, K. Nicholas, R.P. Freckleton, N.V.C Polunin. Coral Reef Cascades and the Indirect Effects of Predator Removal by Exploitation. [https://doi.org/10.1111/j.1461-0248.2004.00593.x]. Ecology Letters **7**, 410–162004. "

23 Mora, Camilo, O. Aburto-Oropeza, A. Ayala-Bocos, P.M. Ayotte, S. Banks, A.G. Bauman, M. Beger. Global Human Footprint on the Linkage between Biodiversity and Ecosystem Functioning in Reef Fishes. [https://doi.org/10.1371/journal.pbio.1000606]. PLoS Biology **9** (2011)

24 D.F.R. Cleary, A.R.M. Polónia, W. Renema, B.W. Hoeksema, P.G. Rachello-Dolmen, R.G. Moolenbeek, A. Budiyanto. Variation in the Composition of Corals, Fishes, Sponges, Echinoderms, Ascidians, Molluscs, Foraminifera and Macroalgae across a Pronounced in-to-Offshore Environmental Gradient in the Jakarta Bay–Thousand

Islands Coral Reef Complex. [https://doi.org/10.1016/j.marpolbul.2016.04.042]. Marine Pollution Bulletin. **110,** 701–17 (2016)

25 C.A. Sánchez-Caballero, J.M. Borges-Souza, G.De La Cruz-Agüero, S.C.A. Ferse. Links between Fish Community Structure and Habitat Complexity of a Rocky Reef in the Gulf of California Threatened by Development: Implications for Mitigation Measures. [https://doi.org/10.1016/j.ocecoaman.2016.12.013]. Ocean and Coastal Management **137**, 96–106 (2017)

26 S.J. Campbell, and S. T. Pardede. Reef Fish Structure and Cascading Effects in Response to Artisanal Fishing Pressure." [https://doi.org/10.1016/j.fishres.2005.12.015]. Fisheries Research. **79**, 75–83 (2006)

27 Green, L. Alison, D.R Bellwood. Monitoring Functional Groups of Herbivorous Reef Fishes as Indicators of Coral Reef Resilience A Practical Guide for Coral Reef Managers in the Asia Pacifi c Region.. [http://cmsdata.iucn.org/downloads/resilience_herbivorous_monitoring.pdf]. Science,. **7**, 70 (2009)

28 Hughes, Terence P., Maria J. Rodrigues, David R. Bellwood, Daniela Ceccarelli, Ove Hoegh-Guldberg, Laurence McCook, Natalie Moltschaniwskyj, Morgan S. Pratchett, Robert S. Steneck, and Bette Willis. Phase Shifts, Herbivory, and the Resilience of Coral Reefs to Climate Change.. https://doi.org/10.1016/j.cub.2006.12.049. Current Biology. **17**, 360–65 (2007)

29 D.R.Bellwood, T.P. Hughes, C. Folke, M Nystrom, M. Nyström. Confrontign the Coral Reef Crisis. [https://doi.org/10.1038/nature02691]. Nature **429** (June) 827–33 (2004)

30 Russ, G R. Grazer Biomass Correlates More Strongly with Production than with Biomass of Algal Turfs on a Coral Reef. [https://doi.org/10.1007/s00338-003-0286-5]. Coral Reefs. **22**,63–67 (2003)

31 Jompa, Jamaluddin, L.J. McCook. Effects of Competition and Herbivory on Interactions between a Hard Coral and a Brown Alga. https://doi.org/10.1016/S0022-0981(02)00040-0. Journal of Experimental Marine Biology and Ecology. **271**, 25–39 (2002)

32 Miller, J. Robert, D.C. Reed, M.A. Brzezinski. Community Structure and Productivity of Subtidal Turf and Foliose Algal Assemblages.[https://doi.org/10.3354/meps08131]. Marine Ecology Progress Series. **388**, 1–11 (2009)

33 Syms, Craig, G.P. Jones. Soft Corals Exert No Direct Effects on Coral Reef Fish Assemblages. [https://doi.org/10.1007/s004420000617]. Oecologia. 127, 560–71 (2001)

34 T.R. McClanahan, B.A. Cokos, E. Sala. Algal Growth and Species Composition under Experimental Control of Herbivory, Phosphorus and Coral Abundance in Glovers Reef, Belize. [https://doi.org/10.1016/S0025-326X(02)00051-6]. Marine Pollution Bulletin. **44**, 441–51 (2002)

35 P.L. Jokiel, K.S. Rodgers, E.K. Brown, J.C. Kenyon, G. Aeby, W.R. Smith, F. Farrell. Comparison of Methods Used to Estimate Coral Cover in the Hawaiian Islands. . [https://doi.org/10.7717/peerj.954]. *PeerJ.* **3**: e954 (2015)

36 J.S. Hannak, S. Kompatscher, M. Stachowitsch, J. Herler. Snorkelling and Trampling

in Shallow-Water Fringing Reefs: Risk Assessment and Proposed Management Strategy. https://doi.org/10.1016/j.jenvman.2011.06.012. Journal of Environmental Management. **92**, 2723–33 (2011)

37 R.E. Grizzle, K.M. Ward, R.M.S. AlShihi, J.A. Burt. Current Status of Coral Reefs in the United Arab Emirates: Distribution, Extent, and Community Structure with Implications for Management. https://doi.org/10.1016/j.marpolbul.2015.10.005. Marine Pollution Bulletin. **105**, 515–23 (2016)

38 C. Sherman, W. Schmidt, R. Appeldoorn, Y. Hutchinson, H. Ruiz, M. Nemeth, I. Bejarano, J. J.Cruz Motta, and H. Xu. Sediment Dynamics and Their Potential Influence on Insular-Slope Mesophotic Coral Ecosystems. [https://doi.org/10.1016/j.csr.2016.09.012]. Continental Shelf Research. **129**, 1–9 (2016)

39 K.E. Kohler, S. M. Gill. Coral Point Count with Excel Extensions (CPCe): A Visual Basic Program for the Determination of Coral and Substrate Coverage Using Random Point Count Methodology. [https://doi.org/10.1016/j.cageo.2005.11.009]. Computers and Geosciences. **32**, 1259–69 (2006)

40 N.V.C. Polunin, D.W. Klumpp. Ecological Correlates of Foraging Periodicity in Herbivorous Reef Fishes of the Coral Sea. [https://doi.org/10.1016/0022-0981(89)90121-4]. Journal of Experimental Marine Biology and Ecology **126**, 1–20 (1989)

41 S.M. Lewis, P.C. Wainwright. Herbivore Abundance and Grazing Intensity on a Caribbean Coral Reef. https://doi.org/10.1016/0022-0981(85)90206-0. Journal of Experimental Marine Biology and Ecology. **87**, 215–28 (1985)

42 Rotjan, Randi D., and Sara M. Lewis. Parrotfish Abundance and Selective Corallivory on a Belizean Coral Reef. [https://doi.org/10.1016/j.jembe.2006.03.015]. Journal of Experimental Marine Biology and Ecology. **335**, 292–301 (2006)

43 Wen, C.K. Chang, K.S.Chen, H.J. Hsieh, C.M. Hsu, C.A. Chen. High Coral Cover and Subsequent High Fish Richness on Mature Breakwaters in Taiwan. [https://doi.org/10.1016/j.marpolbul.2013.04.031]. Marine Pollution Bulletin **72**, 55–63 (2013)

44 Pinault, Mathieu, Claire Bissery, Gilles Gassiole, Hélène Magalon, Jean Pascal Quod, and René Galzin. Fish Community Structure in Relation to Environmental Variation in Coastal Volcanic Habitats. [https://doi.org/10.1016/j.jembe.2014.06.005]. Journal of Experimental Marine Biology and Ecology. **460**, 62–71 (2014)

45 Williams, D.McB. Patterns.Processes in the Distribution of Coral Reef Fishes. The Ecology of Fishes on Coral Reefs. [https://doi.org/10.1016/B978-0-08-092551-6.50021-0]. *Academic Press, Inc* (1991)

46 N. Jayaprabha, S. Purusothaman, M. Srinivasan. Biodiversity of Coral Reef Associated Fishes along Southeast Coast of India. [https://doi.org/10.1016/j.rsma.2017.12.010]. Regional Studies in Marine Science. **18**, 97–105 (2018)

47 EK. Eastwood, D.G. Clary, D.J. Melnick. Coral Reef Health and Management on the Verge of a Tourism Boom: A Case Study from Miches, Dominican Republic." [https://doi.org/10.1016/j.ocecoaman.2017.01.023]. Ocean and Coastal Management **138**, 192–204 (2017)

48 Lortie, J. Christopher, J. Lau, M.J Lajeunesse. Graphical Presentation of Results. https://doi.org/10.1515/9781400846184-023. Handbook of Meta-Analysis in Ecology and Evolution. 339–47 (2013)

49 B.L. Gilby, T. Stevens. Meta-Analysis Indicates Habitat-Specific Alterations to Primary Producer and Herbivore Communities in Marine Protected Areas. [https://doi.org/10.1016/j.gecco.2014.10.005]. Global Ecology and Conservation. **2**, 289–99 (2014)

50 Burgess, Scott C., Kate Osborne, and M. Julian Caley. Similar Regional Effects among Local Habitats on the Structure of Tropical Reef Fish and Coral Communities. [https://doi.org/10.1111/j.1466-8238.2009.00516.x.]. Global Ecology and Biogeography. **19**, 363–75 (2010)

51 X. Jiang, W.X. Lu, H.Q. Zhao, Q.C. Yang, Z. P. Yang. Potential Ecological Risk Assessment and Prediction of Soil Heavy-Metal Pollution around Coal Gangue Dump. [https://doi.org/10.5194/nhess-14-1599-2014]. Natural Hazards and Earth System Sciences. **14**, 1599–1610 (2014)

52 Y. Sagawa, S. Armand, A. Lubbeke, P. Hoffmeyer, D. Fritschy, D. Suva, K. Turcot. Associations between Gait and Clinical Parameters in Patients with Severe Knee Osteoarthritis: A Multiple Correspondence Analysis. Clinical Biomechanics. **28** (3): 299–305. https://doi.org/10.1016/j.clinbiomech.2013.01.008 (2013)

53 J.O. Bell, R. Galzin. Influence of Live Coral Cover on Coral-Reef Fish Communities. [https://doi.org/10.3354/meps015265]. Marine Ecology Progress Series **15**, 265–74 (1984)

Potential *Harmful Algal Blooms* (HABs) in Segara Anakan Lagoon, Central Java, Indonesia

Rose Dewi[1, 3*], *Muhammad* Zainuri[2,4], *Sutrisno* Anggoro[1], *Tjahjo* Winanto[3], and *Hadi* Endrawati[5]

[1]Coastal Resources Management Doctoral Program, Fisheries and Marine Science Department, Diponegoro University.
[2]Departement of Oceanography, Fisheries and Marine Science Department, Diponegoro University.
[3]Fisheries and Marine Science Department, Jenderal Soedirman University.
[4] Centre of Excellent Mitigation of Natural Disaster and Coastal Rehabilitation, Diponegoro University.
[5]Fisheries and Marine Science Department,Diponegoro University.

Abstract. Segara Anakan Lagoon (SAL), located along side of soutern coast on western part of Central Java 108°46'–109° 05'E; 7° 34'–7° 48'S. SAL is necessasrily ecosistem as nursery ground. This has been becoming important research relating to primary productivity. The existance of antropogenical activities around the area, is changing inrush input into lagoon. This is going to be worried about giving influences in physical, chemistry, and biological factors of the water and causing deflation in rate value of primer productivity. Rainy and dry season's variability with important impacts on the phytoplankton community structure, abundance and dynamics. Furthermore, the number of coastal ecosystems with identified eutrophication symptoms is increasing worldwide due to the increasing anthropogenic pressures. One of [1]the first symptoms of eutrophication is enhanced phytoplankton biomass. Generally, Fitoplanton is important biological indicators in the process. However, high abundance of biomass phytoplankton could give harmful effect toward lagoon, they could produce toxic substances that will be accumulated, it can be endangering .The aim of the research is to determine and calculate potential species of *Harmful Algal Blooms* (HABs) in SAL with spastial and temporal approach. The spatial approach is done at seven different stations of ecological characteristics SAL, with representations: natural factors and the presence of anthropogenic activities. On the temporal approach (*time series*) for a year, it refers to the *monsoon* wind pattern, which are called season

*Corresponding author : +6281327929555 | rose.83unsoed@gmail.com

(western, the transitional season 1, eastern, and the transitional season 2).The results of laboratory tests are discussed descriptively. Phytoplankton sample were taken vertically using plankton nets, meshsize 25 μm. The result showed that SAL have been found and consisted of 82 species from 5 division phytoplankton: *Chrysophyta, Chlorophyta, Cyanophyta, Pyrophyta, and Euglenophyta*. 20 species from 9 genus were identified potential as HABs, with four genus of them were known having toxin and endangering for human (*Nitzschia, Oscillatoria, Anabaena* dan *Protoperidinium*).Whether, five genus do not produce toxins but giving deflation Oxygen of waters with anoxia condition (*Chaetoceros, Coscinodiscus, Rhizosolenia, Thalassiosira, Thalassiotrix*).The highest of HABS abundance is dominated during the transitional season II. On the spatial approach, antropoegenic characteristics station contribute to the whole height of HABs. Influence of rainfall, spatially, antropogenical pressure, and hydrodinamical watres causing enhancement of HABs potency.

1 INTRODUCTION

Geographically, Segara Anakan Lagoon (SAL), locating on the ordinate of 7°35"- 7°46" S and 108°45" - 109°01" E, covering 14.221,8 ha, in the southern coastal area in Java island, at Kampung Laut Sub-district, Cilacap Regency. The area is accounted from a bay, the mouth of the lagoon facing to east, there is Nusa Kambangan Island as a barrier, separating the SAL from the Indian Ocean. The existence of shape changing up to now, It has occured through various level in environmental changing. SAL is an area that does not stand alone, but there are affecting various factors in ecological changes of the region [4, 7].

Lagoon is greatly affected, freshwater flow entering in high quantity with suspended sediment, mainly from Citanduy river [15]. S.Citanduy has contribution for catchment area of 3.520 km2 or supplying for 80% of the water flow to the SAL [14]. The river flow brings the supply with organic materials continuously, so it is to trigger the consolidating process of coastal sediments, as a source of nutrients and forming the physical structure of sediments. Condition causing most of the volume of water bodies from SAL, is replaced with mangrove forests, then it converted its use into various anthropogenic activities [4, 9].

Supported by continuous aquatic dynamics, this is causing physical contact complexity, aquatic chemistry as well as microbiology triggering for the availability of accumulation of concentration of macronutrient elements. The accumulation of sedimentation mainly occurs in the western part of the lagoon, while the eastern part of the lagoon is dominated by hydrodynamics of the Indian Ocean [16].

The annually high sedimentation in the lagoon, results in macronutrient accumulation, especially Nitrogen (N) and Phosphate (P), being glued on sediment particles, as the process of natural response. If the condition occurs continuously, it is worried of causing the phenomenon of "eutrophication". The initial impacted biota indicator of aquatic eutrophication is phytoplankton, indicated by adaptation or tolerating the condition in water instability [18, 3, 34]. High consentration of N and P affecting in the content of chlorophyll-a contained in phytoplankton will further impact on the abundance of phytoplankton [10, 16].

It is further explained that in macronutrient utilization process, there are two categories of phytoplankton namely: productive phytoplankton (non-toxic), functioning as natural food or as primary producer in food chain, so it can show the level of water productivity. Condition reflecting in the high primary productivity of waters depends so much on the ability of waters to synthesize inorganic materials into organic materials through photosynthesis by phytoplankton as primary producers [25, 18]. On the other hand, there is a type of phytoplankton being capable of producing toxin (HABs) *Harmful Algal Blooms*. Generically the term HABs refers to high levels of phytoplankton growth with toxin causing anoxia, ie decreased oxygen loss in water body, mass mortality of fish and resulting in negative biomagnification processes in the food chain. So it is perceived as a harmful [22].

The influence of natural factors (input of freshwater flow, sedimentation) and anthropogenic activity of land use change, supported by fluctuation of environmental change of monsoon. Ecological conditions are worried of supporting the occurrence of eutrophication phenomena that affect the composition and abundance (*blooming*) phytoplankton, especially potentially known as HABs, affecting the aquatic organisms and decreased dissolved oxygen content (anoxia) and decreased the average value of primary productivity. This makes identification of the type of HABs phytoplankton in the SAL area, becomes very important for further study, especially in lagoon waters as nursery ground areas.

2 MATERIAL and METHOD

The analysis of the composition and abundance of phytoplanton was done in-situ temporally (time series) for a year, based on the representation of monsoon winds [22], that based on Meteorological and Oceanographic data, Indonesian waters are affected by the monsoon wind, consisting of:

- **M1 : West Season** : December – February (rainy season, February)
- **M2 : Transition Season 1**: March – May (rainy to dry season, May).
- **M3 : East Season** : June – August (dry season, July)
- **M4 : Transition Season II**: September-November (dry to rainy season, Oktober).
- **M5 : West Season** : December – February (rainy season, December).

The reference is used as an in-situ retrieval time of data of 7 times spatial repetition with different ecological characteristics in the SAL region. It is aimed to evaluate the ecological influencing (response) effect of natural factor impacts, at stations (2, 3, 4, 6) and affected anthropogenical activities of the community around the lagoon at the station (1, 5, 7). Further characteristics of the spatial approach are presented in (Table 1).

Tabel 1. The Characteristic of Spatial approach on sampling of Phytoplankton at the area of Segara Anakan, Cilacap Sub-district.

	Characteristics	Ordinate	
S1	TPI dan Majingklak port	108°48'02.4"BT	07°40'27.6"LS
S2	The border of gate (west plawangan) of SAL directly bordering with Indian Ocean	108°46'56.7"BT	07°41'59.0"LS
S3	The mouth of Cikonde and Cimeneng river	108°49'47.9"BT	07°40'34.6"LS
S4	The area of Mangrove	108°51'36.5"BT	07°41'44.9"LS
S5	Kampung Laut settlement	108°52'14.0"BT	07°42'19.5"LS
S6	Conservative tourism "Mina" "Ujung Alang, district of Kampung Laut	108°52'45.4"BT	07°42'55.4"LS
S7	Cultivating place by using floating net	108°48'56.0"BT	07°41'01.0"LS

- Phytoplankon sampling was done vertically, using plankton-net of 25 μm by filtering 100 liters of water, on the surface of lagoon waters with a depth of 0.5 to 1 m [1].
- Preservation of samples by using formalin up to 4% [21], as many as 10 drops. Identification was using a binocular light microscope with magnification of 40 times. The calculation is done in 20 times, the field of view and 3 times of repetition
- Stages of identification of productive and HaBs phytoplankton is classified by [20, 28, 33] and classification of phytoplankton type of HABs refers to [27, 29, 30, 32]. Calculation of phytoplankton abundance with the formula based on Lackey Drop Microtransect Counting method [1] :

$$K = F \times N \quad (1)$$

with

$$F = \frac{Q_1}{Q_2} \times \frac{V_1}{V_2} \times \frac{1}{P} \times \frac{1}{W} \quad (2)$$

Description :
K : Abundance of Phyoplankton (indv/liter)
N : Total of phytoplankton
F : Total of plankton individu /Liter
Q_1 : Weidth of covering glass 18x 18 (324mm^2)
Q_2 : Weidth of visibility (1,11279 mm^2)
V_1 : Water volume in container bottle (100 ml)
V_2 : Water volume under container bottle (0,05ml)
P : Total of observed Visibility (20 times)

 https://doi.org/10.1051/e3sconf/20184704010

W : Netted water volume (100 liter)

- Measurement of water quality parameters of dissolved oxygen content (DO) in situ, is to support the study, carried out by using *Water Quality Checker* (WQC) (Merk Toa DKK-Japan : Type WQC-22A), by inserting (dipping) the device on the surface of the water, recording after the tool shows a constant number.

3 RESULT AND DISCUSSION

Phytoplankton was obtained from the identification in 82 species of Segara Anakan from 5 (five) divisions: 54 species of Chrysophyta, 13 species of Chlorophyta, 8 species of Cyanophyta, 3 species of Pyrophyta, and 4 species of Euglenophyta are spatially distributed. These types of phytoplankton are from the Chrysophyta Division, Bacillariophceae class (diatoms) dominating the entire station and season. Diatoms are an important species (adaptive) in phytoplankton populations and are key to identification of plankton [6]. Phytoplankton Bacillariophyceae can produce cell division quickly, with time intervals of every four hours. In the increase of macronutrient nutrient concentration, diatoms are able to reproduce three times in 24 hours [5, 11, 28].

There are variations of the phytoplankton of freshwater, estuarine and marine waters identified during the spatio-temporal approach. According to [32] is suspected due to the hydrodynamic factors of the waters: the direction and velocity of currents, which affect the composition and abundance of plankton. These factors play a role in the process of migration and the spread of plankton, the passive organism of its movement.

Decent potential phytoplankton as (HABs) is 20 species (9 genera) of all identified phytoplankton. The variability of SAL phytoplankton showed there are distributed HABs species and in spatial (stationary) or seasonal (temporal) periods during the study. The results of the research that at waters with semi-covered system as lagoon, will be identifying species with the type of HABs [8].

Table.2 The measurement of Phytoplankton abundance and dissolved oxygen content at the area of Segara Anakan Lagoon Cilacap Sub-district

The Abundance of Phytoplankton (indv/ L)				Dissolved Oxygen (mg/ L)
Spatial Approach (S)		Temporal Approach (M)		
S1	8355,60 ± 3049,46	M1	6873,43 ± 2296,90	2,46 ± 1.09
S2	4474,80 ± 3408,25	M2	9956,57 ± 5300,23	3,23 ± 2.01
S3	6058,80 ± 1860,57	M3	6618,86 ± 3650,68	2,03 ± 1.43
S4	8712,00 ± 6444,89	M4	5685,43 ± 2075,74	2,90 ± 0.74
S5	7286,40 ± 4137,68	M5	4129,71 ± 3139,88	5,40 ± 2.45
S6	5940,00 ± 4356,00			
S7	5742,00 ± 1945,04			

In the nine genera of HABs identified, there are four toxic genera and make human sick of, such as *Nitzschia, Oscillatoria, Anabaena* and *Protoperidinium*. While there are non-toxic species that are included to damaging phytoplankton in ecosystem, obstructing fish's gill function with its sharpened and chained morphology. Moreover, some of them create insuffenciency of oxygen in waters.

Types of phytoplankton are: *Chaetoceros, Coscinodiscus, Rhizosolenia, Thalassiosira, Thalassiotrix*

The results of the other research stated that there are phytoplankton generally, being able to cause HABs such as *Chaetoceros, Pseudo-Nitzschia, Trichodesmium erytrhraeum (Oscillatoria*) and *Anabaena* [15]. Types of *Pseudo-Nitzschia* multiseries diatoms and *Anabaena* are included to phytoplankton that can produce poisoned. The types of toxic phytoplankton are such as *Nitzschia, Protoperidinium, Anabaena, Mycrocystis, Oscillatoria, Alexandrium* and *Dinophisis*, there are also non-toxic species that can have a detrimental effect in the waters of *Bacteriastrum, Chaetoceros, Coscinodiscus, Skeletonema, Ceratium, Rhizosolenia, Thalassiosira, Thalassiotrix* and *Noctiluca* [2].

The types of phytoplankton are a type of tropical phytoplankton, so they likely can be found in Indonesian waters [28]. Several occurrences of HABs in Indonesian waters are including found in July 1996, there was a blooming type *Oscillatoria* in Ambon Bay [30]. Lastest event which is blooming is *Coscinodiscus*, sp. causing mass mortality in the Bay of Jakarta (Ancol Waters) at the end of November 2015.

The group of potential HABs phytoplankton in the SAL come together and the types of *Chaetoceros*, sp., *Rhizosolenia* and *Coscinodiscus*, sp. can be found mostly in both coastal and estuary waters [5]. Although, those three types of HABs phytoplankton are a type of diatom plankton living in marine waters [33]. However, it is suspected that the hydrodynamic factors in the waters producing the types of marine phytoplankton (ocean of Indonesia) could be distributed into the SAL.

Table. 2 The Type and Abundance of potential Phytoplankton as HABs (indv/ L) in the area of Segara Anakan lagoon, regency of Cilacap.

No.	Types of Phytoplankton	West Season (February) (M1)	Transitional season I (May) (M2)	East Season (July) (M3)	Transitional season II (October) (M4)	West season (December) (M5)
1	*Anabaena hallensis*	0	0	0	198-3960	0-198
2	*Oscillatoria princeps*	198-792	990-2376	396-3564	594-1584	0-792
3	*Chaetoceros,* sp.	0 – 198	0	0	0	0
4	*Coscinodiscus linetus*	198-594	0	0-594	396-990	0-198
5	*Nitzschia acicularis*	0	594-1188	198-594	0	0-396
6	*Nitzschia actinastroides*	198-396	0	0	0	0
7	*Nitzschia apectabilis*	0 – 198	0	0	0	0
8	*Nitzschia closteria*	0	0	0	0-594	0
9	*Nitzschia longissima*	0	0	0	396-1188	0

10	*Nitzschia nyassensis*	0 – 198	0	0	0	0-198
11	*Nitzschia philiphinarum*	198-396	0	0	0	0
12	*Nitzschia ricta*	0	198-990	0	0	0
13	*Nitzschia subrostrata*	0 – 198	0	0	0	0
14	*Nitzschia vermicularis*	198-396	1782-3366	198-594	0 – 198	0
15	*Nitzschia zigma*	396 – 594	0	0	0	0
16	*Rhizosolenia,* sp.	0	0	0	0 – 198	0
17	*Rhizosolenia robusta*	0	0	0	0	0-594
18	*Thalassiosira rotula*	0 – 198	0	0	0	0
19	*Thalassiotrix frauenfeldii*	0 – 198	0	0	0	0
20	*Peridinium,* sp.	198-792	0-594	0	0	198-396
	Jumlah Spesies	**13**	**5**	**4**	**7**	**7**

Adaptation by *Coscinodiscus*, sp. is easily found in the whole research station (S) with morphologically forming large round. In other hand, *Thalassiosira* and *Thalassiotrix* have good adaptation in the environment, colonizing with long arranged chain. The type of *Chaetoceros*, sp. has a very quick response to added nutrients in the waters and potentially create the dominant group in the nutrification competition [17]. The result in the identification, which is obtained in 11 species of *Nitzschia* in SAL region is one of diatom that can produce toxin beside *Chaetoceros*, sp [33].

In spatial approach, the types of high phytoplankton (HABs) are existing in stations with anthropogenic characteristics, with much more dominated by phytoplankton being capable of adapting into environmental conditions with the highest organic material, with containing of 20 species of HABs (Table 2). According to [27] that 50% of (N) inorganic dissolved nitrogen derive from agricultural activities, while 70% of total phosphates come from a variety of domestic activities. The condition resulted in phosphate concentration (P) larger in lagoon than open sea [36].

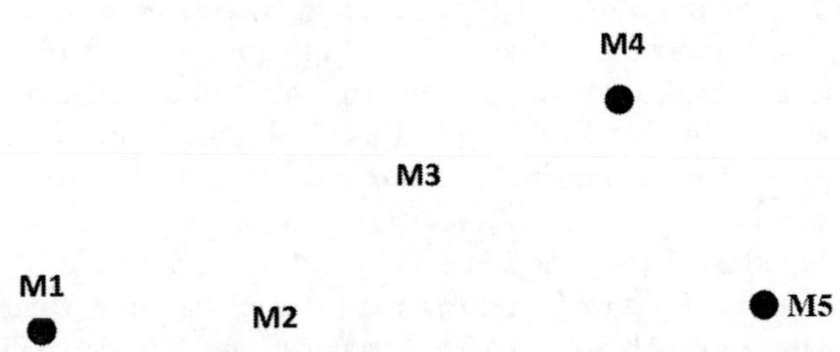

 https://doi.org/10.1051/e3sconf/20184704010

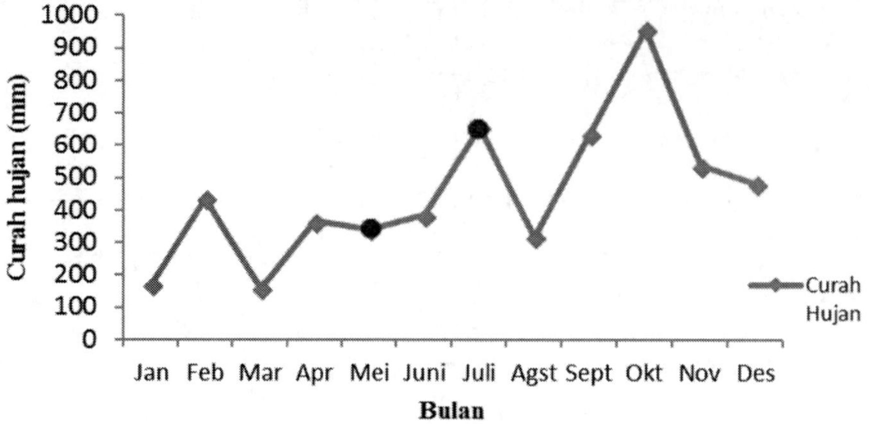

Fig 1. The averaged rainfall in regency of Cilacap during 2016.

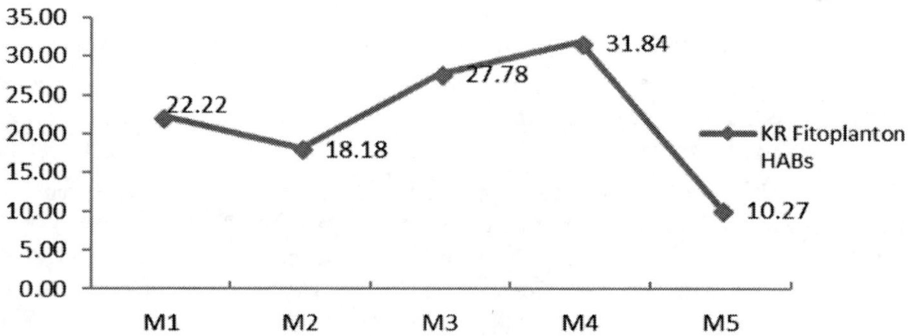

Fig 2. Relative abundance of (%KR) potential Phytoplankton as HABs pada at the area of lagoon of Segara anakan Kabupaten Cilacap.

Description : **(M1)** : West Season (February); **(M2)** : Transitional season I (May); **(M3)** : east season (July); **(M4)** : Transitional Season II (October); **(M5)** : west season (December).

In the temporal approach, season (M1) with abundance of (6873,43 ± 2296,90) is a potential season having the highest composition of phytoplankton HABs (Table 2). Allegedly this season is a transitional season between the waiting phase until the mechanism of flushing time, ie macronutrient flushing into the sea waters (freshwater change) and is the transition from west season to come into the transition I. Flushing time, is a time required by waters, to have freshwater rinsing in the lagoon, is mainly influenced by high rainfall and river flow [12]. So, it resulted in the abundance of distributed phytoplankton into the waters of the sea (Indian Ocean). In the subsequent season, flushing time is coming along with increased rainfall is accompanied by the increase in the relative abundance (KR) of potential phytoplankton as HABs. It is shown from the calculation of % KR phytoplankton HABs (Figures 2 and 3).

In season (M3) with 658 mm rainfall, which is supporting the abundance of phytoplankton of (6618,86 ± 3650,68) indv / L, has KR HABs potential of 27.78% of total phytoplankton. Furthermore, the increased rainfall reached 968 mm. at

(M4), with phytoplankton abundance only at the range of (5685.43 ± 2075.74) invd / L, but it supports the highest KR HABs of 31.84% of total phytoplankton. Furthermore, the decreased rainfall reaching 483 mm, went with the decrease of phytoplankton abundance to (4129,71 ± 3139,88) indv / L and KR HABs was only about 10,27%.

The result of the research showed that the high potency of phytoplankton HABs in the SAL region, occurred in the rainy season. Eutrophication is possible to happen in the rainy season and the rainfall affected runoffs subsequently give contribution to the spatial-temporal distribution patterns of types of phytoplankton, especially HABs in tropics [24, 25]. In other hand, in the dry season, there is an increase in sea water temperature producing rising nutrients from the bottom to the water column above it. The main source of eutrophication phenomena comes from wastes on the land, supported by seasonal changes, resulting in blooming of phytoplankton

The dynamics of phytoplankton abundance are affected by the ecological conditions of the waters, and the community of it, is controlled by dominanting species, showing the strength of the species against other species (*blooming*). These conditions bring important change to the ecological community [23]. But so far, there is no clear threshold regarding to quantity of phytoplankton at the time of the blooming. Further that the condition of phytoplankton blooming, can occur when the quantity of abundance can be much more than the averaged quantity of phytoplankton per month [19].

The excessive macronutrient accumulation in waters (eutrophication), derived from wastes on the land and seasonal changes leading to blooming of phytoplankton [24]. This is due to phytoplankton being able to show the tolerance of aquatic instability, by responding quickly to the entrance of pollutants in aquatic bodies [8, 15]. Explosive population of blooming of phytoplankton can happen in non-toxic (productive) phytoplankon as well as on phytoplankton species which are capable of producing toxin HABs [22].

The result of measurement of dissolved oxygen content (DO) with a temporal approach (time series) is ranged from 2.03 to 5.40 mg / L with an annual average of 3.20 mg / L (Table 2). The low DO at (M4) is due to the relative abundance of HABs of (31.84%). The use of macronutrients by potentially toxin-containing phytoplankton (HABs), can lead to anoxia, ie the decreased and even dissolved oxygen (DO) loss in water bodies, thus, negatively impacting degradation of waters. Another impact from phytoplankton blooming is also shown after mass death, which will be described by bacteria, this process requires high oxygen consumption, resulting in dissolved oxygen concentrations depleted, resulting in conditions (anoxia) or oxygen depletion in waters [19, 22]

4 Conclusion

What to be paying attentiont for the relative abundance (KR) of HABs in the Area of Segara Anakan Lagoon area during the study period, averaged 22.28% of the total phytoplankton abundance, with proportions: (M1) 22.22% (13 species); (M2) 18.18% (5 species); (M3): 27.78% (4 species); (M4): 31.84% (7 species); (M5) 10.27% (7 species) of the total identified phytoplankton each season. Patterns show with increasing rainfall will be accompanied by increased relative abundance of phytoplankton potentially as HABs, and vice versa. Spatially, stations with anthropogenic activity characteristics also support high phytoplankton potential HABs. This indicates that the whole research station contains many organic materials that can be transformed into nutrients, which are then fully utilized by potentially HABs as phytoplankton. However, it should be noted that the dominance of phytoplankton group capable of adaptation to environmental conditions with the content high of organic material macronutrients, will contribute to the occurrence of eutrophication phenomenon

References

1. American Public Healt Assosciatin (APHA). Standart methods for the examination of water and waste water. 21th Edition. Edited by: Andrew. D Eaton, Lenore.S Clesceri, Eugene.W Rice, Arnold. E Greenberg. Centennial Edition (2005).
2. Aryawati, R., D.G Bengen, T. Prartomo, H. Zulkifli. Harmful Algal in Banyuasin Coastal Water, South Sumatra. Biosaintifika. *Journal of Biology and Biology Education*. UNNES. **8**(2) (2016).
3. Ayuningsih, M.S, I.B. Hendrart and P.W. Purnomo. Distribution and Abundance of Phytoplankton and Chorophyll-a in the Sekumbu Bay JeparaRegency: Relationship with Nitrate and Phosphate Content in Water. *Diponegoro Journal Of Maquares. Manajement Of Aquatic Resources*. **3**(2): 138-147 (2014).
4. Ardli, E.R., and M. Wolff. Land use and land cover change affecting habitat distribution at Segara Anakan lagoon, Cilacap, Indonesia. *Reg. Environmental Change*. 9: 235-243 (2009[b]).
5. Arinardi, O.H., A.B. Sutomo., S.A Yusuf., Trimaningsih, Asnaryanti, dan Riyono S.H. Kisaran Kelimpahan dan Komposisi Plankton Predominan di Perairan Kawasan Timur Indonesia. LIPI - Pusat Penelitian Oseanografi, Jakarta. 140 hlm (1997).
6. Basmi, J. Planktonologi: Bioekologi Plankton Algae. Fakultas Perikanan dan Ilmu Kelautan, Institut Pertanian Bogor (1999[a]).
7. Carolita, I, E. Parwati, B. Trisakti, T. Kartika dan G.Nugroho. Pendekatan prediksi perubahan lingkungan di Kawasan Perairan Segara Anakan. *Pertemuan Ilmiah Tahunan MAPIN XIV*. Pemanfaatan efektif penginderaan jauh untuk peningktan kesejahteraan bangsa. Institut Teknologi Sepuluh Nopember, Surabaya (2005).
8. Coutinho, M.T., A. C. Brito., P. Pereira., A.S. Gonçalves and M.T Moita. A phytoplankton tool for water quality assessment in semi-enclosed coastal lagoons: Open vs closed regimes. *Journal Estuarine, Coastal and Shelf Science*. 110: 134-146 (2012).
9. Dewi, R., M. Zainuri., S. Anggoro., dan T. Winanto. Analisis Perubahan Lahan Kawasan Laguna Segara Anakan Selama Periode Waktu (1978 – 2016) Menggunakan Satelit Landsat Multitemporal. Jurnal Omni Akuatika. 12 (3) : 144-14, 2016. Research Article. ISSN: 1858-3873 print/ 2476-9347 online (2016).

10. Dewi, R., M. Zainuri., S. Anggoro., T. Winanto.,and Endrawati, H. Characteristic and The distributtion of Spasio – Temporal Macronutrient in the Lagoon area of Segara Anakan. International Journal of Marine and Aquatic Resource Conservation and Co-existence. Research Article 2 (1): 51-57, October 2017. http://ejournal.undip.ac.id/index.php/ijmarcc (2017[a]).

11. Dewi, R., M. Zainuri., S. Anggoro., dan T. Winanto. Analisis Komposisi dan Kelimpahan Fitoplankton Secara Spasio-Temporal Kawasan Laguna Segara Anakan. Prosiding Seminar Nasional Tahunan XIV. Hasil Penelitian Perikanan dan Kelautan, 22 Juli 2017. Jilid II MSP. Departemen Perikanan, Fakultas Pertanian, UGM (2017[b]).

12. Dyer, K.R. Estuaries: A physical Introduction (2nd Edition), John Wiley and Sons, New York, 195 pp (2000).

13. Hallegraeff,G,M. Harmfull Algal Blooms : A Global Overview. In Hallegraeff,G,N, D.M Anderson, A.D Cambella, Editor. Manual on Harmfull Marine Microalgae. Franch: UNESCO Publishing.p. 1-22 (1995[a]).

14. Holterman P, Burchard H, and Jennerjahn T.C. *Hydrodynamic of The Segara Anakan lagoon.* Regional Environmental Change **9** (4), 245 – 258 (2009).

15. Jennerjahn, T.C.,_Holtermann., I. Pohlenga and B. Nasir. Environmental Conditions in The Segara Anakan Lagoon, Java, Indonesia. Synopsis of Ecological and Socio- Economic Aspect of Tropical Coastal Ecosystem With Special Reference to Segara Anakan (2007[a]).

16. Jennerjahn, T.C., B. Nasir, and I. Pohlenga. Spatio-temporal variation of dissolved inorganic nutrients related to hydrodynamics and land use in the mangrove-fringed Segara Anakan Lagoon, Java, Indonesia. *Reg Environ Change* 9:259–274 (2009).

17. Lagus, A., J. Suomela., G. Weithoff., K. Heikkila., H. Helminen., dan J. Sipura. Species- specific Differences in Phytoplankton Responses to N and P Enrichment and The N : P ratio in The Archipelago Sea, Northern Baltic Sea. J. Plankt. Res. 26 : 779-798. Doi : 10.1093/plankt/fbh070 (2004).

18. Livingston, R.J. Eutrophication Processes in Coastal Systems. Origin and Succession of Plankton Blooms and Effect on Secondary Production in Gulf Coast Estuaries.. CRC Press LLC. Marine Science Series Boca Raton. Printed in the United State of America. ISBN. 0-8493-9062-1. Pp 327 (2001).

19. Mulyasari, R. Peranginangin, Th. D. Suryaningrum, dan A. Sari. Penelitian Mengenai Keberadaan Biotoksin pada Biota dan Lingkungan Perairan Teluk Jakarta. *Jurnal Penelitian Perikanan Indonesia*. **9(5)**: 39-64 (2003).

20. Newell, G.E and R.C Newell. Marine Plankton, A Practical Guide. Hutchinson of London. 244 hal (1993).

21. Nontji, A. Tiada Kehidupan di Bumi Tanpa Keberadaan Plankton. LIPI-Pusat Penelitian Oseanografi, Jakarta (2006).

22. Nontji, A. Plankton Laut. Jakarta: Lembaga Ilmu Pengetahuan Indonesia. LIPI Press. Jakarta. ISBN: 978-979-799-085-5. 331 Hlm (2008).

23. Odum, E.P. Fundamental of Ecology. (3rd edition).W.B. Saunders Co (1971[a]).

24. Panggabean, L.M.G. "Red Tide" di Indonesia.: Perlukah Diwaspadai? *Oseana*. **19**(1):33-38 (1994).
25. Parslow, J.J. Hunter and A. Davidson. Estuarine Eutrophication Model. Final Report Project E6 National River Health Program Water Services Association of Australian Melbourne Australian. CSIRO Marine Research. Hobarth, Tasmania (2008).
26. Pednekar, S.M, S.G.P, Matondkar and V. Kerkar. Spatio Temporal Distribution of Harmfull Alga Florain The Tropical Estuarine Complex of Goa, India. The Scientific World. 2012: 11. P.doi : 10.1100/2012/592676 (2012).
27. Pintado, J. G., M.M. Mena, G.G. Barbéra., J. Albaladejo and V.M. Castillo. Anthropogenic Nutrient Sources and Loads From a Mediterranean Catchment into a Coastal Lagoon: Mar Menor, Spain. Science of the Enviroment. 373:220-239 (2007).
28. Praseno, D.P dan Sugestiningsih. *Red tide di Perairan Indonesia*. Pusat Penelitian dan Pengembangan Oseanografi-LIPI. Jakarta. 82 hlm. 2-34 (2000).
29. Sachlan, M. Planktonologi. Fakultas Perikanan dan Ilmu Kelautan Universitas Diponegoro (1982).
30. Sidabutar, T. Red tide phenomenon which can result in losses to the fisheries sector. Proceedings of the Annual National Seminar II Research and Marine Fisheries. Department of Fisheries and Marine Resources, Faculty of Agriculture, University of Gajah Mada, 114-126 (2006[a]).
31. Sidharta, B. R. The Current Status of Research on Harmfull Algal Bloom (HAB) in Indonesia. *Jurnal of Coastal Development*. Februari 2005, **8**(2): 75-88. ISSN: 1410-5217 (2005).
32. Suryanti. Kajian Tingkat Saprobitas di Muara Sungai Morodemak pada saat Pasang dan Surut. Universitas Diponegoro. Semarang. *Jurnal Saintek Perikanan*. **4**(1)2: 76-83, (2008).
33. Sze, P. A Biology of The Algae. Second Edition. Wm. C. Brown Publishers. ISBN 0-697-01373-1. 259 pp (1993).
34. Tomas, C,R,. *Identifiying Marine Phytoplankton*. Academic Press, United States Of America.ISBN 0-12-693018-X(Pbk:Alk.paper) (1997).
35. Vadrucci, M.R., E. Stanca., C. Mazzioti, N. Ungaro and A. Basset. Ability of Phytoplankton trait sensitivity to tighlight antropogenic pressures in Mediteranean lagoons:A size spectra sensitivity index (ISS-phyto). Ecological Indicators. 34: 113-125 (2013).
36. Zaaboub, N., A. Ounis., M. A Helali., B. Béjaoui., A.I. Lillebø., E.F. Silva and L. Aleya. Phosphorus Speciation in sediments and assessment of nutrient exchange at the water-sediment interface in a Mediterranean lagoon: Implications for management and restoration. *Journal Ecological Engineering*. 73: 115-125 (2014).

TRANSMISSION LOSS ESTIMATION OF UNDERWATER SOUND BASED ON THE NOISE INTENSITY EMMITED BY MV. PENGAYOMAN IV IN TANJUNG INTAN CRUISE LINE, CILACAP

Lesa Triwahyanti[1], *Ajeng* Sekar Cyndana[1], *Yessy* Hurly Sefnianti[1], *Ratna* Juita Sari[1], Amron *Amron*[1*]

[1] Marine Science Department, Jenderal Soedirman University, Kampus Unsoed Karangwangkal, Purwokerto 53122 Indonesia

Abstract The water is an imperfect acoustic medium, acoustic energy can lost due to absorption and scattering thus resulting in transmission loss. The value of transmission loss in water is an important thing to know, for example in the application of active acoustic. This study aim is to determine the value of transmission loss in Tanjung Intan's cruise line, Cilacap. The acquisition of vessel's sound data was obtained by hydrophone and vessel data (distance) by CCTV. Both sound data and vessel data were connected to personal computer to record the audio and visual signal synchronously for three consecutive days. Results showed that vessel characteristics such as intensity, frequency and interval duration have a value of 43.35 (dB), 6613.54 (Hz) and 216.15 (ms) respectively. While the value of water absorption coefficient has a range of 500-900 dB/km, the estimated transmission loss value was between 5-11 dB/km.

Corresponding author : twoone_brond@scientist..com

1 INTRODUCTION

Indonesia is the biggest archipelago in the world which is largely surrounded by oceans. This allows the development of technology to optimize the potential of Indonesia in the field of shipping, fisheries, military, and another matters related to Indonesia's potential in maritime field. One of them is the development of underwater communication. The ability to communicate effectively in underwater is important for marine research, oceanographers, commercial marine operators, oil industry, and defense. Underwater communication method can use acoustic or electromagnetic method. Both of these methods propagate waves through the medium and will be diffused, reflected and absorbed by the water. However, the acoustic method has less absorption than electromagnetic so can reach longer distances in the water medium [1].

The water medium is an imperfect acoustic medium. When the sound spreads through the water, acoustic energy can be lost by absorption and scattering [1]. Urick [2] adds that acoustic wave propagation will through the process of loss of intensity caused by geometrical spreading loss and absorption loss. This is known as propagation loss (transmission loss) which is one of the sonar parameters that express a quantitative value of some phenomena related to the propagation of acoustic waves at sea.

According to Duncan and Parsons [3], calculating transmission loss is complicated by its dependence on many different factors, including the horizontal separation of the source and receiver, their positions in the water column, the temperature and salinity profiles of the water column, the water depth, the slope and roughness of the seabed, the geology of the seabed, and the roughness of the sea surface. Whereas according to Burnett [4], transmission loss is influenced by environmental and non-environmental factors. Environmental factors that affect such as sea conditions, surface temperatures, depth profiles of temperature and also daily and seasonal geographic variations. While non-environmental factors such as transmitter and receiver depth, transmission range and frequency. Hence the stochastic process characterizing transmission loss is implicitly indexed by the corresponding transmission and receiver depths and transmission range.

The value of transmission loss in water is important to know, for example in active acoustic application. According to study conducted by Lermusiaux *et al.* [5], the prediction of transmission loss values can be used to determine the geoacoustic model of the type and thickness of the sediment. In addition, it can also be used to find out the bathymetry and condition of the water column. Transmission loss can also be used to know the dynamics of the tides in the water column. Meanwhile, according to Jones *et al.* [6], acoustic transmission modeling can be used to view the presence of water column features.

2 MATERIALS AND METHODS

The research was conducted on 22-25 January 2018 for 3x24 hours. Sampling was conducted on MV. Perkasa 7 that leaning on the dock of Pandu Station, Cilacap Central Java (7044'30'' S 108059'49" T). MV.Pengayoman IV which is the object in this study sailed in the Tanjung Intan cruise line, Cilacap. The method used observation method by recording the sound intensity of MV.Pengayoman IV used hydrophone and ship activity used CCTV.

https://doi.org/10.1051/e3sconf/20184704011

2.1 Data Acquisition

Data acquisition has been done for 3x24 hours. Underwater sounds was recorded used a Hydrophone Sea Phone SQ26-H1 was at depth of 1-2 meters and ship activity as support data was recorded used CCTV (**Fig. 1**).

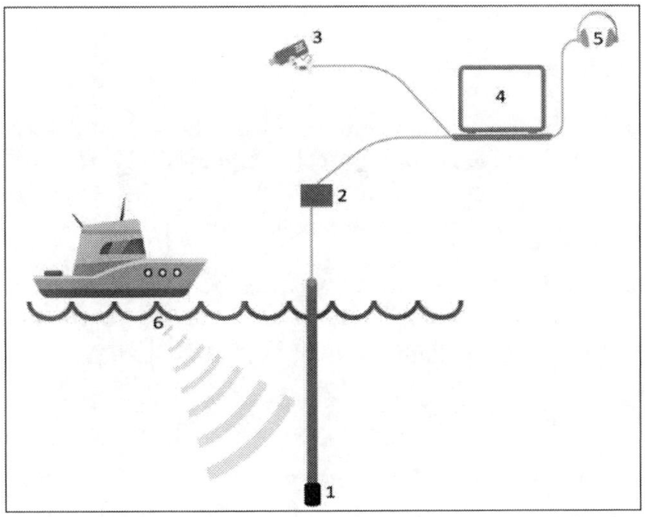

Fig 1. Instrumentation preparation design

2.2. Data Processing

2.2.1 Measurement of Vessel Distance

The distance of the vessel to the hydrophone was measured by projecting the vessel's position on CCTV into the Google Earth Software. The observed vessels were sought the position and distance to the hydrophone so can be seen the difference of resulted sound intensity. To simplify the positioning of the vessel, used fixed objects around the cruise line to be projected as buoy, wijayapura ferry port and cement mining **(Fig. 2)**.

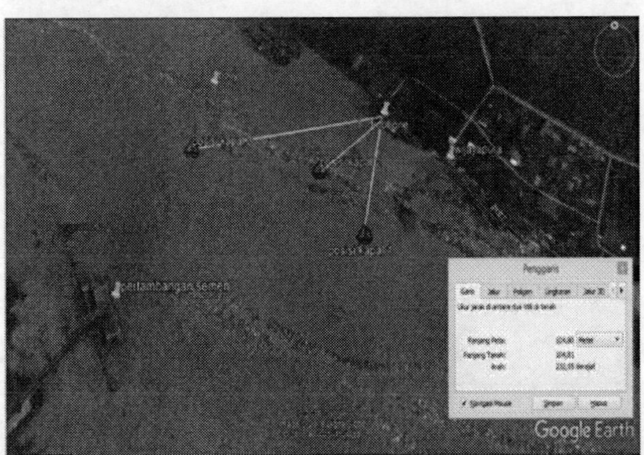

Fig 2. Measuring Vessel distance from hydrophone

2.2.2 Data Processing of Vessel Noise Characteristics

Passive acoustic systems will generate a lot of raw data. The recorded data contains the sound required in the study (sound of the vessel) and not required (natural sounds such as wind, rain, waves or mining activity around the study site). Therefore, prior to data analysis, sound processing was required to separate the required sound and not required by Wavelab 6.1 software. Stages performed were Envelope, Spectrum, and Sound Characteristic determination.

2.2.3 Transmission Losses Value Calculation

Absorption loss and geometric spreading loss on the acoustic wave propagation will give the transmission loss value according to the equation [2] which can be expressed in equation (1):

$$TL = 20 \log R + \alpha R \quad (1)$$

Where the R factor (Distance) is entered because the α is usual expressed in dB / km (or dB / kyard). Since the absorption coefficient value (α) has not been obtained, the TL value is calculated using ordinary least square estimation (equation.2):

$$Y = a - bx \quad (2)$$

Where **Y** is a representation of the intensity at X distance, **a** is a representation of the initial intensity and **bx** is the value of transmission loss.

3 RESULTS AND DISCUSSION

3.1 Vessel Noise Characteristic

MV. Pengayoman IV is one type of ship crossing ferry or also called Ro-Ro ships. Roll-on/roll-off (RORO or Ro-Ro) ships are vessels designed to carry wheeled cargo, such as cars, trucks, and motorcycle that are driven on and off the ship on their. MV. Pengayoman IV has a specification length of 21 m, speed 4km / hour and capacity 127 GT. Dussardier [7] stated that the use of Horse Power 3-3.5 times the Gross Tonnage of the vessel. The vessel produces noise characteristics as in table 1. This result was obtained by recording distance between 35-80 meters from hydrophone.

Table 1. Average result of noise characteristic of MV.Pengayoman IV .

Ship Name	Ship Size	Intensit y (-dB)	Frequenc y (Hz)	Interval Duratio n (ms)
Kapal Pengayoman IV	127 GT	43,35	6613,54	216.15

3.2 Coeficient Absorption

When acoustic waves propagate in water, some of the sound energy will be absorbed by water and converted into heat [1]. This absorption occurs through two main mechanisms including the viscosity of the seawater itself and the molecular relaxation. According to Urick [2], molecular relaxation is a major cause of absorption loss at acoustic frequencies below 100 kHz. At high frequencies (more than 500 kHz), acoustic pressure changes too quickly to avoid molecular relaxation, and absorption of acoustic energy is absent. Here is the graph of coefficient of absorption based on frequency:

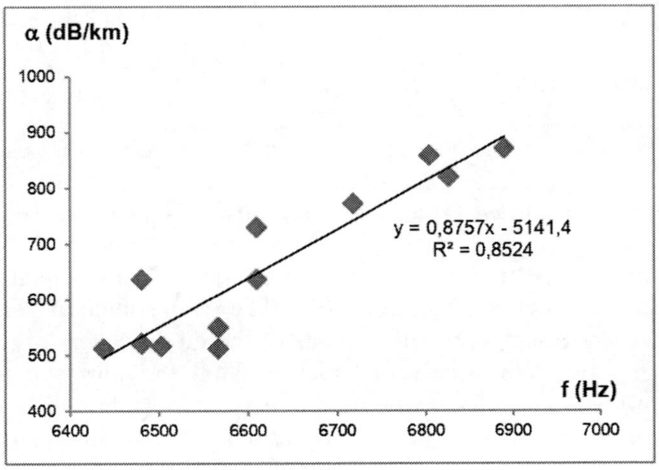

Fig 3. Graph of frequency to absorption coefficient value

The amount of energy absorbed (α) will depend on various factors. In other words α is a function of the sound frequency, pressure, temperature, and density of the water [8]. The value of the absorption coefficient will increase as the frequency. The frequency range obtained was 6400-6800 Hz and the absorption coefficient was 500-900 dB / km.

3.3 Transmission Loss Estimation

Transmission loss is generally influenced by non-environmental factors such as transmitter and receiver depth, transmission range and frequency. Therefore, the stochastic process that controls transmission loss is implicitly indexed by the transmission distance, the depth of the receiver and the transmission range [4]. The estimated transmision loss value that was affected by the transmission distance of 35-80 m ranges from 4.80 to 11.02 dB/km **(Fig. 4)**.

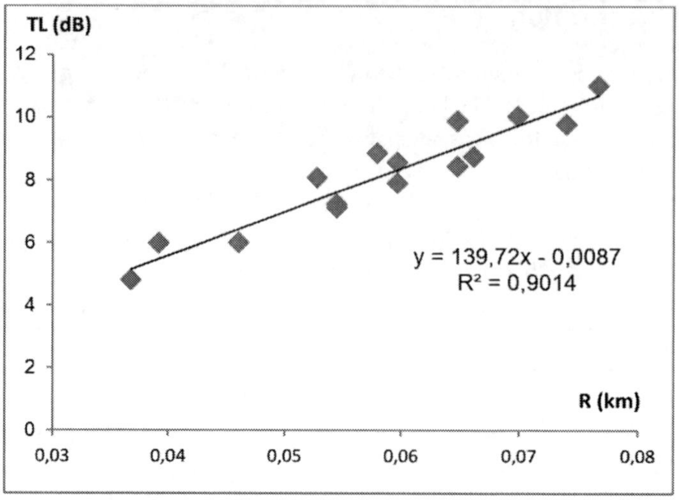

Fig 4. The value of transmission loss based on the distance traveled by sound waves

Various activities on the surface of the waters by various commercial vessels produce a significant contribution to underwater noise [9]. The main source of vessel noise is when the vessel speed is low because the components of the diesel generator (SSDG) donate all the noise from the vessel. When the vessel speed is high, the noise source comes from the dominant drive of the vessel. Basically every vessel has a vibration source. The biggest vibrations occur in the engine room. This is because the engine room has main engine that works as the main driver of the vessel. Vibration on the vessel produces varying noise levels according to the distance from the vibration source and the attenuation that inhibits the propagation of noise from the vibration source [10].

Based on the research that has been done, MV. Pengayoman IV resulted noise characteristic of intensity 43.35 (dB) , frequency (6613.54 Hz) and interval duration 216.15 miliseconds. While previous research conducted by Lesage and Barrete [11] on ferry ship with specification length 80 meter, speed 28 km / hour, size 2173 GT and the power 2000 HP was produced noise intensity characteristic equal to 40-75 (-dB) and frequency 175-6000 Hz. According to Ross [17] anthropogenic sound as in the shipping activity has a frequency reaching 10 KHz. Changes of noise characteristics can be used as a reference in the transmission process in the water.

According to Preisig [12], acoustic propagation through seawater is characterized by a significant frequency of absorption and relatively slow speed propagation. The value of the absorption coefficient will increase along with the increase in frequency **(Fig 3)**. As pressure waves propagate through water, the transfer of thermal energy associated with chemical relaxation processes drives the absorption of sound by seawater salts [13].

The frequency is proportional to the absorption coefficient. Conceptually, higher frequencies involve faster particle velocities and thus higher friction losses. In seawater the relaxation of certain compounds can cause additional losses owing. For each compound, there is a critical 'relaxation' frequency below which the losses occur. The induced pressure reduction of molecules to ions which takes a certain time to complete is called relaxation. At high frequencies, there is no energy that absorbed because sound pressure cycles too quick for the reduction to occur and thus by this process. The magnesium sulphate relaxation dominates the absorption in the frequency range 2–500 kHz, and at lower frequencies there is a further loss associated with boric acid. The frequency is the main determinant of the absorption, but α also depends on the water temperature and salinity[1].

The limiting factor which determines the highest useful frequency for the detection of targets at a given range is rapid increase of absorption with frequency. The received signal is below the noise level and cannot be detected because the absorption loss is so great. While the absorption in freshwater is relatively small at low frequencies, since the molecular relaxation loss is absent [1]. Coefficient absorption is one of the transmission loss parameters.

The accumulated decrease in acoustic intensity as an acoustic pressure wave propagates outwards from a source can be defined as transmission loss. The effects of a variety of propagation phenomena in the sea is summarizes by transmission loss. Adding the effects of geometrical spreading, absorption and scattering, then transmission loss can estimated [14].

Chitre [15] assumes that the energy intensity at each wave surface point will be reduced as the square of the distance traveled by the waves and transmission loss increases according to the square of the distance. Based on the study conducted by Duncan *et* a.l [16] in 10 km distance, range of transmission loss between 75-90 (dB/km). While in 80 km distance, range of transmission loss will increase between 110-120 (dB/km). In **Fig. 4**, the increased transmission loss was proportional to the increase of distance. The statement is justified by [4], transmission loss is implicitly caused by the transmission distance, the depth of the receiver and the transmission range.

4. CONCLUSION

The characteristics of ship noise consist of intensity, frequency, and duration of intervals. Changes in noise characteristics can be used as a guide in determining the value of transmission loss. The absorption coefficient is one of the parameters that can affect the transimission loss. The value of the absorption coefficient increases with increasing frequency, and the value of transmission loss increases with distance.

Authors gratefully acknowledge to Pelindo III, Pandu Station and the Capt of MV. Perkasa 07 for the permits, places and facilities provided in

REFERENCE

1 E.J. Simmonds, D.N. Maclennan. *Fisheries Acoustics,2nd edition*. (Blackwell Science, Oxford,UK, 2005)

2 R.J. Urick. *Principles of Underwater Sound*. 3rd Edition. (McGraw-Hill, New York, 1983)

3 A.J. Duncan, M.C.G. Parsons. *Proceedings of Acoustic 2011* (Gold Coast, Aus. Acous.Soc, 2011)

4 T.D. Burnett. *The Caracterization of Variability in Transmission Loss in The Ocean*. (Naval Postgraduate School, California, 1971)

5 P.F.J. Lermusiaux, X. Jinshan, F.C. Chi, J. Sen, Y.C. Linus, J.Y. Yiing. IEEE, **35,** 895-916 (2011)

6 A.D. Jones, Z. Alex, V.G. Michael. *Proceedings of Acoustics 2013* (Victor Harbor, Australian Acoustical Society, 2013)

7 A. Dussardier. *Consideration on Dissels. Fishing Boat of the World II.* (Jan-olof Traung London, Fishing News (Book) Ltd, 1960)

8 J. Widodo. Oseana. **14,** 81-92 (1989)

9 M F. McKenna, D. Ross, S.M . Wiggins, J.A. Hildebrand. America. J. Acous. Soc. **131** 92-103 (2012)

10 C.S. Clay, H. Medwin. *Accoustical Oceanoghraphy: Principles and Aplications*. (John Wiley and Sons, New York, 1998)

11 V. Lesage, C. Barrete. Mar. Mammal Sci, **15,** 65-84 (1999)

12 J. Preisig. *Mobile Computing and Communications Review*. **11**, 2-10 (2006)

13 I.J. Doonan, R.F. Coombs, S. McClatchie. J.Mar.Sci. **60,** 1047-1055 (2003)

14 M.C. Domingo, R. Prior. Computer Communication. **31**, 1227–1238 (2008)

15 M. Chitre. *Underwater Acuostic Communications in Warm Shallow Waters Channels. Ph.D.* Thesis. (Electrical and Computer Engineering. National University of Singapore, Singapore, 2006)

16 A.J. Duncan, A. N. Gavrilov, R. D. McCauley, and M. Parnum America. J. Acous. Soc.**134**, 201-216 (2013)

17 D. Ross. *Mechanics of Underwater Noise*. (Pergamon, New York, 1976)

The Slope Dynamic of Cilacap Backshore During Transition Season

Florencius EkoDwi Haryono[1*], *Lazuardy* Nalendra[1], *Mukti* Trenggono[1], *Amron* Amron[1], Hartoyo[1]

[1]Marine Science Department. Faculty of Fisheries and Marine Science.Jenderal Soedirman University, Purwokerto

Abstract. Indonesia as one oftropical and archipelago countriesis daily illuminated by the sunlight.The integration of sunlight and earth pseudo-movement cause a dry and rainy seasons. The transition from one to another seasonoccur twice a year, which isMarch-May (first transition season) and September-November (second transition season). This research aim was to analyze the foreshore slope dynamic atCilacapcoast during the transitionseason. Data of slope dynamic were collected bi-weekly from March to May 2017 at 16 sites, where the distance of each site was 2 miles.The angle of slope was measured perpendicular to the shoreline using Theodolite,*i.e.* at the boundary of coastal vegetation, center site and foreshore waterline at low tide. The backshoreslope at four sites were increase at the second month and at the end of the transition seasonwas decline. The changes of foreshore slope at other siteswere decrease. The monthly dynamic of foreshore slope in the transitionseason underwent a narrow of change and different condition.The highest and lowest slope of the coast occurred on April (8%) and May (5.2%) consecutively. The transition season wasmostly influenced by the changes of wind pattern thatwouldaffect the ocean wave and the impacttobackshore slope was vary.The backshore slope changes were shown that more slopingat the end of the season. The transition season had an impact onbackshore slope and statistically, the monthly changes of coast slope weresignificant difference [$p_{(0.05)}$].

1 Introduction

The position of Indonesia lies around the equator always gets sunshine every day and included as a tropical climate country. The influence of the condition has cause two seasons, the rainy season and the dry season. The transition from one to another season occurs twice a year, which areMarch-May (first transition season) and September-November (second transition season). However, inter annual fluctuations are not only locally but also remotely controlled by regional- and global-scale climate variability modes [10]. The fundamental difference between two seasons can be impact to winddirection. Wind direction in monsoon comes from the Northwest (West wind) and in the dry season is dominated by the Southeast wind. Wind direction during the transition season is unpredictable, at the end of the first transition season the wind direction turns to the Southeast wind. Unpredictable wind direction and wind power have affected to direction and force of wave. The meeting area between land and sea is know as a beach.

The continuous influence of wind, waves and tides has led to coastal dynamics [10]. Oceanic current and wave affect the changes of coastal morphology. The tide as a long wave with wavelength can reach one half of the equator (Kalay, 2008). Those condition has impact to influencethe process of sediment dispersal. The key role in process of sediment dispersal is influence of tidal plays [1]. Wind, water, ice, and gravity are the energies which capable of transporting and sinking the sediment [6]. Furthermore, the highly number of erosion along the coast causingthe vulnerability of coastal [4].

The process of coastal dynamic has displayed along shore that impact to the specification of coastal morphology. Rivers support the main source of coastal material and coastal plains supply. Theestuarine system effectively traps the particles originating from the watershed and dominated by fluvial sources, particles of a more oceanic provenance [2]. On the other side, sea-level rise has played a fundamental role in the development of the low-lying coastal environment [14]. The difference betweenthe response of natural and human interventions to coastal as backdrop of researched of coastal dynamic. The importance is evaluated to provide direction for future research that would help to form a way out of the present situation and sustainable planning for this coastal. The objective of the

Corresponding author :marine_2807@yahoo.com

study was to analyze the dynamics of coastal backshore in the March-May as the transition season in the Southern coast of Cilacapdistrict, Central Java

2 Material and method

The data collection is to measure the slope of the backshore of Cilacap district. Data of backshore slope was collected bi-weekly from March to May 2017 at 16 sites, and the distance of each site was 2 miles. The angle of slope was measured perpendicular to the shoreline using Theodolite. The slope backshore data of each location is taken on three points *i.e.* on each site at the boundary of coastal vegetation, center site and foreshore waterline at low tide.Furthermore, data were analyzed descriptive and statistical p $_{(0.05)}$as well as coastal slope mapping.

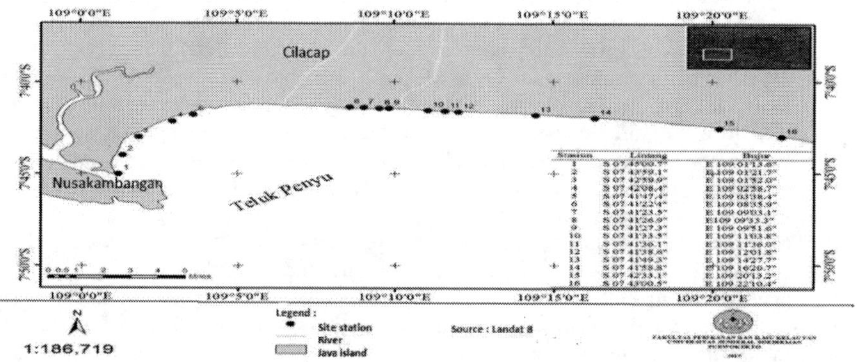

Fig. 1. Sites of study

3 Result and discussion

The Southern Coast of Java Islandis an integral part of the permanent influence of the tropical season. The seasonis experienced the influence of seasonal winds that remain in the rainy and dry seasons.These conditions affect the oceanographic parameters of the Indonesian oceans, which further affects to the coastal of Cilacap. The blow of seasonal winds fix and stronger influence to the ocean wave. The influence of winds at transitional season is more dynamic. The influence can be impact to the changes of wind direction.These conditions indirectly affect the dynamics of Cilacapcoast morphology. Cilacap beach is a`gulfwaters that stretches from East to West [Fig. 1]. Coastal conditions in the eastern part are open water and directly affected oceanographic conditions of Indonesian ocean. The Nusakambangan Island protects the Western coast.The direction and strength of the wind blow vary during the transition season. Those conditionsaffect the fixed waves from the Indonesian oceans, wind-driven currents and waves.Additionally,the blow of wind affectsthe constant influence of currents and tidal waves. All these effects take place in an integrated manner and the result cause to coastal morphological changes. The slope of the backshorethat part of coastal morphological parameters undergoes change during the transition season.

The dynamics of the slope of the coastal plains was obtained to change during the transitional season with the changes was synchronizedwiththe influence of direction and wind energy. (Fig. 1).

Corresponding author :marine_2807@yahoo.com

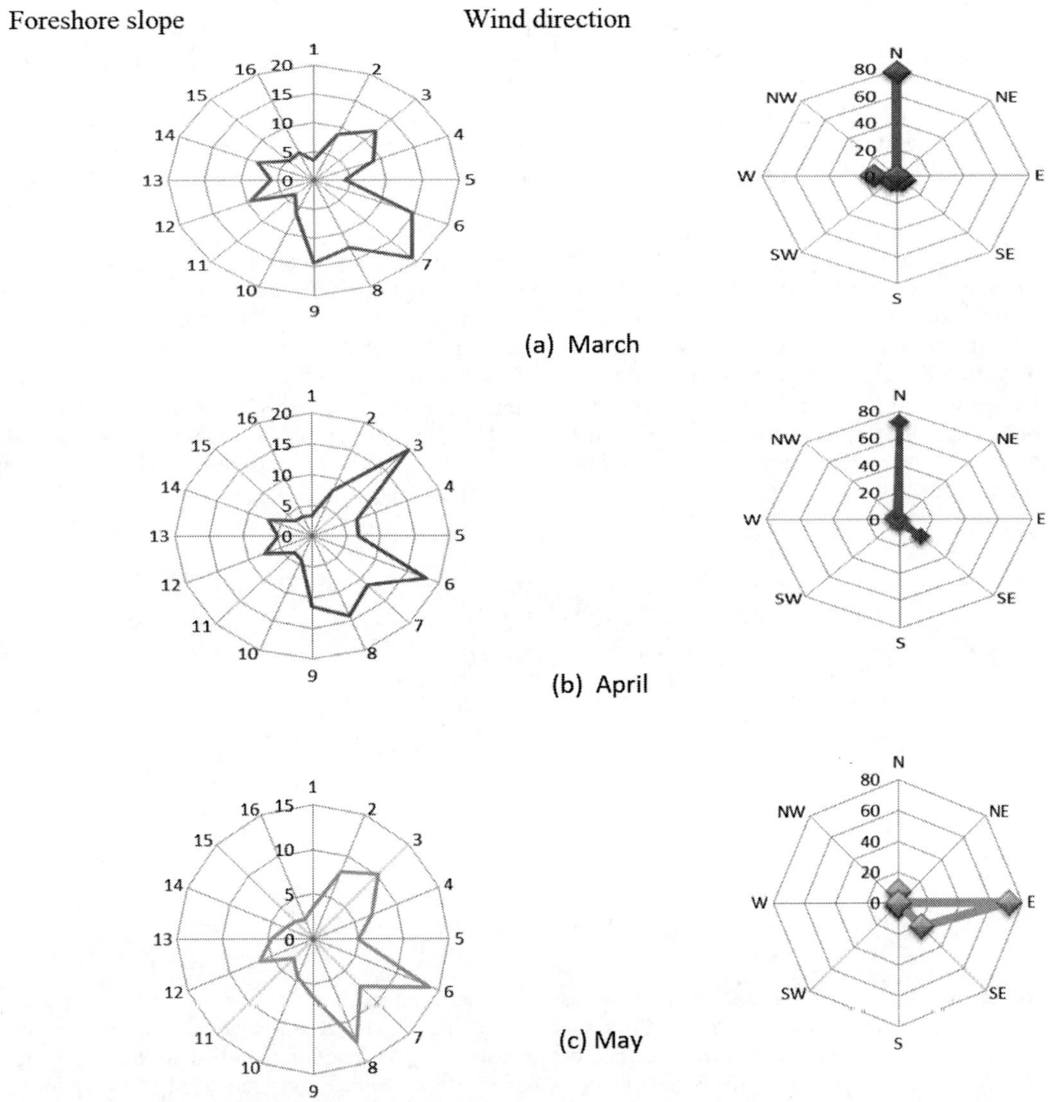

Fig. 2. Degree of backshore slope dynamics of Cilacap coast at each station

The change of backshore slope in March to April was highly increasedon 5 stations, namely the station 3, 6, 7, 8 and 9. Station 3 was sheltered beach with the existence of Nusakambanganisland from the direct influence of oceanographic of Indonesian ocean. Location of the station was Eastern of the Ocean Fishery Port of Cilacap (PPSC). The station 6, 7, 8and 9on the Southern of Serayu riverestuary which is open beach from the oceanography of Indonesian ocean. The backshore slope on stations 3, 6 and 8 were increase and the slope on station 7 and 9 were decrease. The backshore slope on 5 stations was decrease generally. The shrinking slope of the coastal plains indicates that the coastal plain is getting sloping.

Station 3 was obtained as greatest increment of tilt among other stations. The station condition was closely related to the construction of ship channel that built into the sea. The construction has caused a deflection of tidal currents, flowing along the coast from East to West and turn down the construction of the path to the South. At the location, accumulation the style of tidal period causesthe sediment to be moved to the sea. The final result was increaseto the backshore slope. Position of station 6 and 8 wasin the two of river estuaries. Impact of those locationwas increase the backshore slope. The condition is related to the direct influence of the impact of Indonesian ocean, shape of river estuary and tide and low tides. The transition season conditions still cause rain which

Corresponding author :marine_2807@yahoo.com

sedimentation on the coast, Based on the increasing slope showing the effect of the season on moving sediment according to the pattern of tidal current and low tide to the nearby station. Station 5 and 7 receive sediment and the smaller the angle of tilt.

Direction of wind in March is dominated fromSouth and blow weak West wind. April is dominated from Southand began to blow weak of Southeast wind. Wind-induced wave as wind driven current patterns. The current was moving sediments to coast and the impact causing the sediment become accumulate on shore. The upper layerof water is influenced by wind conditions [11]. The influence of the sediment displacement to beach gets sediment delivery and the impact of that to the angle of the backshore slope become smaller.

The backshore slope from April to May was decrease on each station, except on station 2, 4 and 5. The decrease inclination as a result of the weakening of South winds force that turns anEast and Eastern winds at the end of the transition season.The change of wind direction affectsdirection and wave energy. The location of station 2, 4 and 5 are located near toseaward coastal buildings, namely station 2 located at breakwater, station 4 and 5 at Western and Eastern Electric Power. Seaward jetting have affect the tidal currents. Based on the structure of the electric power causechange of tidal current and receding current direction. The direction of the tidal stream had previously been down the coast turned into a sea. In these region the energy of the current moves the coastal sediments to seaward and causes the increasing backshore slope. Monthly dynamic of foreshore slope in each site (Figure 3) wasshow fluctuate.

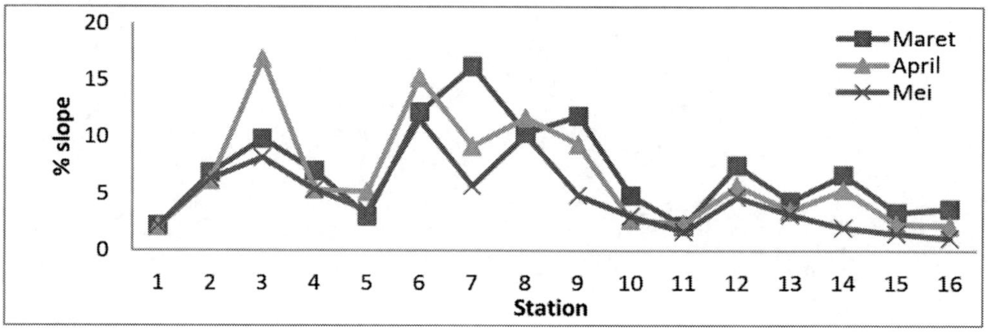

Fig. 3. Monthly backshore slope dynamics

The pattern of tidal current along the coast has moved coastal sediments on the East coast and becomes increasingly gentle at the end of the season. Wind direction of the season is dominated by the South wind at the beginning of the season and the East wind is getting stronger at the end of the season. Such conditions affect to the height of the wave, which the stronger wind causing the higher waves. The resultant style of seasonal wind of the Indonesian ocean wind and the coastal morphology of Cilacap as gulf, integration of both conditionsbecome specific impact to the backshore slope of Cilacap coastal plain, especially the movement of coastal sediment transport at the backshore.

Wind direction at the end of the transition seasonon May was dominated of East wind and wind power become stronger. The wind affects the wave height directly and the wave becomes higher. These wave energy moving toward the shore moved the sediments at sea to shore and the slope of the coastal plains became even flatter, further effects to coastal morphology [12]. Statistically of monthly backshore slope dynamic [Kruskal Wallis $p<0.05$] was obtained significant difference.

4 Conclusion

The dynamics of Cilacap backshore slope foreshore in transition season (March-May) was obtained greater in March to April and progressively flatter at the end of the season [May]. The seasonal winds affect to coastal slope and affect the width of the coastal structure. Based on the conditions need an attention to stakeholder to the direction of tidal currents and wind driven current that support the change of the backshore slope of the coastal plain of Cilacap.

Corresponding author :marine_2807@yahoo.com

References

1. Akter, J., Sarker, M.H., Popescu, I. and Roelvink, D.. *Journal of Coastal Research,* **32(5),** 1212–1226 (2016)
2. Alexander, C.R. and Windom, H.L., *Journal of Coastal Research*: **33**, Issue 3: 548-563 (2017)
3. Anggraeni, S.K., A. Satriadi., A.D.S.P. Agus. *Journal Of Oceanography*. **5**, No: 3, Hal: 390-397 (2016)
4. Boruff, B. J.; Emrich, C., and Cutter, S. L. *Journalof Coastal Research,* **21(5),** 932–942 (2005)
5. Fadika, U., A. Rifai., B. Rochaddi. *Journal Of Oceanography*. **3**, No: 3. 429-437 (2016)
6. Friedman, G. M. & Sanders, J. E. John wiley & Sons, Inc, 792pp (1978)
7. Kalay, D.E. *Mastes Tesis.*SekolahPascasarjana, Institut Pertanian Bogor. Bogor (2008)
8. Komar P. D. CRC HanbookOf Coastal Procesess and Erosion.CRC Press, Inc. Boca Raton, Florida (1983)
9. Loupatty, G. *Jurnal Barekeng*. Vol. 7(1):19-22 (2013)
10. Manta, G.; Barreiro, M.; Ortega, L., and Defeo, O., *Journal of Coastal Research*. **33**, Issue 3: 531-536 (2017)
11. McPhaden, & S. P. Hayes.*.J. Geosphys*. Res. **96**: 3331-3342 (1991)
12. Nontji, A.. PenerbitDjambatan. Jakarta (2005)
13. Scott, T., G. Masselink, P. Russell. School of Marine Science and Engineering, University of Plymouth, Drake Circus, Plymouth, United Kingdom. No. 10, p:1-20 (2011)
14. Vankoningsveld, M.; Mulder, J. P. M.; Stive, M. J. F.; Vandervalk, L. and Vanderweck, A.W. *Journal of Coastal Research,* 24(2),367–379 (2008)

Corresponding author :marine_2807@yahoo.com

Interaction of oceanography patterns towards the abundance of phytoplankton, zooplankton and ichthyoplankton in teluk penyu waters of Cilacap

Florencius Eko Dwi Haryono[1*], *Rose* Dewi[1] , *Taufik* Budhi Pramono[2], *Rifki* Ahda Sumantri[1], *Tri* Nur Cahyo[1], *Dewi* Wisudyanti[1]

[1]Marine Science Department. Faculty of Fisheries and Marine Science. Jenderal Soedirman University, Purwokerto-Banyumas
[2]Aquaculture Study Programe of Jenderal Soedirman University. Purwokerto, Banyumas

Abstract. Coastal waters of Teluk Penyu-Cilacap district is an enlarging area for many species of fish larvae. The waters is directly influenced by the physical, chemical and biological of Indian Ocean patterns. Fish communities inhabit in gulf waters to take advantage of the high productivity in the coastal waters to support their livelihoods. The gulf area is usually associate with other productive ecosystems *i.e.* rivers that empties into it and mangrove forests. The study aim is to analyze the influence of oceanography pattern to the existence of ichthyoplankton [fish larvae] inhabit at Teluk Penyu waters of Cilacap district. Sampling is conducted monthly from April - June 2009 at 3 sites. The sites were set 1 mile of distance from coastline at Teluk Penyu Cilacap, PPSC and RSPC waters. Fish larvae were collected using larvae net mesh size 0.5 mm and 75 cm mouth diameter and phytoplankton and zooplankton are collected using plankton net . Larvae net was towed on the surface of the waters using a motorized boat with a speed 1 knot. Data were analyzed statistically based on Principal Components Analysis/PCA. 30 species of phytoplankton and 25 species of zooplankton species were identified from the waters. The species of phytoplankton and zooplankton were obtained difference species at each site. The highest abundance of phytoplankton were species of *Chaetoceros* and the abundance of zooplankton from the highest subsequence are species of *Coelosphaerium dubium* and *Platydorina*. The highest of relative abundance of ichthyoplankton in Teluk Penyu tourism waters was obtained 41,463% of the crawfish [*Crumrnophthalmus* spp.], followed by pheasant fish [*Auriglobus nefastus*] 8.53667%, and swordfish (*Chirocentrus dorab* spp.) 2.439%, and fish beloso [*Tumbilic saurida*] and gulamah fish (*Johnius dussumieri*) which are 1.2195%. The abundance of phytoplankton and zooplankton associated to ichthyoplankton abundance. The Southern of Teluk Penyu gulf waters was highest abundance area of ichthyoplankton.

1 Introduction

Bay, lagoon, and mangrove forests are part of coastal ecosystems that have a very important role to fish survival. Those areas are care for several important species such as fish and shrimp [23]. The fish community utilizes the high productivity in the coastal areas to support their lives. This area usually associated with other productive ecosystems such as rivers that lead to it and mangrove forests [7]. The estuary adjacent coastal waters is also an important support for fisheries activities [2].

Ecosystem of Teluk Penyu waters at Cilacap district is a very dynamic area. These waters are heavily influenced by the physical, chemical, and biological conditions of the Indian Ocean *i.e.* the current, wind, tidal, wave, and salinity and DO, pH, nitrate, nitrite and biological conditions. These ecosystem also inseparable influence of Segara Anakan mangrove ecosystem which gives a lot of nutrient inputs that affect to the fertility of the waters, and support the life of various species of fish and shrimp. The ecological function of Teluk Penyu coastal waters is enlarging area for many species of fish larvae that supported of the enrichment of nutrients. The high diversity of fish larvae in the waters illustrates the high support of the ecosystem towards the continuity of fishery activities. The main objective of fisheries management utilize the existing of fishery resources maximally while maintaining the sustainability of the resource that is being exploited. In order to preserve fishery resources, information on biological and ecological aspects of fishery resources is needed, especially in the case of the species, abundance and distribution of fish larvae in Teluk Penyu.

The larvae are the stage or level of fish after the eggs hatch until the egg yolk is absorbed [6]. The uses term of larvae which refer to the egg yolk larvae and post larvae for young fish between larvae and juvenile stadia. Stadia larvae is terminated when the egg yolks have been absorbed [21]. Divide the phases of fish life into egg phase, larval pre flexion, larval flexion, larval post flexion, larval transition, juvenile and adult [13]. The newly,

*Author email : marine_2807@yahoo.com

 https://doi.org/10.1051/e3sconf/20184705002

hatched larvae are marked by (a) the presence of an egg yolk sac located at the bottom of the front and (b) a fingerless fin that surrounds the body of the larvae, from the back, tail, to the bottom just behind the anus. In some types of fish the boundary of the circumference fins are located just behind the yolk sac [20]. The development of larvae in the outline is divided into 2 stages namely pro larvae and post larvae. Distinguish the pro larvae has an egg yolk sac, transparent body with several pigment tails whose function is unknown. The pectoral and tail fins are present but are not yet perfectly shaped and most of the newly released pro larvae from this eggshell have no apparent belly fins but only a bulge shape. Respiratory and circulatory systems have not been perfect. The mouth and the jaw have not developed and the intestine is still a straight tube. The food comes from the remaining egg yolks that have not been absorbed. The post larvae period is the larval period from the disappearance of the yolk sacs to the formation of new organs or the completion of the extent of the existing organs so that at the end of the post larvae morphologically already have the form almost the same as the parent [6].

Some taxonomic properties that can be used to identify fish are (a) various structures or body parts, such as eyes, head, body shape, stomach and fins; (b) the order of occurrence of its fins and position, photophores and bone elements; (c) larval size; (d) pigmentation and (e) significant signs such as swollen fin folds, elongated and altered fins, choker pricks, thorns on pre-operculum and others [20], [13], [26].

The existence of planktonic larvae is greatly influenced by complex and simultaneous environmental factors, affecting the distribution, live pass rate and fish larvae abundance. Environmental factors affecting the life of fish larvae are numerous and do not work independently, although it is often the case that one factor is more dominant than the other and the factor can be a limiting factor in larval life [20].

The distribution of ichthyoplankton in Segara Anakan Estuary, Cilacap, is strongly influenced by turbidity factor. The abundance of *Engraulis* larvae (anchovies) in Segara Anakan is significantly associated with salinity and temperature, while the *Stolephorus* larvae abundance is associated with abundance of zooplankton. The results also show that the ichthyoplankton community in Segara Anakan, Cilacap is more abundant in the interior of estuaries that have smaller environmental variations [17]. The distribution of fish larvae is related to oceanographic conditions of waters and seasons, also strongly influenced by the hydrologic cycle [16]. The peak abundance of fish larvae in Segara Anakan mangrove waters mostly in November and the lowest abundance in January and related to sea level [tide condition] [17]. In St Estuary, South Africa, the distribution and abundance of ichthyoplankton is strongly influenced by the Current Agulhas [10]. The point out the importance of current influences on the distribution of ichthyoplankton in coastal waters of Southeast Australia [10]. Meanwhile, fish distribution in the Gulf of Florida follows salinity gradients from estuaries [15]. The description above showing the distribution of ichthyoplankton somewhere is inseparable from the environmental factors that encompass the environment.

3 Material and Methods

Gulf of Teluk Penyu is a coastal ecosystem located in the South of Java Island, precisely in Cilacap district, Central Java. Oceanographic conditions of these ecosystem are strongly influenced by the Indian Ocean such as currents, tides, temperatures and salinity. In addition, Teluk Penyu ecosystem is also influenced by Segara Anakan mangrove ecosystem and several rivers that empties in the vicinity such as the Donan, Kembangkuning, Serayu, Kali Sabuk, Mandiraja and Bodo. These waters is also inseparable from industrial activities such as oil refinery Pertamina UPDN IV, Holcim cement and three electric power plan. Anthropogenic activities of household activities that exist to the waters.

Sampling is done at 3 sites perpendicular to the coastline which distance to coastline 1 mill. Distance between site stations about 2 mills. First Site 1 is located at Teluk Penyu turism beach, 2nd at coastal waters of Cilacap Ocean Fishing Port (PPSC) and 3th site is located in coastal waters of Pertamina Cilacap Hospital (RSPC). Sampling was conducted from April to June 2009.

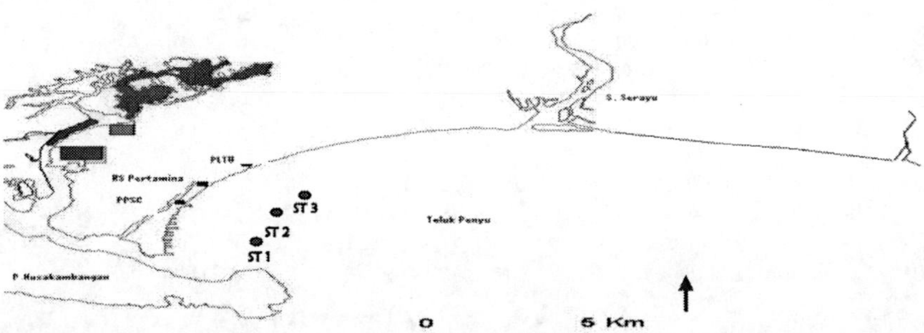

*Author email : marine_2807@yahoo.com

Fig. 1. Sampling site stations

Sampling was done monthly at each sites. Fish larvae are collected using of meshsize 0,5 mm and 75 cm mouth diameter of net larva. The net was towed horizontally 1-2 m under sea surface with boat speed 1 knot for about 30 minutes. Fish larvae were preserved with 4% of formalin.

The identification of ichthyoplankton is based on [3], [4], [14], [11], [13]. Identification of larvae was done up to the most probable of taxon. The larva was divided by stage larva, post larva or juvenil.

The biological, physical and chemical variables was observed *i.e.* salinity, temperature, turbidity, brightness, dissolved oxygen, pH, nutrients (nitrate, nitrite, orthophosphate), phytoplankton and zooplankton. Sampling of water for the measurements of variables was performed at each station along sampling of fish larvae.

The composition of the ichthyoplankton species identification unit is describe the richness of ichthyoplankton in Teluk Penyu gulf waters. The composition each site is relatively expressed in percentage (%) as the ichthyoplankton proportion of species from the total catch d in each site. The ichthyoplankton abundance is defined as the number of one species per station.

Determination of distribution of water biophysical-chemical variables among observation were analyzed multivariate analysis approach *i.e.* Principal Components Analysis (PCA) based on [9], [1]. PCA is a descriptive statistical method for presenting in graphical form, the maximum information contained in a data matrix. The data matrix consists of observation stations as individual of larvae (rows) and aquatic biophysical-chemical variables as a quantitative variable (column). Data variables have not same unit of measurement, and before the PCA, data is normalized through centralization and reducing. PCA results are not realized from the initial variable value, but from synthetic indices obtained from linear combinations of initial variable values [12], [1].

Relationship of Ichthyoplankton abundance to aquatic biophysics variables is obtained with multiple regression analysis. The Stepwise method is used to know the relationship between biophysical-chemical variables with ichthyoplankton abundance. Multiple regression analysis was used in accordance with the procedure [9]. The Stepwise method is either a method of multiple regression analysis which begins by the independent variables (in this case environmental parameters) which have the strongest correlation with the dependent variable (larval abundance). Then whenever other free variables are included, testing is carried out by inserting or removing the previous free variable [23]. The PCA and CA calculation process is done by XLSTAT version 5.0 software. Software SPSS version 11.0 is used to calculate multiple regression analysis.

4 Results and discussion

4.1 Ichthyoplankton Abundance

Distance of observation station in waters was 1 mile from coastal waterline, started from Teluk Penyu tourism beach (TP), Cilacap Ocean Fishing Port (PPSC) and Pertamina Hospital Cilacap (RSPC). The distance between site from Teluk Penyu to PPSC was 1 mile, while the distance between RSPC was 2 miles. Relative Abundance of Inchtyoplankton at each observation site station [see Table 1]

Table 1. Relative abundance of Ichthyoplankton at site stations

No	Spesies	Scientific Name	T. Penyu	PPSC	RSPC
1	Sidat	*Anguila marmorata*	0	0	1.2195
2	Gulamah	*Johnius dussumieri*	1.2195	0	0
3	Selar	*Crumrnophthal mus*	41.463	31.70	0
4	Condro	*Tylosurus crocodilus*	0	0	1.2195
5	Belut laut	*Anguila* sp	0	1.2195	0
6	Remang	*Congresox talabon*	0	1.2195	0
7	Pedang	*Chirocentrus dorab*	2.439	0	0
8	Beloso	*Saurida tumbil*	1.2195	0	0
9	Ikan Buntal	*Auriglobus nefastus*	8.53667	2.439	0
10	U. Windu	*Penaeus monodon*	0	0	1.2195

*Author email : marine_2807@yahoo.com

11	Lobster	*Pannulirus sp*	0	3.659	3.000
12	U. Galah	*Macrobranchiu m sp*	0	0	0

The highest relative abundance was obtained in Teluk Penyu tourism waters 41.463% of the crawfish [*Crumrnophthalmus* sp.], followed by pheasant fish (*Auriglobus nefastus*) 8.53667%, swordfish [*Chirocentrus dorab* sp] 2.439%, and beloso [*Tumbilic saurida*] and gulamah [*Johnius dussumieri*] which were 1.2195% [see Fig. 1]

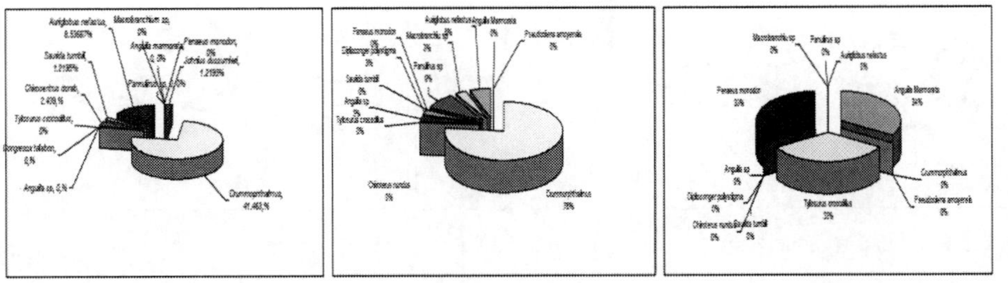

(a) (b) (c)

Fig. 1. Relative Abundance of fish larvae (a) Teluk Penyu tourism of Cilacap waters, (b) PPSC waters and (c) RSPC waters.

The relative abundance of the highest fish larvae species at PPSC station as well as 31,70% of *Crumrnophthalmus* sp. Whereas relative abundance in RSPC waters were obtained 3 larvae species. Relative abundance in the waters is obtained sea eel [*Anguilla* sp], condro fish [*Tylosurus crocodilus*] which are respectively 1.2195%. Those species above were not found in Teluk Penyu and PPSC stations.

4.2 Plankton Abundance

The existence of plankton at sites station in this case includes phytoplankton and zooplankton were obtain difference of abundance [see Table 3]. The highest abundance of phytoplankton in Turism Penyu gulf waters subsequence were obtain *Rattulus rattus* [35,952,875 ind.100 l^{-1}], *Astrionella japanica* [21,628., 38 ind.100 l^{-1}], and *Chaetocheros lorenzianum* [18832.664 ind.100 l^{-1}]. The abundance of phytoplankton in PPSC waters was dominated by *Chaetocheros lorenzianum* [177,756,5 ind.100 l^{-1}]. *Astrionella japanica* [42.774,65 ind.100 l^{-1}] and *Nitzschia lorenziana* [1,536,096 ind.100 l^{-1}]. While the abundance in RS. Pertamina Cilacap waters with the highest order was obtained *Astrionella japanica* [31,348,2328 ind.100 l^{-1}], *Chaetocheras lorenzianum* [10,068,3795 ind.100 l^{-1}], and *Nitzschia lorenziana* [8.26,679 ind.100 l^{-1}].

The highest abundance of zooplankton [see Table 4] in Teluk Penyu tourist site was obtained *Chaetoceros* sp [9276.1437 ind.100 l^{-1}] and subsequent *Coelosphaerium dubium* [378.594 ind.100 l^{-1}] and *Platydorina* 198.895 ind.100 l^{-1}]. Abundance in PPSC with the highest order was *Coelosphaerium dubium* [407,38 ind.100 l^{-1}], *Microcystus airruginosa* Kutz [93,452 ind.100 l^{-1}], and *Bacillaria paradoxa* [28.7494 ind.100 l^{-1}]. While the zooplankton zipper in Pertamina waters with the highest order is *Chaetoceros* sp. [9910.226 ind.100 l^{-1}], *Coelosphaerium dubium* [558,057 ind.100 l^{-1}], and *Rabdonela* sp [161.7556 ind.100 l^{-1}].

Table 3. Phytoplankton abundance

No.	Genera [Phytoplankton]	bundance (ind.100 l^{-1})		
		TP	PPSC	RSPC
1	*strionella japanica*	21628.038	42774.650	31348.232
2	*iddulphia heteroceros*	256.316	98.239	165.350
3	*anthocamptus, sedang kopulatus*	0	7.182	0
4	*athypna ungulata*	0	7.182	0
5	*haetoceras leave*	93.452	187.837	50.317
6	*haetoceras lorenzianum*	18832.664	177756.500	10068.379
7	*haetoceras pseudocurvisetum*	237.242	424.166	100.655
8	*iatoma*	4.74796	0	0
9	*ytilium sol*	83.876	59.892	57.520
10	*udorina wallichii*	0	50.317	0
11	*emialus indicus*	0	7.182	0
12	*emidiscus hardmanianus*	2.394	14.364	0
13	*yalotinea disiliensis*	7.182	21.567	21.567
14	*mnadia lenticularis*	2.393	43.135	0
15	*ngbya spirulinuidae*	0	35.953	0
16	*itzschia*	498.445	0	447.524
17	*itzschia closterium*	14.385	93.452	0
18	*itzschia lorenziana*	1174.129	1536.096	826.679
19	*itzschia seratia*	0	7.182	0
20	*scillatoria*	14.366	79.087	14.385
21	*cillatoria limmosa*	0	226.458	0
22	*eridinium*	107.837	74.278	122.223

Table 4. Zooplankton abundance

Genera [*Zooplankton*]	Abundance (ind.100 l^{-1})		
	TP	PPSC	RSPC
Bacillaria paradoxa	0	28.749	28.748
Bacteriastrum varians	50.316	16.757	7.182
Brachionus calcyflorus	7.182	14.363	0
cepepode	4.787	0	0
Ceratium hirudinella	105.443	0	0
Ceratium tripos	105.443	0	0
Coelosphaerium dubium	378.594	407.38	558.057
Cyclops fimbriatus	26.377	21.545	43.134
Cyclops fuscus sedang copulatis	19.173	7.182	7.182
Cyclops strenus (cepepoda)	0	7.182	0
Dinoflagelata	0	7.182	0
Meta -nauptilus dari cyplops strenus	11.9699	14.385	14.385
Microcystus airruginosa Kutz	40.740	93.452	14.385
Nauplius	98.239	26.355	157.269
Nauplius cepepode	21.567	0	0
Nauplius cumtocamtus staphylinus	0	7.182	0
Nauplius cyclops srenuus	2.393	14.385	0
Nauplius dari diaptomus [illegible]	[illegible]	[illegible]	[illegible]

4.3 Ichthyoplankton abundance and biology, physics and chemistry oceanography

Ichthyoplankton abundance for each observation station can be known as follows:

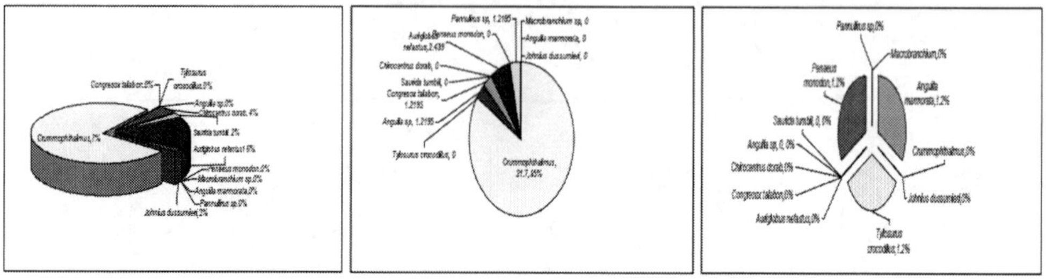

Fig. 2. Ichthyoplankton Relative Abundance (a) Tourism Teluk Penyu (b) PPSC (c) RSPC.

Relative Abundance of individual larvae of ichthyoplankton at each captured site is compared to the total number of larvae. The highest of relative abundance of fish larvae in Teluk Penyu tourism waters was obtained 41.5% of crawfish larvae [*Crumrnophthalmus* sp], then puffer fish (*Auriglobus nefastus* sp) 8,5% and swordfish larvae [*Chirocentrus dorab*] 2,4%. Relative abundance of fish larvae in PPSC waters also with the highest number was obtained 31.7% of *Crumrnophthalmus* sp, followed by 2.4% *Aurivlobus nefastus* sp and sea eel larvae [*Anguilla* sp], ghost (*Congresox talabon*) and lobster larvae [*Panullirus* sp] was 1.22%. While the relative abundance of fish larvae in RSPC was obtained 1.22% of eel [*Anguila marmorata*], condro [*Tylosurus crocodilus*].

4.4 Ichthyoplankton dynamics and biological, physical and chemical parameters

The dynamic of water chemistry-biophysical parameters at each observation station can be seen as follow :

Table 2. Dynamic of biophysic-chemis and phyto-zooplankton abundance.

	Oceanographic Parameters										
Stasion	A	B	C	D	E	F	G	H	I	J	K
T.Penyu	30.7	0.34	30.3	38.3	3.97	7.8	3.05	0.05	0.64	21628.04	9276.1
PPSC	31.7	0.57	30.3	30.3	5.23	8	1.61	0.11	0.07	177756.5	407.38
Pertamina	31	0.77	31	31	5.67	8	1.9	0.09	0.08	31348.23	9910.2

A [Temperature °C], B [current velocity ms^{-1}], C [Salinity ppt], D [Clerarity], E [DO], F [pH], G [Nitrat], H [nitrit], I [Posphat]. J [Phytoplankton abundance (ind.100 l^{-1})], K [Zooplankton abundance (Ind.100 l^{-1})]

Teluk Penyu gulf waters are generally semi-closed areas. Biological and oceanographic conditions are influenced by fluctuations pattern of tidal currents of the Indonesian ocean and geographical of coastal conditions. Tidal current type of Cilacap waters is semi diurnal which two springs and neaps daily. The coastline stretching from the boundary of Cilacap at Eastern border curving to form a circular arc towards the Western boundary and leading to South, thus the pattern of tidal currents along the coast from the East to the West and South along the Southern boundary. Such conditions result in pairs of organic materials accumulate in the Southern boundary. These indication are known the dissolves of organic material in the waters on sampling time, with organic soluble material more in the direction of in the South of the curved boundary, namely the mouth Tanjung Intan ships cannel or East Plawangan. Those conditions cause the waters of South part Teluk Penyu waters relatively more fertile than the waters in the Eastern. Indicator of the condition is known by the abundance of phytoplankton and zooplankton higher than the abundance PPSC and RS.PC site stations at Eastern [Table 2]. Association of the abundance of ichthyoplankton at each observation decrease in the Eastward direction, and the high abundance is obtained in the Teluk Penyu tourism waters.

*Author email : marine_2807@yahoo.com

5 Conclusion

The fertility of coastal waters is influenced by the hydrologic cycle and geographical conditions, furthermore influence the abundance of phytoplankton and zooplankton. Those abundance associated the ichthyoplankton abundance. Southern of Teluk Penyu gulf waters was highest abundance area of ichthyoplankton.

References

1. Bengen, D. G. Fakultas Perikanan dan Ilmu Kelautan IPB. 88 pp (2000)
2. Blaber, S.J.M., D.P. Cyrus, J.J. Albaret, C.V. Ching, J.W.Day, M. Elliot, M.S. Fonseca, D.E. Hoss, J. Orensanz, I.C. Potter, and W. Silvert. *Journal of Marine Science*. **57** : 590-602 (2000)
3. Delsman, H. C. Treubia. VIII : 109-254 (1926)
4. Delsman, H. C. Treubia. XIV : 109-254 (1932)
5. Dinas Kelautan dan Perikanan Kabupaten Cilacap. Data Produksi Perikanan Kabupaten Cilacap. 234 pp (2002)
6. Effendie, M.I. Yayasan Pustaka Nusatama, Yogyakarta (1997)
7. Ekau, W, P. Westhaus-Ekau and C. Medeiros. *Archive of Fishery Marine Research*. **47**(2/3) : 183-200 (1999)
8. Gray, C.A and A.G. Miskiewicz. Coastal and Shelf Science. **50** : 549-570 (2000)
9. Hair, J.F., R.E. Anderson, R.L. Tatham and W.C. Fifth Edition. Prentice-Hall International, Inc, London. 432 pp (1998)
10. Harris, S.A., D.P. Cyrus, and L.E. Beckley. *Estuarine, Coastal and Shelf Science*. **49** : 789-811 (1999)
11. Jayaseelan, M.J.P. UNESCO Publishing. 193 pp (1998)
12. Legendre, P. and L. Legendre. Elsevier Science, Amsterdam (1998)
13. Leis, J. M and B.M. Carson-Ewart (eds). Brill, Leiden. 850 pp (2000)
14. Leis, J.M and Rennis, D.S. New South Wales University Press, Sydney. 269 pp (1983)
15. Ley, J.A., C.C. McIvor and C.L. *Estuariane, Coastal and Shelf Science*. **48** : 701-823. 1999.
16. Nurdawati, Syarifah. *Jurnal Penelitian Perikanan Indonesia*. **13** (2) (2007)
17. Nursid, M. *Jurnal Ilmu-Ilmu Perairan dan Perikanan Indonesia*. **14** (1), 45 – 51 (2007)
18. Nursid, M. Tesis. Program Pascasarjana Institut Pertanian Bogor, Bogor. (tidak dipublikasikan). 63 pp (2002)
19. Okiyama, M. (ed). Tokay University Press, Tokyo. 1154 pp (1989)
20. Romomihtarto, K dan S. Juwana. Plankton Pusat Penelitian dan Pengembangan Oseanologi Lembaga Ilmu Pengetahuan Indonesia. Jakarta (1998)
21. Russel, F.S. The 456 pp (1976)
22. Sanchez-Velasco, L., C. Flores-Coto and B. Shirasago. *Estuariane, Coastal and Shelf Science*. **43** : 707-721 (1996)
23. Santoso, S. PT. Elex Media Komputindo, Jakarta. 432 pp (2000)
24. Tomigama, O. Fisheries Science. **66** (3) : 442 – 451 (2000)
25. Unesco Technical Papers in Marine Science. Report of the CICAR Ichthyoplankton Warkshop. Mexico City. 42 pp (1975)
26. Westhaus-Ekau, P. Series Course on The Sea and Its Resources. Jenderal Soedirman University, Faculty of Biology, Purwokerto. 54 pp (2002)

*Author email : marine_2807@yahoo.com

LOCAL FISHERMAN RESISTANCE IN KELAPA KAMPIT COASTAL AREA TOWARD OFFSHORE MINING ACTIVITIES

Ibrahim[1] ,*Dwi* Haryadi[2], *Nanang* Wahyudin[3]
[1]Political Science Department, Universitas Bangka Belitung, Indonesia
[2]Law Department, Universitas Bangka Belitung, Indonesia
[3]Management Department, Universitas Bangka Belitung, Indonesia

Abstract*Tin mining activities in Belitung Island have lasted since colonialism era to this moment. In 1990s tin exploitation in Belitung was stopped as it was assessed as no longer productive. In 1999, however, the mining had massively recommenced and at the same time, tourism and fishery were being actively developed as the new economic power. Recently, tin exploitation has shifted into offshore, and since then the conflicts began to appear. Rejection actions were continuously performed by fishermen through anti offshore mining audience, demonstration and campaign in social media. One of the rejections was when production suction vessel entered Pering waters, KelapaKampit East Belitung. At the urging of local fisherman group, written agreement between East and West Belitung head regencies, as well as the special committee recommendation of East Belitung house of representative, finally the suction vessel moved out from Pering waters. This study aimed to analyze fishermen's and stakeholders' knowledge, attitude, and expectationtoward offshore mining activities. Their point of views are needed in order to obtain complete perspective regarding the positive and negative sides, whether there is the meeting point among various interests and how long this conflicts will last. It is a mixed method research between qualitative and quantitative one. Quantitative approach employed questionnaire and SPSS 20 was used to analyze the results. Data collection technique used non-probability sampling with specific characteristic, fisherman community in Pering coastal area in which the amount applied Solvin theory. Qualitative approach was used to explore various* *phenomenons developing in society through interview with stakeholders related to offshore mining issues. Qualitative research was applied to explore phenomenon developing on site, either historically or contextually. Primary data was collected through interview with stakeholders in East Belitung related to the major issues raised. Based on the questionnaire results on fishermen and the interview with the related stakeholders, there were some findings. First, fishermen's knowledge about offshore mining was very limited. Second, most fishermen were agreed that offshore mining damaged ecosystem, reduced catching and disturbed tourism. Third, the expectation if offshore mining keep operating, fishermen will continuously fight for it. In stakeholder and social society level; though there is one supporting offshore mining, so far however, the majority still put their side on fishermen, tourism, and environment's interest and thus refusing offshore mining.*

Corresponding Author:iim_babel@yahoo.com

1 Introduction

Belitung Island is well known as Tin Island as that of Bangka in which they have experienced tin glory from time to time [1]. There is no exact number on when tin was first found and started to mine in Belitung. Based on J.J. van Sevenhoven report from Holland, in his visit in 1823 to Belitung, he had found tin nail and thus it can be said if tin has been mined at least in 17th century. In 1850, Prince Hendrik Baron van Tuyll van Serooskerkenorganizedly submitted proposal to open tin mining in Belitung and was granted on 23 March 1852. In 1852, this business switched to *Billiton Maatschappij*that established subsidiary *GemeenschappelijkeMijnbouwMaatschappij*(GMP) [2].

Other history document regarding tin finding in Belitung could also be quoted from the journal of John F. London in 1853 stating that "short journey for about 14 days for us was very important as the land we studied, and although it was shallow, it was found tin seeds on top layer and thus it could be concluded that this island part is as rich as the first mining business though it was performed as an experiment" [3].

In Belitung, tin glory in the past can be seen through the historical buildings available at this moment, in Manggar, TanjungPandan, and Gantung. Tin exploitation in the past ended in 1990. PT Timah suggested that tin production in Belitung had reduced and thus it must be moved to Bangka, and there would be company employees' reduction as a result of this policy. In 1999, non-conventional, massive mining activity as a result of central government policy was also occurred in Belitung Island. Hundreds of illegal mining popped up as well as leaved hundreds of ponds damaging environment [4,5]

As onshore tin deposit had been minimized, meanwhile the potential of offshore tin deposit was huge, and thus mining companies started to shift their exploitation area from onshore to offshore. Moreover, at this moment, it's already available more progress suction vessel technology reaching offshore tin mining compared to previous generation as the dredger. It is the offshore mining activity and its impact that have been a sensitive issue in Belitung since several years ago to this moment. While in another side, Belitung has started to develop the potential of fishery and marine tourism. This potential has been supported by the phenomenal of 'LaskarPelangi' novel by Andrea Hirata to abroad. Both sectors have been considered as more promising, sustainable, and environment friendly. As the consequence, mining sectors started to leave behind as it damaged the environment, non-sustainable, and the presence of dark experience in the past regarding tin employee rationalization in Belitung.

Apart from suction vessel and dredger, there have been also floating tin mining with the number of around 2,500. PT. Timah itself operates 11 dredgers. Some suction vessels operate by PT. Timah counterparts. All offshore mining activities have thrown away their hundred tons waste as sand and mud and thus it becomes sediment covering coral reef, algae habitat, and other marine biota. Bangka Belitung University study results suggested that 50% of coral reef in Bangka waters is damaged resulted from sedimentation. Offshore mining mostly operates in the area 5 miles from the beach. It is the small fisherman catching area. Daily fishermen are the most affected by marine ecosystem resulted from this mining.

The conflict of interest among miner, fisherman, and tourism as a result of offshore mining has repeatedly occurred not only in Belitung but also in Bangka. In 2015, in PT. Timah socialization related to production suction vessel operation planning on Matras beach, there were more rejections from fishermen as it was considered as damaging. Still in the same year, thousands of fishermen came to Governor Office to refuse suction vessel since fishermen catching dramatically dropped. In West Bangka, Belo Laut citizen also refused suction vessel and production suction pontoon operation. The fisherman also held demonstration in PT. Timah Office, Belinyu marine unit to refuse offshore mining in Kelabat. Offshore mining rejection action by fishermen recently occurred was conducted by South Bangka fishermen who held the demonstration in Local Police Office and PT. Timah.

In Belitung, October 2012, fisherman community came to Belitung Head Regency Office to sign the offshore mining exploration permission revocation. In local government level itself, there had been an agreement between East and West Belitung Head Regency regarding offshore mining that was then followed up by delivering a letter to Governor. RustamEffendie, a senior tourism actor, regretted offshore mining, considering that tourism has been proved to create employment. In the grass root level, people put their signature to support offshore mining rejection in Belitung Island in NyiurMelambai beach, Manggar East Belitung.

One of offshore mining rejections by fisherman occurred in Pering waters, Mayang village, KelapaKampit that became the focus in this study. They felt threatened by the existence of suction vessel. Many efforts have been conducted by local fishermen, starting from dialogue, demonstration, and even to the extreme effort such as burning the suction vessel. In addition, it collectively raised resistance effort through the appearance of FORTTAL (Forum Rakyat Tolak Tambang Timbang/Tin Mining Resistance People Forum) supporting people's struggle. East Belitung Local Government itself, in accordance to their political promise during their last campaign, assertively refused offshore mining. Belitung House of Representatives even created a special committee for it. This paper was eager to explore how actually local fishermen community resistance was, particularly in Mayang village in which its coastal area has become the IUP coordinate point and there was also rejection on vessel suction that would be operated. There are three basic items measuredin this paper: local concept knowledge, attitude, and expectation of fishermen and stakeholders as well as social society on offshore mining activity.

2 Methodology

This study applied mixed method between qualitative and quantitative study. Though different, both type of this study can fill each other [6]. This study type was used in considering of the wider information needed to formulate various phenomenon develop in the field, either historically or contextually. Therefore it will observe, take a close look and directly interact with natural setting [7]. Primary data was collected through interview with stakeholders in East Belitung related to the issues raised, such as Marine and Fishery Services, All Indonesian Fisherman Association (HimpunanNelayanSeluruh Indonesia/HNSI), Indonesian Marine and Fishery Institution, Tin Mining Resistance People Forum, and coastal village apparatus.

Quantitative approach was employed to measure fisherman community perspective, especially in Mayang village reacting on suction vessel presence. Based on KelapaKampit in Number of 2017, there were 195 Mayang people working as fisherman, and by using Solvin theory, it could be included 67 fisherman respondents being able to represent the population. However in execution, it was included 70 respondents. Questionnaire with question was organized in likert scale through descriptive approach measuring three variables: knowledge, attitude, and expectation toward offshore mining. SPSS 2.0 was applied to analyze questionnaire result data. Data collection technique was performed by non-probability sampling system and quota sampling with special characteristic, fisherman community.

3 Discussion

The discussion with fisherman resistance toward offshore mining began from descriptive analysis into substance. Fisherman was first described regarding age, gender, and others to see the fisherman profile that would be the major object of this study. Its perception was then about knowledge, attitude and their expectation toward offshore mining. Lastly, it was delivered description on stakeholder and local social society point of view.

3.1. Fisherman Profile

Fisherman profile was a short description regarding age, gender, population status, education level, length to be fisherman, and have they ever mined. This profile was important to know and understand fisherman character that would influence its perception on the knowledge of its context, attitude, and future expectation.

Table 1. Age

		Frequency	Percent	Valid Percent	Cumulative Percent
Valid	<21 yo	1	1.4	1.4	1.4
	21-30 yo	13	18.6	18.6	20.0
	31-40 yo	20	28.6	28.6	48.6
	41-50 yo	21	30.0	30.0	78.6
	>50 yo	15	21.4	21.4	100.0
	Total	70	100.0	100.0	

Table 2. Gender

		Frequency	Percent	Valid Percent	Cumulative Percent
Valid	Male	70	100.0	100.0	100.0

Table 3. Education Level

		Frequency	Percent	Valid Percent	Cumulative Percent
Valid	Elementary School	21	30.0	30.0	30.0
	Junior High School	19	27.1	27.1	57.1
	Senior High School	30	42.9	42.9	100.0
	Total	70	100.0	100.0	

Table 4. Population

		Frequency	Percent	Valid Percent	Cumulative Percent
Valid	Local	66	94.3	94.3	94.3
	New comer	4	5.7	5.7	100.0
	Total	70	100.0	100.0	

https://doi.org/10.1051/e3sconf/20184705003

Table 5. Length to be Fisherman

		Frequency	Percent	Valid Percent	Cumulative Percent
Valid	<10 yo	7	10.0	10.0	10.0
	10-20 yo	36	51.4	51.4	61.4
	21-30 yo	19	27.1	27.1	88.6
	31-40 yo	5	7.1	7.1	95.7
	>40 yo	3	4.3	4.3	100.0
	Total	70	100.0	100.0	

Table 6. Having Ever Activated as Miner

		Frequency	Percent	Valid Percent	Cumulative Percent
Valid	Yes	7	10.0	10.0	10.0
	No	63	90.0	90.0	100.0
	Total	70	100.0	100.0	

Based on Table 1 above, it could be seen that most fishermen becoming respondents were between 31 and<50 years old, 88% or 56 fishermen of total 70 respondents. There were only 18% or 13 fishermen were between 21 and 30 years old. Looking at the age side, the fishermen were mostly married with family economy burden. Based on Table 2, fisherman gender was all male who became the family backbone. Within education context, based on Table 3, 43% or 30 fishermen were senior high school graduation. There were then 21 and 19 fishermen respectively graduating from elementary school and junior high school. With maximum senior high school graduation, even there were junior high school and elementary school certificate holder, it made them having minimum skill to look for other profession. Hence, it was normal for them to refuse suction vessel, since they considered that it would disturb their only economical source.

From the demographic aspect, whether they were as local people or new comer. Based on Table 4, most fishermen of 66 people (94%) were local people who were born or have already long settled in Mayang village. There were only 4 fishermen as new comer. They were dominantly local people, indirectly having onward and inward connection with their land of birth or the place they lived and grew up. Local people felt to have responsibility to preserve their marine quality, their beach beauty as well as wealth of fish resources and others that during this time had been enjoyed since they were born.

Based on Table 5, there were 36 people (51%) having profession as fisherman for years. There were 19 people (27%) as fisherman at the age of 21 – 30 years old. The rest were working for less than 10 years (7 people), and in contrast, there were 3 people working as fisherman for more than 40 years. This data showed that most respondent profession was fisherman for ten of years who made ocean as the family economy resource. It means that it's not easy to shift their profession into other sectors, or just giving compensation from their mining production. Thus, fisherman was not only profession but also the obligation attached to them to preserve the ocean. Meanwhile, the question as they have ever mined during fishing, based on Table 6, of 70 fishermen, 63 (90%) stated that they had never mined, only 7 (10%) people who had ever mined, and that's the onshore mining not the offshore one. Meaning that in path with the length they mined (Table 5), majority fishermen in Mayang village truly made fisherman as their major profession and family economy source, although mining activity was sometimes tempting. In Kampak West Bangka coast, for example, research result showed the average earnings per month from mining could be twice more compared to catching fish [8].

3.2 Context Knowledge

Table 7. Offshore Mining Permitted as Regulation

Alternative	Frequency	Valid Percent
Totally do not Know	24	34.3
Do not Know	44	62.9
Know	2	2.9
Total	70	100.0

Table 8. Very Big Contribution of Offshore Mining

Alternative		Frequency	Valid Percent
Valid	Totally do not Know	20	28.6
	Do not Know	48	68.6
	Know Much	2	2.9
	Total	70	100.0

Table 9. Village People Depend on Ocean

Alternative		Frequency	Valid Percent
Valid	Do not Know	17	24.3
	Know	12	17.1
	Know Much	41	58.6
	Total	70	100.0

Table 10. Some People's Group Supporting Offshore Mining

Alternative		Frequency	Valid Percent
Valid	Know	7	10.0
	Know Much	63	90.0
	Total	70	100.0

Based on Table 7, fisherman community had minimum knowledge and understanding that offshore mining could be permitted if fulfilling the requirements. There were 24 (34%) fishermen who stated that they totally didn't know and 44 (63%) fishermen didn't know. There were only 2 fishermen knowing it. This very limited knowledge had become the high cause factor of fisherman resistance on offshore mining activity. The fisherman knowledge itself regarding the great offshore mining contribution for the territory was illustrated in Table 8. There were 20 fishermen stated totally not knowing, 48 not knowing, and only 2 fishermen stated knowing it. The high percentage of fishermen who didn't know it indicated the minimum information about offshore mining contribution. Based on Table 9, it could be seen that majority fishermen, 63 (90%), were very dependent on ocean in terms of their livelihood. This data was in accordance with fisherman profile as the length to be fisherman and the very limited number for them who ever worked as miner. Thus, local people were truly having profession as fisherman and their livelihood.

The presence of offshore mining becoming pros cons issue had indeed emerged the difference on attitude and opinion in the community itself. Based on Table 10, fishermen knew and understood that there were other community groups also supporting offshore mining. There were 41 (58%) fishermen answering knowing much, and 12 fishermen suggested to know it. This data suggested that fishermen were actually aware if there were some local people beyond them supporting offshore mining and also conducting some efforts to make it happen.

3.3 Miner Attitude

Table 11.Community is Involved in Offshore Mining Permission Socialization

Alternative		Frequency	Valid Percent
Valid	Agree	17	24.3
	Totally Agree	53	75.7
	Total	70	100.0

Table 12. Offshore Mining Damaged Coral Reef

Alternative		Frequency	Valid Percent
Valid	Do Not Agree	1	1.4
	Agree	2	2.9
	Totally Agree	67	95.7
	Total	70	100.0

Table 13. Offshore Mining Reducing Fisherman Catching

Alternative		Frequency	Valid Percent
Valid	Do Not Agree	1	1.4
	Agree	2	2.9
	Totally Agree	67	95.7
	Total	70	100.0

Table 14. Offshore Mining Disturbing Marine Tourism Development

Alternative		Frequency	Valid Percent
Valid	Agree	9	12.9
	Totally Agree	61	87.1
	Total	70	100.0

Table 15. Offshore Mining is Difficult being Side by Side with Fisherman Activity

Alternative		Frequency	Valid Percent
Valid	Agree	11	15.7
	Totally Agree	59	84.3
	Total	70	100.0

Table 16. Offshore Mining Only Benefit Small Number of People

Alternative		Frequency	Valid Percent
Valid	Do Not Agree	1	1.4
	Agree	4	5.7
	Totally Agree	65	92.9
	Total	70	100.0

Table 17. Offshore Mining is Banned and Dealt With Firmly

Alternative		Frequency	Valid Percent
Valid	Do Not Agree	1	1.4
	Totally Agree	69	98.6
	Total	70	100.0

The question in these indicators wanted to explore as how actually fisherman attitude toward the issues related to offshore mining. Based on Table 11, there were 53 (75%) fishermen who were totally agree if in the offshore mining permission proposal process, they were involved in socialization and thus it could be explained about its mining process, what efforts performed to prevent the impacts. Therefore, fisherman community was actually available to get involved and dialogued. Likewise, within the dialogue the company needed to hear fisherman opinion and input in order to guarantee the good and proper offshore mining, and thus there wouldn't be any conflict of interest between fisherman and miner.

One of offshore mining impacts is damaging coral reef, and it was justified by fisherman based on Table 12. There were 67 (97%) fishermen stating that offshore mining was truly damaging coral reef. This got along with previous research performed by Coral Reef Exploration Team in several points throughout offshore mining in Bangka. Most fishermen also stated that offshore mining had impact on the fishermen catching reduction. On Table 13, there were 67 (97%) fishermen suggested that offshore mining diminishing their catching. In accordance with the survey result, research by Nurcahya and Agustina [9] that offshore mining had reduced water quality as total dissolved solid increased and pH decreased; the changing on sea floor had caused the changing of flora, fauna, and plankton diversity and the increased of coral reef and fish mortality index related to them. The number of fish caught in offshore mining location was also decreased.

In terms of tourism development disturbance, in Table 14, there were 61 (87%) fishermen who were totally agree with it. The rest 13% said agree. It means that most fishermen were agree that offshore mining would definitely disturb marine tourism development that had become the major potential in Belitung. Related to tourism development, the greatest challenge for government and society were that are we able to shift economy backbone from mining basis to tourism basis, and can we become an example for developing a good tourism without waiting for the ending of mining activity [10]. "LaskarPelang" booming must be the tourism revival moment, and thus it must be supported by every development sector [11].

Fishermen society negative response toward offshore mining above in turn came to the perspective that offshore mining was difficult to get along with fisherman activity and tourism development. This could be seen from Table 15 that there were 59 (84%) fishermen suggested totally agree, and the rest suggested disagree. This majority opinion illustrated the rejection toward offshore mining activity. The rejection attitude, apart from impact factor, was also caused by offshore mining production that was only enjoyed by some people. Based on Table 16, 65 (93%) stated such thing. Along this time, tin economy had capitalist nuance in which the capital owner gained the most profit [12]. Whereas in UNCLOS 1982, the fisherman within archipelago country has traditional fishery rights [13]. It means that fisherman is actually the ocean owner and thus their interest must be prioritized. Finally, in accordance with Table 17, 69 (98.6%) fishermen suggested that the mining should be forbidden and firmly dealt.

3.4 Miner's Expectation

Table 18. The Attitude if Offshore Mining Keep Existing

Alternative		Frequency	Valid Percent
Valid	Protest Action	34	48.6
	Massively Resist	36	51.4
	Total	70	100.0

Table 19. The Form of Resistance Performed

Alternative		Frequency	Valid Percent
Valid	Dialogue	7	10.0
	Demonstration	46	65.7
	Violence if needed	17	24.3
	Total	70	100.0

Expectation indicator aimed to directly find out fisherman attitude related to the offshore mining existence. Related to the expectation if offshore mining keep existing, the data was rather balance in which 34 (48.6%) fishermen chose to have protest action and there were 36 (51.4%) fishermen choosing to massively oppose. Both data showed that fisherman would not keep silent and kept making resistance with various approaches, either through protest action or massively resist. Toward the choice of massively resist, there were more than 46 (65,7 %) fishermen choosing to make demonstration. There were only 7 (10 %) fishermen choosing to have dialogue and there were at least 17 (24,3 %) fishermen who would commit violence if needed. Here, it could be seen the effort to resist that would be performed by fisherman community if offshore mining remain permitted. This must certainly get attention from local government and all related parties in making decision in association with offshore mining in order that no party will be lost or even victimize. The environment conflict occurred between fisherman and mining company has two character, they are open and latent. Open character is conducted by protest action, dialogue until demonstration, while the latent one occurred as the mining keeps operating and the society obtains incomparable compensation to things they have experienced. Furthermore, latent conflict can also be happened between fisherman and the organization they work in, such as HNSI. Therefore, the fishermen can be then split, some believe to HSNI and some other doesn't believe much on HSNI [14].

3.5 Greatest Expectation

Fisherman expectation could also be seen from their greatest hope that could be classified into several offshore mining rejection augmentations: economy, environment, and tourism aspect. In economy aspect, some fishermen stated that marine is their life source in which their dependency on it is very high, as the only family income source, and mining will definitely diminish revenue and thus their life will be getting more difficult. The statement from one of the fishermen, Mh submitted: "Our life is greatly dependent on ocean and thus it must be preserved and free from mining." As was then stated: "If there is suction vessel, there must be followed by small offshore miner. Ocean will be more damaged and our livelihood will be more difficult." In the environment aspect, fisherman suggested that offshore mining would damage coral reef, hence making it difficult to catch fish. Therefore, they refused suction vessel. For fisherman, ocean is the legacy for their children and grand children later, thus it must not be ever damaged. Af said: "Marine damaged by suction vessel is not worth it with its result, thus we must preserve it." Bd said: "Mining waste will cover coral reef as the home for fish." The research result in South Bangka showed that the content of Pb and Fe within ocean water in tin mining area was higher than that of outside it and the quality threshold for it was 0.08 mg/L. This substance is absorbed by squid body and on the higher level, it will be poisonous and influence the stock deposit and in turn have impact on squid fishing area.

There was Pb heavy metal bioaccumulation within squid gill, 1.771 mg/L [15].

Third, the fishermen's greatest expectation related to offshore mining rejection was the big dream on the tourism progress as the future sustainable mainstay sector. The sustainable development principal itself does need the environment law awareness [16] and this had been possessed by the fishermen. Dh, for example, conveyed that offshore mining would disturb marine tourism. Yl then stated to make ocean as marine tourism. There was also Fe who asserted that Belitung ocean is famous with marine tourism, but not with offshore mining. In addition to the three expectations above, some fishermen also explicitlyexpressed their hope to the government in order to be more concern on tourism sector and minor fishermen fate rather than thinking about offshore mining in that its destruction would be difficult to overcome. Although law politic of natural resources domination is firmly asserted in constitution; however, their management resulted in injustice and nationally mining exploitation didn't also deliver prosperous to society [17]

3.6 Offshore Mining, Between Pros and Cons

Offshore mining rejection in East Belitung, especially by fishermen in KelapaKampit reaping pros and cons was actually not only disharmony among fishermen, tourism actors, and mining companies. However it's more than that, pros cons attitude, supporting and not supporting, agree and disagree also involved many parties either formally or informally, politically or non-politically. Politically, East and West Belitung Head Regency had agreed together to reject offshore mining. On the urging of fisherman union, East Belitung House of Representatives finally made Special Committee in association to the rejection of fisherman society in Pering waters Mayang Village in terms of KM Kamilah suction vessel presence that already operated in its IUP OP coordinate. The Special Committee issued several recommendations: first, asking local government to evaluate IUP and to Province Government to revoke and order to pull Kamilah suction vessel out of East Belitung water; second, in issuing permission in East Belitung Regency, Province Government should involve local government; third, asking Province Government and Provincial House of Representatives in discussing RZWP3K Local Regulation Meeting especially for East Belitung Regency to involve and ask input from stakeholder and community elements in East Belitung Regency; fourth, asking to Province Government through East Belitung Head Regency in order that there wouldn't be any mining activity in East Belitung ocean territory before Local Regulation regarding Zones Planning on Coastal Area and Small Islands was set; and the last, Special Committee agreed to refuse KIP presence and reject offshore mining in East Belitung water territory. This Special Committee recommendation showed House of Representatives partiality on fisherman community and tourism development.

Apart of Head Regency and House of Representatives, East Belitung Marine and Fishery Agency also agreed that offshore tin mining would impact on ocean ecosystem damage and fisherman catching area reduction. Zulfi who works on PNKPWP in DKP stated that the technology owned by KIP had not been able to minimize ocean ecosystem damage, and thus reducing fisherman catching area. No matter how big the offshore mining will be, it will always be non-sustainable in nature. Therefore, it will be better to prioritize tourism and marine fishery sector development that is sustainable in nature. Do not only calculate the economical profit, but it must also count the ecological lost value. East Belitung itself has proposed zero offshore mining to province and up to this moment this local regulation draft regarding zones hasn't been finalized.

Offshore mining problem from the perspective of East Belitung vice chief of All Indonesian Fisherman Association, MulyaKaryadi, had different opinion. According to him, maritime resource was not merely fish, shrimp, squid, and others, but also mining materials; moreover, if the minerals were as tin metal and the mineralcontaminant widely needed in industrial world. Therefore, the potential must be maximized in order to accelerate local development. It must be admitted that recently local government needs lots of fund, rather than striving to invite uncertain investors, it is better to permit offshore tin mining to operate as the revenue obtained is already obvious. The concept will be asking 10% revenue sharing and thus it will be expected around 20 trillion rupiah. The fund will accelerate development in all sectors, such as fish processing industry, tourism industry, and other programs supporting fisherman's and tourism actors' prosperity. There will always impact as the consequence of mining; however, the solution must be found out, but not just reject without solution. For example: tailing should not directly throw away to ocean without processing. We have ways or recent technology and environmental friendly, hence among offshore mining, tourism, and fisherman activities can run in harmony.

While from the village apparatus side, Mentawak head village, BapakDaerobi whose some of his citizens have profession as fisherman, seeing this offshore mining rejection as a result of several things. First, the benefit could not be enjoyed by people in general. Second, past disappointment when the tin was stated as run out so that there was employees cut off. Third, they were not willing if their ocean was destroyed as the ocean potential such as squid, and fish, and most of them were small fisherman. Fourth, as this time local people along with government are focusing on tourism program, and tourism aware group formation. There are then the compensation offered that will not influence people's rejection, since the community has already been comfortable with the ocean condition that has become their revenue expectation.

Indonesian Marine and Fishery Institution of East Belitung led by MukhaediMachwari stated that as an institution, they definitely rejected offshore mining, even we along with fisherman and local government have made letter to central government, since offshore mining will damage ocean ecosystem, fish breeding, turbid and muddy water as well as other damages. However, personally, I have other opinion. By seeing the minimum tourism infrastructure, the economy development remaining dependent on tin, and local government fund limitation, therefore for the sake of larger benefit and fisherman interest, tin mining should be permitted with special control and restrictions and responsibility; started from its processing, area permission, environmental friendly technology application, supervision and much involving local people, compensation improvement for local government and local people minimum 10%, as well as special compensation for directly impact society. The presence of offshore mining in East Belitung has produced social society power. One of them is People Forum for Refusing Offshore Mining led by Mr. Albani. This institution was founded as a result of citizen desire (fisherman, employee, and general people) to firmly refuse offshore mining. It doesn't believe on guarantee that offshore mining will not damaging. Dredge/suction vessel operates in fisherman fishing area must be disturbing much. The rejection was also caused by past disappointment of many former tin company employees. The compensation offered will not be comparable with the ocean damaged caused. Meanwhile many people depends their life on ocean. We are willing to support if there is really available the environmental friendly technology, however it should be shown and socialized to citizen. Environmental or ocean friendly suction vessel technology availability can actualize what was stated by Sudrajat [18] as good and green practice mining.

4 Conclusion

Resistance or fisherman rejection in Pering waters toward offshore mining has been very great. There were several reasons: first, the minimum socialization on offshore mining operation, the contribution and technology employed, as well as the efforts to overcome the impact; second, offshore mining, according to fishermen would destroy coral reef, diminish fisherman catching area and in turn decrease their income that depended a lot on ocean; three, offshore mining would disturb tourism development that had just begun; fourth, the offshore mining product would not be comparable with the marine damage caused, non-sustainable and only enjoyed by a small number of people; fifth, there hadn't been any environmental friendly offshore mining technology, including the proper reclamation pattern. While local government, house of representative, and the stakeholder, as well as social society, mostly had similar opinion with the fishermen to refuse offshore mining.

Thank you to the Directorate of Research and Community Service that has funded this research in the scheme of 2018 Competency Grant.

Reference

1 Bahari U. 2015.*Bangka Tin Magnet from Time to Time.*In Bangka History Seething. Bangka: Bangka Regency Culture and Tourism Agency

2 Sujitno S. 2007.*The Impact of Indonesian Tin Presence along the History*. Jakarta: Cempaka Publishing, Jakarta.

3 Kurniawan W. 2016.*The Last 'Kulek' (An Introduction to SawangGantong Tribe History).*Manggar: LPMP Air Mata Air

4 Haryadi D.2014.*Illegal Tin Mining (Life Time Problem?)*, In: After 14 years (The Journey Note of Bangka Belitung Islands Province). Yogyakarta: Imperium

5 Haryadi D.2015.*Criminogen Factor of Illegal Tin Mining in Bangka Belitung*. Law Problems Journal UNDIP Vol 44 No 1.

6 Sumanto 2014.*Research Method Theory and Application, Psichology, Education, Business Economy, and Social*. Yogyakarta: Center for Academic Publishing Service (CAPS)

7 Sugiyono 2009.*Quantitative and Qualitative Method and R&D*. Bandung: Alfabeta

8 Alpharesy MA. Anna Z. Yustiati A. 2012. *Revenue Analysis and Labor Fisherman Household Expenses Pattern in Kampak Coastal Area, West Bangka Regency.*Marine Fishery Journal UNPAD Vol 3, No 1.

9 Nurcahya E. Agustina F. 2015.*Managing the socio-economic impact of tin mining in Bangka Island, Indonesia - preparation for closure.* Conference: The 10th International Conference on Mine Closure, At Vancouver June 2015.

10 Megawandi Y. 2014.*Tourism Role in Bangka Belitung Islands Economy*. In: After 14 Years (The Journey Note of Bangka Belitung Islands Province). Yogyakarta: Imperium

11 Zukhri N. 2015.*LaskarPelangi Tourism Brandisization.* Imperium: Yogyakarta

12 Ibrahim. 2013. *Tin Chaos and the Stammered of Ideology Pancasila*. Imperium: Yogyakarta

13 Solihin A. 2010.*Marine and Fishery Law Politics (Issue, Problem, and Critical Policy Review).* Bandung: NuansaAulia

14 Sari YTW.Satria A. *Fisherman Access toward Coastal Resources in Mining Area.*Tran discipline Journal of Sociology, Communication, and Human Ecology. Vol 05. No 03

15 Febrianto A. Baskoro MS. Simbolon D. Haluan J. Mustarudin.*The Impact of Tin Mining Activities on Squid (Uroteuthischinensis) Fishing Ground In South Bangka.* International Journal of Sciences: Basic and Applied Research (IJSBAR) (2015) Volume 23, No 1, pp 283-293.

16 Utomo S. Haryadi D. 2013.*Mineral and Coal Mining Law*. Yogyakarta: Danadiyaksa.

17 Rosadi O. 2012.*Law Quo Vadis, Ecology and Social Justice*. Yogyakarta: Thafa Media.

18 Sudrajat N. 2010.*Indonesian Mining Theory and Practice according to Law*. Yogyakarta: PustakaYustisia

Effects of el nino on distribution of chlorophyll-a and sea surface temperature in northen to southern sunda strait

Mukti Trenggono[1*], *Amron* Amron[1], *Wanda* Avia Pasha[1], and *Damar* Lazuardy Rolian[1]

[1] Fisheris and Marine Science Faculty, Jenderal Soedirman University, Kampus Unsoed Karangwangkal, Purwokerto 53122 Indonesia

Abstract. Sunda Strait is an important passage for the Java Sea water to flow into the Indian Ocean. There is Java sea in North close to Karimata Strait and Eastern Indian Ocean in South part. Headed from north to south of Sunda Strait, has a high primary productivity that signifies the fertility of water. The strong El Nino (SOI) in 2015, hypothesized to affect variability values of a-chlorophyll content and sea surface temperature in this area. The research aims to know the distribution pattern of chlorophyll-a, and sea surface temperature due to El Nino effect on transition season I (March, April and May 2015). The relationship of both with SOI was analyzed by simple linear correlation analysis. The results showed that the distribution of chlorophyll-a in Northen Sunda Strait is not affected by El Nino but comes from the mouth of the Musi River. The pattern of sea surface temperature distribution from Northern to Southern of Sunda Strait showed in this area affected by El Nino so that the temperature is cooler. The relationship of chlorophyll-a with SOI has a negative moderate correlation (-0.532), indicating that chlorophyll-a in this waters have the direction opposite to SOI and sea surface temperature with SOI has a strong positive correlation (of 0.959).

1 Introduction

The Sunda Strait is located between the islands of Indonesia are among the territorial waters of the island of Sumatra and Java, shown in Figure 1. Astronomical position in the Sunda Strait is located at coordinate 2-5 ^{0} South latitude and 107-104 ^{0} East longitude [1]. The water body of the Sunda Strait is one of the areas that was essential in the activity of fishing. The dynamic water body conditions of the Sunda Strait influenced by the flow from the Java Sea and the Indian Ocean so that the mixing of the two water masses affect[1] the abundance and productivity of pelagic fisheries in the region. Most of the mass of the water heading to the Indian Ocean through the Sunda Strait.

* *Corresponding author: mukti.trenggono79@gmail.com*

Information on variability spatial temperature and chlorophyll-a sea level can be used to simplify the management and utilization of fishery resources as the basis for the suspect and determine the waters the potential for fishing ground. Sea surface temperature (SST) is the water temperature close to the surface of the sea. The water temperature of the sea mainly in the surface layer is heavily dependent on the amount of light receives from the Sun. The areas that receive the most sunlight is an area that is located at a latitude of 0° , therefore, the temperature of the sea water is highest at the equator [2]. Seawater surface temperature usually ranges from 27 °C – 29 °C in the tropics and 15 °C – 20 °C in subtropical areas. This temperature is decreasing regularly according to depth. The seawater temperature is relatively constant between 2°C – 4°C at a depth of more than 1000 m [3]. Sea surface temperature forms a zone based on the latitude. Getting closer to the Equator (low latitude) the temperature will increase in contrast to high latitude, the temperature decreased.

The crucial part is the nutrient for the growth of phytoplankton in the sea as well as the distribution of nutrients can be caused by the existence of a tropical storm on the surface, the upwelling and downwelling transfer [4]. The concentration of chlorophyll-a in waters of the beach and coastline are higher due to due to supply nutrient supply through run-off of the River from the Mainland, whereas low concentrations of chlorophyll-a in waters off the coast because of the lack of supply of nutrients from the mainland directly [5]. However, in some places still found the concentration of chlorophyll-a high enough, though far from the Mainland. The situation caused by the presence of upwelling. Upwelling or water mass ascent is the process of the accession of the mass of water from the deeper layers to the layer above or head to the surface [6].

Climate variability and change on the coast and the sea can have an impact on the uncertainty of the time and location of the occurrence of a potential fishery catch. Seasonal and annual climate variability in Indonesia is affected by the monsoon and ENSO [7]. Monsoon climate affects Indonesia through the movement of a point after the culmination of the Sun which resulted in Indonesia experienced a rainy season and a dry season [8].

The main factors that also affect the variability of climate in Indonesia are the ENSO (El Nino Southern Oscillation). ENSO is atmospheric sea interaction are centered in the region of Equatorial Pacific Ocean [8] (Aldrian, 2008) which cause global climate anomalies [9] Give ENSO effect on the symptoms of the condition of the sea in Indonesia that is becoming cooler in warmer El Nino and La Nina year [8]. ENSO consists of three normal events, namely the phenomenon of El Nino and La Nina. Division criteria on each depending on the Southern Oscillation Index (SOI) from April (0) to March (+ 1). The aims of this study are to determine the distribution pattern of chlorophyll-a, and sea surface temperature due to El Nino effect on transition season I (March, April and May 2015).

2 Material And Method

2.1 Satellite Data

Sea Surface Temperature Data and Chlorophyll-*a* were downloaded from https://oceancolor.gsfc.nasa.gov with spatial focus 4 km using Aqua Modis level 3 monthly and Chlorophyll-*a* data using OCI algorithm. Aqua/Terra Satellite has 36 bands and the satellite can orbit in entire earth every 1 – 2 days, Aqua tracks over the South to North and Terra tracks over North and South [10].

2.2 Methods

https://doi.org/10.1051/e3sconf/20184705004

The method used is the assimilation of the modeling result by using the Ocean Data View 4 software for spatial variability which analyzed by descriptive and statistic using correlation to find out the influence of ENSO on the Sea Surface Temperatures against different parts of the North and the South of Sunda Strait using Minitab statistical software 16. SPL data and chlorophyll-a were gained using MODIS image recording in March, April, and May of the year 2015. The image data processed using the *Ocean Data View* 4 and SeaDas 7.4. The coordinates point when cropping the image were from 2^0 North; 100^0 West; 8 ^{0} South; 110^0 East.

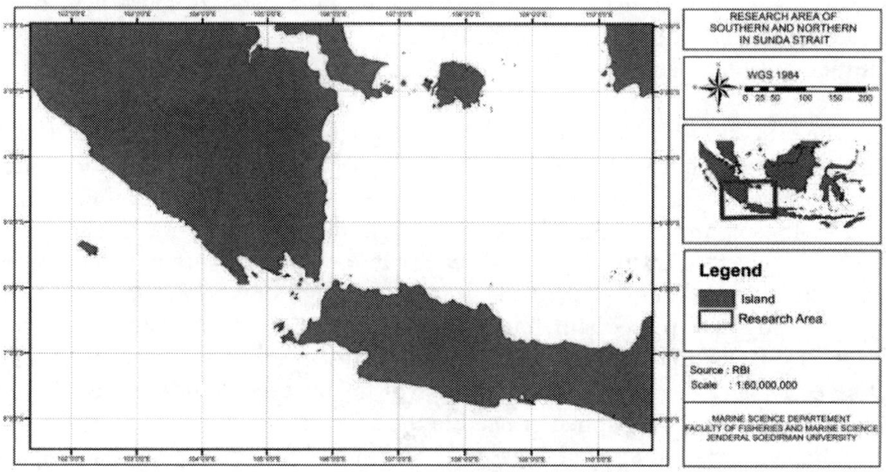

Fig. 1. Location of Research.

3 Result and Discussion

3.1 Distribution of Sea Surface Temperature

https://doi.org/10.1051/e3sconf/20184705004

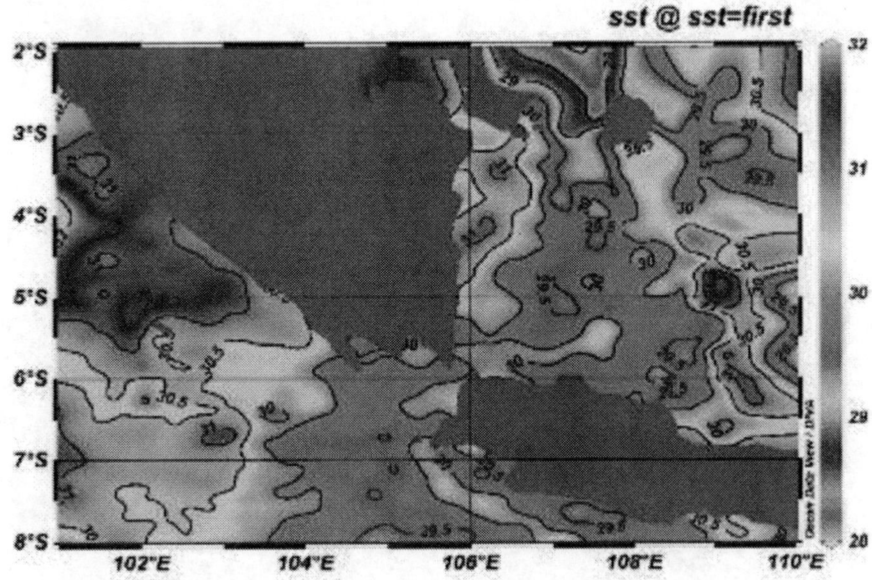

(a)

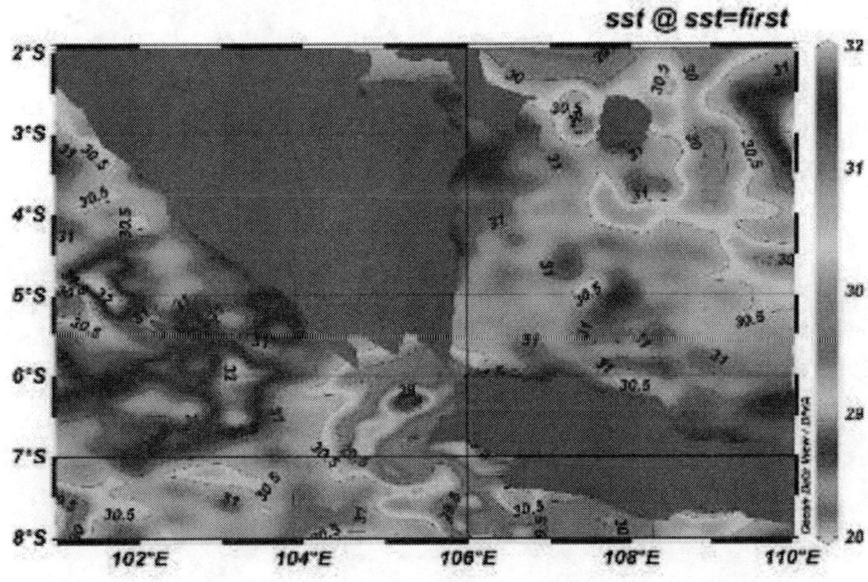

(b)

https://doi.org/10.1051/e3sconf/20184705004

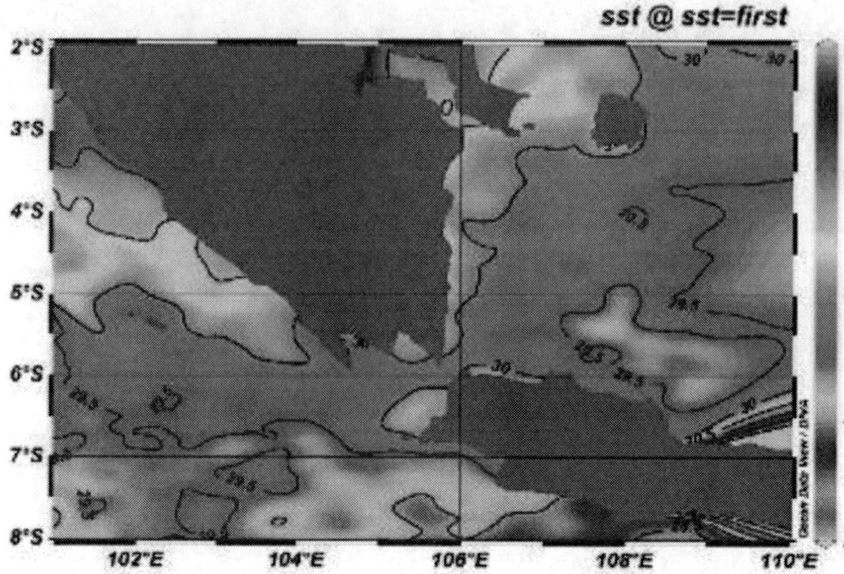

(c)

Fig. 2. Spatial distribution of Sea Surface Temperature 2015 in the Sunda Strait (a) March, (b) April, (c) May.

The results of the analysis show the temperature in March, April and May 2015 through the image (2a-c) has been changed. Surface temperature in March (Figure 2.a) shows the presence of mixing temperature on the northern part of the Sunda Strait ranges from 29.5-31.5^{0}C in territorial waters near the Karimata Strait. The sea surface temperature on Sunda Strait which located between Java and Sumatra island have same tend to be stable until Hindia Ocean valued 30^{0}C (Figure 2.a). On April 2015 (Figure 2.b) the waters of the southern part of tend to increase from March 2015, the magnitude of temperature in South of Sumatra sea range between 30^0 C – 31.5^0 C, in general, the water's temperature is normal. In Figure 2.c May 2015 showed a lower temperature difference i.e. amounting to 29.5 ^{0}C in both parts the waters spread out equally were the North and the South is relatively the same.

3.2 Distribution of Chlorophyll- *a*

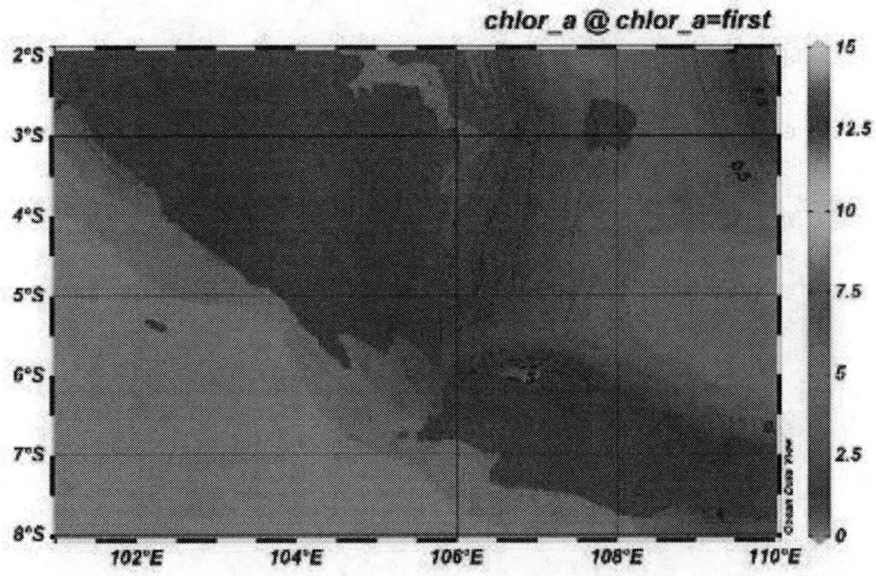

(a)

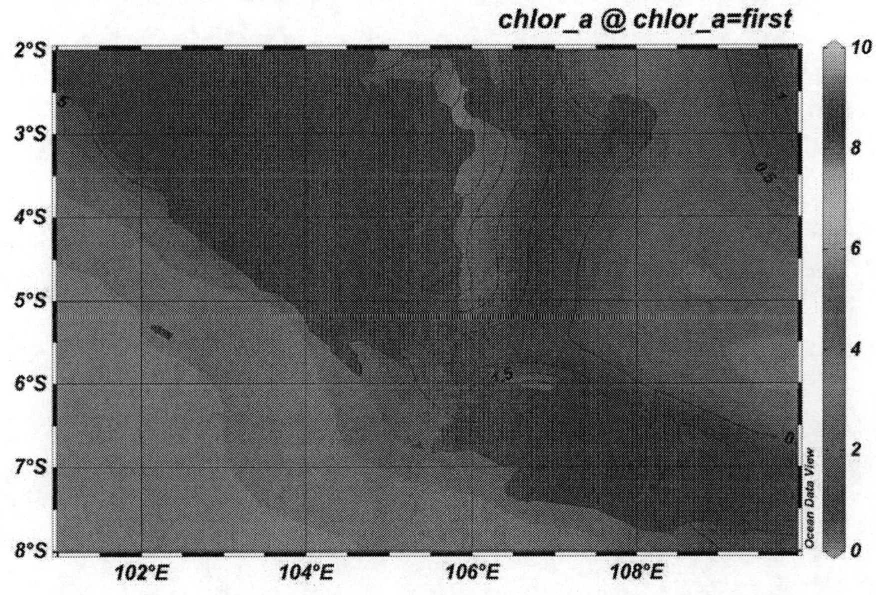

(b)

https://doi.org/10.1051/e3sconf/20184705004

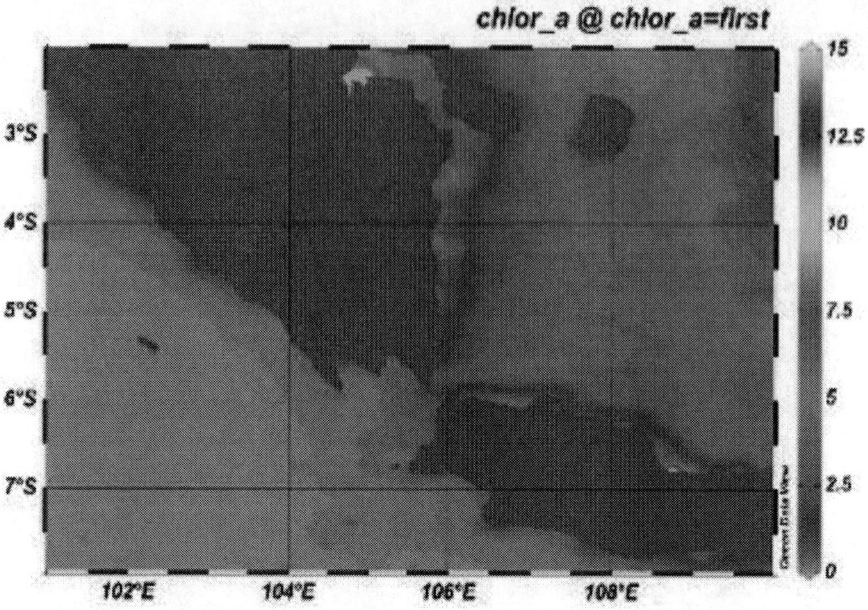

(c)

Fig.3. Spatial distribution of chlorophyll-a in 2015 the Sunda Strait (a) March, (b) April, (c) May

In Figure 3.a looks that chlorophyll in March in the northern part of the Sunda Strait has chlorophyll range 1.5-5 mg/m³ and the southern part of the Sunda Strait has the value ranges of chlorophyll 0-1.5 mg/m³. In Figure 3.b looks that chlorophyll in April in the northern part of the Sunda Strait has chlorophyll range 1-6 mg/m³ and the Southern part of the Sunda Strait has the value ranges of chlorophyll 0-1.5 mg/m³. In Figure 3.c look that chlorophyll in May has increased from the previous month, in the northern part of the Sunda Strait has chlorophyll range 1.5-9 mg/m³ and the southern part of the Sunda Strait has the value ranges of chlorophyll 0-1.5 mg/m³.

3.3 The Correlation Between Sea Surface Temperature With SOI

Table 1. *Pearson* correlation Sea Surface Temperature between with SOI.

Pearson correlation of SST and SOI	0.959
P-Value	0.183

Based on the coefficient of the correlation between sea surface temperatures with SOI years 2015 of 0.959 positive relationships with directions as well as the level of the relationship is very strong. The value of the p-value of the year 2011 of 0.183 which is greater than 0.05, so it can be said that there is no significant correlation in 2015 between sea surface temperatures with the SOI. The correlation between sea surface temperature by SOI in the year 2015 indicates the direction of the relationship is positive, so the temperature of the sea surface with the SOI has value and the same direction. Sea surface temperature and SOI in the year 2015 decline.

3.4 The Correlation Chlorophyll- *a* With SOI

Table 2. *Pearson* correlation Chlorophyll-a between with SOI.

Pearson correlation of CHL-A and SOI	-0.532
P-Value	0.643

Based on the coefficient of correlation between Chlorophyll- *a* with SOI year 2015 of 0.532 negative relationships with directions as well as the level of strong relationships. The value of the p-value of the year 2011 of 0.643 which is greater than 0.05, so it can be said that there is no significant correlation in the year 2015 with Chlorophyll- *a* between SOI. Correlation with Chlorophyll- *a* SOI in the year 2015 indicates the direction of the relationship is negative, so the Chlorophyll- *a* with SOI has value and opposite direction. If the value of the Chlorophyll- *a* is an increase, then the value of the SOI will go down. If the value of the Chlorophyll- *a* has decreased, then the value of the SOI will rise.

On the territory of the western part of Sumatra occurred high differences in temperature with the southern part near the island of Java. The high temperature in the western part of Sumatra occurred due to the influence of the monsoon. There is a relationship between the variability of monsoon and sea surface temperature [11]. Both North and South area of Sunda Straits to Karimata straits were the warmest month phenomenon and a lower temperature difference indicating the presence of upwelling [10]. The highest temperature areas exist in the territorial waters of Karimata Strait with temperature 30^0C state occurs due to the input of warm water from the river which modified before gets the mass output from the ITF (Indonesian Throughflow) [12].

Spatial distribution of chlorophyll in March, April and may seem that chlorophyll concentration values in the northern part of the Sunda Strait relative high in this case because the coastal areas located around the estuary of the Musi River, which resulted in a sufficient intake of nutrient substances the height of land. It is also reinforced by [13] the spread of chlorophyll-a concentration, in general, is relatively lower in the central part of the waters and the ever increasing towards the coastal waters. Movement of water masses to the East is causing the vacancy and will be replaced by water masses from a lower layer which contain cooler nutrient-rich. According to [14], stating that the concentration of chlorophyll-a can be caused by a high occurrence of nutrient-enriched on the surface layer of the waters through the various processes of the dynamics of water masses, including upwelling, vertical mixing of water masses and the mass movement of water.

Spatial distribution of chlorophyll on March, April and may seem that chlorophyll concentration value in the South of Sunda Strait relative low. This was a result of low water masses input in South of Sunda Strait which contains high nutrients from the coastal. On the Southwestern part of Java, the concentration of chlorophyll-a height that occurred in the East and the transition Seasons II is caused by the occurrence of upwelling along the southern coast of Java, at a time when Southeast Monsoon [15].

https://doi.org/10.1051/e3sconf/20184705004

CONCLUSION

The pattern of sea surface temperature distribution from Northern to Southern of Sunda Strait showed in this area affected by El Nino so that the temperature is cooler. The results showed that the distribution of chlorophyll-a in Northen Sunda Strait is not affected by El Nino but comes from the mouth of the Musi River. The relationship between chlorophyll-a with SOI has a negative moderate correlation (-0.532), indicating that chlorophyll-a in this waters have the direction opposite to SOI in contrast to the relationship between sea surface temperature with SOI which has a strong positive correlation (of 0.959).

Reference

1. J.T. Potemra, P.W. Hacker, O. Melnichenko, and N. Maximenko. J. G. *Satellite estimate of freshwater exchange between the Indonesian Seas and the Indian Ocean via the Sunda Strait*, **121,** 5098-5111 (2016).

2. P.K. Weyl. Oceanography. *An Introduction to The Marine Environment*. John Wiley & Sons Inc. New York (1970).

3. C.A.M. King. *Introduction to Coastal Oceanography*. McGrow Hill. New York (1963).

4. Zhao, Hui. and Ling Tang. J. G. *Effect of 1998 El Nino on the distribution of phytoplankton*. **112** (2007)

5. Nybakken, J.W. *Biologi Laut: Suatu Pendekatan Ekologis*. Gramedia. Jakarta (1992)

6. H. V. Sverdrup, M. Johnson and R. H. Fleming. *The Oceans, Their Physics, Chemistry and General Biology*. Prentice-Hall Inc. New York, NY (1942)

7. Z. Sitompul and E. Nurjani. J.B.I. *Pengaruh El Nino Southern Oscillation (ENSO) terhadap Curah Hujan Musiman dan Tahunan di Indonesia*. **2**, 1 (2013).

8. E. Aldrian, *Meteorologi Laut Indonesia*. Jakarta : Badan Meteorologi dan Geofisika (2008).

9. K. E, Trenberth, and J. M. Caron. J. C. *The Southern Oscillation Revisited: Sea Level Pressures, Surface Temperatures and Precipitation*. **13**, 4358 – 4365 (2000).

10. D W. Kusuma Murdinanto, A. Aden, L Y. Sukresno B, Jatisworo D. Hanintyo R. IOP Conference Series: Earth and Environmental Science **98** (2017)

11. Z . Kunarso, M. Ario, M. Raden, P. Bayu, Harmon. IOP conf. Series: Earth and Environmental Science. *Impact of Monsoon to Aquatic Productivity and Fish Landing at Pesarawan Regency waters*. **116** (2017).

12. R. Susanto, D. Zexun, Wei, R. T. Adi, Fan Bin, Li Suhijang, Fang Guohong. Acta Oceanol. Sin. *Observations of the Karimata Strait througflow from December 2007 to November 2008*. **32**, 1-6 (2013) .

13. B. Nababan, K. Simamora. J.I.T.K.T. *Variability Of Chlorophyll-A Concentration And Sea Surface Temperature Of Natuna Waters*. **4** (2012).

14. D. J, Pugesehan. J. Agroforestri. *Analisis Klorofil-A Fitoplankton (Produktivitas Primer) di Perairan Pantai Netsepa Kabupaten Maluku Tengah*. Politeknik Perdamaian Halmahera. Tobelo. **4**, 272- 278 (2010).

15. R. D. Susanto, A. L. Gordon, Q. Zheng. *Upwelling along the coasts of Java and Sumatra and its Relation to ENSO.* Geophysical Research Letters, **28,** 1599-1602 (2001).

Genetic mutation in mangrove *Acanthus ilifolicius* base on DNA Barcode (rbcL and matK gen) in the different environment change in coastal Cilacap, Central Java, Indonesia

[1*] R. Taufan Harisam, [1]Florencius Eko Dwi Haryono, [1]Bintang Marhaeni, [1]Norman Arie Prayogo, [1]Petrus Hary Tjahja Soedibya

[1*] Fisheries and Marine Science Faculty, Jenderal Soedirman University, Indonesia

Abstract. mangroves are salt-tolerant forest ecosystems of tropical and subtropical intertidal regions. They are among most productive, diverse, biologically important ecosystem and inclined toward the threatened system. In recent years, DNA barcoding using plastid markers *rbcL*and matKhas been suggested as an effective method to enrich traditional taxonomic expertise for rapid species identification, mutation genetic and biodiversity inventories. This research use survey method and descriptive qualitative analysis in the laboratory.This research aimed to determine the mutation DNA of plant barcoding standard*A. Ilifolicius* based gene *rbcL* and matK compare with species in the different location. Total DNA was isolated and successfully amplified by the Polymerase Chain Reaction (PCR) using primers based on the gene *rbcL*and matK. The results of sequencing long DNA fragments showed 760 bp are amplified by the forward primer and bp were 760 bp amplified by the primer for reverse. This study indicated that had been a mutation spesies in contaminated mangroves compared with uncontaminated mangroves.

1 Introduction

The coastal wetlands of the tropics and the subtropics of the world are characterised by the presence of a unique group of plant species, the mangroves. Mangroves are exist along the sheltered inter-tidal coastline, in the margin between the land and sea in tropical and subtropical areas. This ecosystem endowed with productive wetland having flora and fauna adapted to local environment such as fluctuated salinity, anoxic condition, and water level [31]. They are most productive and biologically important ecosystems of the world which provide goods and services to human society in marine and coastal systems [25]. They have unique features such as aerial breathing roots, extensive supporting roots, buttresses, salt-excreting leaves and viviparous propagules [11],[34].

The genus *Acanthus* has mangroves species i.e. *A. Ilifolicius* in mangroves all over the world. The genus *Acanthus* L. belonging to the family Acanthaceae is an Old World genus native of tropics and subtropics with about 30 species. It is often distinguished from the related genera by spiny leaves, spicate terminal inflorescences, two bracteoles and uniform anthers [11]. Four species *A. ilifolicius*, *A. Ilicifolius* L., *A. volubilis* Wall. and *A. xiamenensis* are known from mangrove communities and are classified as true mangrove species [28]. Of these *A. xiamenensis* is endemic to China and all the other species are common in Indo West Pacific (IWP) region. However, the taxonomical identity of *A. xiamenensis* in China is not clear; treated *A. xiamenensis* and *A.ilicifolius* as the same species [37]. Today they are threatened by high rates of anthropogenic disturbance, including habitat destruction, pollution, fragmentation, and changes in oceanic and estuarine environments due to climate change [36], [1], [10].

DNA barcoding appears to be a promising approach for taxonomic identification, characterization, and discovery of newer species, facilitating biodiversity studies [17]. It helps researchers to appreciate genetic and evolutionary associations by collection of molecular, morphological, and distributional data [17]. DNA barcoding is currently used effective tool that enables rapid and accurate identification of plant [22]. The Consortium for the Barcode of Life (CBOL) recommended *rbcL* + *matK* as the core barcode. However, these core barcode further combined with the *psb*A-*trn*H intergenic non-coding spacer region whichimproved discrimination power of core barcode. The noncoding intergenic region *psb*A-*trn*H exhibits high rates of insertion/deletion and sequence divergence [20]. These features make *trnH-psbA* highly suitable candidate plant barcode for species resolution. Later on, the nuclear ribosomal internal transcribed spacer (ITS) region considered as supplementary barcode, though China Plant Barcode of Life claimed ITS region had higher

*)Corresponding author : taufan.ltd@gmail.com

discriminatory power than plastid core barcodes [3], [18]. The ITS region has some limitations which prevent it from being a core barcode such as incomplete concerted evolution, fungal contamination and difficulties of amplification and sequencing [18]. Plastid gene large subunit of the ribulose-bisphosphate carboxylase gene (*rbcL*) is of 1350 bp in length and choice for DNA barcoding [4]. The maturase gene *matK* is about 1500 bp long and located within the *trnK* gene encoding the tRNALys (UUU). Substitution rate of the *matK* gene is highest among the plastid genes [17]. Plastid gene *matK* can discriminate more than 90 % of species in the Orchidaceae but less than 49 % inthe nutmeg family [20], [27]. In another case, identification of 92 species from 32 genera using the *matK* barcode could achieve a success rate of 56 % [12]. However, a recent study of the flora of Canada revealed 93 % success in species identification with *rbcL* and *matK*, while the addition of the trnH-psbA intergenic spacer achieved discrimination up to 95 % [2]. The species discrimination was lower (<50 %) for *rbcL* + *matK* combination in the study of tropical tree species in French Guiana [13]. Lower discriminationwere reported in closest and complex taxa of *Lysimachia*, *Ficus*, *Holcoglossum* and *Curcuma* using *rbcL*and *matK* [39], [41], [22], [5]. The lowest discriminatory power was observed in closely related groups of *Lysimachia* with *rbcL* (26.5–38.1 %), followed by *matK* (55.9–60.8 %) and combinations of core barcodes (*rbcL* + *matK*) had discriminationof 47.1–60.8 % [41]. Delineating mangrove species from putative hybridsusing morphological characters are always questionable. Putative hybrids were reported within the major generaof *Rhizophora*, *Sonneratia* and *Lumnitzera* and recentl in *Bruguiera* [10]. In the present study, we assessed mutation mangrove species using plastid coding loci *rbcL* and *matK* mangroves from coastal Cilacap, Central Java, Indonesia. This is our first steptowards DNA barcoding of mangroves based on plastid genes. Our study might be helpful towards mangrove conservation.

2 Methods

2.1 Sample collection

In the present study, leaf samples of 20 mangrove *A. Ilifolicius* were collected fromcoastal Cilacap, located on located on the west coast of Cilacap with geographical latitude of 8°35' S - 8°48' S and longitude of 108°46' E - 109°03' E.

2.2 DNA extraction

High content of phenolics, latex, mucilage, secondary polysaccharides and metabolites in these plants make it a difficult system for nucleic acid and protein isolation from mangrove plants. Cetyl-trimethyl ammonium bromide (CTAB) protocol for DNA extraction from mangroves was modified. Leaf tissue was pulverized in liquid nitrogen and pulverized leaf sample (0.2 g) were mixed with CTAB buffer (20 mM EDTA; 1.4 M NaCl; 2 % PVP-30; 1 % β-mercaptoethanol; 10 % SDS and 10 mg/ml proteinase K). The suspension was incubated at 60 °C for 60 min with gentle mixing and centrifuged at 14,000 rpm for 10 min at room temperature with equal volume of chloroform: isoamyl alcohol(24:1). The aqueous phase was transferred to a new tube and DNA was precipitated with 0.6 volume of cold isopropanol (−20 °C) and chilled 7.5 M ammonium acetate followed by storing at −20 °C for 1 h. The precipitated DNA was centrifuged at 14,000 rpm for 10 min at 4 °C followed by washing with 70 % ethanol. DNA was finally dissolved in TE buffer (10 mM Tris–HCl, 1 mM Na2EDTA, pH 8.0) and its quantity and quality was confirmed by agarose gel electrophoresis and nanodrop (Thermo Scientific, USA).

2.3 PCR and sequencing

Amplification of plastid genes (*rbcL* and *matK*) was carried out in 50-μl reaction mixture containing 10–20 ng of template DNA, 200 μM of dNTPs, 0.1 μM of each primers and 1 unit of Taq DNA polymerase (Thermo Scientific, USA). The reaction mixture was amplified in Bio-Rad (T100 model) thermal cycler with temperature profile for *rbcL* (94 °C for 4 min; 35 cycles of 94 °C for 30 s, 55 °C for 30 s, 72 °C for 1 min; repeated for 35 cycles, final extension 72 °C for 10 min) and for *matK* (94 °C for 1 min; 35 cycles of 94 °C for 30 s, 50 °C for40 s, 72 °C for 40 s; repeated for 37 cycles, final extension72 °C for 5 min). The amplified products were separatedby agarose gel (1.2 %) electrophoresis and stainedwith ethidium bromide [35]. Two pair of universal primers *rbcL* (*rbcLa*_F and *rbcLa*_R) and *matK*_390f and *matK*_1326r were used for the amplification purpose [19], [5]. PCR amplification was performed using forward and reverse primers [3]. Primers used for DNA amplification of *rbcL* and matK gene were *rbcLa*_ f : 5'-ATGTCACCACAAACZAGAGACTAAAGC-3'and rbcL724 -r : 5'-"GTAAAATCAAGT CCACCRCG" -3'; *matK_390-f* : 5'- "CGATCTATTCATTCAATATTTC"-3' and matK_1326-*r*: 5' TCTAGCACACGAAAGTCGAAGT-3'.

*)Corresponding author : taufan.ltd@gmail.com

3 Result and discusion

3.1 Genetic Mutations of Mangrove *A. Ilifoliciu s*using Barcode DNA based on rbcL and matK Gen

The sequencing result of *rbcL* produced an average of 580 bp without any insertion, deletion and stop codon, whereas *matK* sequencing produced 712 bp with few insertion and deletions in the form of gaps without stop codon. Sequence alignment of *rbcL*and matK gene of *A. ilifolicius*using Multalin (available at http://multalin.toulouse.inra.fr/multalin) showing different number and position of nucleotide (Picture 1 and Picture 2).

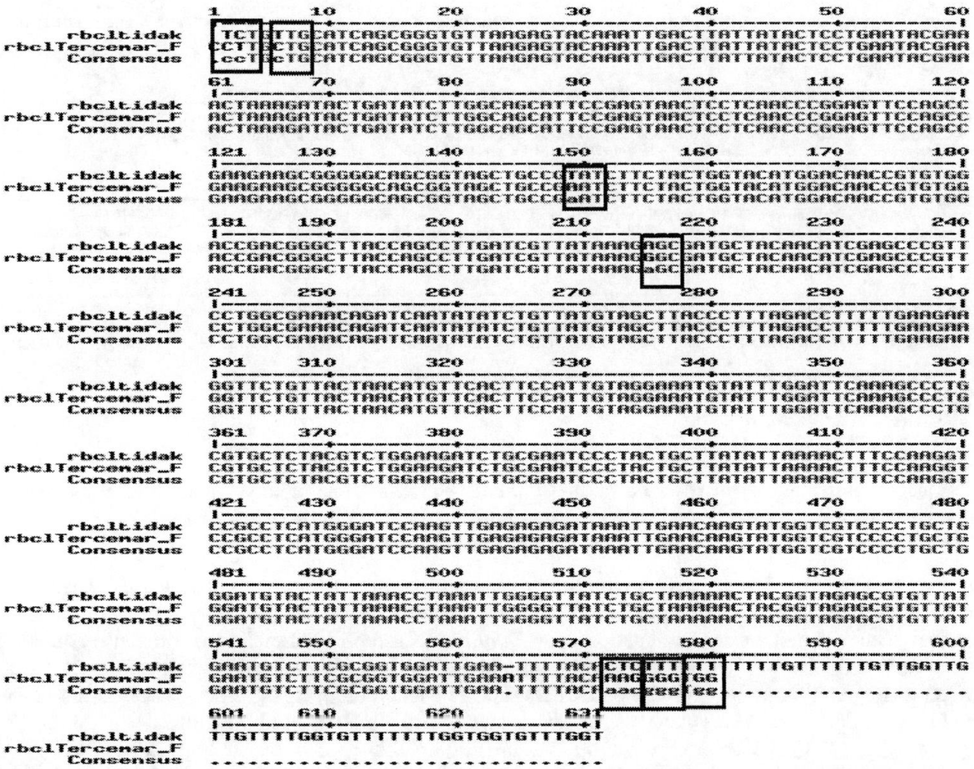

Fig. 1. Sequence alignment of rbcL gene of *A. ilifolicius*between contaminated mangroves compared with uncontaminated mangrovesusing Multalin (available at http://multalin.toulouse.inra.fr/multalin)

*)Corresponding author : taufan.ltd@gmail.com

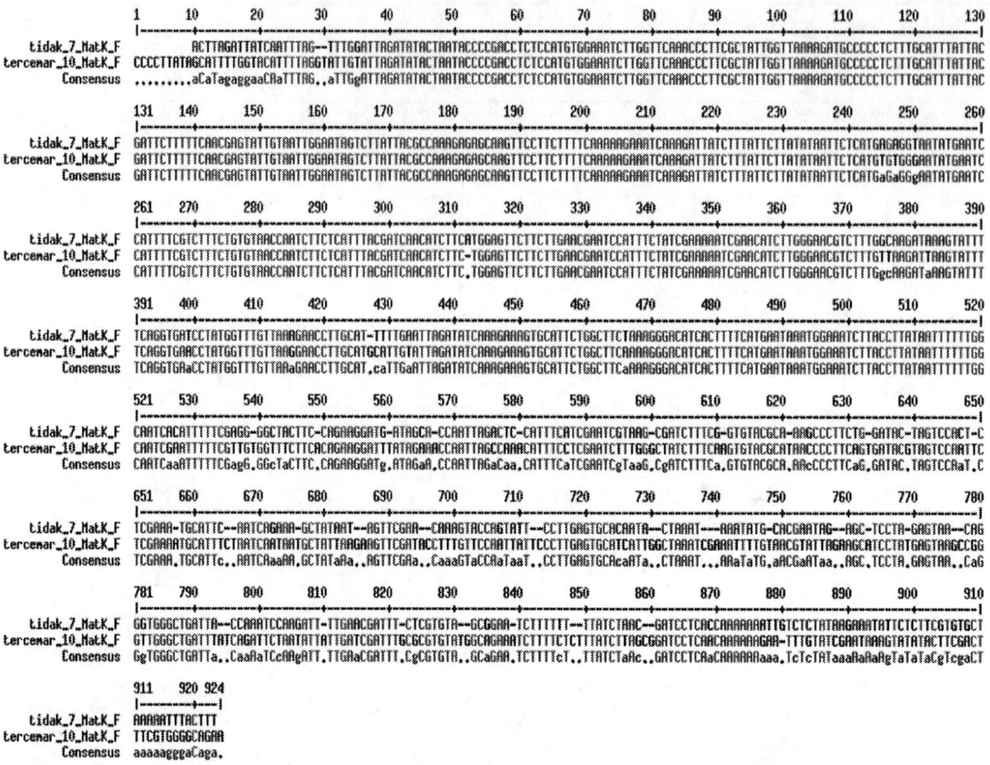

Fig. 2. Sequence alignment of matK gene of *A. Ilifolicius* between contaminated mangroves compared with uncontaminated mangrovesusing Multalin (available at http://multalin.toulouse.inra.fr/multalin)

Position of different nucleotides in species with the same or a different number of nucleotide difference was not the same in rbcL and matK at uncontaminated and contaminated location. The different of both nucleotides indicated that occure of mutation. The mutation of *A. Ilifolicius* based on rbcL gene has 7 mutation point. The mutation occurs at the locus 1-10 bp, 150 bp,215 bp and 570-500 bp there are TCT-CTT, GTT-GCT, GTA-GAA, GAG-GGG, ACT-AAA, CTT-GGG dan TTT-GTG. The mutation of *A. Ilifolicius* based on matK genes has 108 point mutation as seen in sequence alignment result. There are 10 bp, 12 bp, 14 bp, 15 bp, 16 bp, 17 bp, 18 bp, 19 bp, 20 bp, 23 bp, 31 bp, 35 bp, 245 bp, 247 bp, 250 bp, 375 bp, 376 bp, 382 bp, 399 bp, 415 bp, 428 bp, 429 bp,433 bp, 466 bp, 526 bp, 527 bp, 536 bp, 537 bp, 542 bp, 544 bp, 559 bp, 565 bp, 576 bp, 587 bp, 595 bp, 597 bp, 598 bp, 602 bp, 610 bp, 624 bp, 631 bp, 647 bp, 664 bp, 672 bp, 673 bp, 682 bp, 684 bp, 694 bp, 698 bp, 699 bp, 700 bp, 703 bp, 707 bp, 709 bp, 725 bp, 726 bp, 729 bp, 743 bp, 745 bp, 749 bp, 753 bp,756 bp, 757 bp,779 bp, 782 bp, 793 bp, 797 bp, 798 bp, 800 bp, 803 bp, 806 bp, 815 bp, 824 bp, 835 bp, 846 bp, 856 bp, 858 bp, 869 bp, 877 bp, 878 bp, 879 bp, 882 bp, 884 bp,888 bp, 889 bp, 890 bp, 892 bp, 894 bp, 896 bp, 898 bp, 900 bp, 902 bp, 904 bp, 906 bp, 907 bp, 908 bp, 911 bp, 912 bp, 913 bp, 914 bp, 915 bp, 916 bp, 917 bp, 918 bp, 919 bp, 921 bp, 922 bp and 923 bp.

To the best of our knowledge, current research is the first attempt of performing genetic mutation in mangrove *A. ilifolicius* based on DNA barcoding from Cilacap, Central Java, Indonesia using plastid core markers *rbcL* and *matK*. The rbcL barcoding genes has differentiation 7 nucleotide sequences nucleotide that is TCT→CTT, GTT→GCT, GTA→GAA, GAG→GGG, ACT→AAA, CTT→GGG dan TTT→GTG. The rbcL barcoding genes has differentiation 108 nucleotide sequence. The change of nucleotide bases caused converts the encoded amino acid caused by a point mutation [17]. The Point mutation is involves mutation the replacement of one base pairs (bases substitution), the one base at the DNA sequences replaced with different base. The mutation is the permanent change wich heredity in the genome (gene and nucleotid sequence) in some organism [1]. The gene coding caused of change nucleotid sequence without changed of amino acid (synonym mutation) [33], [34], [31] The change of nucleotide base in rbcL and matK gene maybe caused of long term anthropogenic pollution [40].

*)Corresponding author : taufan.ltd@gmail.com

5 Conclusions

DNA barcoding can be used to detection of mutation in *Acanthus ilifolicius*. The mutation of *A. Ilifolicius* based on rbcL and matK gene has 7 and 108 mutation point. The change of nucleotide bases caused by a point mutation. The change of nucleotide base in rbcL and matK gene maybe caused of long term anthropogenic pollution.

This project was funded by BLU (Junior Lecture Research) Jenderal Soedirman University.

References

1. Alongi D.M. Environ Conserv. 29: 331-349 (2002)
2. Burgess KS, Fazekas AJ, Kesanakurti PR, Graham SW, Husband BC, Newmaster SG, Percy DM, Hajibabaei M, Barrett SCH. **20**:8–12 (2011)
3. CBOL Plant Working Group. PNAS USA **106**:12794–12797 (2009)
4. Chase MW. Ann Missouri Bot Gar **80**:528–580 (1993)
5. Chen J, Zhao J, Erickson DL, Xia N, Kress WJ. Mol Ecol Resour 15:337–348 (2015)
6. Corpet F. *Nucleic Acids Res*16:10881-10890 (1988)
7. Dereeper A, Guignon V, Blanc G, Audic S, Buffet S, Chevenet F, Dufayard J-F, Guindon S, Lefort V, Lescot M, Claverie JM, Gascuel O. *Nucleic Acids Res* **36**:465-469 (2008)
8. Dereeper A, Guignon V, Blanc G, Audic S, Buffet S, Chevenet F, Dufayard J-F, Guindon S, Lefort V, Lescot M, Claverie JM, Gascuel O. *Nucleic Acids Res* **36**:465-469 (2008)
9. Dong W, Liu J, Yu J, Wang L, Zhou S. *PLoS ONE* 7:e35071 (2012)
10. Duke N, Meynecke J, Dittmann S, Ellison A, Anger K, Berger U, Cannicci S, Diele K, Ewel K, Field C. Science. **317**: 41-42 (2007)
11. Duke NC. University of Queensland and Norman C. Duke, Brisbane. (2006)
12. Fazekas AJ, Burgess KS, Kesanakurti PR, Graham SW, Newmaster SG, Husband BC, Percy DM, Hajibabaei M, Barrett SCH. PLoS ONE **3**:e2802. (2008)
13. Gonzalez MA, Baraloto C, Engel J, Mori SA, Pétronelli P. PLoS ONE **4**:e7483 (2009)
14. Hebert PDN, Cywinska A, Ball SL, deWaard JR. Proceedings of the Royal Society of London, Series B 270: 313–321 (2003)
15. Hebert PDN, Gregory TR. Systematic Biology **54**: 852–859 (2005)
16. Hebert PDN, Ratnasingham S, DeWaard JR. Proceedings of the Royal Society B Biological Sciences 270:S96–S99 (2003b)
17. Hilu KW, Borsch T, Muller K. Am J Bot **90**:1758–1776 (2003)
18. Hollingsworth ML, Clark A, Forrest LL. Mol Ecol Resour **9**:439–457 (2009)
19. Kress J, Erickson DL. PLoS ONE **2**:e508 (2007)
20. Kress WJ, Erickson DL, Swenson NG. PLoS ONE **5**:e15409. (2010)
21. Kuzmina ML, Johnson KL, Barron HR, Hebert PDN. BMC Ecol **12**:25 (2012)
22. Li HQ, Chen JY, Wang S, Xiong SZ. Mol Ecol Resour **12**:783–790 (2012)
23. Li X, Yang Y, Henry RJ, Rossetto M, Wang Y, Chen S. Biol Rev **90**:157–166 (2015)
24. Lissy KP and Simon TK. J. Mari. Biol. Ass. India. **43** :222-227 (2011)
25. Little DP, Stevenson DW. Cladistics **23**:1–21 (2007)
26. Manojit Bhattacharya, Ashish Ranjan Sharma, Bidhan Chandra Patra, Garima Sharma, Eun-Min Seo, Ju-Suk Nam, Chiranjib Chakraborty & Sang-Soo Lee. Mitochondrial DNA Part A, **27**:4, 2744-2752 (2011)
27. Newmaster SG, Fazekas AJ, Steeves RAD, Janovec J. Mol Ecol Resour 8:480–490 (2008)
28. Polidoro BA, Carpenter KE, Collins L, Duke NC, Ellison AM. PLoS ONE 5: e10095 (2010)
29. Purushothaman N, Newsmaster SG, Ragupathy S, Stalin N, Suresh D, Arunraj DR, Gnanasekaran G, Vassou SL, Narasimhan D, Parani M. *Genet Mol Res* **13**:2959-2968 (2014)
30. Purushothaman N, Newsmaster SG, Ragupathy S, Stalin N, Suresh D, Arunraj DR, Gnanasekaran G, Vassou SL, Narasimhan D, Parani M. *Genet Mol Res* **13**:2959-2968 (2014)
31. Ragavan P, Saxena A, Mohan P.M, Jayaraj R.S.C, Ravichandran K. J.Biodiversitas.Vol. 16 (2), page 225-237 (2005)
32. Ross HA, Murugan S, Li WLS. Syst Biol **57:**216–230 (2008)
33. Saadullah, Khan ZUD, Ashfaq M, Zaib u Nisa. Journal of Biodiversity and Environmental Sciences 8:175_186. (2016)
34. SaddeA.A, Rahul Arvind Jamdade, Kundan Kumar. *SpringerPlus 5:1554* (2016)
35. Sambrook J, Fritsch EF, Maniatis T. pp 5.1–5.17 (1989)
36. Valiela I, Bowen J.L, York J.K. Bioscience. 51: 807-815 (2011)

[*)]Corresponding author : taufan.ltd@gmail.com

37. Wang WQ, Wang M. The Mangroves of China. Science Press, Beijing [Chinese] (2007)
38. Widyastuti A, Yani E, Nasution KE, Rochmatino. Biodiversitas. **19** (1) 246-252 (2011)
39. Xiang XG, Hu H, Wang W, Jin XH. Mol Ecol Resour **11**:1012–1021 (2011)
40. Xu, S., He Z.,Zhang, Z., Guo, Z., Guo, W., Lyu, H., Li, L., Yang, M., Du, Z., Huang, Y.,Zhou, R., Zhong, C., Boufford, D.E., Lerdau, M., Wu, C.i., Duke, N.C., Shi, S. National Science Review. (2017)
41. Zhang CY, Wang FY, Yan HF, Hao G, Hu CM, Ge XJ. Mol Ecol Resour **12**:98–108 (2012)

The pattern of coastline change based on the characteristics of sediment and coastal slope in Pangenan coast of Cirebon, West Java

Suhendra[1], *Amron* Amron[1,*], and Endang *Hilmi*[2]

[1]Marine Science Department, Jenderal Soedirman University 53122 Purwokerto, Indonesia
[2]Aquatic Resources Management Department, Jenderal Soedirman University 53122 Purwokerto, Indonesia

Abstract. The coastline changes depend on stability of coastal bodies, such as characteristics of sediment and coastal slope. This research aimed to determine the relationship between the rate of coastline changes with the characteristics of sediment and coastal slope. The coastline changes were analysed by Landsat satellite images in 1991, 1999 and 2017 used End Point Rate (EPR) method on Digital Shoreline Analysis System (DSAS) module. Sediment characteristics (grain size and sediment statistics mean, sorting, skewness and kurtosis) were analysed by dry sieves and hydrometer with graphical method. The coastal slope used ASTER DEM (Digital Elevation Model) data analysis. The results showed that coastline changes at Pangenan coast of Cirebon in 1991-1999, 1999-2017 and 1991-2017 had average accretion rate respectively was 10.72 m/year, 7.25 m/year, 8.97 m/year and average abrasion rate was -12.49 m/year, -9.67 m/year, -6.70 m/year. The sediments were dominated by coarse silt, that had characteristics, were very well sorted, very fine skewed and platykurtic. The coastal slope was categorized as flat. The conclusion of this research was the rate of coastline change had not significant correlation with sediment and coastal slope.

1 Introduction

The coastline is a feature line between the land and sea boundaries. The position is not fixed each time, always change continuously based on the coastal condition [1]. A process of abrasion and accretion along the coast can characterize coastline changes. Coastal abrasion occurs when a coast is reduced by sediment causing coastline to be retreat. The accretion occurs when a coastal sediment addition [2]. The dynamics of coastline changes generally occur on the coast of Indonesia, including Pangenan coast. Pangenan coast is located in the east coast of Cirebon Regency, which is generally dominated by the accretion process [3, 4]. However, coastal abrasion occurred in some coastal areas of Cirebon [3].

The main factor-causing coastline changes that is the wave. Waves force the coastal bodies and caused abrasion [5]. However, the power of wave influence in causing coastline

* Corresponding author: twoone_brond@technologist.com

changes depends on stability of coastal bodies, such as characteristics of sediment and coastal slope. Beaches with flat slope will give a faster effect to the waves in causing coastline changes than the steep beaches [6]. Flat beaches have finer sedimentary characteristics compared to steep beaches [2]. So research on coastline changes associated with the characteristics of sediment and coastal slope is very relevant to do.

The aims of this research are: 1) to determine the rate of change of coastline at Pangenan coast of Cirebon, and 2) to determine the relationship between the rate of coastline change with sediment characteristic and coastal slope at Pangenan coast of Cirebon.

2 Methods

The study was conducted based on survey method [7]. The method consists of 3 (three) stages namely the determination of sampling point, data collection and data processing.

2.1 Study site and sampling point

This research was conducted at Pangenan coast, Cirebon, West Java with coordinates in 6°45'00"-6°48'00"S and 108°38'00"-108°41'00"E. Sampling point was determined by 9 points divided into 3 different areas along the study sites using systematic random sampling.

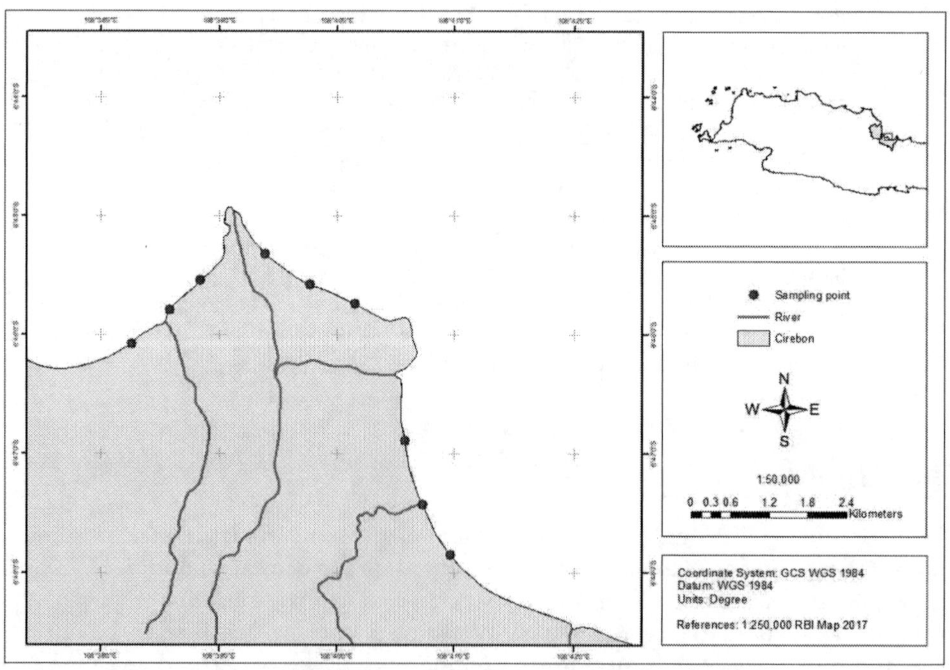

Fig 1. Map of study site and sampling points in Pangenan coast of Cirebon.

2.2 Data collection

Landsat satellite image data of Pangenan Cirebon consisted of images of 1991, 1999 and 2017. Coastal slope data were obtained based on estimates using DEM data. The DEM data used ASTER Global DEM V2 with 30 × 30 meters spatial resolution. Both of the data obtained from https://earthexplorer.usgs.gov. Sediment samples was carried out at each point by using a core sampler [7].

Table 1. Landsat image data type used for coastline change analysis.

Year	Satellite/sensor	Acquisition		Spatial resolution
		(dd-mm-yyyy)	(hh:mm:ss)	
1991	Landsat-5/TM	05-07-1991	02:47:00	30 m
1999	Landsat-7/+ETM	05-09-1999	02:47:14	30 m
2017	Landsat-8/OLI	25-05-2017	02:53:51	30 m

2.3 Data processing

2.3.1 Coastline data

Landsat images in 1991, 1991 and 2017 were first processed each time using ER Mapper software which includes: geometric correction, radiometric correction, band stacking and image enhancement [8]. Land and water (sea) boundaries were determined using Band-Ratio method.

Coastal delineation was done by visual interpretation method [9, 10]. The delineation result of the coastline was performed tidal correction [11].

Coastline data in 1991, 1999 and 2017 were overlayed, so it might be determine the suspected abrasion and accretion area. The coastline changes were analyzed using the Digital Shoreline Analysis System (DSAS) module with the End Point Rate (EPR) method contained in ArcGIS [12]. Then, the data was divided into 9 zones spatially, representing the research station.

2.3.2 Sediment samples

Sediment samples analyzed grain size using standard ASTM D 1140-00 and D 422-63 and calculated sediment statistics based on graphical method [13] include Mean size (Mz), Sorting (δ1), Skewness (Sk_i) and Kurtosis (K_G).

2.3.3 Coastal slope data

Coastal slope data derived from ASTER GDEM V2 data, processed in ArcGIS software to obtain the average of coastal slope in the study area. The average slopes were classified as flat (<5°), moderate (5°-30°) and steep (>30°) [14]. Then, the data was divided into 9 zones spatially, representing the research station.

3 Results and discussion

3.1 The rate of coastline change

The rate of coastline change in the study, consisted of accretion and abrasion rate. Accretion rate in Pangenan coast of Cirebon based on Table 3, in 1991-1999 occured in almost all zones except zone 6 and 7 which occured abrasion rates. The average accretion rate in that years was 10.72 m/year and the average abrasion rate was -12.49 m/year. In 1999-2017 and 1991-2017, accretion rate occured in same zone (i.e zone 1, 2, 3, 4) with average respectively was 7.25 m/year and 8.97 m/year, also abrasion rate occured in same zone (i.e zone 5, 6, 7, 8, 9) with average respectively was -9.67 m/year and -6.70 m/year. The highest accretion rate in the range 1991-1999 occurred in zone 4 with 15.96 m/year accretion and in the range 1999-2017 and 1991-2017 occurred in zone 1 with 11.36 m/year and 10.46 m/year accretion. Whereas the highest abrasion rate in all year ranges occurred at zone 6, with -19.79 m/year, -22.33 m/year and -22.53 m/year abrasion in respectively.

The tendency of the accreted coast was located in the gulf. Whereas the abrasioned coast was located in the promontory [15]. Based on Figure 3, it can be seen that the coastline of zones 1, 2, 3 and 4 from 1991 to 2017 was located in the gulf area which resulted in the accretion process in this area. Whereas the coastline in zones 5 and 6 was located in the promontory, so that in this area occurs the process of abrasion.

Coastlines in zones 7, 8 and 9 which were located in gulf occuring accretion in 1991-1999, but in 2017 occuring abrasion. It was due to the coastline in 2017 at the area had different effects compared to the other coastline zones. Based on field observation, the coastal seawall had been established, resulting in abnormal coastline changes.

The accretion and abrasion rate in the period between 1991-1999 and 1999-2017 generally decreased. The decreasing in the rate of change of this coastline based on field observation might be caused by changes in coastal morphology with the presence of coastal protective building. In addition, weak oceanographic pressure is also likely to cause a decrease in the rate of coastline change [16, 17].

Table 2. The rate of accretion/abrasion in Pangenan coast of Cirebon.

Zone	Accretion/abrasion (m/year)		
	1991-1999	1999-2017	1991-2017
1	8.52	11.36	10.46
2	11.71	8.26	9.35
3	14.59	8.32	10.30
4	15.96	1.05	5.76
5	2.34	-4.36	-2.24
6	-19.79	-22.33	-21.53
7	-5.18	-6.41	-6.19
8	8.84	-7.84	-2.58
9	13.05	-7.40	-0.95
Average accretion rate	10.72	7.25	8.97
Average abrasion rate	-12.49	-9.67	-6.70

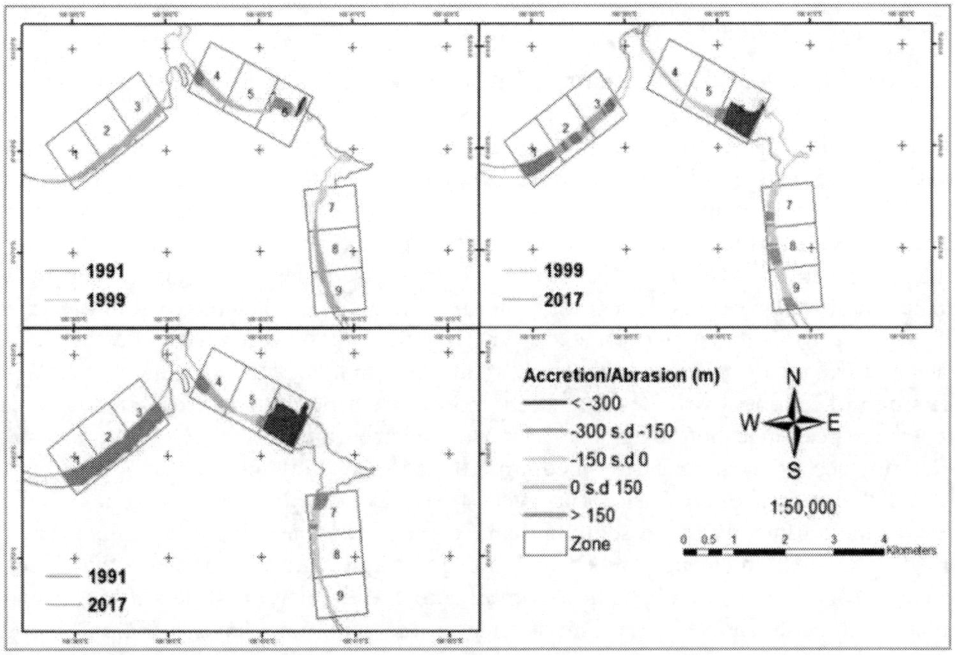

Fig 2. Map of accretion/abrasion in Pangenan coast of Cirebon.

3.2 The characteristics of sediment

3.2.1 Grain size

Based on Table 4, the coarse silt dominated the sediment type in all stations, with ranging from 29.96 to 70.68%. Whereas the least type of sediment is the granules, with ranging from 0 to 0.81%. If the sediment samples of each station are combined and averaged, then the order of sediment dominance in Pangenan of Cirebon consists of 52.06% coarse silt, 26.11% fine silt, 7.5% fine sand, 6.06% very fine sand, 5.99% medium silt, 0.86% coarse sand, 0.74% clay, 0.5% very fine silt and 0.18% granule.

Table 2. Grain size percentage of coastal sediment type in Pangenan coast of Cirebon.

Sediment type	Percentage (%)								
	St. 1	St. 2	St. 3	St. 4	St. 5	St. 6	St. 7	St. 8	St. 9
Granule	0	0	0.06	0.81	0	0.71	0.02	0	0
Coarse sand	0.07	0.08	0.73	1.25	0.35	5.13	0.12	0.03	0.01
Fine sand	0.06	0	0.33	23.87	22.4	20.38	0.2	0.13	0.09
Very fine sand	0.14	0.19	0	20.43	24.89	8.2	0.31	0.21	0.16
Coarse silt	53.91	48.19	57.82	32.45	29.96	35.7	69.8	70.68	70
Medium silt	4.47	10.08	6.66	4.23	3.2	6.41	6.33	6.2	6.37
Fine silt	36.88	38.09	31.07	16.95	19.2	23.49	23.21	22.75	23.36
Very fine silt	2.24	1.12	1.11	0	0	0	0	0	0
Clay	2.24	2.24	2.22	0	0	0	0	0	0

3.2.2 Sediment statistics

Based on the calculation of sediment statistics, the mean size of each station can be seen in Table 5. The mean size produced using unit phi (ϕ), Then, classified based on Udden-Wenworth scale. The mean size at stations 1, 2 and 3 belong to medium silt. The mean size at stations 4, 5 and 6 belong to medium sand to very fine sand. And the mean size at stations 7, 8 and 9 belong to coarse silt. Other sediment statistic data such as sorting coefficient, skewness and kurtosis can be seen in Table 6.

The larger mean size indicates the finer types of sediment, the smaller sorting value indicates the better sorted sediment, and the more negatively skewness value indicates the more presence of a large particle size, whereas the more positively indicates the more presence of a small particle size [18]. And the smaller kurtosis value indicates the distribution of sediment is flatter against the normal distribution and vice versa if the larger indicates the peak distribution against the normal distribution [19].

Dominance of mean sizes for medium silt and coarse silt in each of 3 stasions, very well sorted sediment in all stasions and very fine skewed sediment in 4 stasions indicated there was the uniformity of the sediment, an excess of fine particles and dominance of fine sediment type in Pangenan, Cirebon. The sorting value can also indicate the deposition type, the characteristic of deposition current and the deposition time [20]. Because the analyzed sediments showed very fine sorting, so the sediments have sufficient deposition time due to the weak water movement conditions and tend to be stable at the sampling site [18].

The accreted beaches in the study sites tend to be dominated by the average of silt sediments, whereas the abrasioned beaches are dominated by the average of sand sediments. This result contrast with research by [21] who reported that the abrasioned area in Egyptian Nile Delta had a finer characteristic of coastal sediments compared with the accreted area. The difference is probably caused by the presence of river mouths at study sites that contribute to high sedimentation. As reported in the [22] study that there was a significant relationship between coastline changes with sediment transport and runoff.

Table 3. Mean size sediment in Pangenan coast of Cirebon.

Station	Mean	Classification
1	5.07	Medium silt
2	5.16	Medium silt
3	5.21	Medium silt
4	3.77	Medium sand
5	3.88	Very fine sand
6	3.95	Very fine sand
7	4.82	Coarse sand
8	4.84	Coarse sand
9	4.82	Coarse sand

Table 4. Mean size sediment in Pangenan coast of Cirebon.

Station	**Sorting**		**Skewness**		**Kurtosis**	
	δ1	Classification	Sk_1	Classification	K_G	Classification
1	5.07	Medium silt	-0.28	fine skewed	0.65	very platykurtic
2	5.16	Medium silt	-0.08	symmetrical	0.68	platykurtic
3	5.21	Medium silt	-0.47	very fine skewed	0.53	very platykurtic
4	3.77	Medium sand	0.09	symmetrical	1.06	mesokurtic
5	3.88	Very fine sand	0	symmetrical	1.11	leptokurtic
6	3.95	Very fine sand	0.26	coarse skewed	0.89	platykurtic
7	4.82	Coarse sand	-0.40	very fine skewed	0.88	platykurtic
8	4.84	Coarse sand	-0.39	very fine skewed	0.91	mesokurtic
9	4.82	Coarse sand	-0.41	very fine skewed	0.88	platykurtic

3.2 Coastal slope

The slope value based on DEM data analysis had a average slope in study site was 2.46°. The lowest coastal slope was in zone 7 with 1.38° and highest in zone 5 with 4.19°. All zones were classified as flat beaches (Table 7).

Based on the average range of slope in Table 7, it showed that Pangenan coast of Cirebon was classified as flat beach. The accreted beaches had slope ranging from 1.99° to 3.83°, whereas the abrasioned beaches were ranging from 1.38° to 4.19°. The results was related to the statement of [3] that the beaches in Cirebon have a slope ranging from 2° to 4°. Flat beaches have high vulnerability to coastline changes [23, 6].

Table 3. Coastal slope in Pangenan coast of Cirebon.

Zone	**Coastal slope**	**Classification**
1	2.2	Flat
2	3.08	Flat
3	1.99	Flat
4	3.83	Flat
5	4.19	Flat
6	2.19	Flat
7	1.38	Flat
8	1.49	Flat
9	1.83	Flat

3.3 Correlation

The correlation between the rate of coastline changes with the characteristics of sediment and coastal slope in Pangenan coast of Cirebon based on correlation analysis, showed poor correlation and not significant. The value of significance between the rate of coastline change in 1991-1999, 1999-2017 and 1991-2017 with mean, sorting, skewness and kurtosis and coastal slope value was > 0.05 which means there is no significant relationship between each variable. However, based on description analysis showed the tendency was accreted coastline had finer sediment and flatter slope compared to abrasion coastline.

4 Conclusions

Based on the discussion described above, it can be concluded that:
1) The dynamic rates of coastline change at Pangenan coast of Cirebon in 1991-1999, 1999-2017 and 1991-2017 respectively had the average accretion rate was 10.72 m/year, 7.25 m/year, 8.97 m/year and the average abrasion rate was -12.49 m/year, -9.67 m/year, -6.70 m/year.
2) The relation of coastline change rates, sediment characteristic and coastal slope of Pangenan coast of Cirebon based on correlation analysis had no significant relationship. However, the result tendency was accreted coastline had finer sediment and flatter slope compared to abrasion coastline

References

1. M. K. Ghosh, L. Kumar, C. Roy, ISPRS J. of Photogramm. and Rem. Sens., **101**, 137-144 (2015)
2. B. Triatmodjo, *Teknik Pantai* (1999)
3. P. Astjario, F. Harkins, *J. Geologi Kelautan,* **3**, 19-26 (2005)
4. P. Raharjo, F. Novico. *Buletin Geologi Tata Lingkungan,* **22**, 115-127 (2012)
5. P. D. Komar, *Beach Processes and Sedimentation*, (1976)
6. J. Yin, Z. Yin, J. Wang, S. Xu, J. of Coast. Conserv., **16**: 123–133 (2012)
7. C. Rachmani, Rifardi, M. Galib, J. Online Mahasiswa (JOM) Bidang Perikanan dan Ilmu Kelautan, **4** (2017)
8. R. Azizul, Rifardi, M. Galib, Int. J. of Science and Research (IJSR), **6**, 1945-1948 (2015)
9. G. Winarso, S. Budiman, Judijanto, *22nd Asian Conference on Remote Sensing*, 1-5 (2001)
10. F. Kasim, A. Salam, J. Ilmiah Perikanan dan Kelautan, **3**, 160-167 (2015)
11. Sakka, *Model Perubahan Garis Pantai di Sekitar Delta Sungai Jeneberang, Makassar, Sulawesi Selatan*, Disertasi, Sekolah Pasca Sarjana, Institut Pertanian Bogor, (2012)
12. E. R. Thieler, E. A. Himmelstoss, J. L. Zichichi, A. Ergul, *The Digital Shoreline Analysis System (DSAS) Version 4.0 - An ArcGIS extension for calculating shoreline change (ver. 4.4, July 2017)*. U.S. Geological Survei, Reston. https://pubs.er.usgs.gov/publication/ofr20081278 (2017)
13. R. L. Folk, W. C. Ward, J. of Sedim. Petrol., **27**, 3-26 (1957)

14. NOAA, *Environmental Sensitivity Index Guidlines Version 3.0. NOAA Technical Memorandum NOS OR&R 11* (2002)
15. M. Arief, G. Winarso, T. Prayogo, J. Penginderaan Jauh, 8, 71-80 (2011)
16. K. M. Dewidar, O. E. Frihy, J. of Oceanog. and Marine Science, **1**, 028-039 (2010)
17. B. Deepika, K. Avinash, K. S. Jayappa, Int. J. of Environ. Science and Tech., **11**, 395–416 (2013)
18. A. Bayhaqi, C. M. Dungga, Depik, **4**, 153-159 (2015)
19. S. N. Bramha, A. K. Mohanty, M. K. Samantara, S. N. Panigrahi, K. K. Satpathy, Indian J. of Geo Marine Sciences, **46**, 1562-1574 (2017)
20. T. Solahuddin, E. Triarso, R. A. Troa, J. Segara, **2**, 42-48 (2006)
21. O. E. Frihy, P. D. Komar, Marine Geo., 115, 253-261 (1993)
22. B.-L. Cui, X.-Y. Li, Geomorph., **127**, 32–40 (2011)
23. T. S. Kumar, R. S. Mahendra, S. Nayak, K. Radhakrishnan, K. C. Sahu, J. of Coast. Research, **26**, 523-534 (2010)

The sinking ship policy to the perpetrator of illegal, unregulated and unreported fishing in criminal law perspective

Agus Raharjo*, *Tedi* Sudrajat[2], *Rahadi* Wasi Bintoro[3], and *Yusuf* Saefudin[4]

[1]Department of Criminal Law, Faculty of Law, Jenderal Soedirman University, Purwokerto – Indonesia
[2]Department of Administrative Law, Faculty of Law, Jenderal Soedirman University, Purwokerto – Indonesia
[3]Department of Procedure Law, Faculty of Law, Jenderal Soedirman University, Purwokerto – Indonesia
[4]Indonesian Teachers' Association of Victimization Jl. Prof. dr. HR Boenyaim 708 Purwokerto – Indonesia

Abstract The sinking ship policy by Minister Susi Pudjiastuti to the perpetrator of illegal, unregulated and unreported (IUU) fishing in the territorial sea of Indonesia raises the pros and cons from many perspectives. The policy background and policy legal consequences in criminal law perspective are the problems that will be discussed in this article. The research method is normative and empirical research. The results showed that the policy comes since the previous Government let perpetrators of the theft of fish in the sea of Indonesia. This caused losses for Governments as well as the fishermen who technically cannot compete in terms of fishing technology. The sinking ship policy also made due to the slowness of criminal justice cases against illegal, unregulated and unreported fishing in Indonesia which obviously does not comply with the principle of quick, simple and low cost justice. However, this policy is problematic in criminal law perspective, especially against the application of presumption of innocence principle and the proof process in criminal justice. Therefore, there needs to be laws that can be used as a guide and protector for this policy. Thus there is no problem in the future and instead become a breakthrough in fisheries law enforcement..

1 Introduction

There is a paradox that can be found in this country. On the one hand, we should be proud because Indonesia is one of the mega-biodiversity countries [1], where some of its territories are included in the biodiversity hotspots of Sundaland (West Nusantara) and Wallacea [1, 2], others still have natural ecosystems with high biodiversity of Sahulland (East Nusantara), and become the center of the world's coral triangle [3, 4, 5, 6].

[1] Corresponding author: agus.raharjo007@gmail.com

However, on the other hand, we should be saddened by the existing conditions, in which Indonesia became one of the countries with the highest extinction rate of biodiversity in the world. Natural habitat conversion is a major cause of biodiversity loss - illegal deforestation and conversion to oil palm plantations, whereas in the ocean biodiversity is lost due to the uncontrolled IUU Fishing [7] and a major threat to the sustainability of marine food resources [8].

Indonesia as an archipelagic country has approximately 5.8 million km^2 of marine sea (75% of the total area of Indonesia), consisting of 0.3 million km^2 territorial sea sea, 2.8 million km^2 of marine sea of the archipelago, and 2.7 million km^2 Exclusive Economic Zone of Indonesia (ZEEI). Based on data from the Geospatial Information Agency (BIG), Indonesia's coastline reaches 99,093 km [9]. The potential of fish resources in the area is estimated at 6.4 million tons per year [10], should be the driving force for economic development and welfare of the community, especially fishermen. However, the life of fishermen in Indonesia identic with poverty [11].

The limitation of fishing gear and vessel are technical reasons which are not the only one reason, but the illegal fishing activities of foreign and domestic fishermen that cause great potential were not explored to the maximum potential. Illegal fishing poses a threat to global food security [12], as the contribution of this activity reaches 30% of the total world catch [13, 14]. Indonesia's economic losses due to illegal fishing according to FAO reach Rp 30 trillion per year [15], but the loss number shown by the Ministry of Marine Affairs and Fisheries is even more fantastic, that is 20 million US dollars every year. This is only economic loss mentioned, not to mention losses due to marine ecosystems damage that could affect the sustainability of potential fishing in Indonesia [16].

Based on the background of unexplored potentials, unbridle illegal fishing, and especially changes in government policy since 2014, which emphasizes the vision of Indonesia as a maritime country, so the steps towards improvements in law enforcement are conducted. One of the most controversial policies is the ship sinking caught doing illegal fishing in the ocean of Indonesia. Actually, this policy is not a new thing, because it is regulated in Article 69 Verse (4) of Law no. 45 of 2009, stated that in carrying out its functions, the fishery investigator and/or supervisor may take special action in the form of burning and/or drowning of a foreign-flagged fishing ship on the basis of sufficient preliminary evidence.

In the period of 2009-2012 the number of ship that were excoriated only 37 ships were drowned [17]. However, since 2014-2016, the government has drowned 151 ships (18, 16].

This policy caused controversy, both on a practical and theoretical level. On a practical level, this policy is widely criticized by neighbour countries and foreign fishermen who have reduced their fish catch, as well as by domestic fishermen who do not comply with the rules of fishing in Indonesia's teritory. At the theoretical level, the sinking of ships without the judicial process does invite problems, let alone the vessel can serve as evidence in the trial. Based on Article 46 Verse (2) of the Criminal Procedure Code, evidence can be returned to the rightful, deprived of the state, or seized for destruction, or seized for evidence of another matter. This can be done if the trial is completed. These issues will be discussed in this paper as the hope of clarifying the legal position of the illegal fishing ship drowning in Indonesian criminal law.

2 Research Methods

The approach method used to answer the problem is the normative and empirical approach. In the normative approach, the law is seen as building a system of norms/rules that should support each other and not against in a system. The empirical approach is used to assess the facts, both in the form of factors causing law enforcement in the field of fisheries, the value of losses due to illegal fishing and policy implementation. Research data form is in the primary, secondary and tertiary legal material obtained through literature study which analyzed with qualitative analysis.

3 Causative Factors and Value Losses Due to IUU Fishing

In general, the vastness of Indonesian sea and the richness embodied in the sea is an attractiveness for anyone to use them, not least for criminals. Illegal fishing is a generic term for calling them, but as the time running, this term is not limited to "thieves" alone, extends to those who do not report the catch or fish in places that are prohibited by the authorities. Understanding Illegal, Unreported and Unregulated Fishing based on International Plan of Action (IPOA) which was initiated by FAO, script on IUU Fishing, are as follows [19, 20]:

(1) Illegal fishing is an illegal fishing activity which:
  a. Conducted by national ships or foreign ships in sea under the jurisdiction of 1 (one) country, without the consent of that country, or contrary to its laws and regulations;
  b. Conducted by ships flying the flag of a member country of a regional fisheries management organization but acting contrary to the conservation and management provisions established by that regional organization and binding the country, or other relevant international law provisions, or;
  c. Violate the provisions of national law or other international obligations, including those committed by states in cooperation with a related regional fisheries management organization

(2) Unreported fishing is a fishing activity that:
  a. Not reported, or deliberately reported by giving incorrect data to the authorities of the relevant national authorities, which are contrary to applicable laws or regulations in that country, or;
  b. Conducted within the territory competence of a regional fisheries management organization, where such activity is not reported or misreported, thus contrary to the reporting procedures of the organization.

(3) Unregulated fishing is a fishing activity carried out:
  a. In territories under the management of a regional fisheries management organization, by ship without nationality, or by ship flying the flag of a non-member country of the organization, or by a fishing entity, in a manner which is neither appropriate nor contrary to the provisions conservation and management measures of the organization, or;
  b. In the territory or on fish stocks that have no management and conservation arrangements, where such activities are carried out in a manner contrary to state responsibility under international law on marine biological resource conservation.

The extent of IUU Fishing can actually be a safety net for Indonesia to secure and utilize its marine wealth. Current government policies that pay more attention to the development of maritime axis is also a driver for law enforcement in Indonesian sea. There are 3 (three) Fisheries Management Areas (FMA) that are often victims of illegal fishing practices: FMA of South China Sea, there is decrease in stock density and demersal fish species size; Arafura Sea FMA there is a decrease in stock and weight of shrimp; and FMA of the Indian Ocean, a decrease in the number of catch and weight of tuna [21]. IUU Fishing is a factor that causes Indonesia can not maximally utilize its marine wealth. The rise of IUU Fishing is caused by many factors that influence it.

There are 7 factors causing illegal fishing in Indonesia, namely: (1) world fish demand is increasing, while world supply is decreasing, overdemand occurs especially for certain marine fish such as tuna, it is imendorong fishery world hunting fish anywhere by legal or illegal; (2) the price disparity of fresh fish intact in other countries than in Indonesia is quite high; (3) fishing ground in other countries is already exhausted, while in Indonesia is still promising, whereas they must maintain the fish supply for their consumption and must maintain the production of processing in the country; (4) the Indonesian sea is very wide and open, on the other hand supervision capability is still very limited; (5) fish processing system in the form of licensing system is open, the limitation is only limited to fishing gear, so it is less suitable if faced with factual condition of Indonesia geography, especially ZEE Indonesia bordered with another countries' sea; (6) the limited number of monitoring facilities and infrastructure and human resources supervision; and (7) the perception and cooperation of law enforcement apparatus in handling of criminal case of fishery still not solid, especially in terms of understanding legal action and commitment of supervising ship operation in ZEE [8, 7]

Due to the rampant illegal fishing, caused not only losses in terms of economy but also in terms of ecology. Supervision and Control of Marine and Fishery Resources (Pengawasan dan Pengendalian Sumber Daya Kelautan dan Perikanan/P2SDKP) discloses, some of the economic losses due to IUU fishing are (1) fishery levies paid by the tariff of Indonesian ship; (2) fuel subsidies enjoyed by unauthorized foreign ships; and (3) stolen fish production (volume and value). This loss is the impact of illegal fishing because of the direct loss of catch value that should be accepted by every coastal state [22, 21, 23, 24].

If an example is taken from one part of the Indonesian sea, Arafura Sea, it appears that the losses due to illegal fishing in the sea reached US $ 4 billion. When it calculated losses due to illegal fishing in the sea Arafura since 2001 - 2013, then the value can reach a fantastic amount, which is 0.52 Quadrillion rupiah (520 trillion rupiah, assuming exchange rate $ 1 = Rp 10,000) [8]. Economic losses can be calculated mathematically, but there are other disadvantages that are more important and can not be judged materially but are strongly related to the nation's self-esteem is the destruction of the image of Indonesia in the international because it is considered as unmanageable fisheries [25].

There are other problems in the management of marine resources, they are: (1) the use of net that violates the rules; (2) illegal transport of cargo in the middle of the sea, since the volume of cargo is not recorded at the port and as a result is not taxed; (3) the hunting of protected marine animals, where Indonesia is the world's largest shark kill, although this is against the rule. A total of 73 million sharks are killed every year just for the fins only. Hundreds of thousands of tons of sharks are captured by 7 countries alone. The main shark finers are Hongkong and China [26]. Despite protection laws and marine conservation areas, more efforts are needed to stop the slaughter of sharks [7]. Based on studies from UCSB and Research and Development Center Marine and Fisheries, if exploitation continues, fish biomass in Indonesian sea will drop to 80% by 2035 [27].

These issues are what the government wants to solve, one of them is the issuance of the ship's sinking policy. The value of large losses due to illegal fishing, the impact that is generated for both the state economy and the decline in the welfare of fishermen become a consideration of this controversial policy. The hope for sovereignty in the sea of the archipelago and the development of maritime axis which is the government's policy today is a strategic step to realize the welfare of the people, especially fishermen.

4 Criminal Law Perspective on Sinking IUU Fishing Shipper

There are 4 (four) aspects of the study related to the criminal law, namely the actions (criminal acts), people who commit crimes, punishment, and victims of crime. When referring to Law no. 45 of 2009 on Amendment to Law no. 31 of 2004 on Fisheries, these four aspects already exist in the arrangement, namely: the first aspect of the forbidden act is in Article 84 - Article 105; the subject of criminal offenses in the form of persons and corporations, although in corporate matters still contain controversy regarding the non-criminalization of corporations when committing a fishery crime (Article 101); the main types of punishment are two, namely prison within a certain time and a fine; and about the victims, of course the country and people of Indonesia (especially fishermen).

When referring to some research results, there are various ways of foreign fishing ship to conduct illegal fishing activities in the Arafura Sea, among them by falsifying permits and numbers on the hull of their ship, doing illegal fishing techniques in Indonesia, using Indonesian flags, arresting the night are, using some crew members of Indonesian citizenship, and so forth. Of the various types of illegal fishing by foreign fishing ship in the Arafura Sea, at most is to use the way of falsifying permits and writing the number of fake ships on board. This method is mostly found in the Arafura Sea, in addition to stealing fish, this ship can benefit refueling fuel (at government subsidized prices) and other supplies at the Fishery Port of Indonesia [8]. Such modus operandi is covered by article 84 until article 105.

Sinking vessel policy is actually stated in Article 69 verse (4) of Law no. 45 Year 2009, but the implementation until 2015 is not effective. The previous Government let perpetrators of the theft of fish in the sea of Indonesia. This caused losses for Governments as well as the fishermen who technically cannot compete in terms of fishing technology. The sinking ship policy also made due to the slowness of criminal justice cases against illegal, unregulated and unreported fishing in Indonesia which obviously does not comply with the principle of quick, simple and low cost justice. Therefore, the new government then set up a Task Force on Eradication of Illegal Fishing (Satgas 115) approved by President Joko Widodo through Presidential Decree no. 115 T 2015. Task Force 115 is a one roof enforcement system consisting of elements of the Navy, Police, Marine Security Agency (Badan Keamanan Laut/BAKAMLA) and the Attorney General, facilitating coordination, encouraging synergies, and implementing facilitation functions in combating illegal fishing [27].

The fishery law is a special law or special criminal law, which has a specificity compared to the general criminal law concerning the conduct, arrangement, legal subject, punishment, or judicial and procedural law. Based on such matters, then, pursuant to the perspective of criminal law, the sinking of ships or any action existing or regulated in such special legislation is allowed and valid, even against the common criminal law (material or formal). The problem that arises is the act of sinking the ship as it is legitimate so it is not considered a vigilante action.

When referring to Article 69 verse (4), the act of sinking the vessel may be carried out by the investigator on the basis of "sufficient initial evidence". When referring to a general definition in the criminal procedural law concerning such words, Article 17 of the Criminal Procedure Code provides that an arrest warrant is made against a person alleged to have committed a crime on the basis of sufficient evidence of the matter. The explanation of this article confirms that a sufficient initial proof is a proof of the beginning to suspect a criminal offense.

The development of the meaning of the word came after the decision of the Constitutional Court No. 21/PUU-XII/2014 declare unconstitutional conditional on the phrase "preliminary evidence", "sufficient preliminary evidence", and "sufficient evidence" in Article 1 number 14, Article 17 and Article 21 verse (1) Criminal Procedure Code, evidence in accordance with Article 184 Criminal Prosedure Code (witness testimony, expert statement, letter, guidance, and description of defendant). The Constitutional Court considers the minimum requirement of two evidence and examination of the candidate for the transparency and protection of a person's rights so that before a person is designated as a suspect he has been able to give a balanced account. This avoids any arbitrary action by the investigator, especially in determining sufficient preliminary evidence.

What is described in the Criminal Prosedure and the decisions of the Constitutional Court is related to the determination of status or action on the subject of law (human) in the investigation, but what is required as "sufficient initial evidence" in Article 69 verse (4) of Law no. 45 Year 2009 is on the object or tool used for crime (ship) which incidentally is not a legal subject. This can be seen from the explanation of the article, that is, "sufficient initial evidence" is the preliminary evidence to suspect a criminal offense in the field of fisheries by foreign-flagged fishing ship, for example, foreign-flagged fishing boats do not have Fishing Permit (Surat Ijin Penangkapan Ikan/SIPI) and Fishing Vessel License (Surat Ijin Kapal Pengangkut Ikan/SIKPI), as well as apparent fishing and/or transporting fish when entering the fishery management area of the Republic of Indonesia. The word in italics indicates that the perpetrator of a fishery crime is a ship, not a human being. The new thing, in this case, is to establish the status of the object or tool of crime (to be drowned, seized, or seized by the state) used the standard to establish the status of the legal subject in the criminal procedure law; and the establishment of a new legal subject, ie objects (ships) as criminals. Can these judgments be judged equally by establishing equal positions between humans and things?

It seems that lawmakers use the precautionary principle when using the requirements for the determination of legal subject status (human and corporate) on the determination of action against ship used for illegal fishing. At this point, the argument is acceptable, but the ship is not a mere human object, then a fishery law is directed to an object or a human being. Ships are only tools, which, when used for crime, fall into the category of evidence. Under Article 76A, this evidence may be destroyed, auctioned or handed over to a group of fishermen or cooperatives for use. At this point, the principle of the benefit of evidence in the ship's sinking policy is not achieved. Destruction of evidence can only be done if the goods are really dangerous if used by the public, such as narcotics, psychotropics, liquor, and others after obtaining approval from the local district court chairman. The question is what is the danger of a ship if it has been seized or taken away by the state?

At the practical level, policy brings significant change. Recorded until October 2017 has been drowned 317 ships and 90 other ships waiting for execution. Minister Susi Pudjiastuti claims this policy raises fish stocks by nearly 10 million tons per year (2015) to 12.54 million tons per year (2016). National fish consumption rose to 7kp/capita. As a result, the exchange rate of fishermen also increased because the fish in the local market is easily accessible and the price is not too expensive [28].

In the making policies should be based on strong theoretical studies and followed by accurate implementation. Criminal law in the field of fisheries is included in a special criminal law that may create rules that deviate or conflict with the general criminal law. However, such deviations should be followed by strong arguments, including if they are related to the basic principles of criminal law. This explanation is important in criminal law in order to be a material for the development of theory, and in practical terms, it can be translated correctly so that there is no gap between theory and practice.

5 IUU Fishing Prevention and Countermeasures

The prevention and control of IUU Fishing is not an easy thing, considering the position of Indonesian sea bordered on 10 countries so that cooperation between countries becomes an important issue. However, bilateral and regional cooperation has not been able to overcome the problem of illegal fishing in Indonesian sea [29]. It still takes more than a political step to realizing sovereignty in Indonesian sea.

Some policies that have been issued to achieve this are the moratorium on the issuance of licenses for foreign fishing ship for six months; Ministry of Marine Affairs and Fisheries Regulation no. 57 of 2014 which prohibits transshipment or sale of fish in the middle of the sea [30]; there is also Presidential Regulation no. 44 of 2016 on a negative list of foreign investments stating that fishing is closed to foreigners [27]; and some bilateral and regional cooperation with some neighboring countries.

There are several examples of research that can be taken into account by preventing and tackling IUU Fishing, such as charting. The baganisasi policy in addition to empowering existing fishermen is expected to reduce the activity of border water territorial violations by fishing ship from other countries. Thus the charting program will not only improve the economy of the border community but also be able to overcome the border problems of the region with neighboring countries [31]. The results of Solihin et.al [20] show that there are problems in eradicating illegal fishing in Sebatik Island, Kab, Nunukan, East Kalimantan due to low sea patrols and weak coordination of law enforcement apparatus. The charting policy has a positive impact, because: (a) impedes the entry of foreign fishermen; (b) shows the control of sea by RI, and (c) Other State Naval Ships entering the territorial sea of Sebatik Island and its surroundings.

Some international provisions can also be adopted to realize water sovereignty from the threat of illegal fishing. According to rule 19 in Safety of Life at Sea (SOLAS) V in 2002, the International Matrition Organization (IMP) determined that any ship weighing more than 300 GT (Gross tonnage) are required to have their Automatic Identification System (AIS) transmitters. This rule requires each of the ship to continue communicating AIS signals unless they are in a location where there are regulations prohibiting communication on such information [32].

International regulations require every ship above 300 GT to send a data using AIS to be an opportunity to detect ships doing fish theft. Indonesia's ability to detect AIS signals from LAPAN-A2 / Orari satellites enlarges the opportunity. The detection process is done by analyzing the travel data from the AIS system. The types of fish thefts that can be detected by this algorithm are trans-shipment, tiger trawls, territorial zone violations, non-reporting violations, violation of capture areas, and violations do not activate the AIS transmitter [33].

The prevention and handling of illegal fishing are not enough only by the issuance of policies, the application of research results or the adoption of international arrangements on fishing rules, the use of fishing gear or the utilization of technology. Prevention and mitigation of illegal fishing is the joint responsibility of all elements of the nation. There needs to be the involvement of the most relevant parties, the fishermen as the front guard in sea supervision where they rely on and depend on marine resources. Therefore, it is necessary to raise awareness about the important role of fishermen in safeguarding Indonesian sea through skills education and training not only in fishing capability but also in maintaining the sovereignty of fisheries in the Indonesian sea.

The empowerment of local wisdom in the prevention of illegal fishing is also worth considering because it is proven to be implemented in several areas [34, 35]. More than that is to prevent abuse of power from the law enforcement apparatus or the so-called "unscrupulous", in order not to use its power and authority to seek its own advantage.

6 Conclusion

Natural wealth contained in the Indonesian ocean is an attraction for anyone who wants to use it. Illegal fishing is a way of exploiting marine wealth that is not legally and ecologically destructive. The ship's sinking policy is an attempt to prevent and cope with illegal fishing which has caused much controversy. Practically this policy brings the benefits of decreasing the number of illegal fishing in Indonesian sea, increasing the number of fisheries production and fishermen welfare. On the other hand, this policy from the side of the criminal law contains issues particularly related to the phrase "sufficient initial evidence" as a condition of vessel sinking, the determination of the vessel as a legal subject (the perpetrator of the crime), and the actual value of the vessel's benefit is more useful than drowned. Criminal law in the field of the fishery is a special criminal law that may deviate from the general criminal law, but the deviation should be followed by a complete and accurate explanation in order to avoid debate and misinterpretation.

The researcher would like to express gratitude to some of the parties who has helped researcher during looking for data and finishing this paper. First and foremost, the researcher express thank you to Chairman of research institutes and community service, Dean of Faculty of Law, Jenderal Soedirman University. Also, thanks for my student, Luthfi Kalbuadi and Imas Puji Nuryanti, which has been helpful in searching data. Hopefully the help given will receive a reply from Allah SWT.

Reference

1. R.A. Mittermeier, P.R Gil, M. J. Pilgrim, T. Brooks, C.G. Mittermeier, J. Maloreux, G.A.B. da Fonseca, P.A. Seligmann, and H. Ford. *Hotspots Revisited: Earth's Biologically Richest and Most Endangered Terrestrial Ecoregions*. (Conservation International, New York, 2005)
2. L.J. Gorenflo, S. Romaine, R.A. Mittermeier, K. Walker-Painemilla. Proc Natl Acad Sci USA **109**, 8032-8037 (2012)
3. J.E.N. Veron. *Coral of the World*. (3 Volumes). (Australian Institute of Marine Science. Townsville, Queensland, 2000)
4. C. M. Roberts, C.J. McClean, J.E.N. Veron, J.P. Hawkins. G.R Allen, D.E. McAllister, C.G. Mittermeier, F.W. Schueler, M. Spalding, F. Wells, C.T.B. Werner. Science. **292,** 1280-1284 (2002)
5. TNC. *Coral triangle facts, figures and calculations: Part II: Patterns of Biodiversity and Endemism*. (The Nature Conservancy, Arlington VA, 2008)
6. UNEP-WCMC, WorldFish Center, WRI, TNC. 2010. *Global distribution of warm-water coral reef, compiled from multiple sources including the Millennium Coral Reef Maping Project.*
7. Sutarno, A. D. Setyawan. *Prosiding Seminar Nasional Masyarakat Biodiversity Indonesia*, **1**, 1-13 (2015)
8. P. Rahardjo. *Pandangan Ilmiah pada Workshop on Parameters and Indicators of Habitats to be Expressed in Map of Trawl Fishing Gear Management in Arafura Sea, Kerjasama Ditjen Perikanan Tangkap dengan Food Agriculture Organization (FAO),* Royal Hotel Bogor, 1-18 (2013)
9. G. Samantham. *National Geographic*, **21**(2013)

10. R. Dahuri. *Cetak Biru Pembangunan Kelautan dan Perikanan Menuju Indonesia Yang Maju, Adil-Makmur, dan Berdaulat*. (PKSPL-IPB, Bogor. 2010)
11. E. Sutrisno. Jurnal Dinamika Hukum. **14**, 1-12 (2014)
12. Agnew. Estimating the Worldwide Extent of Illegal Fishing. J. Plosone. **4**, 1 1-8 (2009)
13. A. Solihin. J. Mar. Fisheries, **1**, 113-210 (2010)
14. A. Solihin. *Politik Hukum Kelautan dan Perikanan*. (Bandung, Penerbit Nuansa Aulia, 2010)
15. P2SDKP. *Refleksi 2007 dan Outlook 2008: Pengawasan dan Pengendalian Sumberdaya Kelutan dan Perikanan.* (Jakarta, PSDKP, 2008)
16. M. R. Isnurhadi. Sekuritisasi Illegal. J. Hub. Int. **10**, 13-27 (2017)
17. DetikFinance. *RI sudah Tenggelamkan 38 Kapal Asing Illegal.* (, 2016)
18. D.H. Agung. *Susi telah Tenggelamkan 151 Kapal Pencuri Ikan*. Tirto.id, http://tirto.id/susi-tenggelamkan-151-kapal-pencuri-ikan-ED, accessed Maret 21, 2018.
19. A. Gustina. J. Dinamika Hukum. **14**, 340-349 (2014)
20. A. Solihin, S. Koeshendrajana, and F. Y. Arthatiani. *Paper on Seminar Nasional Tahunan IX Hasil Penelitian Perikanan dan Kelautan*, (Yogyakarta, 1-18, 2012)
21. A. Solihin, M. Imron, dan A. Wahyono. Buletin PSP. **20,** 205-211 (2012)
22. Marine Resource Assessment Group Ltd (MRAG). *Review of Impact of Illegal, Unreported and Unregulated Fishng on Developing Countries. Final Report.*(London, MRAG)
23. R. Yulianti and M. Ikhwan. J. Dinamika Hukum **18,** 29-36 (2018)
24. E. Sutrisno, and R. Apriani. J. Dinamika Hukum **17**, 266-272 (2017)
25. G.A. Wagey, S. Nurhakim, V.P.H. Nikijuluw, Badrudin, T.J Pitcher. *A Study of Illegal, Unreported and Unregulated (IUU) Fishing in the Arafura Sea, Indonesia*. (Research Center for Capture Fisheries, Agency of Marine and Fisheries Research, Jakarta, 2009)
26. Save Our Seas. How many sharks are caught each year? Save Our Seas Foundation. 2013.http://saveourseas. com/articles/how_many_sharks_are_caught_each_year, access on January 15, 2015.
27. Marta, and M. Fajar. Susi dan Tiga Pilar yang Tak Sekadar Mitos. *Kompas*. September 07, 2017. Accessed Oktober 10, 2017. http://ekonomi.kompas.com/read/2017/09/07/ 051546726/susi-dan-tiga-pilar-yang-tak-sekadar-mitos.
28. Haruskah Susi Pudjiastuti Berhenti Menenggelamkan Kapal? http://www.dw.com/id/ haruskah-susi-pudjiastuti-berhenti-menenggelamkan-kapal/a-42077235.
29. S. V. Muhammad. Politica **3**, 59-85 (2012)
30. P. A. N. Indra. Tenggelamkan Kapal Lagi, Bu Susi? Tirto.id, Agustus 04, 2016. http//tirto.id/ tenggelamkan-kapal-lagi-bu-susi-bw5E, accessed Maret 21, 2018
31. Koesrianti. J. Mimbar Hukum **20**, 193-410 (2008)
32. IMO. SOLAS 1974: Brief History – List of amendments to date and where to find them. (2013) http:/www.imo.org/ en/ KnowledgeCentre/ Reference AndArchives/ HisoryofSOLAS/Documents. Access date June 20, 2017.

33. E.M. Husni, M. Riksa, R.S. Andanawari, and R. H. Triharjanto. Jurnal Teknologi Dirgantara. **14**, 81-90 (2016)

34. J. M. Monteiro. J. Dinamika Hukum. **17,** 83-88 (2017)

35. Adwani, Mahfud, and Rosmawati. Jurnal Dinamika Hukum **16**, 1-7 (2016)

Corporate criminal liability towards the doer of illegal, unregulated and unreported fishing in indonesia

Rahadi Wasi Bintoro[1,*], *Agus* Raharjo[2] , and *Tedi* Sudrajat[3]

[1]Department of Procedural Law Faculty of Law Jenderal soedirman University, Purwokerto, Indonesia
[2]Department of Criminal Law Faculty of Law Jenderal soedirman University, Purwokerto, Indonesia
3Department of Administrative Law Faculty of Law Jenderal soedirman University, Purwokerto, Indonesia

Abstract. The seas in Indonesia are so wide and its natural wealth (fish) just can be explored and exploited by the ships which are owned by corporation. Those ships are potential to do illegal, unregulated and unreported (IUU) fishing. Even though corporation might commit a crime but the corporation can not be asked yet for the criminal liability. Nevertheless, criminal liability is charged to corporation's staff. This research is included as normative research based on the study of law material. Stipulation of Article 101 Law Number 45 year 2009 jo Law Number 31 year 2004 on fishery, state that the claim and sanction in fishery criminal case which done by corporation will be given to corporation's staff in form of fine penalty and added third of the given penalty. This is a reverse of criminal law because the other fields make corporation as the doer and can be asked for their responsibility. The theory development on criminal corporation liability seems does not followed by law maker. Consequently, there is no deterrent effect towards the doer of fishery criminal corporation cases. It needs law amendment and insert the corporation criminal liability clause, so there is a visible deterrent effect for the doer.

1 Introduction

Indonesia is an archipelago, with a vast territory and it contains a lot of natural sources, either on land or in the sea. As an archipelago, Indonesia has a wealth marine which is priceless. This is a gift from the Almighty God who should be thanked and used properly for the prosperity of all Indonesian people.[1]

The strategic value of Indonesia as a rich archipelago on fish and natural resources has affected the region to has a potential conflict. Various kind of violation are often perpetrated and continued to evolve as if there were no fear or awkwardness to commit

* Corresponding author: rahadiwasibintoro@gmail.com

further violations. This is urging the government to enforce the law against any act that is considered to violate and harm the nationality.

The fishery sector is one of the sectors that has the greatest potential in optimizing the state revenues, on the other hand it can also be a source of reduction to the state revenues if companies or corporations commit the illegal action, Illegal practices, Unreported, Unregulated (IUU) Fishing practices that intended only to increase the assets of the corporation's own wealth.

The example of fish catching without giving any concern to the environmental sustainability is like it done using explosives, moreover it is likely to be done by foreign corporations. Furthermore, this Illegal mode, Unreported, Unregulated (IUU) Fishing has even developed no longer perform desperate action by entering the territorial waters of Indonesia. Nowadays there are many indications of Indonesian-flagged ships fishing in Indonesian waters but actually taken to the high seas, then trans- shipment to foreign ships. [2] At present, more data show 60-70% of captains and crew of foreign fishing vessels are recidivists in the ship's home country. [3] The total of drowned-thieves' fishing boat from 2014 to April 2017 reached until 317 ships. The number of fishing vessels captured by Satgas 115 to June 2017 was 294 fishing boats with the details of 116 Indonesian ships and 66 foreign ships. The foreign ship consists of 6 Malaysian flagged vessels, Vietnam 54 units, Philippines 5 units, and Taiwan 1 unit. [4]

This is a challenge for the law enforcement in the field of fisheries, given the violations in the field of fisheries, such as illegal fishing, still continually occur, even when the law enforcement done firmly. Unfortunately, the occurrence of Illegal, Unreported, Unregulated (IUU) Fishing show that the purpose of crime, namely the deterrent effect [5] for the perpetrator itself and others, has not been achieved. Moreover, illegal fishing tends to be done by corporations. Therefore this article will discuss the liability of corporations on the criminal acts in fisheries.

2. Research Method

This paper is a part of legal research and to answer the problem, the approach method used is a concept and legislation approach. Therefore, the primary source of legal material is the legislation relating to the object of this study, they are Law relating on Illegal, Unreported, Unregulated (IUU) Fishing, and it is supported by secondary legal materials, in the form of literature for instance books and scientific journals. After the source of legal material is collected, then qualitative analysis is conducted by describing the data quality in the form of a regular, coherent, logical, non-overlapping and effective sentence, thus it is facilitating the interpretation of the data.

3. Research And Discussion

3.1. The Development of Corporate Criminal Liability Arrangement

Crime can be identified by the occurrence of harm (harm), which then resulted the criminal liability. [6] Discussion of responsibility begins to appeal when the word 'accountability' is attached to the legal subject of the corporation, remembered that the Indonesian Criminal Code (KUHP) considered a subject of criminal law is only an individual in a natural biological connotation (natuurlijk persoon).

The principle of corporate liability is not regulated in general criminal law (Kitab Undang-undang Hukum Pidana/ KUHP), but is spread in certain criminal law. The unfamiliarity of the principle of corporate responsibility in the KUHP is due to the fact that the subject of a crime known in the KUHP is a person in a natural biological connotation (natuurlijke persoon). In addition, the KUHP also still adheres to the principle of sociates delinquere non potest where legal entities are considered as the one who is unable to commit a crime. [7] Thus, the fictional thinking about the nature of legal entities (rechspersoonlijkheid) is not applied in the field of criminal law. [8]

However, in the doctrine of criminal law, the application of the principle of corporate responsibility has undergone rapidly in line with the rise of corporate crime itself. Initially, corporations have not been acknowledged as perpetrators of a criminal offense, therefore responsibility for criminal offenses is charged to the board of corporations. Furthermore, the corporation is recognized as a criminal offender, while responsibility for criminal acts is still charged to its board, as adopted in Law no. 12 year 1952 on firearms. In recent developments, beside as an agent, the corporation may also be held liable for a crime. Legislation that embraces this model is Law no. 7 Year 1955 on Economic Crime, Law no. 6 Year 1984 on Post, Law no. 23 Year 1997 on the Environment, Law no. 31 Year 1999 jo Law no. 20 Year 2001 on Corruption, and Law no. 15 Year 2002 on Money Laundering Crime. In later development, the principle of corporate responsibility is widely adopted in legislation, such as: Law no. 5 Year 1984 on Industry, Law no. 8 Year 1995 concerning Capital Market, Law no. 5 Year 1997 on Psychotropic, Law no. 22 Year 1997 on Narcotics, Law no. 23 Year 1997 on Environmental Management, Law no. 5 Year 1999 concerning Prohibition of Monopolistic Practices and Unfair Business Competition, Law no. 8 Year 1999 on Consumer Protection, and Law no. 20 Year 2001 on the Eradication of Corruption.

Corporate crime represents a greater disadvantage if compared to individual crime or commonly referred to as conventional / traditional crime. This is caused by the characteristics of corporate crime itself, among others: First, the crime is difficult to be seen (low visibility), because it is usually covered by normal and routine work activities, involving professional skills and complex organizational systems; Secondly, the crime is complex (it always relates to lies, deception and theft and is often related to a scientific, technological, financial, legal, organized, and involving many people and runs for years; Third, the occurrence of spreading responsibility (diffusion of responsibility), which is increasingly widespread due to organizational complexity. Fourthly, diffusion of victims, such as pollution and fraud, Fifth, detection and prosecution barriers, as a result of unbalanced professionalism between the law enforcement apparatus and the perpetrators of the crime, the sixth, the ambiguous law that often causing the losses in enforcing the law, and the seventh, the dual attitude of the offender's status. [9] Therefore it is not surprising that until 2010 only two corporations that become the defendant and the accused by the AGO u PT Newmont Minahasa Raya in Manado (Decision no. 284 / Pid.B / 2005 / PN. Mdo) and PT Giri Jaladi Wana in Banjarmasin (Decision No. 812 / Pid.Sus / 2010 / PN.BJM). [10]

3.2. The Corporate Accountability in fishery crime in Indonesia

Until now there is no recorded history in Illegal, Unreported, Unregulated (IUU) Fishing where corporations are criminally charged. The criminal tendency is attached to the captain of the ship, for example, is a horrendous case in 2014, the Panaman-flagged ship named MV Haifa with a captain named Zhu Nian Le who performed the activities as a Tramper (fish carrier) ship that containing the mixed fish belonged to PT Avona Nian Lestari which be exported to China, on December 24th, 2014 the MV Haifa's ship anchored in Wanam Papua which belongs to Indonesia territorial waters. The supervisor of fishery work unit of PSDKP (Marine Resources and Fishery Inspector) Kaimana in Wanam Papua together with the related institution then do inspection of the document and the completeness of ship. They examined various ship completeness including Letters Feasible Operations (SLO) and Sailing Approval Letter (SPB).

From the inspection, The Main Naval Base (Lantamal) IX Ambon then issued a warrant for the commencement of investigation (SPDP) received by the High Prosecutor of Ambon on January 9, 2015 by letter no. B / 21/1/2015. Subsequently, the file was received by the prosecutor team on February 12, 2015 and based on the results, the file had been completed (P-21) on February 13 2015 and on 23 February 2015 the delivery of evidence and the suspect to the Prosecutor by the investigator happened. The proof consists of Panamanian-flagged MV Haifa's ship with a weight of 4,306 GT (Gross Ton), ship document along with frozen mixed fish as much as 800,658 kg and frozen shrimp as much as 100,044 kg. Subsequently the case was handed over to the Fishery Court at the Ambon District Court on February 24, 2015. During the trial process, the MV Haifa 's ship was carrying out fishing activities in Indonesian national territory with SIKPI-NA license 20.14.0001.02.42482 valid until February 6, 2015 issued by Marine and Fisheries Ministry (KKP) and the ship also bagged SPB (Letter of Consent Sailing) No.BB.4 / 63/22 / XII / KUPP.KMN-2014 date December 19, 2014.

It was found that there was a violation conducted by the ship; First, the ship is not bagging SLO (Letters Worthy of Operation). Second, the ID Transmitter VMSNo.4958945 based on PSDKP monitor results from 22-24 December 2014 was inactive. Third, the vessels of this MV Haifa transporting 800,658 kg of mixed fish and 100,044 kg of frozen shrimp, among the 800,658 kg of mixed fish found 15,000 kg of shark type fish lonjor and hammerhead shark which include to the protected species of fish and are prohibited from being arrested, traded, and exported. Then based on these violations the Public Prosecutor charged the defendant, in this case is the ship captain, with the Article 100 Jo Article 7 Paragraph (2) Letter M of Law Number 31 Year 2004 Jo of Law Number 45 Year 2009 about Fisheries. Further, punish the defendant by fine of 200 Million Rupiahs with a six month subsidy of confinement.

The example above proves that although the Law has ruled that corporations could be criminally convicted, unfortunately, the enforcement of Illegal, Unreported, Unregulated (IUU) Fishing law has not been able to touch the corporation. In the doctrine of criminal law, the corporate criminal liability has essentially five theories, namely: [11] First , identification theory, commonly referred to as direct corporate criminal liability or direct criminal responsibility of corporate. According to this theory, corporations can do a number of offenses directly through a board that is closely related to the corporation, acting for and on behalf of the corporation so that it is viewed as the company itself. They are not a substitute and therefore , corporations' liability is not personal liability. Second , strict liability. Strict liability is defined as strict liability by law. So accountable corporations solely based on sound laws - laws regardless of who make a mistake. In strict liability elements to mistake does not need to be proven. Third , vicarious liability. The doctrine of accountability substitute, more emphasis on accountability of the corporate board as the agent of the act of the corporation. This teaching departs from the doctrine of the respondent superior, based on the employment principle and the delegation principle. This doctrine is exceptional individual accountability adopted in criminal law based on adugium nemo punitur pro alieno delicto (no one punished for the actions of others).Thus the vicarious liability in it contains a principle of substitute liability for action criminal committed by others. Fourth , the aggregation theory that states that criminal liability can charged to legal entities if deeds is done by a number of people who meet the element of offense which is between one with the others being assembled and not standing alone. Fifth , corporate culture model doctrine or cultural model of work. This teaching focuses on express and implied legal entity policy affect the workings of the legal entity. Legal entities can be accounted for criminally if any person's actions have a rational basis that the legal entity authorize or permit the act to take place.

In addition to these theories, Sutan Remy Sjahhdeini [12] introduced another theory called the combined teachings of a combined teachings identification, teachings of intra vires, teachings of interconnection functions, teachings of benefits, the teachings of legal entity and the teachings of aggregation. According to this theory, to be an accountable corporation as the subject of offense, it must be fulfilled six important elements. First, a crime must be committed by directing mind (controlling personnel) of the corporation, well done alone or ordered by him to do by others (doctrine of identification). Secondly , the offense is committed in order intent and purpose of corporation (intra vires). Third , the offense is committed by the offender or above command of the giver of orders in the course of his duties in the corporation. Fourth, the crime is committed to provide benefits for corporation.Fifth, the perpetrator or the giver of the order have no reason for justification or reason forgiving to be relieved of liability criminal. Sixth , actus reus and mens rea not must be on one person but mens rea must there is on directing mind (personnel penggendali) corporation, whereas its actus reus can performed by one or more others (doctrine of aggregation).

Law Number 45 Year 2009 about Change on Law Number 31 Year 2004 about Fishery on Article 1 Number 14 which states "Everybody is an individual or corporation" also on Article 1 Number 15 which defines corporation as stated that " Corporation is a group of people and/or wealth which is organized well whether it is legal corporation or not". Thus it is clear that Law Number 45 Year 2009 about Change on Law Number 31 Year 2004 about Fishery has adopted corporation as law subject other than human, so it is clear that on the Law, the principle "Universitas elinguere non potest" has been abandoned. Therefore, if a corporation does criminal act on fishery, it can be charged on criminal liability.

The model development on corporate criminal liability can be divided into four those are: First, corporate administrator as the maker and organizer who criminally responsible (first phase of corporate liability development); Second, corporation as the maker, yet the administrator must criminally responsible (second phase of corporate liability development); Thirdly, corporation as the maker, and also must be criminally responsible (third phase of corporate liability development); Fourthly, administrator and corporation, both of them as the doer of criminal act and must criminally carry the responsibility.

There is corporate criminal liability system that is enacted on Law Number 45 Year 2009 about Change on Law Number 31 Year 2004 about Fishery which corresponds with the development of corporate liability in the second phase. That is the criminal liability will be charged to administrator, this is stated on Article 101 which is:

> "In criminal act as mentioned in Article 84 paragraph 1, Article 85, Article 86, Article 87, Article 88, Article 89, Article 90, Article 91, Article 92, Article 93, Article 94, Article 95, and Article 96 done by corporation, prosecution and criminal sanctions imposed on its administrator and the penalty will be added one third of the criminal sanction imposed."

That rule is only burden the responsibility to an individual or corporate administrator for law actions that have been done by a corporation. This thing will cause the court ignoring the status of legal entity from the corporation and burden the responsibility to private parties and perpetrators of the company (corporation administrator) by disregarding the principle of limited liability of corporations as legal entities. Although corporation can do direct criminal act through the administrators. In this case, it can be identified as an act by the company or the corporation itself. Thus administrator's action is seen as corporation's action. Therefore, criminal liability can be truly burdened to the corporation if only that criminal action is done by someone who is the directing mind of the corporation. Directing mind is an action, behavior or policy created by members of the board or organ of the company/corporation or manager who decides the direction, activity, operational of the corporation. The theory can also be called as "alter ego" theory. [13] Moreover, based on "doctrine of aggregation" that if there is a group of people who has done criminal act but these people act for and on behalf of the corporation or for the sake of the corporation then the corporation can be charged the criminal liability.

The flaw of law enforcement officers in taking decisive action towards the crime that is done by corporation is very worrisome, since the crime impacts from it are very big. [14]

The regulation that is on Article 101 Fishery Law suffered a setback, because positioning the corporation as the perpetrator of illegal, unreported, unregulated (IUU) fishing criminal act yet the corporation itself is not criminally charged but the punishment is shifted and charged towards the administrator. The setback of subject of fishing criminal act as stated in Article 101 Fishery Law obviously cause problem and challenge to recover the state's asset from illegal, unreported, unregulated (IUU) fishing criminal act that is obtained and hid by the corporation administrator, either foreign perpetrator or Indonesian. Besides, giving the punishment to corporation administrator will not guarantee that they will not do it again in the future. In fact, many of corporation parties cover on dummy company which deliberately they built to protect the main corporation. Besides that, the magnitude of the loss value which is caused by the act of illegal, unreported, unregulated (IUU) fishing reaches 240 trillion rupiah per year. This still has not been as a consideration which makes Indonesia as if not serious yet to handle this criminal act, this is caused because of one the pillars of law enforcement that is juridical aspect is still fragile.

Related to a corporation criminal liability, in Indonesia there is a different regulation in one law to another about corporation as the offender in natural resources criminal action. In Article 78 paragraph 14 Law Number 41 Year 19999 jo Law Number 19 Year 2004 about forestry states that if the criminal act is done by and/or on behalf of legal entity/business entity, the prosecution or criminal sanction will be charged on the administrator. The same thing is also regulated on Article 101 Law Number 45 Year 2009 about Fishery, that the criminal act that is done by a corporation, prosecution or criminal sanction will be charged on the administrator. Thus, law on forestry and law on fishery have the same concept on formulating corporation as the offender

Different regulation about corporation as the offender can be seen on Article 112 Law Number 39 Year 2014 about plantation. It is clearly stated that criminal act that is done by corporation then the corporation will be punished and given penalty adding one third. Meanwhile in Article 116 paragraph 1 Law Number 32 Year 2009 about environment, it has its characteristics in formulating corporation criminal liability. If the environment criminal is done by or on behalf of business entity, then punishment will be charged on the company and/or someone who gave the order to do that crime. Same with environmental law, Article 163 on Law Number 4 Year 2009 about Mineral and Coal, states that in criminal act that is done legal entity, besides imprisonment and penalty to the administrator, the punishment is also charged on the legal entity in form of penalty.

Each law has different formula to regulate the corporation as the offender First, in fishery and forestry law has the same formulation that if corporation as maker—responsible organizer – is sentenced. Second, plantation law has its own formula, if corporation as maker – corporation who will be sentenced or responsible. Thirdly, environmental law adds new thing, if corporation as maker – corporation and the person who gave the order will be sentenced. Fourthly, mineral and coal law makes more simple formulation, corporation as maker – the administrator and corporation will be sentenced. Therefore it needs a reformation on the corporation criminal liability regulation through penal policy.

This condition potentially causes overlapping of authority which is fragmentary,[15] "centric"and/or "sectoral egoism" and it does not show as a unity of criminal law enforcement system. Moreover, each sub-system has created "internal rules" in the form of circulars, decree and so on whose purpose is to harmonize and/ or to promote the harmonization of general rules.[16] in context of corporation criminal liability needs a reform in criminal law [17] which its main focus is to do an improvement towards the corporation criminal liability. The improvement can start from formulation stage, the application to execution stage and the last purpose is for the social defense and social welfare.

4 Conclusion

Based on the analysis it can be concluded that liability towards illegal, unreported, unregulated (IUU) fishing acts have not come to corporation, because the liability is limited towards the administrator only. This is a setback, because on certain laws, corporation has been able to be burdened of the criminal liability. The difference on set concept of corporation liability makes the legal uncertainty therefore legal reform is needed in form of "criminal corporation liability", the application to the stage of execution.

References

[1] Ridwan. Jurnal Dinamika Hukum, **12** (2012)

[2] S Pudjiastuti. *(Satgas 115)*, Monday,July 10 (2017)

[3] H Lubis. SK.Perestasi Reformasi, No.479 year 16 , date December 4 2015, p.6, kol. 1-4.

[4] S M Sari. (http://industri.bisnis.com/read/20170711/99/670163/susi-pudjiastuti-ini-modus-baru-penangkapan-ikan-ilegal, retrieved from January 13 2018)

[5] Sulistyanta and H Malo. Jurnal Dinamika Hukum, **17** (2017)

[6] H Gross, *A Theory of Criminal Justice* (Oxford University Press, New York, 1979)

[7] M Ali and H Amrani. (Raja Grafindo Persada, Jakarta, 2015)

[8] T M Sirait. Jurnal Dinamika Hukum, **17** (2017)

[9] Setiyono. (Averros Press, Malang, 2002)

[10] EOS Hiariej. (Cahya Atma Pustaka, Yogyakarta, 2014)

[11] A Pinto & M Evans. (Thomson Sweet and Maxwell, London, 2008);

[12] SR Sjadeini. (Grafiti Pers, Jakarta, 2006)

[13] *Ibid.*, p. 70

[14] RT Prasetyo, U Ma'ruf, A Mashdurohatun. Jurnal Khaira Ummah, **12** (2017)

[15] H Nugroho. Jurnal Dinamika hukum, **13** (2013)

[16] Sulistyanta. Jurnal Dinamika Hukum, **13** (2013)

[17] SW Kim. Jurnal Dinamika Hukum, **13** (2013)

Harmonization of regulation in water territorial management becoming a fair economic benefit distribution towards regional autonomy

Tedi Sudrajat[1*], *Agus* Raharjo[2], *Rahadi* Wasi Bintoro[3], and *Yusuf* Saefudin[4]

[1] Department of Administrative law, Faculty of Law, Jenderal Soedirman University, Purwokerto - Indonesia
[2] Department of Criminal law, Faculty of Law, Jenderal Soedirman University, Purwokerto - Indonesia
[3] Department of Procedural law, Faculty of Law, Jenderal Soedirman University, Purwokerto - Indonesia
[4] Faculty of Law, Pekalongan University, Pekalongan - Indonesia

Abstract. The extent of Indonesian territorial waters along with its natural wealth inside brings economic benefit, yet on the other side it invites problems. Especially with the existence of regional autonomy, the competition to obtain the economic benefit causes fiercer competition between regions. The competition causes the division emergence of marine areas which implicates towards the fate of fishermen. This relates to their catchment area, imposition of income tax, and technical restriction on fishing. This research used normative approach by emphasizing the comparative study of water territorial arrangement in various regions. Based on the research, in autonomy region which has marine water, they regulate the object very detail and there are some which exploit their area. The detailed and thorough regulation with its practice sometimes cause the territorial waters dispute among the regions. It causes a confusion for the government or fishermen in obtaining the economic benefit of their own water. Hence, in level of regulation, it needs a legal harmonization between autonomy regions in utilization of water territorial. In practical level it often needs coordination to create fair economic benefit for the stakeholders.

1. Introduction

Based on article 18 of the 1945 Constitution of the Republic of Indonesia (1945 constitution), it is emphasized that "The Unitary State of the Republic of Indonesia is divided into provinces and provinces divided into districts and cities, with each province, district and city having local government, which regulated by law". Substantively, Article 18 of the 1945 Constitution also mandates the division concept of power vertically, which is a distribution of power between the National or Central government with other lower government units.[1] Implementation of government policies on the implementation of governance in the regions, then further elaborated through the concept of autonomy as described in Law No. 23 Year 2014 concerning Regional Governance .

According to the Regional Governance Law, autonomy is implemented to manage Governmental Affairs and the interests of local communities. In relation to the distribution of its affairs, the Regional Governance Law divides Government Affairs consisting of absolute

* Corresponding author ; tedi.unsoed@gmail.com

governmental affairs, and concurrent government affairs. Absolute governmental affairs fully entrusted by the Central Government. The concurrent governmental affairs is a Government Affair which is divided between the Central Government and the Provincial Region and the Regency/Town Region. Concurrent government affair which is submitted to the region as the basis for the implementation of Regional Autonomy, and then carried out significantly in the area.[2]

In relation between law and regional autonomy, local government is given the authority to arrange and manage. The arranging authority is implemented through the establishment of Rules and Regulations Regional Heads. The managing authority carried out the peace, order and prosperity for the people of the region, including taking care of the water territory in its area.[3] Under the Regional Governance Act, the conservation management authority in the territorial waters of a government affairs'choice which resulted in that the fisheries management authority, marine and space arrangement in the territorial waters become the right from the local government.

Observing the settings, and correlating within the context of autonomy, there is authority in the area of exploration, exploitation, conservation and management territorial of waters exist at 4 (four) miles calculated from the baseline area region.[4] Observing the object of the arrangement, then in the region can manage the territorial of waters for the greatest importance of the local community. But the factual is that the problems that often occur in coastal areas and oceans are conflict of resource utilization and spatial use, which is only seen by the Regional Governance from the economic aspect through the exploitation of water resources, taxes and user charges. The gap between the interests of autonomous functions, water management and economic empowerment of the region has created the precepts chaos in the regional autonomy. Therefore, this paper is going to examine the issue of management arrangements harmonious territorial waters in the vertical and horizontal implications fairly to the existence of regional autonomy.

2. Research methods

The method used is Normative Juridical approach which use law and analytical approach. In order to achieve legal harmonization, the evaluation and analysis of laws and regulations related to the management of territorial waters, either vertically or horizontally are conducted. As for the activity of analyzing regulatory issues concerning territorial waters in regional autonomy, the authors used normative qualitative analysis through grammatical interpretive model and systematic interpretation. The results of the analysis are presented comprehensively, all inclusive and systematic using deductive logic.

3. Discussion

a. Identification and Inventory Management of Territorial Water Regulation

In order to realize Indonesia as a legal state, the Government is obliged to carry out the development of national law which is carried out in a sustainable, planned, integrated and comprehensive manner in a national legal system. Development of sustainable national law is expected to guarantee the protection of the rights and obligations of all Indonesian people to meet the community's need for legislation that protects the community. The efforts made to create legal development is to harmonize the law.

Evaluation and analysis of regulations on the management of the territorial waters are conducted in the framework of regulatory harmonization, or commonly known by synchronizing either vertically or horizontally. Regulatory harmonization efforts are based on the theory of legislative formation. Based on the stuffenbau theory (general theory of hierarchy of legislation Hans Kelsen), legislation can be grouped generally into 4 (four) levels, namely: First, the provision that contains the basic norm (*grundnorm*) which is Law; Second, the legislative provisions that describe the basic norms that is Law; Third, the provisions established by the government as the implementation rule of Law are the implementing legislation, and Fourth, the organic provisions to operationalize in detail the Implementing Legislation are among others: Presidential Regulation, Ministerial Regulation, and Local Regulation.[5] As the reinforcement of Hans Kelsen's theory, it is able to be matched with the theory of Algemeine Rechtslehre (Hans Nawiasky). Based on Nawiasky theory this legal norm consists of First, *Staatsfundamentalnorm* (State Fundamental Norms); Second, *Staatsgrundgesetz* (Basic Rules/Principles of State); Third, *Formell Gesetz* (Formal Law); Fourth, *Verordnung and Autonome Satzung* (Rules of Execution and Autonomous Rules).[6]

To create the harmonization, it is necessary to compare the regulations relating to the study object, that is the management of territorial waters which becomes the authority of the region, in the form as follows:

1. Law Number 25 Year 2004 concerning National Development Planning System;
2. Law Number 31 Year 2004 concerning Fisheries as amended by Act Number 45 Year 2009;
3. Law No. 17 Year 2007 concerning the National Long-Term Development Plan of 2005-2025
4. Law Number 26 Year 2007 concerning Spatial Planning;
5. Law Number 27 Year 2007 concerning the Management of Coastal Areas and Small Islands as amended by Law Number 1 Year 2014;
6. Law Number 10 Year 2009 concerning Tourism
7. Law Number 23 Year 2009 concerning Environmental Protection and Management;
8. Law Number 23 Year 2014 concerning Regional Government as amended by Act Number 9 of 2015;
9. Law Number 32 Year 2014 concerning Marine, and
10. Other organic regulations

b. Regulatory Harmonization Efforts

In the context of regional autonomy, the preparation of Regional Regulations shall be used as legal means in the first instance, policy instruments for implementing regional autonomy and assistance tasks; Secondly, as an effort to accommodate the specificity and diversity of the region and channel the aspirations of the people in the regions; and Third, as a tool of development in improving the welfare of the region. Regional Regulation is not merely a regulation of the implementation of the law above it, but more than that as a policy that is able to absorb and accommodate special conditions for regional self-reliance (*zelfstandigheid*) and aspirations of local citizens.

In case of making a regional regulation, the regulations made by the legislators must be based on the principle of establishment of the rules. Broadly speaking, the principle is a general proposition expressed in general terms without suggesting specific ways of its implementation applied to a series of deeds to be a good guide to the action. Principle is a fundamental assumption and consideration that is the basis of social behavior. Thus, the principles of law are not concrete rules of law, but they are the basic or general thoughts which are the background of the "positive law" contained in and in every legal system incarnate in the laws and regulations.

In formulating the rules, there are principles in the formulation of good state regulations (*beginselen van behoorlijke regelgeving*). The principle of the establishment of good legislation is formulated in Law Number 12 Year 2011 concerning the Establishment of Regulations particularly Articles 5 and 6 which define as follows:

1. Clarity of purpose;
2. Appropriate institutional or organizers;
3. Conformity between type and material of charge;
4. Can be implemented;
5. Utility and usability;
6. Clarity of formulation; and
7. Openness.

Meanwhile, the principles that should be contained in the material of Legislation are formulated in Article 6 Paragraph (1) of Law Number 12 Year 2011 concerning the Establishment of Regulations, the content of the Laws and Regulations contains the following principles:

1. Aegis. Any material on the content of legislation should provide the protection to create public peace.
2. Humanity. Any material on the content of legislation should reflect the protection and respect of human rights and the prestige and dignity of every citizen proportionately.
3. Nationality. Any material on the content of legislation should reflect the nature and character of a plural Indonesian nation while maintaining the principle of the Unitary State of the Republic of Indonesia.
4. Kinship. Each content of the statutory matter shall reflect deliberations to reach consensus in any decision-making.
5. Merchandise. Every content of legislation concerns the interests of the entire territory of Indonesia and the material content of legislation made in the region is part of the national legal system based on Pancasila and the 1945 Constitution of the Republic of Indonesia.

6. Bhinneka Tunggal Ika. Any material on the content of legislation should consider the diversity of the population, religion, tribe and class, special conditions of the region and culture in the life of society, nation and state.
7. Justice. Any material on the content of legislation should reflect proportional justice for every citizen.
8. Equal position in law and government. Any material on the content of legislation should not contain things that are distinguishable based on the background, including religion, ethnicity, race, class, gender, or social status.
9. Order and certainty. Any material on the content of legislation should be able to realize order in society through guarantee of legal certainty.
10. Balance, harmony, and harmony. Any material on the content of legislation should reflect balance, harmony, and harmony between individual interests, society and the interests of the nation and state.

Based on the description above, the formation power of special regulation (*regeling*) legislation regenerating has coherence with the authority that encompasses it, in the form of (*attributie van wetgevingsbevoegdheid*) which is the granting of authority to form legislation granted by the Constitution (*groundwet*) or by law (*wet*) to an institution. This attribution of power will create to a new powers. In relation to autonomy, it is affirmed in Article 18 Paragraph (6) of the1945 Constitution that "Regional Government shall have the right to enact regional regulations and other regulations to implement autonomy and assistance tasks. This provision has given the authority of attribution to the local government to form a law.

Therefore, Regional Regulation is one of law products in which the principles of its formulation, application and enforcement must contain the legal values. It must substantially bind in general and be effective in giving the punishment. According to Lawrence M. Friedmann, Punishment is a way to enforce a norm or a regulation.[7] Meanwhile, Legal Punishment is the punishments which are lined up and authorized by law. Every legal punishment contains a statement of law consequences which are some punishments, appointments, or threats

In accordance with the explanation above, the formal principles must be related to the material principle as the base of the regulation substance arrangement. Here are the material principles related to the territorial water management object:

1. Principle of maximizing people's interests as it is stated on Article 33 Constitution 1449 that "the land and the water as well as the natural resources inside them are for the people's prosperity as many as possible".
2. Sustainable Development Principle
   The effort done before a policy stipulated is conducting the Strategic Environmental Assessment (SEA) in every formulation or evaluating some important documents. They are Regional Spatial Planning with its details, the National, Province, and Regent/City Long-Term Development Plan and Mid-Term Development Plan. In addition, the policy, plan, and program which potentially influence the environment are regulated in Article 15 Law Number. 32 Year 2009.
   SEA is basically a mandatory environmental development instrument. It is an obligation that must be obeyed by the policy, plan and program creator. It becomes the base to formulate a development policy, plan or program in regions. It aims to ensure that the sustainable development principle has been a fundamental and integrated to a regional development. This is because SEA consists of assessment like the environment's support and accommodation capacities, the ecosystem service performance, the efficiency of natural resources utilization, the vulnerability level and the adaptation capacity toward the climate change, and the endurance level and potency of the biodiversity. All of them is regulated in Article 15 section (1), Article 17 section (1), and Article 16 Law Number. 32 Year 2009 concerning the Environment Protection and Development.
3. Economy Principle
   The water resources utilization and management in regional autonomy is expected to contribute to the increasing of people's prosperity and the locally generated income so that it can actualize the economic independence and justice.
4. Participative Principle
   The water resources utilization and management will be well-enforced by all related parties participation.
5. Transparency and Accountability Principle

The water resources utilization and management must pay attention on the accountability and transparency aspects in its enforcement.

6. Integration Principle
   The integration principle in the water resources utilization and management is necessary to be conducted by integrating the stakeholders, the planning process, and the enforcement.

Related to the protection of the territorial water management, there are some policies enforced in some regions in order to protect their territorial water:

1. Aceh
   In Aceh territorial water, the Laot (sea) Customary Law forbids any tools or activities that can harm the sea ecosystem like bombing, poisoning, anesthetizing, electrocuting, etc [8] as well as illegal fishing using forbidden equipment. By a modus operandi and motive, the illegal fishing can be categorized as an economic crime. It is obviously regulated in Law Number 31 Year 2004 which was changed into Law Number 45 Year 2009 concerning Fishery. Thus, the illegal fishing is a crime that needs a legal punishment and a specific regulation.[9]
2. East Nusa Tenggara (NTT)
   The Regional Government of NTT released Regional Regulation of NTT Province Number 1 Year 2011 about Spatial Planning of NTT Province which has given law certainty for the coastal areas and the sea allocations as the conservative areas. This regulation has specifically included Sawu Sea into the province spatial pattern as it is stated in Article 25. Some conservative areas have been accommodated in SEA as well. They are Komodo National Park, Pantar Strait Park, Teluk Kupang Marine Nature Park, Teluk Maumere Marine Nature Park, Riung 17 Island Marine Nature Park, etc.[10] In law perspective, these allocations show the NTT government's intention to support the fishery and marine resources sustainability management efforts.

Based on the comparative example above, it should be emphasized that water resource is renewable. If it is not interfered, life will be naturally balanced. In addition, it will be wasted if we do not make use of it. Yet, if the utilization is not balanced with the restoration, it will be degraded and possibly extinct. Therefore, the balancing concept in the water resources management regulation is a resources utilization based on the natural system and carrying capacity. This function can be used by the government to create good governance in managing economic and social resources for development and society.[11] Those concept will create a balancing condition between economic and sustainability development aspects by integrating the legal policy in defending the water areas.

4. Conclution

The regulation formulation must concern on the proper legislation formulation principles as follows: Purpose clarity, proper institution or officer to make the legislation, balanced content type, hierarchy, and materials, applicable, efficiency and usability, formulation clarity, and transparency. These principles are an important guide for the regulation maker to avoid a confusing and multi-interpreted content and to make every legislation is based on the purpose. Beside the principles, other important principles that need to be concerned and included in every legislation content are: Protection, humanity, nationality, kinship, the sense of nusantara, Bhineka Tunggal Ika, justice, equality before the law and government, law order and certainty, balance, harmony, and integration.

Regional Regulation is a kind of legislations and becomes a part of Pancasila-based National Law System. The content of The Regional Regulation is the whole material in case of regional autonomy and co-administration duty, as well as to accommodate the regional specific condition. Related to the water area management, there should be a necessary effort to synchronize the water area management aspects with the integration principles of sustainability development, maximizing people interest, participative, accountability, and transparency. The condition does not eliminate the economic principle in the exploration, the exploitation, the conservation, and the water resources management, instead, it still concern with the sustainability of the regional assets so someday it can be economically beneficial.

References

[1] M. Fauzan, Hukum Pemerintahan Daerah; Kajian Tentang Hubungan Keuangan Antara Pusat dan Daerah, (UII Press, Yogyakarta, 2006)

[2] E. R. Agoes, (2000), Dimanakah Batas-Batas Laut Kita (Department of Marine Affairs and Fisheries, Jakarta 2010)

[3] H. Jalil, Hukum Pemerintah Daerah, (Syiah Kuala University Press, Darussalam Banda Aceh, 2008)

[4] E. Sutrisno, "Implementasi Pengelolaan Sumber Daya Pesisir Berbasis Pengelolaan Wilayah Pesisir Secara Terpadu untuk Kesejahteraan Nelayan", Jurnal Dinamika Hukum, **14** (2014)

[5] Yuliandri, Asas-Asas Pembentukan Peraturan Perundang-undangan Yang Baik Dalam Rangka Pembuatan Undang-Undang Berkelanjutan (Airlangga University, Surabaya, 2007)

[6] Yuliandri, Asas-Asas Pembentukan Peraturan Perundang-undangan Yang Baik Dalam Rangka Pembuatan Undang-Undang Berkelanjutan, (Airlangga University, Surabaya, 2007)

[7] L.M.Friedmann, Sistem Hukum Perspektif Ilmu Sosial, The Legal System; A Social Science Perspective (Nusamedia, Bandung, 2009)

[8] Sulaiman, "Kearifan Tradisional Dalam Pengelolaan Sumber Daya Perikanan di Aceh pada Era Otonomi Khusus", Jurnal Dinamika Hukum, **11** (2011)

[9] Adwani, Mahfud and Rosmawati, "Local Government Role in The Solving of Catching Fish Illegally in Aceh Region", Jurnal Dinamika Hukum, **16** (2016)

[10] J. M. Monteiro, "Local Wisdom Functionalization For Regional Law Efnforcement of Fisheries of Fisheries Management", Jurnal Dinamika Hukum,**17**, (2017)

[11] T. Sudrajat, "Perwujudan Good Governance Melalui Format Reformasi Birokrasi Publik Dalam Perspektif Hukum Administrasi Negara", Jurnal Dinamika Hukum, **9** (2009)

INSTITUTIONAL AND MANAGEMENT ANALYSIS OF THE AUCTION FISH PLACE (AFP) FOR IMPROVING FISHERMEN'S WELFARE IN TEGALSARI AFP, TEGAL CITY, CENTRAL JAVA

***Teuku* Junaidi [1], *Irfina* Fitri Mardani [1], and *Arif* Mahdiana [1]**
[1]**Fisheries and Marine Science University of Jenderal Soedirman , Purwokerto, Central Java, Indonesia**

ABSTRACT Institutional and Management Analysis of The Auction Fish Place (AFP) for Improving Fishermen's Welfare in Tegalsari AFP, Tegal City, Central Java" research aimed to find out the institutional system of Auction Fish Place (AFP) and Evaluation Strategy of Auction Fish Place (AFP) Tegalsari based on its institutional function. This research uses survey method which done by observation and interview. The data used are primary data and secondary data which then analyzed descriptively by using SWOT analysis. The result showed that the institutional and management system in Tegalsari AFP has been running well, from the performance of AFP managers and the tender system that runs smoothly in accordance with applicable regulations. The optimal management strategy of Tegalsari AFP is by improving the existing operational system in AFP, one of them is clean water facility and access to capital for local fisherman.

1. Introduction

Tegal city has one Coastal Fisheries Port (CFC) and three Fish Landing Port (FLP) namely Tegalsari CFC, Tegalsari FLP, Pelabuhan FLP, and Muarareja FLP. Tegalsari FLP is the largest fish landing port. The annual data showed that fish landed at Tegalsari FLP is more than other FLP in Tegal.

Auction Fish Place (AFP) is a public facility build for giving service to the people. AFP's performance is important to be measured for appropriate usage of the facility. AFP's performance can be measured from it is efficiency in auction and facility maintenance. The measurement of AFP performance is also important know customer satisfaction, the customer satisfaction usually keep the AFP active in auction and port economic development [1]

Corresponding Author: teuku_junaidi63@yahoo.com

Auction Fish Place is a formal institution in fisheries sector. Fisherman use this facility to sell their catch. The problem whether AFP is already functioned optimally. AFP rules and system need to be fair for each consumers. Fair AFP will give profit to each of consumers, but if AFP only gives profit to specific consumers, it will lead to fishermen economical downfall [2].

The AFP is built for benefit of fishermen and fisheries cooperation, AFP's aim is to make fishermen prosper and succeed [3] Prosperity is lifestyle and social livelihood, material and spiritual need of safety, decency, and tranquility that give people the ability to fulfill their daily necessity for themselves or community [4]. The sea is rich of resources and having great economic potential that can be used to make fishermen prosper [5]. AFP need to be managed to be optimally functioned as economic development instrument for the fishermen in the vicinity.

Poverty is a state of inability to be able to meet basic needs such as food, clothing, shelter, education, and health. Poverty can be caused by the scarcity of the fulfillment tool basic needs, or difficult access to education and employment. Poverty is a global problem. Some people understand this term subjectively and comparative, while others see it in terms of moral and evaluative, and which others understand it from an established scientific standpoint [6].

SWOT application is how power is able to take profit from the opportunities available, how to overcoming weaknesses that prevent the advantage of opportunities that exists, then how can power be face the existing threat, and the last is how to cope weaknesses that can make the threats real or creating a new threat. SWOT is used to assess strengths and weaknesses from company's resources and opportunities and external challenges faced [7]. Aim and strategies are established with the intent of harnessing internal strength and coping weakness [8].

The aim of this research is to study the institution system and management of AFP Tegalsari to be optimally functioned. The research data will be a source of information to develop and managing AFP, then the fishermen prosperity in Tegal will increased.

2. Material and Method

The Material is Auction Fish Place (AFP) Tegalsari, Tegal. Data sampling using survey method. The first method is Institutional survey for secondary data. The second method is direct sampling at AFP.

Purposive Sampling was used in the research, the AFP sample target are AFP chairman, auctioneer, purchase casher, receive casher, secretary, treasurer, head of business unit, head of security. System Randomized Sampling was used to 10% of total fishermen, and 10% of total fish collectors. The sampling was done in December 2016.

2.1. Data Analysis

SWOT analysis method is used in the research. The method is used to observe working procedure, institution, and system in AFP Tegalsari. SWOT is used to know AFP Strength, Weakness, Opportunity, and Threat.

3. Result and Discussion

3.1.1. Institutional System of AFP Tegalsari

Table1. Managerial access to increase AFP performance to increase fishermen prosperity (n=21)

No.	Description	Percentage
1.	AFP managerial does make fishermen prosper.	85.71 %
2.	AFP managerial does not make fishermen prosper.	14.29 %
Total		100 %

From the table, 85.71 % of respondent stated that AFP managerial does make fishermen prosper. The result showed that AFP Tegalsari institution function is optimally working.

Table 2. Managerial access to balance the fish price in AFP (n=21)

No.	Description	Percentage
1.	AFP managerial does balance the fish price	76.19 %
2.	AFP managerial does not balance the fish price	23.81 %
Total		100 %

From the table, 76.19 % of respondent stated that AFP managerial does balance the fish price in AFP. The result showed that AFP Tegalsari institution function is optimally working.

Table 3. Managerial access to increase AFP performance to increase Locally-generated revenue (n=21)

No.	Description	Percentage
1.	AFP managerial does increase Locally-generated revenue	95.23 %
2.	AFP managerial does not increase Locally-generated revenue	4.77 %
Total		100 %

From the table, 95.23 % of respondent stated that AFP managerial does increase Locally-generated revenue. The result showed that AFP Tegalsari institution function is optimally working.

Table 4. Managerial access to increase AFP performance to do social activity (n=21)

No.	Description	Percentage
1.	Decision taken was based on family decision	14.29 %
2.	Decision taken was not based on family decision	85.71 %
Total		100 %

From the table, 85.71 % of respondent stated that AFP managerial taken decision not based on family decision. The result showed that AFP can take a decision for every social activity followed.

Table 5. Managerial access to do lobbying

No.	Description	Percentage

1.	Managerial does lobbying to stakeholder (Government, Ministry of Fisheries and Marine, Local Government) for solve the problem in AFP	71.43 %
2.	Managerial does not lobbying to stakeholder (Government, Ministry of Fisheries and Marine, Local Government) for solve the problem in AFP	28.57 %
Total		100 %

From the table, 71.43 % of respondent stated that AFP managerial does lobbying to solve the AFP problem.

3.1.2. Managerial Strategy of AFP Tegalsari

Managerial strategy of AFP Tegalsari can be observed from internal factor (strength and weakness) and external factor (threat and opportunity)

a. Internal Factor of AFP managerial strategy can be a power for AFP Tegalsari to become a power to overcome a weakness and upcoming event.

Table 6. IFAS (Internal Factor Analysis Summary) matrix AFP Tegalsari managerial strategy to determine maximum value

No.	**Internal Factors**	**Score**		**Weight**	**Value**	
		Min	Max		Min	Max
1	Good Auction System	1	4	10.7	10.7	42.8
2	Good AFP Service Performance	1	4	10.7	10.7	42.8
3	Good AFP Security	1	4	10.7	10.7	42.8
4	Good Interaction Between employers and customers	1	4	10.7	10.7	42.8
5	Total Fishermen and Fisheries Community	1	4	12.5	12.5	50
6	AFP Geographical Condition	1	4	15	15	60
7	Less Clean Water Facility	1	4	7.1	7.1	28.4
8	AFP Sanitation	1	4	7.1	7.1	28.4
9	Less Capital Access for small fishermen	1	4	7.1	7.1	28.4
10	KPLI Payment Overdue	1	4	8.4	8.4	33.6
Total				100	**100**	**400**

Data and information of strength and weakness of AFP arranged in IFAS matrix and analyzed for AFP Tegalsari institutional system and managerial. The value of internal factor (score x weight), maximum value was obtained. Internal factor criteria:

Very Good: >80% (>320) Maximum value; strength of AFP Tegalsari is very good and dominant
Good: 70-80% (320-280) Maximum value; strength of AFP Tegalsari is good
Above Average: 60-70% (280-240) maximum value; strength of AFP Tegalsari is above average
Average: 50-60% (240-200) maximum value; weakness of AFP Tegalsari is average

 https://doi.org/10.1051/e3sconf/20184706005

Bad: <50% (<200) maximum value; weakness of AFP Tegalsari is dominant

From the IFAS matrix we can determine AFP Tegalsari internal condition. This matrix draw qualitative value of strength and weakness of AFP Tegalsari.

Table 7. IFAS matrix of AFP Tegalsari managerial strategy

No.	**Internal Factors (*Strength*)**	**Score**	**Weight**	**Value**
1.	Good auction operational system	4	10.7	42.8
2.	Good AFP service performance	4	10.7	42.8
3.	Good AFP security	4	10.7	42.8
4.	Good Interaction Between employers and customers	3	10.7	32.1
5.	Total Fishermen and Fisheries Community	3	12.5	37.5
6.	Total Fishermen and Fisheries Community	3	15	45
(Sub-Total)			**70.3**	**243**
No.	**Internal Factors (*Weakness*)**	**Score**	**Weight**	**value**
1.	Less Clean Water Facility	1	7.1	7.1
2.	AFP Sanitation	1	7.1	7.1
3.	Less Capital Access for small fishermen	2	7.1	14.2
4.	KPLI Payment Overdue	2	8.4	16.8
(Sub-total)			**29.7**	**45.2**
Total			**100**	**288.2**

Based on IFAS (table 7), there are six strength factors and four weakness factor in AFP Tegalsari managerial system. AFP location, and total fishermen and fisheries community got highest weight in the strength factors. The weakness of AFP Tegalsari were overdue KPLI payment. Internal aspect analysis resulted in value of 288.2 in AFP Tegalsari. The result showed that AFP Tegalsari internal factor in good condition.

b. External factor consist of opportunity that need to be utilized and threat that need to be overcame in AFP Tegalsari managerial system.

Table 8. External Factor Analysis Summary (EFAS) of AFP Tegalsari strategy to determine maximum value

No.	**External Factor**	**Score**		**Weight**	**Value**	
		Min	Max		Min	Max
1	Good Marketing	1	4	9.9	9,9	39,6
2	Tegal geographical location	1	4	11.4	11.4	45.6
3	High potential in Fisheries	1	4	9.9	9.9	39.6
4	AFP income from Highest PAD	1	4	8.2	8.2	32.8

5	Fish price stability	1	4	12	12	48
6	Water pollution affect fish quality	1	4	10.4	10.4	41.6
7	Bad weather and climate, global warming	1	4	13.1	13.1	52.4
8	Wholesaler dependency	1	4	7.1	7.1	28.4
9	Low on education and experience	1	4	7.6	7.6	30.4
10	Dominant in small fishermen	1	4	10.4	10.4	41.6
Total				**100**	**100**	**400**

Data and information about opportunity and threat of AFP Tegalsari institutional and managerial system analyzed using EFAS. The value (score x weight) for external factors, the criteria were:

Very Good: >80% (>320) maximum value; AFP Tegalsari respond to opportunity very well.
Good: 70-80% (320-280) maximum value; AFP Tegalsari respond to opportunity well.
Above Average: 60-70% (280-240) maximum value; AFP Tegalsari respond to opportunity is above average
Average: 50-60% (240-200) maximum value; AFP Tegalsari not responding to opportunity well.
Bad: <50% (<200) maximum value; AFP Tegalsari respond to opportunity is bad.

EFAS matrix used to determine external factor condition in AFP Tegalsari. EFAS matrix draw qualitative value from opportunity and threat in AFP Tegalsari managerial system.

Table 9. EFAS matrix of AFP Tegalsari managerial strategy.

No.	**External Factor**	**Score**	**Weigh t**	**Value**
(*Opportunity*)				
1.	Good Marketing	4	9.9	39.6
2.	Tegal geographical location	4	11.4	45.6
3.	High potential in Fisheries	3	9.9	29.7
4.	AFP income from Highest PAD	4	8.2	32.8
5.	Fish price stability	3	12	36
(Subtotal)			**51.4**	**183.7**
(*Threath*)				
1.	Water pollution affect fish quality	2	10.4	20.8
2.	Bad weather and climate, global warming	1	13.1	13.1
3.	Wholesaler dependency	2	7.1	14.2
4.	Low on education and experience	2	7.6	15.2
5.	Dominant in small fishermen	2	10.4	20.8
(Subtotal)			**48.6**	**84.1**
Total			**100**	**267.8**

Based on table 9, there are five opportunity and five threat analyzed in AFP Tegalsari external factor. Stable fish price in AFP Tegalsari was the highest opportunity, and the highest threat in AFP Tegalsari was bad weather and climate because of global warming.

External factor analyzed in AFP Tegalsari managerial system resulted in 267.8 in value. This value showed that AFP Tegalsari external factor in good condition, it means AFP Tegalsari can respond to opportunity well.

SWOT matrix analyzed to determine AFP Tegalsari Managerial strategy. SWOT matrix showed a combination between internal and external factors:

Table 10. SWOT matrix of AFP Tegalsari Managerial

IFAS / EFAS	**Strength**	**Weakness**
	1. Good auction operational system 2. Good AFP service performance 3. Good AFP security 4. Good Interaction Between employers and customers 5. Total Fishermen and Fisheries Community 6. Total Fishermen and Fisheries Community	1. Less Clean Water Facility 2. AFP Sanitation 3. Less Capital Access for small fishermen 4. KPLI Payment Overdue
Opportunity 1. Good Marketing 2. Tegal geographical location 3. High potential in Fisheries 4. AFP income from Highest PAD 5. Fish price stability	**SO Strategy:** 1. Increased market share and distribution network of fishery products 2. Increased fishery activities in terms of quantity and capacity of fishery products	**WO Strategy :** 1. Water supply to support TPI operations 2. Improve TPI hygiene 3. Memberikan akses kepada nelayan untuk mendapatkan modal usaha 4. Conduct a policy for timely payment of KPLI
Threat 1. Water pollution affect fish quality 2. Bad weather and climate, global warming 3. Wholesaler dependency 4. Low on education and experience 5. Dominant in small fishermen	**ST Strategy:** 1. Maintaining TPI's operational system and improving TPI infrastructure for market development so that TPI is not only dominated by small fishermen as well as increasing capital in TPI Tegalsari	**WT Strategy:** 1. Improved cleanliness and facilities of TPI to increase comfort in TPI so as to attract buyers 2. Providing capital so that fishermen can still operate

Once the internal and external factors are combined, it will be variously known strategies that suit the environmental conditions of both internal and external environments management of AFP Tegalsari. The strategy has four possibilities:

1) SO Strategy

SO 1: Increased market share and distribution network of fishery products

SO 2: Increased fishery activities in terms of quantity and capacity of fishery products

2) Strategy ST

Keeping AFP's operational system in good condition and improving the infrastructure facilities of AFP agar the development of the market so that AFP is not only dominated by small fishermen as well increased capital in AFP Tegalsari

3) WO Strategy
WO 1: Provision of clean water to support AFP operations
WO 2: Improve AFP hygiene
WO 3: Provide access to fishermen to obtain business capital
WO 4: Conduct a policy for timely payment of KPLI
4) WT Strategy
WT1: Improved cleanliness and facilities AFP to improve comfort at AFP thus attracting buyers
WT 2: Provide capital so that fishermen can still operate

4. Conclusion

The institutional and management system in AFP Tegalsari has been operating functionally, observed from the performance of AFP managers, as well as auction system that runs smoothly accordingly with applicable regulations. AFP Tegalsari management strategy is optimal by improving the existing operational systems in the AFP is one of them is a facility clean water and access to capital for local fishermen, so the welfare of fishermen in TPI Tegalsari region can be improved

References

1. Resti, F.D. Pengukuran Kinerja Pengelolaan Tempat Pelelangan Ikan di PPI Muara Angke, *Skripsi* (Institut Pertanian Bogor, Bogor, 2012)

2. Silalahi, D.G, Aktivitas Kelembagaan Tempat Pelelangan Ikan Sebagai Kelembagaan Ekonomi Masyarakat Nelayan (Studi Kelembagaan Tempat Pelelangan Ikan, Kelurahan Pelabuhanratu, Kecamatan Pelabuhanratu, Kabupaten Sukabumi, Propinsi Jawa Barat), *Skripsi*, (Institut Pertanian Bogor, Bogor, 2006)

3. Pramitasari, S.D, *Analisis Efesiensi TPI (Tempat Pelelangan Ikan Kelas 1, 2 dan 3 di Jawa Tengah dan Pengembangannya Untuk Peningkatan Kesejahteraan Nelayan, Tesis*, (Universitas Diponegoro, Semarang, 2005)

4. Rambe, A, Alokasi Pengeluaran Rumah Tangga dan Tingkat Kesejahteraan (Kasus di Kecamatan Medan, Kota Sumatra Utara, *Tesis*, (Institut Pertanian Bogor, Bogor, 2004)

5. Imron, M, Kemiskinan Dalam Masyarakat Nelayan, Jurnal Masyarakat dan Budaya. Vol **5**, **1**: 63-82, (2003)

6. Retnowati, E, Nelayan Indonesia Dalam Pusaran Kemiskinan Struktural (Perspektif Sosial, Ekonomi, dan Hukum), Jurnal Perspektif, Vol **16**, **3**: 149-159, (2011)

7. Jogiyanto, *Sistem Informasi Strategik Untuk Keunggulan Kompetitif,* (Andi Offset, Yogyakarta, 2005)

8. David, F.R, *Manajemen Strategis (Edisi Sepuluh),* (Salemba Empat, Jakarta, 2006)

Law and Justice for the Oceans: Study on Illegal, Unreported and Unregulated Fishing and Related Crimes

Aryuni Yuliantiningsih*[1,2], *Hartiwiningsih*[1], *Ade* Maman Suherman[2], *and Emmy* Latifah[1]

[1]Law Faculty, Sebelas Maret University, Indonesia
[2]Law Faculty, Jenderal Soedirman University, Indonesia

Abstract. This article examined the ethical and moral dimensions of Illegal, Unreported and Unregulated fishing and related crimes and explores the means to realize justice for the oceans. It needed a new paradigm for addressing the issues. The methods of discussion used statute approach and philosophical approach. Based on the research, it was known that IUU fishing related to other crimes and Transnational Organized Crime, so it was necessary to recognize the concept of transnational organized fisheries crime. The IPOA-IUU fishing and UNCLOS 1982 were not adequate to counter it. In the terms of ethics and morals, IUU fishing had violated the ecosystems ethics, the principles of sustainability, values of respect for human rights and justice for the oceans. Justice for the oceans can be realized with recognition of the ocean rights. It meant, the oceans was recognized had the equal position with humans. The oceans needed space to breathe and to breed for recovery itself. The legal consequences of the recognition of oceans rights, among others, extended the legal standing. It provided sanctions for perpetrators and provided obligations for States to protect the ocean rights.

1 Introduction

World consist of 70 % oceans and all human being depend on it for water supply, foods, weather and air. Almost 50% of population in the world straight depend on the ocean for livelihood, but the human activities drives it contaminated by the pollution, over exploitation, coastal degradation and the climate change, thus causes the ocean in crisis [1].The amount of human need and pressures toward ocean boost the extinction of such species and habitat such as fish, mangrove, seaweed and coral reef. The diversity of marine biota is being threatened, the marine ecosystem gradually lose its endurance. Currently, more than 70% of fish production comes from the stock which gets full exploitation, run over the ability of fish breeding in a sustainable manner [2].

The critical condition of the oceans due to the contamination is from the nature and human acts. The climate change may enlarge the impact of ecosystem damage caused by human activities such as air and water pollution [3].The climate change may also destroy

* Correspondenceauthor: aryuni71@gmail.com

 https://doi.org/10.1051/e3sconf/20184706006

othermarine ecosystem for instance the ecosystem ofsea grass, plankton even mangrove.The increase of rim temperature and unstable weather are the impact of the ecosystem damage thus the marine organism cannot adapt themselves to the change occurred [4].Currently, the ocean is in emergency situation due to the plastics waste. It is estimated that 275 million metric tons (MT) of plastics waste exist in 192 coastal of states with the total 4.8 to 12.7 million MT entering the oceans [5].As well as the damage rate of mangrove is 3-5 times from the deforestation flow and the coral reefs damage in the world has 50% [6].

Other damage of ocean is caused by Illegal Unreported and Unregulated fishing (IUU fishing).According to the report of Food and Agriculture organization (FAO), IUU fishing acts has decreased the world fish stock as much 90,1% and caused the global lose up to 26 million tons of fish in each year[7]. Recently, the acts of IUU Fishing are being the entrance of other crimes and Transnational Organized Crime (TOC). Pursuant to the report of United Nations Office on Drugs and Crime (UNODC) 2011,says that the activity in fishery sector pertain to the transnational crimes are, human trafficking, people smuggling, illicit drug, corruption, and piracy [8]. The relation of illegal fishing activity with TOC and other crimes occur in several states. Environmental Justice Foundation (EJF) and Green Peace have also conducted a study in West Africa, sub Antarctic waters, Samoa, and Thailand. The crews on the ship of IUU fishing got physical and mental persecution, murder, being servitude without sleep or foods, not paid and being the victim of human trafficking [9].

According to the facts above, Indonesian government argues that IUU fishing is not only concerning the stealing fish and simply endanger the food sustainability, but also negatively affect the economy, harm the environment, and violate the human rights. IUU fishing is not solely being national problem as well as being the global enemy thus the international contribution is necessary to solve that.

Human exploitation toward marine resources and human in the fishery business cycle shows that the anthropocentrism ethics simply rooting. Anthropocentrism ethics always exalts that ethics is only for human, not for biotic and abiotic. This ethic then creates the dominated relation of human on nature, which end up on large exploitation for human interest in short time [10]. This anthropocentrism paradigm is necessary to be changed to realize the justice in ocean. The eco-centrism ethics which argue that nature is not an object but have equal position with human being the ideal ethics to be applied. To get over the crisis occurred in the ocean, on June 2016 in the event of United Nations Ocean Conference, Indonesia proposed to admit the Oceans Rights. According to the Ministry of marine and Fishery Indonesia, Susi PudjiAstuti, ocean should be given the rights as human to realize the justice for ocean [11]. This idea is restated at the World Ocean Day in 2017. The acknowledgement of Ocean rights certainly has important legal impact for the ocean and necessary to conducted further reviews. The current paper has two purposes, first is reviewing the dimension of ethics and moral in IUU fishing practices and related crimes, and second is reviewing the ways to realize the justice for ocean as one of the ecosystems should be maintained.

2 The Dimension of Ethics and Moral in IUU Fishing Practices Related to Other Crimes

Fishery is the source of foods, nutrient, income, and the livelihood of hundred million people in the world. Although it has always been recognized that due natural resources may be run out, but until today the fishery resources, treated as will never run out and continually exploited, by not much consider the consequences on the environment. By the higher the demand on fishery product, the states around the world are faced to the doubt of the facts that fishing does in over exploitative way, gradually will decrease the potential of fish resources preservation. The phenomenon of over fishing and habitat damage will be the causes of decrease the fish production supply.

In overcoming the international demand which continually increase, the fishery entrepreneur sometime conducts the IUU fishing by waiving the moral and ethics as well as violates the human rights.

The principle of ethics and consideration on fundamental of human values in fact is the heritage from natural laws doctrine. The values of this universal ethics and moral principles have succeeded to be taught in community. The universal ethics and moral values beside contain the universality and honour as well as are being flexible and eternal. These noble and great values that radiates and its function is to animate the norms of laws or other norms which real and factual applied and bound the international community [26].

Table 1. Dimensions of the ethics of fisheries

Subject	**Objective**
Ecosystem	Ecosystem well-being
Fish stocks	Conservation
Fisheries	Responsible fisheries,sustainable development
Fishers	Safety on board, freedom and well-being, just access
Fishing communities	Eradication of poverty, cultural diversity
Consumers	Right to food, food safety

Source: FAO

FAO have formulated the Code of Conduct for Responsible Fisheries (CCRF) in the 28th conference of Committee on Fisheries (COFI) of FAO in Roma on 31 October 1995, attached on the resolution Number: 4/1995. This governance is the guidance for the national and international efforts to guarantee the implementation of conservation aspect, management and effective development of the aquatic biological resources by ecosystem sustainability and biological diversity. The background of CCRF establishment is one of them because the rising of IUU fishing globally.

The ethics in fishery field is related to the environmental ethics which may be the reference for the states and entrepreneur in utilizing the natural resources, particularly the marine resources. Environmental ethics is the basic morality that gives references in responding all things pertain to the environment as a unity in supporting the human living continuity. Human, natural resources and environment whether biotic or abiotic are the environmental community. All citizens of the environment interact with each other to produce a life. The relation among those environmental unsure should be in justice, as stated by Briant Baxter as ecological justice [12].In the relation between the human to human and human to non human in the context of ecological justice containing two important aspects those are distributive justice on environmental benefits(the justice distribution of environmental benefit) and the environmental justice itself. Similar opinion has been stated by David Scholsberg who says that ecological justice is the implementation of justice concept toward the ecological aspect which the core is the justice in the relation between human to human and human to non human[13].

Discussing the widespread practices of IUU fishing in the world, may be attributed to the growing doctrine in the marine laws history. Legal thinking of Romans against the oceans is based on the doctrine of res communism omnium (the common rights for mankind), which views the oceans use is free or open to everyone. The principle of res communist omniumin addition to the interest of voyage, become the basis for freedom in fishing [14].This doctrine is reinforced by the opinion of Grotius who fought for the principles of freedom in the ocean that is mare liberum which views that marine utilities based on the concept of freedom of the sea [15]. The doctrine of mare liberum has been recognized for long time and accommodated in United Nations Convention on the law The Sea 1982 (UNCLOS). All states have rights to utilize the ocean and the freedom which is recognized internationally as regulated on Article 87 UNCLOS, including the freedom in sailing and fishing.

In its development, the population of fishes which is initially considered to be unlimited then realized that it becomes threatened by the sophisticated fleets and fishing gears. The increase fishing intensity one of them is due to the IUU fishing practices. IUU Fishing is a global issue necessarily to be addressed seriously because it drives the economic losses of states in the world that have high dependence on fisheries [16]. According to the research of Agnew et.al., states that there are 11-26 Mmillion Ton fishes captured by IUU fishing practices, which when converted to be currency can generate the values about US$ 10-30 billion. The values are quite large, given that the global economic contribution of fisheries sector is about $ 217 billion [17]. While FAO states that 90%of the world's fish supply have been over-exploited [18].

Recently, the Illegal Fishing practice has been changed its operation system if compared to the operation system in 1990ies. IUU practices have been a highly sophisticated form of transnational organized crime, with the characteristics of modern ship's moving control and modern fishing gears [19].On the report in 2011, United Nations Office on Drugs and Crime (UNODC) says that the fisheries sector not only undertake the environmental crime but related to the transnational crimes for instance human trafficking, smuggling, drugs trading, corruption and piracy [20]. IUU fishing is also the firth of crimes toward human rights due to the most of the crews used by IUU Fishing ship are the victim of human trafficking and servitude (forced labour). In addition, according to the report of Environmental Justice Foundation (EJF) and Green Peace the relation between IUU fishing and TOC and other crimes occurred in several areas such as West Africa, Sub Antarctic waters, Samoa, and Thailand, and Indonesia. The crews on the ship of IUU fishing got physical and mental persecution, murder, being servitude without sleep and foods, not paid and even being the victim of human trafficking [21].

IUU fishing not only leads to overfishing, but also harms the marine environmental, depletion of fish supply and threatens fisheries sustainability, and the worst threatening the fisherman's living that depend on ocean. The activities of large fishing boat carried out simultaneously with the demand for cheap seafood. There is a correlation between the cheap market price, overfishing, IUU fishing, and the human rights violation, particularly the human trafficking and servitude. As has been occurred in Thailand which has been observed by EJF that international demand for cheap seafood leads to the continue the brutal human trafficking in vulnerable and ruin the whole marine ecosystems. Crime in fishing industry also occurred in Indonesia. On March 2015, Associated Press released its research about the relation of human trafficking on fisheries industry in Benjina, Indonesia. According to the investigation result, found the violations of human rights in the form of child labour, human trafficking, living condition under the standard, physical and mental persecution, works for 18-20 hours per day, works without any fees and fraud committed by PT Pusaka Benjina Resources (PT PBR).

The impact of IUU fishing practices and related crimes have leaded the failure in protecting the marine biodiversity and human rights[22]. Endangered human rights for instance the right to life (provided for Article 6 the International Convention on Civil and

Political Rights), the right to organize, the rights to be free from the persecution and other inhuman or degrading treatment (Art 7 ICCPR), the right to be free from servitude and forced labor (Art.8 ICCPR),the right to work in good condition (Art. 6,7,8,9 of the International Covenant on Economic, Social and Cultural Right/ICESCR), the right to be healthy (Art. 12 ICESCR), the right to get healthy environment (Art. 12.2 ICESCR) and the right to access the foods (Art.11 ICESCR). The organized crime networks are also involved in illegal fishing- whether directly or indirectly in illegal fishing or catching endangered fish species.

The facts about violations of environmental ethics, moral and human rights values pertain to IUU fishing have proven the truth of Garret Hardin's opinion who said that a principle of common ownership has negative impact. In 1968 the United States ecologist, Garrett Hardin, released his article entitled "The Tragedy of the Commons". He argues that the tragedy of common ownership arise when everybody tries to take the natural resources that belongs together for personal interest thus harms other living being. Therefore, the tragedy of common Ownership generally occurs on resources that are public properties [23].This opinion initially favourable to those who use a lot of natural resources, but in the end its availability will be exhausted and will have negative impact for the users and other human, and over exploitation toward ocean will also occur. It causes endangered for a lot of marine species and harms the natural balance.

The perpetrators of IUU fishing have shown a tendency towards anthropocentrism ethics. Anthropocentrism is an ethic which argues that human is the core of system of universe. Anthropocentrism paradigm is considered as a shallow and narrow environmental ethic (shallow environmental ethics). Frit of Capra says that Shallow Ecology is an ecology that put human being over or outer the nature, in dominated position against the nature [24].

Anthropocentrism paradigm has caused the ocean as natural resources particularly fisheries become the human object. It is supported by the principle of freedom in the ocean one of them is freedom of fishing and the principle of common ownership. Certainly it occurs due to the anthropocentrism ethics that quiet rooted. Anthropocentrism ethics is too exalting that ethics are only for human being, not for biotic and abiotic nature. This ethic then creates the relation of human domination against the nature, which ends up on the massive exploitation to nature for short-term human benefits. Illegal fishing practice and other crimes related contrary to the principles of sustainable fisheries and human rights respect.

The main issue of ethic in fisheries widely related to the human ecosystem welfare. According to the ecological justice that there is should be a justice in natural utilization and justice on the natural resources itself. It may be realized in justice based on the principle of sustainable development that is formulated in the justice of intra-generation and intergeneration. Intra-generation justice is aimed to whom live in one generation. It is related to the natural resources distribution fairly applied on national level and/or international level. While inter-generation justice is related to the sustainable fisheries meant that in carrying the natural use, particularly fishing, should consider the future generation so that massive exploitation is not necessary.

The entrepreneur in fisheries should have the ethic of eco-centrism. The eco-centrism paradigm holds that, living creatures and other abiotic objects are intertwined with one to another. The moral responsibility and liability not limited on living creatures, but also applied on all ecological reality. Arne Naess, a Norwegian philosopher, who is one of the figures of eco-centrism paradigm, propose a view known as Deep Ecology. This view is a new ethic which is not human-centred, but centred on all living creatures in relation to address the environmental problems [25].

Respecting the nature is the basic principle for human as the part of universe, who should respect each species in ecological community as well as have moral responsibility to maintain the integrity of ecological community.

3 Laws and Justice for Oceans Related to the IUU Fishing and Other Crimes

Laws and justice cannot be separated and are the fundamental to common life [27]. Common life not only human should be treated fairly but all environments have the same right to be fairly treated. There are international laws regulating the liability of State in conserving the environment, including the Declaration of Stockholm 1972,United Nations Convention on the Law of The Sea1982 The Convention on Biological Diversity (CBD) in 1992 and International Plan of Action- IUU (IPOA-IUU) fishing released by FAO.

Stockholm Declaration considers the public views and principles for assisting human in preservation and improvement the human environment. In 1982, international community has agreed the UNCLOS that provide the marine issues comprehensively. This agreement provides for instance the maritime boundaries, rights and liabilities of State in ocean use, marine resources management, also protection and preservation of ocean. Moreover, there is Convention on Biological Diversity (CBD) in 1992. CBD is trigger due to the concerns about the pace of development, which, in the other hand leads to endanger many species. To stop the severe damage required an agreement governing the biological resources. This convention has three main goals those are, to conserve the biodiversity, to harness sustainable biodiversity, and to ensure fair sharing of benefits from the utilization of genetically sources. Furthermore, in 2001, FAO published IPOA IUU-Fishing. IPOA-IUU fishing is the global action aimed to prevent the fishery resources damage and redevelop the fishery resources which have been or almost extinct, thus the food demand sourced from fisheries for the current and future generation remain guaranteed the availability. Although there are UNCLOS 1982 and other regulations that provide the State's responsibility to conservee the marine resources, maintain the marine resources and environment, not accommodate the right of marine yet because UNCLOS 1982 more set the State's authorities to manage and utilize the marine resources.

Until now, IUU fishing remains be the global and crucial issue to be addressed because have been evolved into transnational and organized crimes so that in the current paper used the term of IUU fishing and related crimes. IUU fishing and related crimes have violated the ecological justice, principle of sustainable and human values so that new concept is required to be addressed. There are two new concepts being fought by Indonesia to address that issue. First, Indonesia is mainstreaming the acknowledgment of fisheries crimes as TOC transnational organized fisheries crime (TOC) in order to easier the international corporation in enforcing the laws [28]. Second, Indonesian government is mainstreaming the concept of ocean rights in United Nation to protect the ocean and realize the healthy oceans.This concept holds that the ocean can be recognized as the subject of laws that has the rights so that the position may be synchronized to human. Ocean requires space for breathing and breeding thus it may be recover.

The concept of ocean rights is inline with the opinion of Cristopher D. Stone who proposed the idea of making the environment as the subject of laws in 1972 on his article entitled; "Should Trees Have Standing? Towards Legal Rights for Natural Object". Stone argues that it is necessary to give legal rights to the natural objects such as forests, oceans, rivers, mountains as natural object which deserves the legal rights. Furthermore, according to Stone, environmental organization that have data and reason to suspect a project/activity will ruin the environment, that organization may propose to the court to be appointed as guardian of that natural object to supervise and manage natural objects against the indication of legal rights violation [29].

Most States in the world including Indonesia acknowledge that become the subjects of laws are human and legal entities, but in its development, environment as the subject of laws has been acknowledged in the constitution in several states of Europe, America, such as Portugal, Franch and Ecuador. [30].For example Article 72 of Ecuador Constitution provides that "Nature has the right to be restored. This restoration shall be apart from the obligation of the State and natural persons or legal entities to compensate individuals and communities that depend on affected natural systems". In the Europe union agreed that the UNECE Convention on Access to Information, Public Participation in Decision on Making and Access to Justice in Environmental Matters (Aarhus Convention). This convention is the first international instrument that regulates the procedure of rights in the context of environment and conceptualizes the access to environmental justice for the perpetrators not States.

The idea promotes the importance of ocean rights for all States, which have ocean and its resources, due to by several things. First, more massive crimes that exploit the marine resources sequent from IUU fishing, marine and coastal pollution due to industrial activities to the trade of marine biota protected. Second, the absence of adequate international legal shade and recognizes the importance of marine rights as the part of widest world ecosystem. Third, the dependency most people in the world toward marine space, ecosystem until the resources content as the transportation line, food and energy supply, as well as other services provider.

The acknowledgement of ocean rights in United Nations certainly is not easy because it must through the long procedures and supported by 2/3 states who attends the UN General Assembly session. The form of acknowledgement may be set forth in a UN resolution. The legal impact if oceans rights recognized by international community for instance: first, the expansion of lawsuit, second, sanctions for violators, and third is the obligation for States to protects those rights.

The expansion of lawsuit may take the form of standing law. It is a legal product that gives individual, group/organization or governmental agencies access to justice as the plaintiffs to demand the restoration of their rights that have been violated. In Law number 32 of 2009 concerning the Environment, access to the rights is granted to several parties, those are, 1. Individual lawsuit (individual); 2. Non Governmental Organization's lawsuit (NGO standing), 3 group representative's lawsuits (Class action) and 4. Government and Local authorities lawsuit and Citizen's lawsuit.

In the context of ocean rights, NGO may perform in the court based on the understanding that NGO is the natural guardian. It is appropriate with the theory proposed by Stone that gives legal right to the natural objects. According to Stone, forests, oceans or rivers as the natural object are deserved to have legal rights and unwise if considered otherwise just because the inanimate characteristic (unable to speak). As the guardian, the court give rightts to NGO to monitor even manage those natural objects. In international laws, the lawsuit is a challenge if associated to the mechanism of overcoming of environmental and marine laws disputes.

The second legal impact is the provisions of sanctions for the violators. The sanctions in environmental laws context are administration, criminal, and civil sanctions. The administrative sanction is important for restoration of damaged or polluted environments and restoration of circumstances. Discussing the imposition of sanction in international laws should through long mechanism. If the violation concerning the issue of maintaining international peace and security then under Article 25 of UN Chartered authorized to impose the sanctions is the UN Security Council. However, if the violation is related to the marine environmental disputes then it may proposed to several international court for instance International Tribunal on The Law of Sea (ITLOS), International Court of Justice and International Arbitration agencies. This mechanism is necessary to be reviewed further because the agencies for addressing this dispute determine that only the State that may be the party to suit the dispute.

The third legal impact is the emergence of State's liability to protect the marine rights. The acknowledgement of ocean rights can be set forth in General Assembly resolution. General Assembly is main part of UN consisting all member States that may take the role in forming the international laws sources of resolution. Although the resolution of General Assembly is not legally binding, it creates the moral liability to the States to agree on it. The resolution also may help creating the international consensus and put the basic for subsequent binding agreement negotiations [33].

4 Conclusion

IUU fishing has become the entrance to other crimes and TOC such as, human trafficking, migrant smuggling, corruption, illicit drugs , servitude and human rights violation thus the term used in the current paper is IUU fishing and related crimes. The fisheries resources exploitation followed by the human rights violation in fisheries business indicates that the economy perpetrators remain use the anthropocentrism paradigm. Viewed from the dimension of ethics and morals, the values of ethics and morals violated are for instance ecocentrism ethics, principle of sustainable, and the values of respecting human rights. The instrument of international laws such as UNCLOS 1982 and IPOA- IUU fishingare not adequate anymore to overcome the IUU fishing and related crimes so that required the criminal court system to address it. Indonesia is mainstreaming the recognition of concept of Transnational Organized Fisheries Crime in UN in order to easier the law enforcement to overcome the IUU fishing and related crimes. The crissis occurred in ocean requirred a solution to realize the justice for the oceans. The solution fought by Indonesian government is the acknowledgement of *ocean rights*. It will give important legal impacts for the ocean, move aside the paradigm of anthropocentrism that put the human as the subject and natural resources as the object become ocean in ecosystem equal to the human that have légal right, enlarge the lawsuit, the State may apply the sanction for the violators al well as determine the liabilities for the States to protect that law suit.

REFERENCE

1 UNESCO, *Healthy Ocean, Healthy People*, (UNESCO, France, 2010). PDF Available onhttp://unesdoc.unesco.org/images/0021/002166/216651e.pdf
2 Southern Ocean Conservation Unit,Aust. Antarctic Mag. **5** (2003)
3 M.A. Santosa, Penegakan HukumTerhadap PenangkapanIkanIlegal OlehKapalIkan Asing di ZEE Indonesia, *Paper has Presented in Focus Group Discussion of Penegakan HukumTerhadap Penangkapan Ikan Ilega lOleh Kapal IkanAsing di ZEE Indonesia Universitas Indonesia,October, 24* (2017)
4 UNODC, *Transnational Organized Crime in Fishing Industry*, (Vienna,2011). PDF Available on:http://www.unodc.org/documents/human-trafficking/Issue_Paper_-_TOC_in_the_Fishing_Industry.pdf
5 B. Islam, Int. J. Chem. Sci.: **11**,3(2013)
6 J. R. Jambeck,et all, Sci. **347**, 6223(2013)
7 FAO, *The State of the World Fisheries and Aquaculture*, (FAO, Rome, 2016). PDF Available onhttp://www.fao.org/3/a-i5555e.pdf

8 Anonim,*Global implication of IUU fishing,* (NIC WP, 2016). PDF Available onhttps://fas.org/irp/nic/fishing.pdf
9 UNODC.*Transnational Organized Crime in Fishing Industry*, (UNDOC, Vienna, 2011)
10 EJF. *All At Sea The Abuse of Human Rights Aboard Illegal Fishing Vessel*. (Environmental Justice Foundation, UK, 2010)
11 EJF, *Pirates and Slaves: How Overfishing in Thailand Fuels Human Trafficking and the Plundering of Our Oceans*, (Environmental Justice Foundation, UK, 2015)
12 A. Satria, *KrisisLautDunia*, (Kompas, Senin, 08 Juni 2015), Available onhttp://pkspl.ipb.ac.id/berita/detail/krisis-laut-dunia, Accessed March 25, 2018.
13 B. Kuncahyo, *Indonesia has urgued the United Nations to Issue Rights of Oceans*, (Antaranews, 11th August 2017), Available on https://en.antaranews.com/news/112232/indonesia-urges-un-to-issue-rights-of-ocean, Accessed Mach 25, 2018
14 B.Baxer, *A Theory of Ecological Justice*(Routledge, London,2005)
15 D. Scholsberg, *Defining Environmental Justice: Theories, Movements, and Nature*,(Oxford University Press, New York, 2007)
16 A. Kiss, Int'l. J. Law, Int'l Comm.**40**,3 (1985)
17 J. Theutenberg, Artic.**37**, 4 (1984)
18 Anonim, *Closing the net:Stopping illegal fishing on the High Seas,*Final report of the Ministerially-led Task Force on IUU Fishing on the High Seas,(IUU Fishing Coordination Unit, UK, 2006), PDF Available onhttps://www.oecd.org/sd-roundtable/papersandpublications/39375276.pdf
19 D. Agnew, et.all, Plos One, **4**, 2(2009)
20 FAO, *The State of the World Fisheries and Aquaculture* (FAO, Rome, 2014), PDF Available on http://www.fao.org/3/a-i5555e.pdf
21 Southern Ocean Conservation Unit, Aust. Antarctic Mag. **5**(2003)
22 UNODC.*Transnational Organized Crime in Fishing Industry*, (UNDOC, Vienna, 2011)
23 EJF. *All At Sea The Abuse of Human Rights Aboard Illegal Fishing Vessel*. (Environmental Justice Foundation, UK, 2010). Available onhttp://ejfoundation.org/report/all-sea
24 Anonim, *Tragedy in the Marine Commons: The Intertwined Exploitation of Ocean Ecosystems and Fisheries Workers*, Human Right at Sea. (2016). Available onhttps://www.ohchr.org/Documents/Issues/Environment/Biodiversity/CoalitionNGOs.pdf (Accessed February 16, 2018)
25 G. Hardin,Sci. **162**, 3859(1968)
26 W. Ambrosius,UWL Undergraduate Research Journal, **VIII** (2005). PDF Available from https://www.uwlax.edu/urc/jur-online/PDF/2005/ambrosius.pdf
27 M.P. Nelson, *Deep Ecology*, Encyclopedia of Environmental Ethics and Philosophy. **207**(2008). PDF Available onhttp://www.uky.edu/ OtherOrgs/AppalFor/Readings/240%20-%20Reading%20-%20Deep %20Ecology.pdf
28 D. Shelton,in Malcom D. Evans (ed). **70**, 166(2014)
29 A. Carty. *Philosophy of International law*, (Edinburg University Press,Edinburg, 2007)
30 C.D. Stone. Southern. Cal. L. Rev.**45**, (1972)

31 H.Tenggono, *Lingkungan Sebagai Subjek Hukum Dalam Mengatasi Permasalahan Lingkungandi Indonesia*. Available on http://www.academia.edu /17549482/ Lingkungan_Sebagai_Subjek_Hukum

32 A. Pramudianto, *HukumLingkunganInternasional* (Raja GrafindoPersada, Jakarta, 2017)

The Roles of the Ministry of Marine and Fishery in Marine Environment Protection from Oil Pollution by Tankers in the Sea in Indonesian Legal System

Elly Kristiani Purwendah,[1] Marsudi Triatmodjo,[2 1] Linda Yanti Sulistiawati[2 2]

[1] Mahasiswa Doktoral Ilmu Hukum, Fakultas Hukum Universitas Gadjah Mada, Yogyakarta, Indonesia
[2 1, 2 2] Fakultas Hukum, Universitas Gadjah Mada Yogyakarta, Indonesia

Aabstract. The Ministry of Marine and Fishery (KKP), as stipulated in the Presidential Regulation Number 2 of 2017 on the Amendment of the Presidential Regulation Number 63 of 2015 on the Ministry of Marine and Fishery, is in charge of organizing government affairs in the field of marine and fisheries to assist the President in organizing the state government. In one of the seven pillars of Indonesia's marine policy, which is in the pillars of the management in the scope and protection of marine environment, one of the major programs of marine environment protection policy is the prevention, mitigation, and recovery of pollution impacts and damage to marine environment. For the achievement of the objectives, the KKP requires a normative support for the protection to marine environment from oil pollution by tankers, in terms of prevention, mitigation, and recovery of pollution impacts and damage to marine environment.

1. Introduction

Indonesia's marine territory which reaches an area of 5.8 million km^2 makes the marine sector potential become invaluable, particularly from the sector of marine natural resource. [1] The potential of marine resources is so important that it is prioritized by Indonesia in the concept of green economy and blue economy that leads to sustainable development as conveyed by the President of the Republic of Indonesia during his speech at the Conference of Rio+20 (the United Nations Conference on Sustainable Development) in Rio de Janeiro, Brazil, on June 20-22, 2012. [2] The marine environment is part of a country's economy. [3] It is estimated that the Indonesia's marine potential reaches US$171 billion per year. In detail, it can be stated as follows; Fisheries: US$ 32,000,000,000/year (IPB, 1997), Coastal Area: US$ 56,000,000,000 per year (ADB 1997) Biotechnology: US$ 40,000,000,000 per year (PKSPL-IPB, 1997), Marine Tourism: US$ 2,000,000,000 per year (DEPBUDPAR, 2000), Petroleum: US$ 21,000,000,000 per year (ESDM 1999), and Marine Transportation:

[1] Correspondence author; elly_kristiani@yahoo.co.id

US$ 20,000,000,000 per year. [4]

The promising condition of the Indonesian sea remains untouched. Hence, the government issued Law Number 32 of 2014 on Marine and the State Gazette of the Republic of Indonesia of 2014 Number 294 in which in Chapter VI on Marine Management Article 14,it states that the Government and the Regional Government in accordance with their authority conduct marine management for the greatest prosperity of the people through the utilization and exploitation of marine resources by using blue economic principles. The blue economy policy adopted by the state certainly causes the consequences for the state to prepare a national legal system related to the state's obligation to develop marine science and inter-state technology in terms of exploration, conservation and processing of marine resources, the protection and preservation to marine environment, marine science research and other activities in marine environment in accordance with the United Nations Convention on the Law of the Sea (UNCLOS). With the aim of accelerating social and economic development for developing countries. The State should strive to create a climate of "economic law" and "favorable law" for the transfer of marine technology that benefits all relevant parties fairly (Article 266 Paragraphs 2 and 3 of UNCLOS).

Furthermore, the most important safeguard in a legal system for the sustainability of its marine resources against exploration,the protection to exploitation activities, and the destruction and pollution of marine environment is stated in Article 193 which states that "states have sovereign rights to exploit their natural resources in line with environmental policies as well as their obligations to protect and preserve marine environment. A good and just legal system that stands for the sustainability and preservation of marine environment is of course a judicial system with an insight on a value of ecological justice that has a perspective on deep ecological concept by prioritizing the marine environment right as the central part of law enforcement system. Marine environmental rights as a central legal system is surely a point of interest that needs to be protected for human needs as the objective of the law is the conflict of human interest. Therefore, the law that is coherent with environmental justice in the concept of blue economy should be regulated in a unity of national legal system in order not in contrary with the interests of economic exploitation which results in the destruction and pollution of marine environment resources. To realize a good and fair legal system and takes sides for the sustainability and preservation of the sea, a legal structure that can play an active role in law enforcement is required.

Some marine areas in Indonesia have certain characteristics, such as the presence of fishermen, ports with their function of sea transportation and trade, and the exploitation of natural resources such as petroleum. This is in accordance with the geographical condition in the Cilacap sea that is vulnerable to oil pollution caused by oil companies.For example, in Cilacap, in one time, there are two pollution cases happened; one from the Single Point Moring (SPM) of PT. Pertamina RU. IV on May 20, 2015, and the second one from the Martha Petrol Ship on May 24, 2015. The oil distribution moves towards the southern island of Nuksakambangan and Teluk Penyu beach and surrounding areas. The pollution certainly caused direct victims which were very disadvantaged economically because fishermen can not catch fish in Cilacap waters.

This paper will discuss how the Ministry of Marine and Fishery plays the role of protecting the interests of marine ecosystem from tanker oil pollution that can lead to decreased quality of marine fisheries and fishermen's welfare. The role is whether the Ministry of Marine and Fishery can help calculate the loss of fishery and the carrying capacity of its ecosystem and how its implementation mechanism.

2The Setting of Marine Environment Protection from Oil Pollution by Tankers

The Indonesian sea has a tangible economic value, such as; fisheries, shipping, offshore mining, health and biodiversity and marine tourism. These economic interests require enhanced marine infrastructures. [5] Various programs in implementing marine policy strategies are required, such as the preparation and development of marine economic data and information base, the creation of a conducive and efficient business investment climate, the development of business world in the national maritime field with international competitiveness, the development of marine economic area in an integrated manner, the blue economic principles in coastal areas, small islands, outer islands and Indonesian waters realistically, optimizing the provision of infrastructure facilities required by the business world and marine business actors particularly fishermen, intensification, extensification and reinforcement of fishery products ranging from pre-production process up to marketing, development of mutually beneficial marine business partnership between small and medium enterprises with big business, and the development of sustainable economic cooperation with marine strategic partner countries. [6]

It became a serious problem when marine economy is superior with all forms of strategic program is implemented. This will be related to the shipping and trading sectors, petroleum and community welfare in the fishermen and fishery sector,and marine environment. Given some of the relevant interests, it will also relate to law enforcement structures. On the one hand, related to the shipping, it will be the authority of the Ministry of Transportation which is, in this case, the Directorate General of Sea Transportation. It will be then related to petroleum with the Ministry of Energy and Natural Resources, and it is lastly related to the welfare of fishermen and fishery communities which will be under the authority of the Ministry of Marine and Fishery. Marine environment as an important matter as the impact of marine economic development will be related to the authority of two major ministries; the Ministry of Environment and Forestry and the Ministry of Marine and Fishery. Problems arise when Marine Law provides protection to traditional fishermen.On the other hand, it also develops the potential of marine and industrial transportation.

It can be seen in the condition of the coastal area of Cilacap which has the characteristics with the presence of the State Owned Company of PT. Pertamina Ru. IV as an oil processing unit and Tanjung Intan Port. Tankers as a means of sea transportation may cause oil pollution in the coastal areas and harm the welfare of fishermen. [7] The three priorities of marine economic development make the Cilacap Sea vulnerable to the pollution of oil sources from tankersgiven the distribution of oil is carried out through sea transportation.

It is argued by John Maddox [8] that pollution will be solved by calculating the price. Pollution is an economic problem: "*we can reduce pollution if we are prepared to pay it*". Then, it must be understood that the problem of pollution is an economic problem, meaning how much we pay by creating anti-pollution prevention tools or indirectly by paying for the losses caused by pollution. H.C. Combs[5] argues that economic changes can solve many environmental problems. According to him, pollution can be controlled economically through legal arrangements, such as the application of fines and taxes on the materials processed or waste.

The Shipping Law provides an understanding of the prevention and countermeasures of pollution from the operation of vessels provided for in Article 227 by stipulating that; each vessel shall prevent and overcome the occurrence of environmental contamination sourced from ships. Operated ships of a certain type and size shall be equipped with the equipment and materials of oil pollution prevention with the approval from the government. Operated ships of a certain type and size shall be accompanied by a pattern of oil pollution control from the ships authorized by the government. The size of the required vessel with an obligation to complete the pollution control shall be further stipulated in the Regulation of the Minister of Transportation of the Republic of Indonesia No. 45 of 2012 on Ship Safety Management, for oil tankers, chemical tankers, and gas carriers with the gross tonnage greater or equal with GT 150 (one hundred and fifty Gross Tonnage).

The Ministry of Marine and Fishery (KKP) is established through the Presidential Regulation No. 63 of 2015 on the Ministry of Marine and Fishery and the State Gazette of the Republic of Indonesia of 2015 Number 111. The discussions will begin with how the Indonesian legal system protects the sea from tanker oil pollution through various regulations. The protection to marine environment is very important to protect the marine ecosystem as Indonesia's marine areas which are very potential for economic growth. Oil pollution is regulated in various regulations in the Indonesian legal system. The regulation of oil pollution in Indonesian law is Law Number 32 of 2014 on Marine and the Supplement to the State Gazette of the Republic of Indonesia Number 5605. In Article 1, point 2, it is stated that the protection to marine environment is a systematic and integrated effort undertaken to conserve marine resources and to prevent the occurrence of pollution and/ or environmental damage in the sea that includes marine conservation, marine pollution control, disaster management, pollution prevention and handling, and damage and disasters. In article paragraph (2), point b, it explains that marine disaster caused by environmental pollution can be oil pollution.

Since the beginning, the KKP considers pollution as a complementary task since the main task of pollution control is within the authority of the Ministry of Environment and Forestry (KLHK). In the Presidential Regulation No. 109/2006 on the Emergency Response of Oil Spill at the Sea made for the sake of quick, precise and coordinated action of shipping activities, oil and natural gas exploitation activities and other activities that contain the risk of accidents and may result in oil spills which can pollute and / or damage the marine environment by mobilizing local, regional and national potentials effectively. The Chairman of the National Disaster Management Team is the Minister of Transportation, and the Minister of Marine and Fishery as the member. The KKP, together with other Ministries (Ministry of Environment and Forestry),is the facilitator representing the government's interest in solving oil pollution problem due to tankers.

In the case of tanker's spill in Cilacap in 2015, the KKP had no active role in facilitating the demands for fish losses against the losses in fishery sector. The sampling and field verification were conducted by KLHK. [9] However, on October 18, 2016, the KKP had a Pollution Handling Team established through the Decree of the Minister of Marine and Fishery of the Republic of Indonesia Number 5/ KEPMEN-KP / 2016 on the Oil Spill Response Team on Marine Resources and Fisheries.

This team was established to overcome and compensate for the oil losses at sea due to marine and natural gas shipping and the concession activities that pollute and damage marine and fishery resources. The team consists of directors, working groups and field personnel. The Steering Committee was tasked to coordinate the implementation of the countermeasures in overcoming the impact of oil spill of marine and fishery resources.

The working group consists of, first,the working group of data-base and data and material collection containing the data-base of marine resources, coastal fishery and Indonesian waters. They also compile environmental sensitivity index maps; calculate the economic value of marine resources,coastal fishery, and Indonesian waters; compile the fixed procedure on material and information collection; conduct the pull-bucket activities in the place of oil spill; process and analyze the data of the results of material and information collection activity; make document and report for loss calculation work report. Second, the working group of loss impact calculation has the duty to compile the fixed procedures to calculate the impact of losses and to calculate the compensation for the impact of oil spillon marine and fishery resources. Third, the working group has the duty on the preparation of the compensation indictment document which contains the standard operating procedure of the indemnity compensation document, the contents of the indemnity document, the submission of documents to the National Command and Control Center for the Operation of Emergency Response of Oil Spill at the Sea (Puskodalnas) at the Ministry of Transportation, to assist the follow-up submitted by PUSKODALNAS until the approval of the compensation, to improve the field officer's ability to handle oil spills. Fourth, the working group of cleaning and rehabilitation mentoring has the duty to develop the operational procedures on coastal biota cleaning and rehabilitation assistance, to mitigate the impact handling of oil spill by cleaning the biota and coast, to provide assistance to the implementation of compensation to the community and the implementation of marine and fishery resource rehabilitation, to monitor and examine the post-occurrence of oil spill and to make documentation and report.

The field officer is in charge of monitoring activities by operating a 24-hour oil spill monitoring system for 7 days a week, receiving reports from the community, submitting reports to monitoring, reporting and indemnification working groups and database working groups, collecting materials and information, taking samples, data and information as the initial evidence after the oil spill, facilitating group work activities in the field and coordinating with relevant agencies. This team was formed only for the implementation of oil spills in Riau Islands and Jakarta.

The KKP has a Standard Operating Procedure (SOP) for calculating the impact of losses in the context of impact mitigation for the prevention of oil spill impact to marine and fishery resources. The identification of the impact of oil spill on the users and utilization of marine and fishery resources based on ecosystem mentions an ecosystem function including coastal areas in some groups, i.e.; regulatory functions in which ecosystems are capable of managing important ecological processes in nature, habitat functions in which ecosystems are home or living places for plants and animals, production function in which ecosystems perform photosynthesis to convert nutrients into energy and subsequently eaten by other living biomass, and information functions in which ecosystems provide reference functions and can serve as a place for relaxation, spiritual and recreational enrichment. The ecosystem functions then form the goods and services which are beneficial and have economic values to humans. [10]Calculating the impact of the oil spill and its loss value according to this SOP will depend on the quantity and intensity of the oil spill volume. If the amount of spill is very high, the oil spill in the sea affects the quality of the surrounding waters which of course will disturb the biota life in it. More significant impacts will be felt when the oil spill reaches the coastal areas as coastal areas contain many important ecosystems and have a direct impact on the decline in the benefits and services of the ecosystem. [11]

The calculation of the total value of oil spill loss impacts consists of three components; i.e.: Compensable value, restoration cost, and assessment cost. Compensable value is anything received in physical or non-physical forms and must be calculated and given to the party or person who is generally an object affected by an oil spill. Compensable value consists of several values which are based on the type of utilization and the utilization of marine and fishery resources in the areas exposed to the oil spill, consisting of; lost economic value from fishing activities (coral reef ecosystem, mangrove and marine), lost economic value from cultivation activity (land, brackish and sea), lost economic value from marine activity, economic value from the damage of coastal ecosystem (mangrove, coral reefs and seagrasses), lost economic value of fishery processing activities, lost economic value from the aesthetics of the coastal environment, lost economic value from cultivation production means (keramba, ponds, etc.), lost economic value from the damage of marine tourism facilities and infrastructures. [12]

In addition to the compensation calculation, the KKP's SOP regulates the cost of recovery as part of pollution control and/ or environmental damage that is carried out in the framework of environmental conservation. The cost of restoration is defined as the costs incurred in recovery efforts to make the environment or parts of it function again as it was.[13] The cost of recovery or *restitutio in integrum* in the lawsuit of unlawful act means the measure of damage or restoration of damage as it was in the original state before the occurrence of an act against the law.In other hand, in the contract, it is intended to be restored to the state of the contract when it is not violated. [14]

The polluted and/ or damaged environments must be restored wherever possible prior to the pollution and/ or damage to the environment. This environmental restoration act applies to the public environment which is the right and authority of the government and the community that includes the rights and powers of individuals and groups of people. The cost of recovery due to oil spill is divided into two; the cost of cleaning up oil spills on coastal and marine areas and the cost of rehabilitation of ecosystem environments (mangroves, coral reefs and seagrasses). [15]

The SOP made by the KKP in 2016 is a serious progress for the KKP in taking into account the sustainability of marine ecosystems as the carrying capacity of captive fishery in coastal areas. Before the SOP was established, the KKP has not played an active role in protecting marine ecosystems as a carrying capacity of captive fishery, in the case of oil pollution by tankers and post-events. In this SOP, the KKP began to formulate a mechanism for handling compensation in the event of pollution and the sustainability of pollution through *restitutio in integrum*. The step forward of the KKP's active role is of course still to be fought since the pollution completion committee is at the minister of transportation, and the KKP is only a member in the emergency oil spill response team at sea.

3 Conclusion

The role of the Ministry of Marine and Fishery in the Marine Environment Protection from Oil Pollution by Sea Tankers in the Indonesian Legal System is still very limited because the Ministry of Marine and Fishery is only a member of the National Team for Emergency Response of Oil Spill at Sea (the Regulation of the President of the Republic Indonesia Number 109 of 2006 on Emergency Response of Oil Spill at Sea). The Ministry of Marine and Fishery only has the authority to map sea losses related to fish, mangroves, coral reefs and seagrass beds. The other environmental losses are the authority of the Ministry of the Environment which is positioned as the vice chairman, while the position of the chairman is the responsibility of the Ministry of Transportation. However, with the preparation of the Standard Operational Procedure (SOP) for the Calculation of Loss Impacts in Order to Overcome the Impact of Oil Spill on Marine Resources and Fisheries, it is expected that the Ministry of Marine and Fishery can play more active role in calculating the loss of fishery and the carrying capacity of the ecosystem, and even the recovery compensation for the oil pollution caused by tankers.

References

1G. R. Lilley. *Prosiding Lokakarya Reformasi Hukum di bidang Pengelolaan Sumber Daya Alam*.(Cetakan I, ICEL, Pustaka Pelajar Offset, 1999)

2Pusat Pendidikan Kelautan dan Perikanan Badan Pengembangan SDM Kelautan dan Perikanan Kementrian Kelautan dan Perikanan, Blue EconomyGoes to Campus, p. 2,4 Kementrian Kelautan dan Perikanan. (2014)

3Albert Gore. Department of States Dispatch, **6**,2 (1995)

4Siregar.*Workshop Forum Rektor Indonesia USU.* (2015)

5Lampiran Kebijakan Kelautan Indonesia, Naskah Akademik Tentang Rancangan Undang-undang Kelautan, Kementrian Koordinator Bidang Kemaritiman Republik Indonesia(2017)

6Ibid

7GregoriusMagnisFinesso, SaatNelayanTerpaksa "Menjala" Minyak, Beta KOMPAS online, Available from http://print.kompas.com/baca/2015/05/29/Saat-Nelayan-Terpaksa-%E2%80%9D-Minyak, Accessed July 17 2015.

8DaudSilalahi. *HukumLingkunganDalamSistemPenegakkanHukumLingkungan Indonesia*(Alumni, Bandung. p. 14.2001)

9Cross, Michael, Hamer, Mick, March.New Scientist.**133,** 3-4 (1992)

10De Groot R.S. M.A. Wilson, R.Mj. Boumans. Ecological Economics**. 41**, 393-408 (2002)

11Pusat Penelitian Sosial Ekonomi Kelautan dan Perikanan Badan Penelitian dan Pengembangan Kelautan dan Perikanan, Kementrian Kelautan dan Perikanan, Kajian Penyusunan Standar Operasional Prosedur (SOP) Perhitungan Dampak Kerugian Dalam Rangka Penanggulangan Dampak Tumpahan Minyak Terhadap Sumber Daya Kelautan dan Perikanan, p.8 (2016)

12 Ibid. p.16

13 M. Yahya Harahap, *Kekuasaan Mahkamah Agung Pemeriksaan Kasasi dan Peninjauan Kembali Perkara Perdata*, Sinar Grafika, Jakarta. p. 5-6 (2008)

14 Henry Campbell Black, Black's Law Dictionary with Pronunciations, Definitions of the Terms and Phrase of American and English Jurisprudence, Ancient and Modern, St. Paul Minn, West Publishing Co. p. 1313(1990).

15Pusat Penelitian Sosial Ekonomi Kelautan dan Perikanan Badan Penelitian dan Pengembangan Kelautan dan Perikanan, Kementrian Kelautan dan Perikanan, Op.Cit. p.17-18

Fishing seasons of fish landed at Sungailiat Archipelago Fishing Port in Bangka Regency

Fauziyah[1*], *Fitri* Agustriani[1], *Desi* Melda Situmorang[1], and *Yuliyanto* Suteja[2]

[1] Marine Science Department, Faculty of Mathematics and Natural Sciences, Sriwijaya University, Indonesia

[2] Marine Science Department, Faculty of Marine Sciences and Fisheries, Udayana University, Indonesia.

Abstract. The fish landed at the Archipelago Fishing Port of Sungailiat is an important aspect for the fishing industry development. One of the factors that influence the sustainability of capture fisheries is optimal fishing operation. Fishing season index can be used to determine an appropriate time in perform of fishing operations. The objective of this research is to determine fishing season pattern of pelagic and demersal fish in the Archipelago Fishing Port of Sungailiat. Time series data on catch and fishing efforts (2006-2015) collected from Sungailiat Archipelago Fishing Port were used to calculate monthly CPUEs and then analysed using moving average method to obtain fishing season index for each month. This results showed that the peak fishing season of eastern little tuna and Barre spanish mackerel occur for 4 months with the best peak season respectively in October and April. The peak fishing season of Yelowstripe Scad, Tile Trevally, Terpedo Scad, sharks and Grouper occur for 5 months with the best peak season respectively in October, May, January, April, and February. The peak fishing season of Marine catfish and Black pomfret occur for 6 months with the best peak season respectively in January and October. While the peak fishing season of Ray occur for 8 months with the best peak season in October.

1. Introduction

The Sungailiat Archipelago Fishing Port (AFP) in Bangka Regency, Bangka Belitung Islands Province has a large potential as a center for the development of capture fisheries industry. Types of fish in the Sungailiat AFP are demersal fish, pelagic fish, crustaceans, and squid. Several species of demersal fish are landed such us eaglerays, ornate threadfin bream, leopard coralgrouper, black pomfret, yellowtail fusilier, marine catfish, red snappers and sharks. Deepbody sardinella, yelowstripe scad, narrow-barred spanish, terpedo scad, kawa-kawa, short bodied mackerel, tile trevally, indo-pacific sailfish, dorab wolf-herring, great barracuda and needlescale queenfish are an example of a landed pelagic fish.

* Corresponding author: siti_fauziyah@yahoo.com

Fishing unit in Sungailiat AFP mostly under 10 GT with fishing gear used by fisherman include handline, drift gillnet, fixed gillnet, surrounding net, mini purse seine, and trap net.

Based on Sungailiat AFP production data for the last 10 years, the production of pelagic and demersal fish resources is fluctiative. The fluctuating production of fish resources is caused by human, availability of fish resources and environment factors. The availability of fish resources factor is strongly influenced by fish season pattern. One approach that can be used to determine the pattern of the fish season is to predict the fishing season pattern that is landed at the fishing port. Fishing season pattern is very useful for fishermen and government. For fishermen it is useful to determine the exact timing for fishing operations [1, 2] so that fishing activities are more planned and efficient [3]. For the government is useful for initial estimates of overfishing [4], more efficient and sustainable fisheries planning [3], utilization and management of fish resources [5, 6, 7], controlling and monitoring the rate of fishing exploitation [2].

Information and research on fishing season pattern that is landed in Sungailiat AFP is poor. Previous studies have only examined the pattern of cuttlefish fishing season [8]. Other studies were related to the technical efficiency of the bottom gillnet [9], the productivity model of the bottom gillnet [10], the finance analysis of bottom gillnet [11], the design characteristics of fishing boat for bottom gillnet [12], the static stability of fishing boat for bottom gillnet [13], bio-engineering of mackerel resources and marketing distribution [14], and bio-economic model of squid fisheries [15].

Based on research that has been done in Sungailiat AFP shows that the study of the fishing season pattern that were landed is very poor so it is difficult to describe the pattern of fish fishing season completely. Therefore, the pattern of the fishing season to be studied covers most of the demersal and pelagic fish landed at the Sungailiat AFP. It is hoped that this will provide a more complete conception of the diversity of fishing season pattern. This study aims to estimate the pattern of fish fishing season that landed at Sungailiat AFP. Information on fish fishing season pattern is needed to know the best time or season to catch the fish, thus reducing the risk of fishing loss. Furthermore, it is expected that the fishing operation will only be done in peak season, so that the optimum catch will be obtained and to keep the sustainability of fish resource [2].

2. Materials and methods

2.1 Study site

This study uses monthly data series of demersal and pelagic fish in 2006-2015. The data was collected by the managers of Sungailiat AFP. The data collected were pelagic fish (Eastern little tuna, Yelowstripe Scad, Barre spanish mackerel, Tile Trevally, and Terpedo Scad) and demersal fish (Marine catfish, Black pomfret, sharks, Ray, and Grouper).

2.2 Data analysis

Data analysis conducted in this research is analysis of catch per unit effort (CPUE), fishing gear standardization, and analysis of fishing season pattern using moving average method.

2.2.1 CPUE

This analysis aims to determine the level of fish resource utilization of each species conducted by several fishing units used by fishermen. The CPUE equation [16, 4] was used as follows:

$$CPUE_i = \frac{c_i}{f_i} \tag{1}$$

Where:

$CPUE_i$	=	the catch per unit of fishing effort in month i (kg/ trip)
c_i	=	the catch of fishing gear in month i (kg)
f_i	=	the effort of the fishing gear in month i (trip)

2.2.2 Standardization of fishing gear

Standardization of fishing gear was done by determining the largest CPUE of each fishing gear and the largest CPUE was used as a standard fishing gear. The stage standardization of fishing gear [17] as follows:

1. Determining the type of fishing gear that can be use as the standardized fishing gear, ie the fishing gear that has the largest CPUE value of each fishing gear.
2. Determine the value of Fishing Power Index (FPI) :

$$FPI_i = \frac{CPUE_i}{CPUE_s} \tag{2}$$

$$FPI_s = \frac{CPUE_s}{CPUE_s} \tag{3}$$

where:

FPI_i	=	Fishing Power Index of the fishing gear i
FPI_s	=	Fishing Power Index of the standardized fishing gear
$CPUE_i$	=	The monthly CPUE of the fishing gear i (kg/ trip)
$CPUE_s$	=	The monthly CPUE of the standardied fishing gear (kg/ trip)

3. Calculating the total value of standard fishing effort.

$$StdEffort_s = FPI_s x f_s \tag{4}$$

$$Std\ Effort_i = FPI_i x f_i \tag{5}$$

$$Std\ Effort_{total} = \sum(FPI_i x f_i) + (FPI_s x f_s) \tag{6}$$

where:

f_i	=	the monthly fishing effort of the fishing gear i (trip)
fs	=	the monthly fishing effort of the standard fishing gear arbitrarily chosen (trip)
$StdEffort_i$	=	the standardized effort of the fishing gear i after standardization
$StdEffort_s$	=	the standardized effort of the standard fishing gear arbitrarily chosen, after standardization
$StdEffort_{total}$	=	the standaridized effort of total fishing gears after standardization

2.2.3 Analysis of fishing season patterns

Estimation of fishing season pattern was calculated per species with moving average method. This analysis was conducted by analyzing time series data on monthly data of catches of 12 species by fishing units that landed in Sungailiat AFP for 10 years. The steps of this analysis [18, 17] were :

1. Create a series of CPUE within 10 years period

$$CPUE_I = n_i \quad (7)$$

where: n_i = CPUE of the i-th order
i = 1,2,3..., 120

2. Create a moving average CPUE for 12 months (MA)

$$MA_i = \frac{1}{12}\left(\sum_{i=i-6}^{i+5} CPUE_i\right) \quad (8)$$

where: MA_i = the 12-month moving average of the i-th order
$CPUE_i$ = CPUE of the i-th order
i = 7,8,9..., n-5

3. Create a centered CPUE moving average (CMA)

$$CMA_i = \frac{1}{2}\left(\sum_{i=i}^{i=1} MA_i\right) \quad (9)$$

where: CMA_i = centered moving average of CPUE in month i
i = 7,8,9..., n-5

4. Create monthly average rasio

$$Am_i = \frac{CPUE_i}{CMA_i} \quad (10)$$

where: Am_i = monthly average rasio in month i
$CPUE_i$ = CPUE in month i
CMA_i = centered moving average of CPUE in month i

5. Create an average value in ixj matrix organized for each month
6. Create average ratio for month i (AAm_i)

$$AAm_i = \frac{1}{n}\left(\sum_{j=1}^{n} Am_{ij}\right) \quad (11)$$

where: AAm_i = average for Am_{ij} i-th month
Am_{ij} = the monthly average ratio in the i x j size matrix
i = 1,2,3,..., 12
j = 1,2,3,..., n

7. The total of monthly average ratios (TAM)

$$TAM = \sum_{i=1}^{12} AAm_i \quad (12)$$

where: TAM = total of monthly average ratios

8. Fishing season Index (FSI)

Ideally, the TAM value is 1200, but many factors cause TAM to not always equal to 1200, therefore the value of the monthly average ratio should be corrected called the Correction Factor (CF)

$$CF = \frac{1200}{TAM} \quad (13)$$

$$FSI_i = AAm_i \ x \ CF \quad (14)$$

Decision criteria:

FSI $\geq$ 100% : peak season
50% $\leq$ FSI < 100% : moderate season
FSI < 50% : low season

3. Result

3.1 Production and CPUE

Fish landed in Sungailiat AFP are generally produced by one or two types of fishing gear (Table 1). When one type of fish was produced by 2 types of fishing gear, then one of them was used as a standard fishing gear

Table 1. Types of fish produced by fishing gear at Sungailiat AFP

Local name	Common name	Habitat	Fishing gear	Standardized fishing gear
Selar	Yelowstripe Scad	P	Handline, Surrounding net	Surrounding net
Kwee	Tile Trevally	P	Handline, trap net	Handline
Tongkol	Eastern little tuna	P	Drift gillnet	Drift gillnet
Tenggiri	Barre spanish mackerel	P	Handline, Drift gillnet	Drift gillnet
Tetengkek	Terpedo Scad	P	Handline, Drift gillnet	Drift gillnet
Manyung	Marine catfish	D	Handline, Fixed gillnet	Fixed gillnet
Cucut	Sharks	D	Handline, Fixed gillnet	Fixed gillnet
Bawal hitam	Black pomfret	D	Surrounding net	Surrounding net
Pari	Ray	D	Handline, Fixed gillnet	Fixed gillnet
Kerapu	Grouper	D	Handline, Trap net	Trap net

P : Pelagic D : Demersal

Total production of 5 pelagic fish species landed in Sungailiat AFP (Figure 1) in 2006-2015 was fluctuated. The lowest production occurred in 2009 and the highest in 2015. In 2015 the production of Eastern Little Tuna, Barre Spanish Mackerel, Tile Trevally, and Terpedo Scad decreased compared to 2014, but the production of Yelowstripe Scad increased dractically. The total production of 5 demersal fish species landed in Sungailiat AFP (Figure 2) increase from 2006 to 2011 and then declined until 2015. The highest total production of demersal fish occurred in 2011 and the lowest in 2015. The highest demersal fish production every year was Ray. The production of Ray was almost annually greater than Yelowstripe Scad, except production in 2013 and 2015. Production of Yelowstripe Scad in 2015 tends to increase, in contrast to Ray that tends to decline.

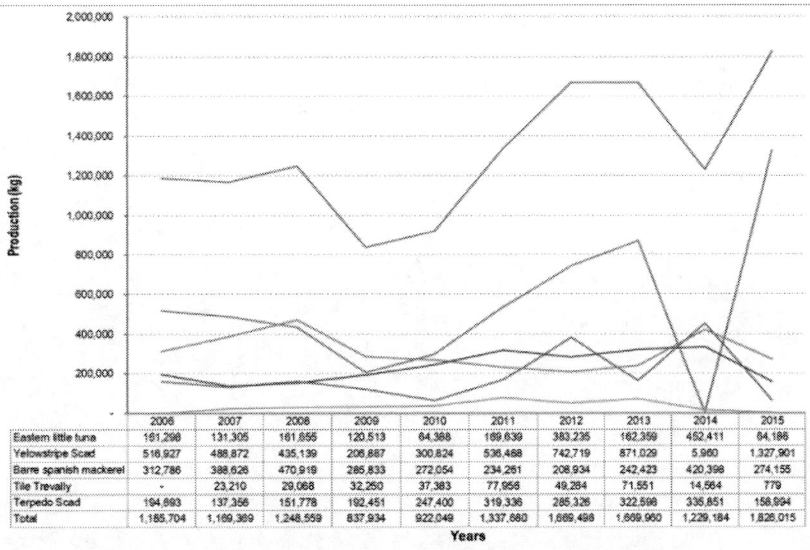

	2006	2007	2008	2009	2010	2011	2012	2013	2014	2015
Eastern little tuna	161,298	131,305	161,655	120,513	64,388	169,639	383,235	162,359	452,411	64,186
Yelowstripe Scad	516,927	488,872	435,139	206,887	300,824	536,488	742,719	871,029	5,960	1,327,901
Barre spanish mackerel	312,786	388,626	470,919	285,833	272,054	234,261	208,934	242,423	420,398	274,155
Tile Trevally	-	23,210	29,068	32,250	37,383	77,956	49,284	71,551	14,564	779
Terpedo Scad	194,693	137,356	151,778	192,451	247,400	319,336	285,326	322,598	335,851	158,994
Total	1,185,704	1,169,369	1,248,559	837,934	922,049	1,337,680	1,669,498	1,669,960	1,229,184	1,826,015

Fig 1. The production of 5 pelagic fish species landed in Sungailiat AFP

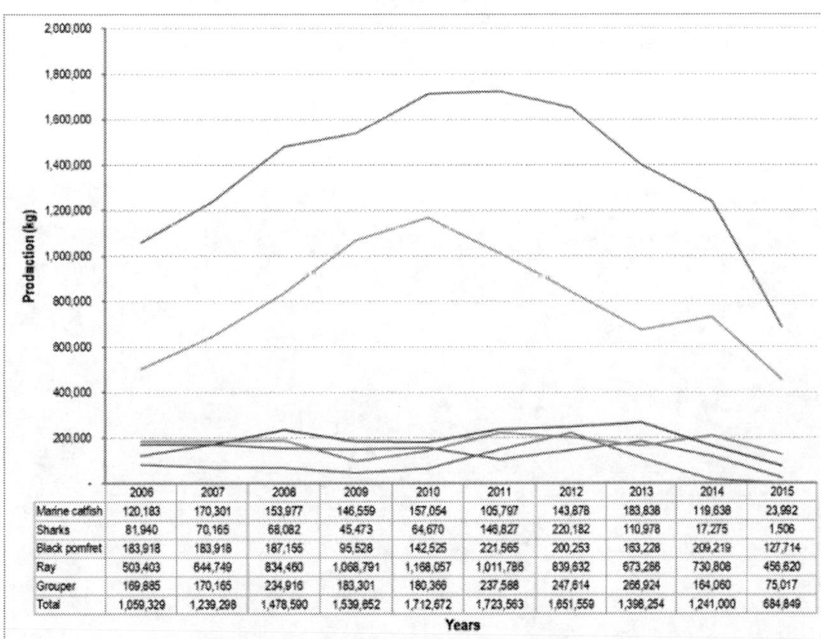

	2006	2007	2008	2009	2010	2011	2012	2013	2014	2015
Marine catfish	120,183	170,301	153,977	146,559	157,054	105,797	143,878	183,838	119,638	23,992
Sharks	81,940	70,165	68,082	45,473	64,670	146,827	220,182	110,978	17,275	1,506
Black pomfret	183,918	183,918	187,155	95,528	142,525	221,565	200,253	163,228	209,219	127,714
Ray	503,403	644,749	834,460	1,068,791	1,168,057	1,011,786	839,632	673,286	730,808	456,620
Grouper	169,885	170,165	234,916	183,301	180,366	237,588	247,614	266,924	164,060	75,017
Total	1,059,329	1,239,298	1,478,590	1,539,652	1,712,672	1,723,563	1,651,559	1,398,254	1,241,000	684,849

Fig2. The production of 5 demersal fish species landed in Sungailiat AFP

Based on the CPUE value (Figure 3), Yelowstripe Scad almost has the highest CPUE value throughout the year compared to other pelagic fish species. In 2006-2009 there was a decline of Yelowstripe Scad CPUE value and then increased until 2015. The same results were also obtained in Yelowstripe Scad production in Sunda Strait in 2006-2013 [19]. The CPUE value of rays (Figure 4) is highest throughout the year when compared to other demersal fish species. CPUE values of Ray were more fluctuated than Yelowstripe Scad. Fish resources (demersal and pelagic) landed on Sungailiat AFP have a fluctuating CPUE value pattern (Figure 3 and figure 4). The increase in CPUE values shows the average increase in catch per trip of the catching unit (improving the performance of the fishing gear used) and vice versa. Decrease in CPUE is one of the over fishing indicators [20].

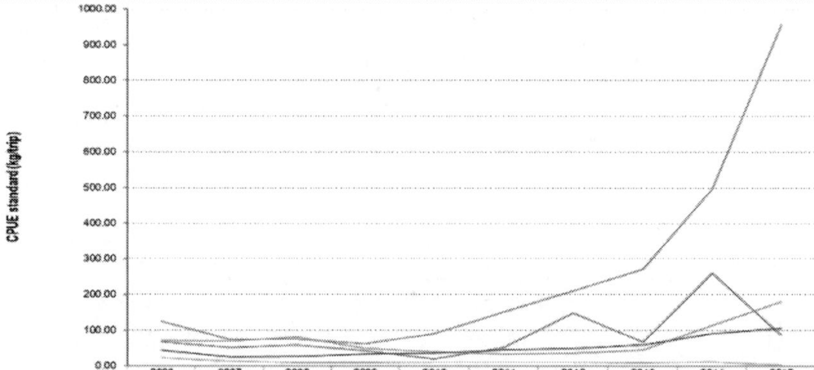

	2006	2007	2008	2009	2010	2011	2012	2013	2014	2015
Eastern little tuna	67.40	51.07	57.96	41.92	19.12	51.47	148.66	66.62	261.51	86.27
Yelowstripe Scad	125.66	75.03	74.43	61.32	90.61	151.29	210.40	272.01	496.70	956.97
Barre spanish mackerel	70.40	69.31	60.98	48.66	36.76	33.23	35.39	44.90	114.80	179.89
Tile Trevally	21.62	12.77	6.95	9.26	9.72	10.47	9.81	6.56	12.22	3.00
Terpedo Scad	43.54	24.31	25.91	32.54	34.99	44.58	47.95	59.30	91.04	107.19

Fig 3. The CPUE of 5 pelagic fish species landed in Sungailiat AFP

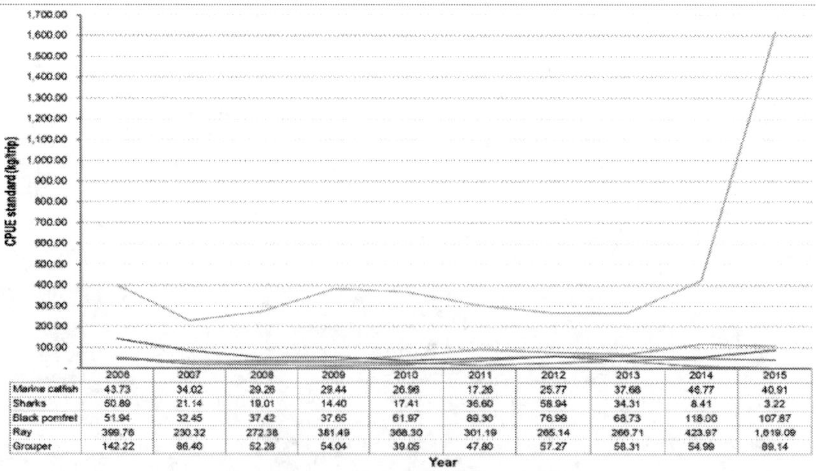

	2006	2007	2008	2009	2010	2011	2012	2013	2014	2015
Marine catfish	43.73	34.02	29.26	29.44	26.98	17.26	25.77	37.68	46.77	40.91
Sharks	50.89	21.14	19.01	14.40	17.41	36.60	58.94	34.31	8.41	3.22
Black pomfret	51.94	32.45	37.42	37.65	61.97	89.30	76.99	68.73	118.00	107.87
Ray	399.76	230.32	272.38	381.49	368.30	301.19	265.14	266.71	423.97	1,619.09
Grouper	142.22	86.40	52.28	54.04	39.05	47.80	57.27	58.31	54.99	89.14

Fig 4. The CPUE of 5 demersal fish species landed in Sungailiat AFP

3.2 Fishing season patterns

Based on the value of fishing season index (Table 2), the peak season of pelagic fishing occurred in first transition season (March - May) and second transitional season (September

- November). In dry season (June-August) and wet season (December-February) were predominantly moderate season.

Table 2. Recapitulation of pelagic fish fishing season

Pelagic Fish	Fishing Season Index											
	Wet Season			First Transition Season			Dry Season			Second Transition Season		
	Dec	Jan	Feb	Mar	Apr	May	Jun	Jul	Aug	Sept	Oct	Nov
Eastern little tuna	75	40^{F}	95	103^{P}	72	45^{F}	59	60	97	125^{P}	229^{P}	199^{P}
Yelowstripe Scad	87	78	102^{P}	99	101^{P}	98	90	94	90	110^{P}	128^{P}	122^{P}
Barre spanish mackerel	70	75	98	134^{P}	141^{P}	97	93	95	80	87	107^{P}	123^{P}
Tile Trevally	94	58	107^{P}	95	103^{P}	162^{P}	89	84	86	85	130^{P}	106^{P}
Terpedo Scad	90	133^{P}	130^{P}	96	125^{P}	101^{P}	93	90	69	75	92	106^{P}

^{F}: low season : moderate season ^{P}: peak season

Peak fishing season of Eastern Little Tuna occurs for 4 months ie March, September, October, and November (most of the 2nd transition season) with the best peak season in October. Similar result also found in Prigri AFP, peak season of tuna occurs in April-June and August-November [21]. At Baron beach (southern waters of Yogyakarta), peak fishing season occurs in July-October with the highest peak occurring in October [22]. In Lempasing Coastal Fishing Port (Lampung Bay), peak catch season occurred in April, July, August, September and October [5].

Peak fishing season of Yelowstripe Scad occurs for 5 months ie February, April, September, October, and November with the best peak season occurring in October. In the Java Sea, fishing season other species Yelowstripe Scad (*Selar crumenophthalmus*) occurred in December [23].

Peak fishing season of Barre Spanish Mackerel occurs for 4 months ie March, April, October, and November with the best peak season occurring in April. These results were different from the fishing season at Lempasing Coastal Fishing Port (Lampung Bay) where the fishing season of Barre Spanish Mackerel occurs in March, May, July, August, and September [5]. In the Java Sea [24], Barre Spanish Mackerel fish abundance in March to June and in October to December throughout the year.

Peak fishing season of Tile Trevally occurs for 5 months ie February, April, May, October, and November with the best peak season occurring in May. Peak fishing season of Terpedo Scad occurs for 5 months ie January, February, April, May and November with the best peak season occurring in January. The previous research in Lampung Bay mentioned that the fishing season of Tile Trevally occurred in January, April, June, September, November, and December [5]. In the dry season, pelagic fish catches were in moderate season. The low season occurs only in Eastern little tuna in January and May. Peak season mostly occurs in first and second transition season II.

The demersal fishing season (Table 3) has a similar pattern to the pelagic fishing season (Table 2). In the dry season and wet season, the captured of the five species of demersal fish were mostly in the moderate season, whereas, during the first and second transition season were mostly in the peak season.

The fishing season peak of Marine Catfish occurs for 6 months during wet and first transition season with the best peak season occurring in April. The similar result also found in Pelabuhanratu (west java), the peak season of Marine Catfish occurred in January, February, September and October [25].

Table 3. Recapitulation of demersal fish fishing season

Demersal Fish	Fishing Season Index											
	First Transition Season			First Transition Season			First Transition Season			First Transition Season		
	Dec	Jan	Feb	Mar	Apr	May	Jun	Jul	Aug	Sept	Oct	Nov
Marine catfish	104^{P}	119^{P}	112^{P}	101^{P}	116^{P}	104^{P}	82	91	93	92	89	96
Sharks	86	65	96	112^{P}	146^{P}	118^{P}	93	91	82	87	117^{P}	107^{P}
Black pomfret	97	105^{P}	125^{P}	118^{P}	76	88	62	76	68	113^{P}	146^{P}	125^{P}
Ray	85	45^{F}	112^{P}	121^{P}	77	109^{P}	115^{P}	104^{P}	90	103^{P}	123^{P}	115^{P}
Grouper	73	81	167^{P}	111^{P}	98	101^{P}	118^{P}	100^{P}	89	82	96	84

^{F}: low season   : moderate season ^{P} : peak season

The fishing season peak Shark occurs for 5 months ie March, April, May, October and November (first and second transition season) with the best peak season occurring in April. This result is slightly different from the fishing season Kejawaan AFP (Cirebon) where the peak season occurs in the east season [26]. In the Pelabuhanratu AFP (Sukabumi), peak season of shark occurred in January, April, May, June July, August, September and December [25]. The fishing season peak Black Pomfret occurs for 6 months ie January to March and September to November with the best peak season occurring in October.

The Ray have the longest peak fishing season (8 months) ie February, March, May, June, July, September, October and November with the peak season occurred in October. The fishing season peak of Grouper occurs for 5 months ie February, March, May, June, and July with the highest peak season occurring in February. The low season occurs only in Ray, which was in January.

Table 2 and Table 3 also show that in December was not the peak season for both pelagic and demersal at Sungailiat AFP. This was thought to be due to the high of wind speed, waves and rainfall [3] it makes fish resources move into the open seas and affect the distribution of fishing season [27]. The movement of these fish will change the distribution and abundance of fish in fishing ground. The decrease of fish abundance in fishing ground will decrease catch per trip [3]. The fishing activities conducted in the west season was ineffective so the fishermen in AFP Sungailiat reduced the trips in those months [8]. During wet season (low season), the fish is still caught by the fishermen although the number of catches per trip is very low [28].

4. Conclusion

The peak fishing season of eastern little tuna and barre spanish mackerel occur for 4 months with the best peak season respectively in October and April. The peak fishing season of yelowstripe scad, tile trevally, terpedo scad, sharks and grouper occur for 5 months with the best peak season respectively in October, May, January, April, and February. The peak fishing season of marine catfish and black pomfret occur for 6 months with the best peak season respectively in January and October. While the peak fishing season of Ray occur for 8 months with the best peak season in October.

We would like to thank UNSRI for the support of DIPA-UNSRI funds in 2016 through competitive scheme and to AFP Sungailiat of Bangka Regency of Bangka Belitung Province

References

1. M. Syahrir, M.S. Baskoro, Darmawan, E. Lubis, E.S. Wiyono, J. Ilmu Per. Trop. **13**, 24-31 (2010)
2. D. Simbolon, B. Wiryawan, P.I. Wahyuningrum, H. Wahyudi, Bul. PSP **19**, 293-307 (2011)
3. J.S. Kekenusa, V.N.R. Warung, D. Hatidja, J. Ilm. Sains **12**, 112-119 (2012)
4. E. Hamka, M. Rais, J. Ipt. PSP **3**, 510-517 (2016)
5. S. Agustina, R. Irnawati, A Susanto, J. Per. Kel. **6**, 74-82 (2016).
6. C.J. Lintang, I.L. Labaro, A.T.R. Telleng, J. ITK Tangkap **1**, 6-9 (2012).
7. Mustasim, J. Airaha **5**, 83-86 (2016)
8. D. Rosalina, W. Adi, D. Martasari, Maspari J. **2**, 26-38 (2011).
9. T. Afridanelly, Fauziyah, F. Agustriani, Maspari J. **2**, 74-76 (2011).
10. Fauziyah, F. Agustriani, T Afridanelly, J. Pen. Sains **14**, 56-60 (2011).
11. T.S. Rahmawati, R Aryawati, F. Agustriani F, Maspari J. **2**, 70-73 (2011).
12. R. Pasaribu, Fauziyah, F. Agustriani, Maspari J. **2**, 54-62 (2011).
13. R. Nopandri, Fauziyah, Rosdirwan, Maspari J. **2**, 63-69 (2011).
14. M.P. Sobari, A. Febrianto, MARITEK **10**, 15-29 (2010).
15. W. Oktariza, B. Wiryawan, M.S. Baskoro, R. Kurnia, S.H. Wisudo, Mar. Fish. **7**, 97-107 (2016).
16. P. Sparre, S.C. Venema, *Introduction to tropical fish stock assessment. Part 1. Manual.* (Rome, FAO, 1998) http://www.fao.org/docrep/w5449e/w5449e01.htm
17. L.I. Azkia, A.D.P. Fitri, I. Triarso, J. Fish. Res. Uti. Man. Tech. **4**, 1-7 (2015).
18. R.I. Wahyu, Zulkarnain, K.P.S. Mara, Bul. PSP **19**, 105-113 (2011).
19. M. Sharfina, M. Boer, Y. Ernawati, Mar. Fish. **5**, 101-108 (2014).
20. R.T. Cahyani, A. Sutrisno, Y Bambang, *Pros. Seminar Nasional Pengelolaan Sumberdaya Alam dan Lingkungan* 378 -383 (2013).
21. A. Wujdi, Suwarso, *Simp. Nas. Pengelolaan Perikanan Tuna Berkelanjutan Bali, 10-11 Desember 2014.*
22. I. Nahib, D. Sutrisno, Globë **12**, 9-20 (2010).
23. U. Chodrijah, T. Hartiati, J. Pen. Per. Indo. **3** (2010)
24. K. Kasim, S. Triharyuni, J. Pen. Per. Indo. **4** (2014)
25. R.S.N. Sulistiawan, D. Martiani, J. Agrosci. **4**, 40-55 (2012).
26. R. Fakhrurrizal, Shutabarat, A, Hartoko, Diponegoro J. Maq. **3**, 71-80 (2014).
27. G. Respondek, J. Groger, J. Floeter, A. Temming, ICES J. Mar. Sci. **71**, 1805-17 (2014).
28. H. Ilhamdi, R. Telussa, D. Ernaningsih, *J. Ilm. Sat. Min. Bah.* **1**, 52-64 (2016).

Prospect on Implementation of National Fish Logistics System: case in Indonesia

Muhamad Azhar[1,2], *Suhartoyo* Suhartoyo[1], *Putut* Suharso[3], *Vivi Endar* Herawati[4], *Nanik* Trihastuti[1]

[1]Faculty of Law, Diponegoro University, Prof. H. Soedarto SH street, Semarang 50275 Indonesia
[2]Center For Administrative Law Studies, Diponegoro University, Prof. H. Soedarto SH street, Semarang 50275 Indonesia
[3]Departement of Library Science, Diponegoro University, Prof. H. Soedarto SH street, Semarang 50275 Indonesia
[4]Department of Aquaculture, Faculty of Fisheries and Marine Science, Diponegoro University, Prof. H. Soedarto SH street, Semarang 50275, Indonesia

Abstract. The policy to realize the sustainable supply of fish (especially marine fish commodities), face the fundamental constraints of seasonal factors, remoteness and characteristic of damaged fish commodities. At the time or harvest season, the production of catch fish can be abundant, but in the famine season, there is the opposite condition. Meanwhile, on the consumer (community) and industry side, fish supply must be available all the time without getting to know the season in constant amount. The research aims to find out the prospect of implementation of national fish logistic system or *Sistem Logistik Ikan Nasional* (SLIN), especially in Indonesia. This research is an empirical juridical research using statutory uproach approach, conceptual aproach, and comparative approach. Then analyzed using synthesis analysis. The results showed that SLIN implemetasi provide great benefits for national fish production, also directly contribute to the increase of national economic development. SLIN Management can run optimally if there is a synergy between all entities involved in each activity from upstream (production side) to downstream (consumption side). In the framework of looking at best practices for upstream downstream supply chain management of fishery commodities as well as in accordance with the mandate of Minister of Maritime Affairs and Fisheries No. 05/Permen-KP/2014 on SLIN. Thus, the Directorate General of Processing and Marketing of Fishery Products sees the importance of better implementation of SLIN. In the Initial Stage by providing several supporting facilities and involving the main operators and supporting operators in the management of selected locations.

Keywords: Policy Implementation; National Fish Logistics System (*Sistem Logistik Ikan Nasional* = SLIN)

Corresponding author: azharundip@gmail.com

1 Introduction

Efforts to realize the sustainable supply of fish (especially marine fish commodities), face the fundamental constraints of seasonal factors, remoteness and characteristic of damaged fish commodities. At harvest time or season, the production of catch fish can be abundant, but in famine season, the opposite condition occurs [1]–[3]. Meanwhile, on the consumer (community) and industry side, fish supply must be available all the time without getting to know the season in constant amount.

Indonesia is a maritime country, which has great potential for fisheries and marine [4], [5]. The development of fish production and consumption for both domestic and overseas (export) has an increasing trend. Fresh fish production has increased by 10% and is also supported by increased production of processed fish. This is also supported by the increasing consumption of fish from year to year with the rate of addition of 3-4% [6] thus indicating the availability of fish in the community has increased. On the other hand fish production, especially from sea catching, is still uncertain because it depends on the season so that the availability of fish both for fresh consumption and for raw materials can not be guaranteed throughout the year.

In relation to the aforementioned matters, one strategy that can be developed is to strengthen upstream downstream connectivity with a coordinated system and mechanism approach involving all relevant stakeholders. The system is a system capable of ensuring continuous availability of fish, both in harvests and famine, to consumers and raw materials processing industries. the system is expected to ensure the stability of fish prices at the market level, encourage the growth of the fishery

*

 https://doi.org/10.1051/e3sconf/20184706009

industry, and maintain the quality of fish through quality assurance system.

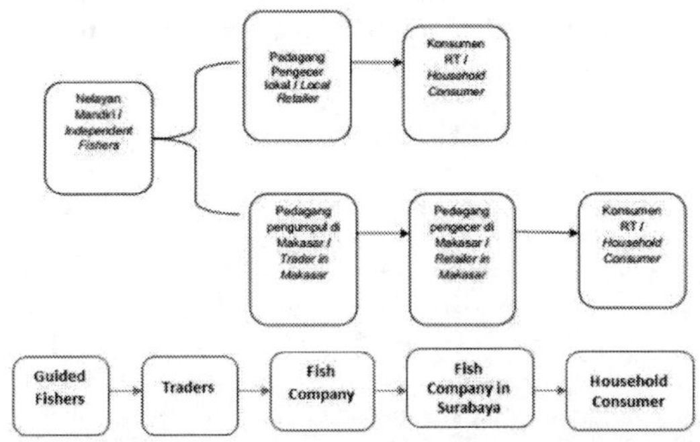

Fig. 1. Line marketing for fish catched in PPS Kendari in 2015 [6]

In addition, inefficient and long distribution channels and inadequate logistics infrastructure are also one of the problems in the provision of fish to meet consumption. The World Bank's publication on the 2010 Logistics Performance Index illustrates that Indonesia ranks 75th out of 155 countries far below the Philippines (number 44) which is geographically similar to Indonesia [7]. It is often said that fisheries trading is the weakest in the chain of economic activity or in the flow of goods from the producer level to the consumer's hand [8]. So the adverse impact of logistics performance in Indonesia is less efficient is reflected in the high price of goods [2], [9] to be paid by consumers in addition to the disruption of competitiveness.

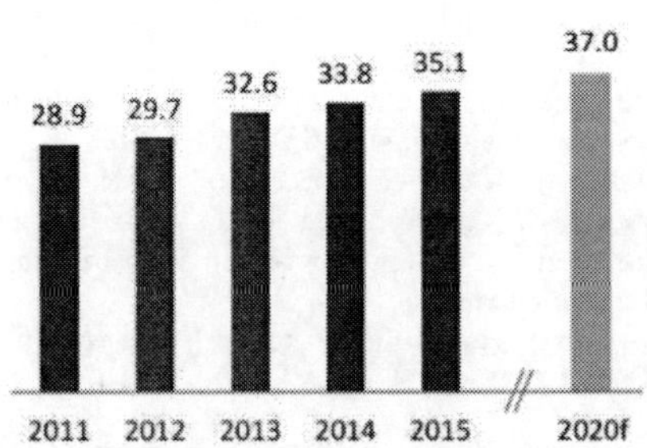

Source: OECD-FAO Agricultural Outlook

Fig. 2. Indonesia's fish consumption per capita (kg/year)

Moving from these problems then the Directorate General of Marketing and Processing of Fishery Products sparked a national policy called SLIN. SLIN is an upstream-downstream supply chain management system for fish and its processed products that aims to stabilize the production of fisheries from upstream to downstream, price control and fulfillment of domestic consumption needs. SLIN must integrate all components from upstream (production/stock procurement), storage to distribution. It is hoped that with this policy the availability of fish can be guaranteed throughout the year and evenly distributed in all provinces in Indonesia but still not incriminating producers and consumers.

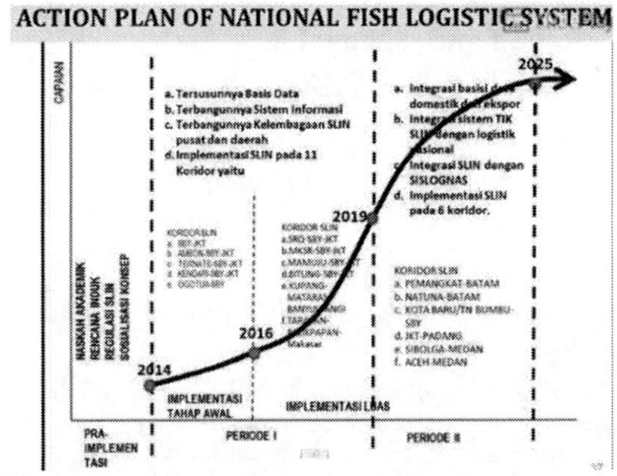

Fig. 3. Action plan of national fish logistic system

Management SLIN can run optimally if there is a synergy between all entities involved in each activity from upstream (production side) to downstream (consumption side). In the framework of looking at best practices for upstream downstream supply chain management of fishery commodities as well as in accordance with the mandate of Minister of Maritime Affairs and Fisheries No. 05/Permen-KP/2014 on SLIN. SLIN implementation of early stage is done in corridor of Kendari - Surabaya/Lamongan - Jakarta in order to fulfill requirement of raw material of pindang industry so it is expected to close the faucet of raw material import.

SLIN implementation has been done since 2014 and will run until 2025. Implementation of the early stages or period I dilukan in order to tersusunya database; establishment of information systems, establishment of central and regional SLIN institutions; and SLIN implemetasi on 11 corridors. While in the second period or the implementation phase is done in the framework of:
Integration of domestic and export databases;
Integration of SLIN ICT systems with national logistics; Integration of SLIN with SISLOGNAS; and SLIN implemetasi on 6 corridors. Easily explanation of the steps can be seen in Figure 3. Thus, this study aims to identify and analyze the prospects for SLIN implementation and also to identify the constraints and challenges faced by the production sector in support of SLIN success.

2 Research methods

This research is a legal research looking from various perspectives. Law is seen as a space for the process of scientific study in order to seek truth. The use of relevant legal research wants to understand the law more thoroughly [10]. In performing implementation analysis, using the method of Regulatory Impact Assessment (RIA) [11], [12]. The need for RIA use arises from the fact that regulation generally has many impacts [13]. It is as difficult to predict without detailed study and consultation with affected parties. The economic approach to regulatory issues also emphasizes the high risk that regulatory costs can outweigh benefits.

3 Result and discussion

The results of the study and discussion will focus on: ***First,*** determination of location and commodity SLIN initial implementation; ***Second***, Determination of SLIN Implementation Operator Initial Stage, and ***Third*** is Provision of Supporting Facilities Infrastructure Utilization.

3.1 Prospects for locating SLIN premises and commodities

Initial SLIN implementation is carried out in corridors that are in real terms a production center and/or collecting center or source of catch fish, especially pelagic fish and connected with distribution centers or market centers to meet the needs of shade industry and other fishery industries.

Determination of location of production center and/or collecting center and distribution center refers to Regulation of Minister of Marine and Fisheries RI number 05/PEMEN-KP/2014 about National Fish Logistic System, with criteria: a) Production center and/or collecting center with target as follows: Available ports for shipping or transshipment ports for fishing vessels; Adequate harbor infrastructure is available; Available port facilities, such as water, electricity and telecommunication facilities; and Available logistics services are held on a regular basis. b) the likelihood of distribution centers with functions as: distribution centers and interconnected with several districts / municipalities which become the main base of the shade industry; logistics services are available on a regular basis; accessible roads are available; there is a sufficient supply of electricity, water and telecommunications.

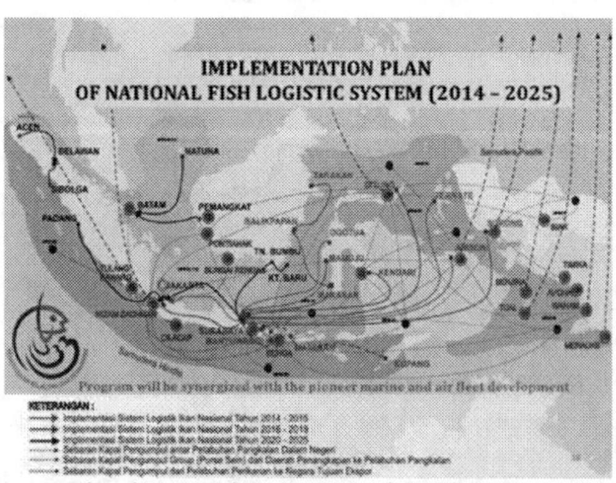

Source: Indonesia Green Infrastructure Summit (IGIS) 2015.

Fig. 4. Implementation Plan

In determining the corridor, consideration includes the potential and production of fisheries, commodity strategic values, local government commitments, regional development related studies and national policy support, such as MP3EI, SISLOGNAS, Fishery Industrialization, Minapolitan, Fishermen Lifestyle Improvement Program. Taking note of the aforementioned matters as well as for the effectiveness of the implementation, SLIN's initial implementation location is in the North Maluku - Sulawesi - Java - Bali corridor and involves business actors.

Fish commodities that become the object of management in SLIN implementation of this early stage is the type of pelagic fish especially for raw materials of industrial pengindangan and other fishery industries. As is known pindang fish is a processed fish that is widely consumed, especially in areas such as Java, Bali and West Nusa Tenggara. This also supported the existence of thousands of processing units of pindang fish spread in the region. Therefore, the provision of pelagic fish as raw material of pindang become very strategic to be managed.

Based on the above description it can be seen that the prospect of determining the location and commodity SLIN initial implementation is done seletif in certain areas that enable the purpose of SLIN utilization can be done optimally.

3.2 Procedure of SLIN implementation of initial implementation SLIN

Early SLIN implementation of SLIN Implementation is carried out openly and accountable and refers to the applicable regulations with the following activity description, Implementation of selection of candidate operator, administrative selection, competency test of operator candidate, and determination of candidate operator. At the stage of selection of prospective operators shall be made by the Selection Committee and Technical Team determined by the decision of the Director General which contains the Selection Committee and the technical team. The selection committee has a membership of at least 5 (five) people and mastered the technical aspects of SLIN and the administration of the management of state property originating from work unit representatives related to early SLIN implementation.

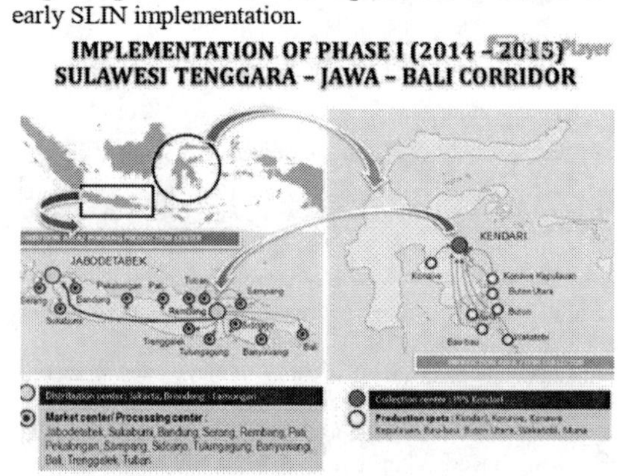

Source: Indonesia Green Infrastructure Summit (IGIS) 2015.

Fig. 6. Implementation of phase I

Administrative Selection is done by the Selection Committee on the things that the prospective operator must meet and the administrative selection method.

Administrative prerequisites that must be fulfilled by the prospective operator, among others: Status of legal entity and permanent position of prospective operators; Organizational structure and stewardship of prospective operators; business experience in the field of fisheries; Financial and technical capabilities of prospective operators; and Operational Plan SLIN implementation activities of the initial stage of prospective operators. While the Administrative Selection Method is carried out with the following stages: 1) The selection committee invites at least 3 (three) prospective operators who are deemed able to complete the administrative requirements documents in the early SLIN implementation; 2) The selection committee shall assess the administrative completeness submitted by the prospective operator; 3) The results of the administrative completeness assessment by the selection committee shall be submitted to the Technical Team for further assessment; 4) The selection committee invites the candidate operator to present the SLIN implementation of Operational Activity Plan of early stage in front of the Technical Team for competency test.

While the stage of each step is competency test operator candidates conducted by inviting prospective operators who have submitted administrative completeness and declared passed by the selection committee to present the operational plan SLIN implementation activities early in front of the Technical Team. The result of the assessment of administrative completeness and suitability of the competency of the candidate operators conducted by the Selection Committee and Technical Team and declared to be passed, is shown to the Director General to be determined. Decision Letter of the Director General on Determination of Operator SLIN Implementation Initial Stage shall include the rights and obligations of the operator.

Operator rights in SLIN implementation are 1). Manage and coordinate the utilization of fishery product processing and marketing facilities provided by the government in the context of SLIN implementation in accordance with prevailing laws and regulations; 2) Propose additional infrastructure facilities in accordance with the duties and functions and availability of the budget required to optimize SLIN implementation; 3) Obtain technical guidance from the government in order to optimize SLIN implementation.

While operator liabilities are 1) Coordinate the involvement of all fishery business actors involved in the implementation of SLIN with the principles of fair and mutually reinforcing partnerships; 2) To accommodate all fisherman catches involved in initial SLIN implementation at a price disekapati the parties; 3) Ensure continuity and quality of availability of shade industry raw material requirements involved in initial SLIN implementation at a price agreed by the parties; 4) Provide maintenance and operational costs for the utilization of fishery product processing and marketing infrastructure; 5) Paying rent for utilization of fishery product processing and marketing facilities in accordance with applicable provisions; and 6) Create and submit quarterly, semiannual and annual reports to the Director General of Processing and Marketing of Fishery Products.

From the above description it can be seen under the preliminary implementation of SLIN implementation preliminary implementation allows can be done selectively and directed. This can be seen from the selection of selection committee selection to the right and obligation of SLIN operator candidates.

4.3 Prospect of provision of utilization of supporting infrastructure

The initial SLIN implementation was intended as a trial of the SLIN conception before being implemented in a more comprehensive concept in accordance with the general definition of SLIN regulated in Candidate of KP Number 05/Permen-KP/2014 on National Fish Logistics System. In connection with this, the Directorate General of Processing and Marketing of Fishery Products, since 2013 has provided several supporting facilities. In addition, the implementation of SLIN implementation in the established locations, also synergize with some infrastructure facilities that have been provided several years earlier. Several types of infrastructure facilities in the implementation of SLIN are as follows: First, marketing infrastructure facilities include Moving Marketing Facilities and other marketing infrastructure, and Second is the facilities of processing facilities such as ice factory, Cold Storage and ABF.

The utilization of infrastructure facilities supporting SLIN implementation of the initial stage by the operator in the framework of the implementation of the main tasks and functions whose status of assets is the property of the state must be leased as regulated in PP. 19 of 2006 on amendment to the government regulation no. 62 of 2002 on Tariff of Non-Tax State Revenue Applicable to the Ministry of Marine Affairs and Fisheries and Regulation of the Minister of Finance No. 96/PMK.06/2007 concerning Procedures for the Implementation, Use, Deletion and Transfer of BMN. The mechanism of rent of state property as such arrangement shall be stipulated as follows:

a. The lease term is adjusted to the Decree of Director General of Processing and Marketing of Fishery Products on the determination of initial SLIN implementation operator.
b. The lease agreement contract may be extended by mechanism and time of renewal in accordance with applicable regulations.
c. Requests for renewal of the lease term by the operator shall be submitted to the Goods Manager no later than 3 (three) months prior to the expiry of the rental period.
d. The goods manager will perform monitoring and evaluation on the implementation of the lease of state property facilities and infrastructure supporting the implementation of other early SLIN.
e. The amount of cold storage rental rate is calculated by using the formula: $T = HD + x$ (where T is the rate of levied, HD is the base price set and x is the price adjustment factor).
f. The basic price tariff for Cold Storage services is Rp. 100 per kg/day.

g. Terms of rental rates and other matters not regulated in PP. 19 of 2006 may refer to Regulation of the Minister of Finance Number 96 / PMK.06 / 2007 or other relevant regulations.

From the above description, the prospect of the provision of supporting infrastructure facilities related to the SLIN implementation is carried out by the Directorate General of Processing and Marketing of Fishery Products. Such support has been conducted since 2013 and has provided several supporting infrastructure facilities. In addition, the implementation of SLIN implementation in the established locations, also synergize with some infrastructure facilities that have been provided several years earlier. Several types of infrastructure facilities in the implementation of SLIN are as follows: First, marketing infrastructure facilities include Moving Marketing Facilities and other marketing infrastructure, and Second is the facilities of processing facilities such as ice factory, Cold Storage and ABF.

4.4 Monitoring, evaluation and reporting

Implementation of the national fish logistics system implementation also utilizes monitoring and evaluation monitoring procedures. This is done in order to detect early deviations that occur. A study conducted in 2015 by Rismutia Hayu Deswati and Muhadjir [6] that the results of the trials were seen by Kendari as upstream areas or suppliers of raw materials pindang still not able to supply full and sustainable. Kendari through KOMIRA as SLIN operator can only supply 65% of the total demand due to several causes including: the lack of ability and competence of fishermen in utilizing science and technology to support the arrest, the absence of government intervention in setting the price for fishermen and the difficulty of obtaining permits.

Implementation of Monitoring, Evaluation, and Reporting SLIN Monitoring is implemented with a view to taking into account the success of the SLIN program. Monitoring is carried out in the context of monitoring, guidance, and evaluation of utilization Initial SLIN implementation support infrastructure. Monitoring time can be carried out before, during, and after SLIN implementation in the early stages. While the implementation of Monitoring is done by the SLIN Implementation Technical Team of the early stage as well as assisting the experts both from internal ministries and external ministries.

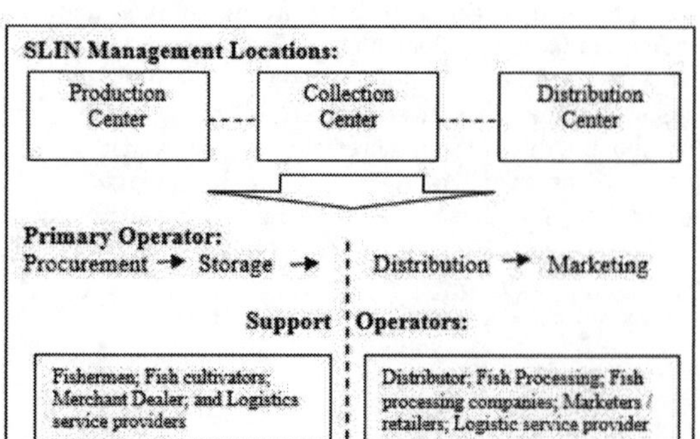

Fig. 7. Operationalization of SLIN management

Implementation Evaluation is done with a target on a particular object object. The evaluation object is emphasized on the suitability and effectiveness of the utilization of infrastructure facilities supporting SLIN implementation in the early stages by referring to the operator's obligations as well as the implementation objectives of early SLIN implementation, among others: 1). Increased percentage of uptake of fishermen products at initial SLIN implementation sites; 2). Control of prices across the supply chain at locations designated as early SLIN implementation areas; 3). The continuity of availability of shade industry materials at initial SLIN implementation sites, and 4). Availability of good quality industrial raw material.

In the final stages is the SLIN program reporting at the agency with the level of authority that exists. Reporting shall be conducted in two ways: first, the Operator is required to submit activity report to the Director General of Processing and Marketing of Fishery Products and forwarded to the relevant Provincial and Regency/City Maritime and Fishery Office, and Secondly, the SLIN implementation report of the initial stage is submitted in writing by the operator quarterly, semiannual and yearly. For operators who do not utilize the facilities or perform their functions in accordance with the obligations of operators which result in delayed implementation of initial SLIN implementation, may be subject to administrative sanctions in the form of revocation of appointment as operator.

From these descriptions it can be seen that Monitoring, evaluation, and reporting in order to know the advantages and kekuarangan program. However, there is also a lack of SLIN implementation. Lack of capability and competence of fishermen in utilizing science and technology to support the capture, the absence of government intervention in determining the price for fishermen and the difficulty of obtaining permits.

5 Conclusion

The prospect of SLIN (national fish logistics system) implementatation has shown a positive trend. the implemetation of SLIN provides substantial benefits to national fish production, directly contributing to the increase of national economic development. SLIN Management can run optimally if there is a synergy between all entities involved in each activity from upstream (production side) to downstream (consumption side). In the framework of looking at best practices for upstream downstream supply chain management of fishery commodities as well as in accordance with the mandate of Minister of Maritime Affairs and Fisheries No. 05/Permen-KP/2014 on SLIN. Thus, the Directorate General of Processing and Marketing of Fishery Products sees the importance of better implementation of SLIN. In the Initial Stage by providing several supporting facilities and involving the main operators and supporting operators in the management of selected locations. Determination of location and commodity SLIN initial implementation is done seletif in certain area of area that enables purpose of SLIN utilization can be done optimally. But there is also a lack of SLIN implementation. Lack of capability and competence of fishermen in utilizing science and technology to support the arrest, there is no government intervention in determining the price for fishermen.

Acknowleddments

The first author would like to thanks to the Dean of Law Faculty (Prof. Dr. Retno Saraswati, S.H., M.Hum), Diponegoro University for his support and kindness.

References

1. M. Nielsen, P. Andersen, L. Ravensbeck, F. Laugesen, D. M. Kristófersson, and H. Ellefsen, "Fisheries management and the value chain: The Northeast Atlantic pelagic fisheries case," *Fish. Res.*, vol. 186, pp. 36–47, 2017.
2. R. M. Rosales *et al.*, "Value chain analysis and small-scale fisheries management," *Mar. Policy*, vol. 83, no. May, pp. 11–21, 2017.
3. F. Jensen, H. Frost, and J. Abildtrup, "Fisheries regulation: A survey of the literature on uncertainty, compliance behavior and asymmetric information," *Mar. Policy*, vol. 81, no. March, pp. 167–178, 2017.
4. Marsetio, "Indonesian Sea Power and Regional Maritime Security Challenges," *J. Marit. Stud. Natl. Integr.*, vol. 1, no. 1, pp. 34–46, 2017.
5. I. N. Putra A., A. Hakim, S. H. Pramono, and A. S. Leksono, "The effect of strategic environment change toward Indonesia maritime security: Threat and opportunity," *Int. J. Appl. Eng. Res.*, vol. 12, no. 16, pp. 6037–6044, 2017.
6. R. H. Deswati and M. Muhadjir, "Support of Production Aspect in National Fish Logistics System (SLIN) in the Kendari City, Southeast Sulawesi (Dukungan aspek produksi dalam sistem logistik ikan nasional (SLIN) di Kota Kendari, Sulawesi Tenggara)," *J. Sos. Ekon. Kelaut. dan Perikan.*, vol. 10, no. 2, pp. 191–202, 2016.
7. B. Hoekman *et al.*, *Connecting to compete 2010: trade logistics in the global economy - the logistics performance index and its indicators*. Washington DC: World Bank, 2010.
8. S. W. Purcell, B. I. Crona, W. Lalavanua, and H. Eriksson, "Distribution of economic returns in small-scale fisheries for international markets: A value-chain analysis," *Mar. Policy*, vol. 86, no. June, pp. 9–16, 2017.
9. M. Laraswati, A. D. Guritno, N. E. Kristanti, and E. Suwondo, "Analysis of logistics cost structure of fish cold supply chain in Java Island," *AIP Conf. Proc.*, vol. 1755, 2016.
10. S. Halliday and P. Schmidt, *Conducting Law and Society Research: Reflections on Methods and Practices*. New York: Cambrige University Press, 2009.
11. A. C. M. Meuwese and S. van Voorst, "Regulatory impact assessment in legal studies," in *Handbook of Regulatory Impact Assessment*, C. A. Dunlop and C. M. Radaelli, Eds. Cheltenham: Edward Elgar, 2016, pp. 21–32.
12. C. Kirkpatrick, D. Parker, and Y.-F. Zhang, "Regulatory Impact Assessment in Developing and Transition Economies: A Survey of Current Practice," *Public Money Manag.*, vol. 24, no. 5, pp. 291–296, 2004. 13. V. Jakupec and M. Kelly, "Development aid: Regulatory Impact Assessment and conditionality," *Impact Assess. Proj. Apprais.*, vol. 34, no. 4, pp. 319–329, 2016.

Practices of Illegal Fishing in Pemalang Region: A Policy Analysis

Wahyuningrat[1)], Tenang Haryanto[2)], Slamet Rosyadi[1*)]

[1)] Faculty of Social and Political Sciences, Indonesia
[2)] Faculty of Law University of Jenderal Soedirman, Indonesia

Abstract. The use of unsustainable fishing gear has resulted in the decline of fish resources and threatens the sustainability of fish resources. This can cause harm to the state and the interests of the fishermen themselves. Although various policies governing fishing gear are already in place, unsustainable fishing practices remain. This, of course, has resulted in overfishing and causing environmental damage. Using the case of fishing practices in Pemalang Region, this study found that fishing gears used by fishermen are prohibited tools because they have negative impacts on the environment. Some fishing gear is quite popular used by fishermen in Pemalang Region are stun, fishing nets, fishing rods and poison. Unfortunately, the use of electric current tool (stun) is found to be more dominant than fishing nets and fishing rods. Similarly, it has developed among fishermen to use toxic methods in fishing. This phenomenon encourages researchers to formulate an effective public policy to control and minimize illegal fishing practices.

1 Introduction

Indonesia as an archipelagic country has huge potential of fish resources and high biodiversity, with Indonesian waters having 27.2 percent of all species of flora and fauna found in the world. Species include 12 percent mammals, 23.8 percent amphibians, 31.8 percent reptiles, 44.7 percent fish, 40 percent molluscs, and 8.6 percent seaweed. The potential of fish resources includes large pelagic fish resources, small pelagic fish resources, penaeid and other crustacean shrimp resources, demersal fish resources, mollusk and cucumber resources, squid, commercial commercial seed resources, coral resources, reef fish consumption resources, ornamental fish resources, turtles, mammals, and seaweed.

The wealth of the sea will not be exhausted forever for the benefit of the welfare of the people of Indonesia when used with attention to sustainability and justice. Fish resources can be utilized as much as possible for the welfare and prosperity of the people of Indonesia. The fact that fish resources have not been able to improve the standard of living is sustainable and just. This is due to overfishing, fish theft and illegal fishing and the use of unsustainable fishing gear.

Unsustainable fishing activities not only threaten the sustainability of coastal and marine natural resource use, but also spur the increase in the number of poor people in the region.

[1]Corresponding Author: slametrosyadi72@gmail.com

The use of unsustainable fishing gear such as the growing cantrang has caused serious problems. The use of fishing gear of this type can damage the marine environment or marine resources because fishing is done with no attention to environmental aspects. In addition it also raises the social problem of conflict between traditional fishermen and machine fishermen. The impact of such environmental degradation is causing the suffering of present and future generations for not being able to consume the much-needed fish for the growth of the human body.

In order for sustainability of resources to be maintained, human (anthrophogenic causes) activities that directly or indirectly potentially undermine the sustainability of fish resources and aquatic ecosystems should be minimized. Given that environmental damage is caused more by human activities as stated by Homer-Dixon [1]. Human activities can cause environmental damage or scarcity of fish resources in three ways: (1) human activities can lead to a decrease in the number and quality of fish resources, especially if resources are exploited at speeds that exceed their recovery power; (2) the decline or scarcity of fish resources is caused by fishing equipment that causes overfishing. With increasing population and declining fish resources will affect the income of the fishermen and cause harm to the state and the interests of the fishermen themselves. This means the utilization of fish resources over the regeneration of the fish itself; and (3) access to unbalanced fish resource environments will also cause many problems. These three factors of human activity can occur individually or in combination.

Human activities that can cause environmental damage also occur in Pemalang District. Most of the fisheries in the waters of Pemalang Regency use an environmentally unfriendly fishing tool that uses stun and toxic materials. It seems that the public has begun to pay less attention to environmental ethics and the principle of justice. If viewed from the anthropocentric perspective of an environmental ethics theory, that human being as the center of the system of the universe. Human beings and their interests are deemed the most decisive in the ecosystem order and in the policies taken in relation to nature, either directly or indirectly. Nature is seen as just objects, tools and means for the fulfillment of human needs and interests. An anthropocentric outlook gives rise to greedy and greedy attitudes and behaviors that cause humans to take their needs from nature without considering conservation of environmental functions. Preservation of environmental functions is not thought of by anthropocentric people.

In the perspective of justice, Santosa and Quina [2] said that there are five principles of justice: (1) the principle of intergenerational equity, every generation of mankind in the world has the right to receive and occupy the earth not in a bad condition due to the actions of previous generations; (2) the principle of intragenerational equity, in which the burden of environmental problems must be shared by the community within a generation; (3) precautionary principle, when there is a threat.

2 Research Method

This research uses several methods, namely:

1. Normative juridical method, conducted through literature study, which examines (mainly) secondary data in the form of legislation relating to Fishing Using Destructive and Inhospitable Materials or Tools. The normative juridical method includes 3 (three) approaches, namely: (a) the legal dogmatic approach (juridical) aims to study and apply legal norms in accordance with applicable laws and regulations relevant to Fishing Arrangements Using Destructive or Destructive Substances and Tools Environmentally Friendly in Pemalang District; (b) the legal theory approach (conceptual), aims to study and apply theories, concepts, opinions, legal teachings, associated with Fishing Using Destructive and Inhospitable Materials

or Tools in Pemalang District; and (c) the philosophical (philosophical) philosophical approach, is to discover and analyze the legal principles that can be used as reference in Fishing Arrangements Using Destructive and Inhospitable Materials or Tools in Pemalang District.

2. The empirical juridical method, or sociolegal, is a study initiated by normative research, followed by in-depth observation to obtain relevant non-legal data.
3. Survey method, is a research method used to find information factually. With this method, it can discuss and analyze a problem that is closely related to Fishing Problems Using Destructive and Inhospitable Materials or Tools.

3 Results and Discussion

Pemalang District is located at coordinates 109 17 '40 "to 109 40' 30" East Longitude and 08 52 '30 "to 07 20' 11" South Latitude. Geographically Kabalang Pemalang is one of the districts located on the north coast of Java Island. While administratively Pemalang District with an area of 111.530 Km2 consists of 14 Districts and 211 Villages and 11 Villages. Based on Law Number 23 Year 2014 on Regional Government, Pemalang District is granted authority of fishery management, in accordance with geographical position with long coastline ± 35 km and width of sea waters 4 miles (1 nautical mile = 1,852 m), so Pemalang Regency has sea an area of 259.28 km. All Pemalang Regency is limited by the north is Java Sea, south of Purbalingga regency, west of Tegal regency, and east of Pekalongan Regency.

Marine fish production in Pemalang Regency reaches 27,507 tons per year with value of Rp 182,444,210.00. In Central Java, Pemalang District ranks fifth after the City of Tegal, Rembang, Pati, and Batang. The catch is quite abundant making Pemalang District as one of Central Java coastal fishing centers. Furthermore, the production of fish cultivation and river fish whose potential is also quite large. For fish cultivation, the area of tambak reaches 1,728 hectares which includes milkfish, shrimp and shrimp soka. Meanwhile Pemalang Regency also has the potential of river fish which is also quite big from the three largest rivers in Pemalang Regency is the river comal, river waluh and river hair.

Most of the population Pemalang livelihood is a fisherman consisting of skipper, pandega, and sideline. While the fish auction places are in Tanjungsari, Asemdoyong, Mojo, Ketapang, and Tasikrejo. Characteristics of residents of Pemalang Regency generally live around the city center (coastal area), where the number of people living in coastal areas reached 57.77% (739,252 souls) of the total population. Fishery is also the largest contributor to Gross Regional Domestic Product (PDRB) Pemalang Regency that reaches 20,036.5 billion rupiah (GDP at Current Prices). In addition, beach tourism namely Widuri Beach became the most favorite place visited by tourists during the last three years. There are 187,538 domestic tourists who come to visit these attractions in 2017.

In addition, the sea for Pemalang Regency is also as a nature reserve area, namely: (1) natural marine reserve area, ie asylum to marine waters, coastal, river estuaries, coral clusters and atolls that have characteristic of diversity and or uniqueness of ecosystem. Marine marine reserve area in Pemalang Regency in the form of river mouth located in District Ulujami; and (2) coastal areas of mangrove forests, ie vegetation areas located on the coast and / or rivers at the mouth of the river that serves as a coastal protection area, is a habitat of certain fauna and also ecological functions. Pemalang Regency is the area that has the widest mangrove area in Pantura which is equal to 10.94%. Furthermore, the beach is also a protected area of marine germplasm located along the coast in Pemalang District namely Petarukan Subdistrict, Taman, Pemalang and Ulujami. Marine germplasm protection in Pemalang Regency is directed to two things: (1) in-situ conservation of germplasm conservation in the form of fish, by establishing and breeding species of fish whose population is limited to germplasm marking, conservation area determination, collection formation , and regulation of germplasm expenditure; and (2) ex-situ conservation, ie the conservation of germplasm of fish outside its habitat which can be done in the form of an aquarium collection container equipped with facilities having certain conditions for storing the germplasm for survival and genetic properties. Next is the protection of the local area that is intended for land use activities that can maintain the sustainability of quantity, quality and provision of water system and smoothness, order regulation and utilization of water from water sources. Local protected areas consist of: (1) the river border area, along the right and left of the river including artificial river / primary irrigation canals which have important benefits for maintaining river functions. The river border area in Pemalang Regency is located in all areas passed by the river that is waluh river and its 15 tributaries pass through Moga, Randudongkal, Bantarbolang, Pemalang and Taman. The hair river consisting of 13 tributaries passing through Moga and Pemalang sub-districts as well as the waluh river along with its 30 tributaries that pass through Belik, Moga, Randudongkal, Bantarbolang, Ampelgading, Comal and Ulujami districts with a length of about 92.5 km; (2) coastal border areas, ie certain areas along the coast, are useful to maintain the sustainability of coastal functions of various activities that may threaten its sustainability. The boundary line criterion is at least 100 meters from the highest tide point towards the land. The coastal border area of Pemalang Regency is located along the coast of Petarukan, Pemalang and Ulujami sub-districts along the 600,272 hectare coastline; and (3) the area around the spring, is the area around the spring that has important benefits for maintaining the function of the spring. The purpose of protection of this area is to protect the springs from cultivation activities that can damage the water quality and physical condition of the surrounding area. The boundary line criterion around the springs is 200 meters around the springs, while the protection of this area is in accordance with existing regulations covering the area within 200 meters around the springs (the area around the springs in the area of Pemalang Regency is 2,722,337 Ha).

Fish resources are very beneficial to people's lives, not least in Kabupaten Pemalang. Besides as a source of protein as well as the main income, especially for the fishermen. Therefore, sustainability and sustainability must be maintained both from its fish resources and its ecosystem. Nevertheless, fishing in Pemalang Regency is currently mostly done with no attention to environmental sustainability. Until the 1990's fishing in Pemalang Regency is still traditional such as installing bubu and fishing. After that fishing is mostly done in ways that are destructive and environmentally unfriendly such as the use of stun guns, explosives (mercury), insect poison and potassium and other chemicals. The results showed that the use of stun fishing equipment by 60 percent, nets 15 percent, fishing rod 15 percent and 10 percent racum.

Meanwhile, modern fishing equipment is destructive and less environmentally friendly today is widely used by fishermen Pemalang District, among others cantrang and payang. Cantrang is one fishing tool in the form of nets. Jar cantrang has a length of 53 meters with a pocket length of 3 meters. The mesh size of cantrang is 0.5 - 4.5 inches. Operation cantrang done at 04.00 am until 13.00 pm. The method of operation of fishing equipment cantrang starting from the stage setting of the decline buoy signs followed by the right strap with the direction of the ship in a circle, then the net is lowered and followed by the left strap to meet the sign buoy. Furthermore the buoy sign was raised to the ship followed by the drawing of the drawn string and the withdrawal was aided with axle. In this process the position of the ship is stopped but the axle machine is still running and this stage is called hauling. Continuous use of cantrang causes extinction of various species of fish and coral reef resources. This is because small fish that have not spawned caught by this tool so as not to have the opportunity to spawn and multiply the species. As a result in a certain period of time, the fish will run out because it did not get natural regeneration.

The next catch is Payang. It is also one of the fishing nets. It is used in Pemalang District that is often called payang gemplo which has a body part that resembles a bag, a wing that blends with a cone-shaped body and a pull strap. Construction of payang net, among others, the overall length up to 200 meters with a pocket length of 7 meters. The size of the mesh (mesh size) is 1 - 20 inches and the pockets are made of warnet nets. Operation of payang is done at 04.00 WIB until 12.00 WIB. The mode of operation starts from the setting stage, where the mark buffer is lowered then followed by the left-handed strap with the direction of the ship's movement in a circle, then the net is lowered and followed by the right-hand strap until it meets the mark buoy. After that the buoy sign is raised to the ship followed by manual drawn drawn rope which usually takes between 9 - 14 people. In this process the ship is still in a state of road but with medium speed. This stage is called hauling. The process works almost the same as cantrang, so the potential destruction of fish resources and ecosystems are also the same as cantrang.

This condition certainly can not be allowed, requires its own policy so that the damages of water environment in Pemalang Regency is not getting worse. The policy can be poured through the Regional Regulation on Prohibition of Fishing Using Destructive and Inhospitable Materials or Equipment. This public policy aims to: (1) protect small fishermen from engaging in activities in the fishing sector; (2) to protect fishermen, especially fishermen located in coastal areas of Pemalang Regency from outside fishermen who use banned fishing gear; (3) protecting the potential of fish resources as a result of the use of prohibited materials or equipment; (4) applying local wisdom as one way to prevent prohibited fishing gear; (5) preserving the environment; and (6) involving the community in the oversight function of the territorial waters of Pemalang Regency.

In such arrangements also need to be affirmed in relation to fishery areas and types of fishing gear that are environmentally friendly and non-environmentally friendly. The fishery areas include sea, river, lake, swamps, rice fields, dams and other waters in Pemalang District. Types of non-destructive and environmentally-friendly fishing gear include gill net, trammel net, fish-folding bubu, crab bubbles, fishing rods, bottom rivers, drifting drifting, tonda fishing lines, and pole and line. Further, other types of destructive and non-environmentally friendly fishing gear include explosives, toxic chemicals, electrically ground devices, trawl, trawl, trawls and trawls drag (seine nets).

In order for the effective arrangement, it is necessary coaching and supervision and community participation. Guidance and supervision may be undertaken by the Regent through the relevant Regional Apparatus Organizations. Coaching and supervision can be made directly to the site and or using technical guidance. The community can participate in assisting local wisdom-based supervision [3]. Hardjasoemantri as described by Santosa and Quina [2], that community participation is needed in decision making related to environment and natural resources.

Community participation provides valuable information and knowledge to decision makers [4, 5]. In addition, public participation will also reduce the likelihood of non-availability of communities to accept decisions that have a significant effect on the formation of public policy [6,7]. In order to empower community participation in supervision and protection, the local government through the relevant Regional Government Organizations can budget the funds through the Regional Revenue and Expenditure Budget. This community participation is intended to encourage the growth and development of community groups of supervisors and conservation groups of fish resources and their ecosystems.

4 Conclusion

Fish resources and ecosystems are needed by the community, both for the fulfillment of nutritional needs and for sources of income. Similarly for local governments, fish resources and ecosystems are expected to increase the GRDP so that per capita society is also getting better. Aquatic resources are also expected to increase tourism potential for both local governments and communities. There are many more benefits that can be drawn from the potential of fish resources along with its ecosystem, so it needs a commitment to maintain its sustainability and sustainability.

In Pemalang Regency has been widely used fishing equipment that can damage and less environmentally friendly, among others, stun, explosives, toxic materials, cantrang and payang. If this is allowed to continue, there will be damage and possibly even the extinction of fish resources and their ecosystems. The sustainability of fish resources and their ecosystem will be disrupted.

These conditions require a policy that can be set forth in a Regional Regulation on Ban on Fishing by Using Destructive and Inhospitable Materials and Tools. The goal is to maintain the sustainability and sustainability of fish resources and their ecosystems. In addition, to empower the community, especially fishermen. Regulatory items may include regulatory objectives, fishery areas, types of eco-friendly and non-environmentally friendly fishing gear, prohibition, guidance and supervision, and community participation.

References

1 Homer-Dixon, T. Environmental, Scarcities and Violent Conflict: Evidence from Cases, *International Security* 19 (1) (Summer), 5 – 44. (1994)

2 Santosa, M. A., & Quina, M. Gerakan Pembaharuan Lingkungan Indonesia dan Perwujudan Tata Kelola Lingkungan yang Baik dalam Negara Demokrasi, *Jurnal Hukum Lingkungan Indonesia*, Vol 01, 23-54. (2014)

3 Sari, M.M.K. Democratic Society in the Local Wisdom: Citizen Participatory Activity in the Environmental Movement. *Advanced Science Letters*, 23 (12), 11666-11669(4). doi:10.1166/asl.2017.10490 (2017)

4 Webler, T., and Tuler, S. Unlocking the Puzzle of Public Participation . *Bulletin of Science, Technology & Society* 22 (3), 179 – 189. doi:10.1177/027676022003002 (2002).

5 Oughton, D. Public Participation - Potential and Pitfalls. *Energy & Environment* 19 (3-4), 485 – 49. doi:10.1260/095830508784641435 (2008)

6 Leitch, V. Securing Planning Permission for Onshore Wind Farms: The Imperativeness of Public Participation. *Environmental Law Review*, 12 (3), 182 – 199. doi:10.1350/enlr.2010.12.3.091. (2010)

7 Li, M., & Zhao, J. Gaining Acceptance by Informing the People? Public Knowledge, Attitudes, and Acceptance of Transportation Policies *Journal of Planning Education and Research.* doi:10.1177/0739456X17732623, (2017)

SUSTAINABILITY OF SEAHORSES: LESSONS LEARNED OF LOCAL WISDOM FROM BINTAN ISLANDS, RIAU ISLANDS PROVINCE

*Fitria*Ulfah[1*], *Afrizal*[2] and *Arief*Pratomo[3]

[1]Social Economic of Fisheries Department, Marine Science and Fisheries Faculty, Raja Ali Haji Maritime University, Tanjung Pinang 29111, Indonesia
[2]Public Administration Department, Social and Political Science Faculty, Raja Ali Haji Maritime University, Tanjung Pinang 29111, Indonesia
[3]Arief Pratomo, Marine Science Department, Marine Science and Fisheries Faculty, Raja Ali Haji Maritime University, Tanjung Pinang 29111, Indonesia

Abstract. Seahorse (*Hippocampus* spp.) has a high economic value in Bintan Island that leads to its exploitation by coastal community intensively. This study was to assess local knowledge on seahorse ecology which was captured across shoreline near the coastal communities. The research was conducted in SebongPereh village Bintan Island. The data were collected by conducting depth interviews and focus group discussion on seahorse fishermen which was selected purposively. The results revealed local fishermen have ecological knowledge related to habitat types, behavior, and monthly variation in abundance, size and reproduction pattern. The result also showed that the existence of the seahorse during the catch season moves around, follow the condition of the seaweed and tidal of sea water. This local knowledge was used by fishermen as a strategy in determining the fishing time and catchment area of a seahorse. Through such local wisdom, local fishing communities established a local institution to manage conversation and sustainability use of seahorse in waters of SebongPereh village.

1 Introduction

One of the coastal areas in Riau Islands Province is Bintan Island. Bintan Island has a large coastal area, with many marine biological resources that have high economic value. One of the most valuable economic resources in Bintan Island is the seahorse (Hippocampus spp.[16], or commonly known as "ondok-ondok" by the local community. The existence of seahorse is mostly found in SebongPereh Village, District, Bintan Regency[14].

SebongPereh village was chosen as study area. Interestingly, in this site, the existence of seahorses was almost repeatedly observed from January to May in each year [14]. In that season, many people are turning the profession into a seahorse catcher due to the higher value of the selling price of dried seahorses that reaches more than 5 million rupiahs/kg. Even when the catch season is over – according to the fishermen – the price of dry seahorse can reach 7 million rupiahs/kg. The high value of the seahorse had driven it catchment by Bintan Island's fishing community intensively. This condition makes the species more vulnerable to overexploitation [14].

However, although the price of seahorses is quite promising, not everyone can take a seahorse in considerable numbers. This is because the experience and the ability to distinguish between a seahorse and sea grass[2], mangroves [15], coral reefs [12], the roots of seaweed or sand[7],[11], becomes the determinant of how many seahorses can be captured. Other skills that are also needed are the diving skills of the fishermen, as well as knowledge of the right time to dive.This condition made not every fisherman can obtain optimal results, even though at that time there were many seahorses in the waters of SebongPereh Village.

Therefore, the knowledge and experience of fishermen in taking the seahorse became one of the keys successes in obtaining a considerable catch.A few studies about local knowledge has been conducted, but with another species such as dugong. Based on that result study, the local knowledge supported the development of appropriately targeted management strategies [17], [6]. This means seahorses could improve the welfare of the community if the fisherman has sufficient knowledge of the management and utilization of the seahorse itself. Such knowledge includes fishing time, fishing gear, catching ways, the condition of seahorse when captured and ecological conditions of seahorse fishing. Based on these conditions and to be better understand the impacts of the harvesting and trade of seahorse requires knowledge of their biology, along with population monitoring and fisheries management [4].So it is necessary to conduct research to obtain information about local knowledge of local communities on seahorses. The aim of this study was to assess the extent of fishing community understanding on seahorse ecology in SebongPereh village which was captured across shoreline near their village.

2 Materials and methods

2.1 Study area

The research was conducted in SebongPereh Village, BintanIsland, and Riau Island Province. The study has been done on February – September 2017.

2.1 Survey design and methodology

Primary data were obtained from in-depth interviews using question guides to key informants[13]. Primary data were also obtained from Focus Group Discussion (FGD) on seahorse fishermen in SebongPereh village. The data was taken in this research consists the number of catches, capture frequency, catch time, fishing gear, the condition of seahorse when it catches, ecological condition of a catchment area, and sale price of a seahorse. While the secondary data obtained from the journal and other relevant references.

The data was collected from 3 key informants and 15 respondents. The technique to determine key informants and respondentsin this study were conducted purposively based on information from the village community. The key informant in this study was the fishermen who got the most catches. While the respondents were those who did the catch in the water sea of SebongPereh. Data analysis is done by using a qualitative approach, by condensing, display data that have been processed, drawing and verifying conclusions[8].

The validity of data is done by comparing data obtained from the key informant with the result of discussion through Focus Group Discussion (FGD) until information is saturated[8], [10].

3 Results and discussion

3.1 Time and catch frequency

According to the fishermen, time of catches the seahorses in SebongPereh village was usually done in January until June, with the highest catches in February and March. In the 4th month, the number of seahorse catches reaches twice more than the 1st month, but lower than in the 3rd month. This information was then evaluated through the score table by making the 1st month (January) as benchmark comparison (h_o). Furthermore, fishermen mentioned that the highest time of peak season is not always same every year, but depends on the seasons

Table 1.Table of comparison of seahorsecatches[a]

Month											
1	2	3	4	5	6	7	8	9	10	11	12[b]
1x	3x	3x	2x	1x	1x[c]	-	-	-	-	-	-

[a]Primary data
[b]Month in a year
[c]Comparison result with the 1stmonth as benchmarks

The fishermen usually catch the seahorsein low tide condition until it will tide between 2-3 hours. In one month, they usually catch the seahorse ± 14 days. They only catch the seahorse in the morning until afternoon.According to the fishermen, in the time toward high tide, the seahorse is relatively easier to be found in the sea than at low tide. It's also easier to found the seahorse at low tide and rainy, compared to when the sea at low tide and sunny. In fishermen perception, at that time, they observed that the seahorse was known to come out from a coral reef, also from the hole of the seabed.According to the fishermen opinion, perhaps it was caused by the temperature and the content of the sea water effects to the presence of the seahorses.

3.2 The number of catches

The amount of the seahorses in one catch was variousdepend on the catch season. In the 1st month (January), fishermen could catch about 20 – 35 seahorses per catch, or around 150 – 350 seahorses in a week, or around 300 – 500 in a month. While in the 2nd and 3rd month, the fishermen could catch the seahorse between in range of 50 – 70 seahorses in once catch, or around 250 – 500 seahorses in a week. But there were also fishermen who can catch up reached 100 seahorsesat once. Probably, the differences of the experience and knowledge among the fishermen resulted in various catch rate individually. When the amount of catch quantified in weight, at the peak season (in the 2nd and 3rd month), the fishermen could get as much as 2 kg – 4 kg dry seahorse. While in another month, the fishermen could get the dry seahorsein range of 1 – 2 kilograms.

3.3 Catchment area: habitat condition

According to the fishermen, the seahorses were found around brown macroalgae, *Sargassum* sp. This seaweed was known by the local community as a "rengkam". Furthermore, it was mentioned that seahorses were also found around the *seagrass* (or known as "setu") although relatively in small amounts. The seahorses more often found in *sargassum* at the edge of lagoon situated up to 200 m towards the beach. Seahorses or commonly known as “ondok-ondok” by local people were also found, likely, around reef crest where the fishermen mentioned it "siring tubir". The fishermen recognized "Siring tubir" is a part before reef slope. Usually around "siring tubir" there were corals, *sargassum*spp, and gorgonian (*Antipathes* spp.) or “akarbahar”. Typically, seahorses that found around the seagrass were generally a type of seahorse that has crown and dark colored (Figure 1).

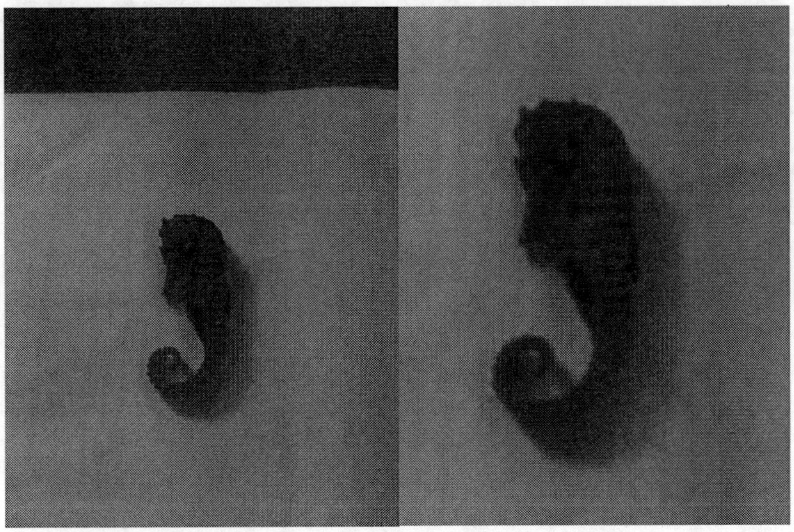

Fig.1. Types of seahorse often found in *seagrass*

Whereas the seahorse that commonly found around *Sargassum*sp were seen as a yellow horse (Figure 2). The seahorses that were often found in seagrass at seawater of SebongPereh were *Hippocampus spinosissmus*, *H. histrik,* and *H. kuda*[17]. While those found around the *Sargassum*sp were *H. barbouri* and *hippocampus comes*. The possibility of seahorses mentioned by fishermen included into these species[4].

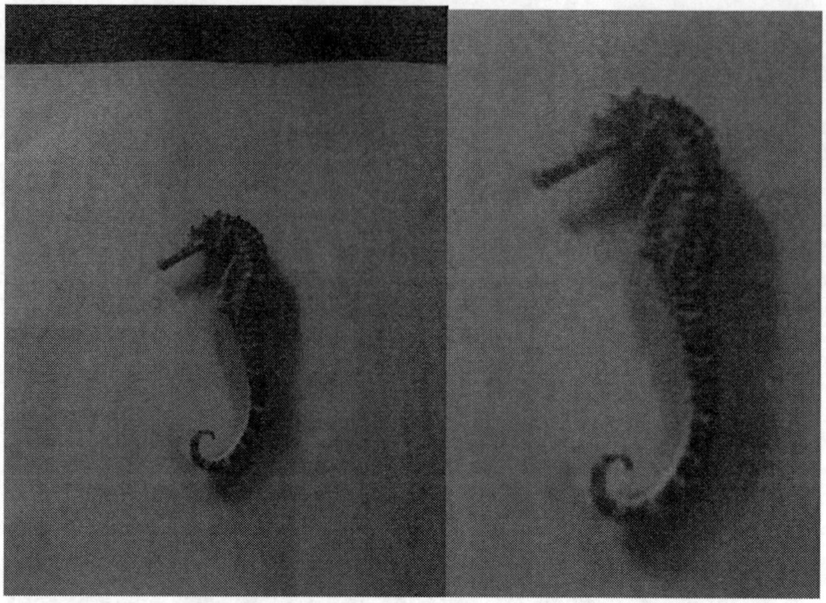

Fig. 2. Types of seahorse found in the *Sargassum*sp

3.4 Monthly variation patterns

Based on the explanation of fishermen, on the first month (January) macroalgae"rengkam" in the lagoon waters are still in abundance and long lifeform condition (Fig. 3). In this month, the sea waves were strong enough, so that seahorses were found relatively tend in the base of themacroalgae.

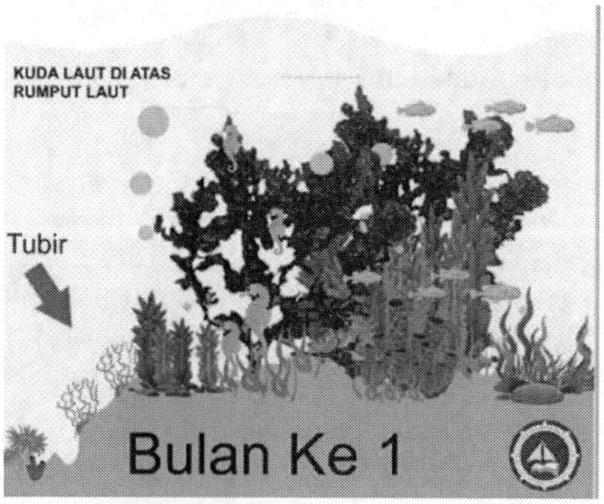

Fig.3. Illustration of seahorses in January

On the 2ndmonth (February), strong wave cause *sargassum*at the base began breakage, so that, there were no leaves. This cause the fishermen have to dive deeper in order to get a seahorse because it commonlyfound in the base of the *sargassum*.Fishermen also said that on the 3rd month (March) some parts of the *sargassum*were fallen, some even begin to rot. In this month the seahorsewas found at the base of the*sargassum*, above it, or in the fallen *sargassum*that has not been rotten yet.

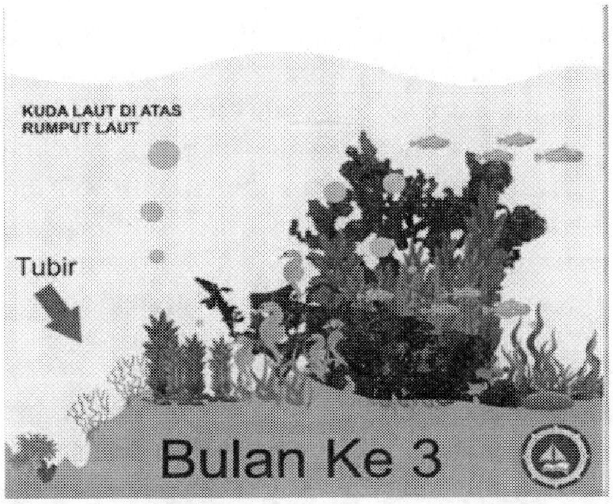

Fig. 4 Illustration of seahorse in March

According to the fishermen, *sargassum*that has fallenmeans it has broken onto the coral reef so that the base and top of the *sargassum* were in parallel position. Broken *sargassum*usuallydoes not drift. It will rot on the rock until it is destroyed. While the driftingsargassum usually is a sargassumthat has broken away because of strong waves.The rotten *sargassum*is a *sargassum* that has been fallen and tendshas blackcolour.

In April the fallen *sargassum* are increasing. In this month, the seahorse could be found at the base, at the end, in a broken *sargassum* but not rotten yet, even in a rotten *sargassum* but not totally decomposed. In May, *sargassum* starts to run out or clear.In this case, most of *Sargassum*has been fallen and rotten, even destroyed. At this time, it began difficult to find the seahorse, but still could found it at the base of *sargassum*, inside the broken *sargassum*, in a rotten *sargassum*, even some of it was found in a coral reef.

Fig. 5. Illustration of seahorse in May

In the 6th month (June), the *sargassums*were getting worn out and clean, increasingly difficulty to find seahorse in SebongPerehVillage waters. This condition continues in the 7th month. In the 9th month, the *sargassum* began to grow again and progressively longer until the 12th month. But in the 12th month, the windwas so strong and so do the seahas waved.This condition made the *sargasssum* "berabuk". "Berabuk" is the local conception about *sargassum* condition when the leaves of *sargassum*were covered up by the substrate.

3.5 Condition and behavior of seahorse

Fishermen mentioned that in the second month (February), they often found seahorses in the pregnant conditions. The number of pregnant seahorses in one catch reaches almost 70% - 80%. It means if the number of seahorses that have been caught reached 30 seahorses, as much as 21 – 24 seahorses were in pregnant conditions. They also could found seahorse in a pregnant condition in another month, but not as much as in the 2nd month. Thisinformation was collected from the fishermen using score table by making the 1st month (January) as benchmark comparison (h_o). Furthermore, in this month fishermen also frequently observed found when the seahorse wastransferring their eggs to each other.

Table 2. Table of comparison the amount of seahorse catches in pregnant condition[s]

Month[b]											
1	2	3	4	5	6	7	8	9	10	11	12
1x	3x	2x	1x	1x	-	-	-	-	-	-	-

[a]Primary data
[b]Month in a year
[c]Comparison result with the 1st month as a benchmarks

3.6 The distribution of seahorse

The fishermen use a hook and sunglasses to getthe seahorse. The hook used to drill down the *sargassum*, or local people say it "memusing".But they use hand to pick the seahorse. While the sunglasses were used to protect the eyes of fishermen when they dived. Furthermore,the fishermen mentioned that theseahorses were sold to collectors who come to SebongPerehvillage. But there were also fishermen who sell directly to the collector which is located in the market town of Tanjungpinang. Then the collectors sell the seahorses to overseas market. But until now, the fishermen do not havequota capture letter of the seahorse.They must have this letter because seahorses are listed as vulnerable or endangered on the 2003 IUCN Red List of Threatened Species [4], [5].Officially,in Indonesia the trading of the seahorse is regulated by the Ministry of Environment and Forestry by issuing quotas of capture and export quota [4].

3.7 Local wisdom: local management institutionalization

SebongPerehvillage seahorse fishermen already have the initiative to form a group of seahorse fishermen. But this group will only be officially instituted in thecoming year (2019). They - encouraged by Government Agent of Coastal and Marine Resource Management (BalaiPengelolaanSumberdayaPesisirdanLaut Padang/BPSPL Padang) – havealso created a marine protected area through Village Regulation[4].The area of conservation is about 15 ha (Fig. 6).

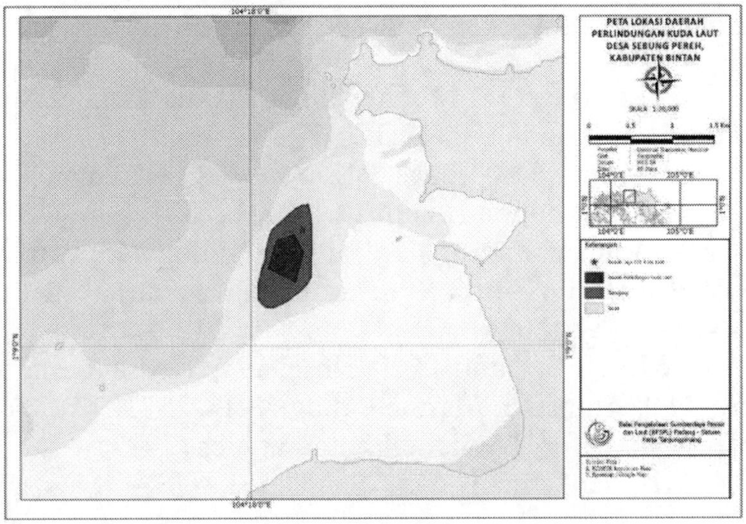

Fig. 6. Marine protected area of seahorse at SebongPerehvillage

As for the provisions on the conservation area is the fisherman is prohibited to make arrests in the conservation area. If a fisherman catches 1 (one) seahorse in a conservation area, then the fisherman is required to return 3 (three) seahorses to the area. Unfortunately, until now there have been no enforcement sanctions given if there are fishermen who violate the rules.In addition to the absence of sanctions for offenders, the SebongPereh Village fisherman has not yet placed supervisors in the conservancy area, even though there are already supervisory huts.

4. Conclusions

The results revealed that based on experience and daily observations, fishermen have quite complex the ecology knowledge to effectively utilize the seahorse as an important source of income from the waters of SebongPereh village. Seahorse fishing and related environmental knowledge can be used to development and implementation of sustainable it uses. Nevertheless, lack of best management practices could lead to the endangered of this types of fisheries. More consultation at interdisciplinary frontiers is needed in order to formulate practical solutions to the core problems of Bintan seahorse fishing community in the future.

The authors wish to thank the KEMENRISTEK DIKTI program for research funding, Dr. Syakti for re-reading the manuscript, community of SebongPereh village, and our students for technical support and research campaigns, Raja Ali Haji Maritime University.

References

1. BalaiPengelolaanSumberdayaPesisirdanLaut Padang [Internet]. Available from :http://bpsplpadang.kkp.go.id/bpspl-padang-dan-

nelayan-bintan-utara-menginisiasi-dpl-kudaluat-seluas-15-hektar (2017).
2. Bell EM, Lockyear JF, McPherson JM, Dale MA. & Vincent ACJ. Environmental Biology of Fishes **67**,35 – 46 (2017).
3. Cullen-Unsworth LC, Jones BL, Seary R, Newman R, Unsworth RKF. Marine Pollution Buletin**11** (007) (2017)
4. Foster SJ, Vincent ACJ.Journal of Fish Biology **65**, 1–61 (2004).
5. Harasti D, Martin-Smith K, Gladstone W. Journal of Fish Biology **85**, 1413 – 1428 (2017).
6. Hashim M, Ito S, Numata S, Hosaka T, Hossain MS, Misbari S, Yahya NN, Ahmad S. Marine Policy **78**:18-25 (2017).
7. Kuiter RH. Seaford: Aquatic Photographics (2017)
8. Miles MB, Huberman AM, Saldana J. USA: Sage Publications (2017)
9. Mulyawan AE, Saokani J. Journal BalikDiwa. **6** (2z0:13 – 19(2017)
10. Moleong LJ. RemajaRosdakarya (2017)
11. Moreau MA, Vincent ACJ. Marine and Freshwater Research **55**, 231 – 239. (2014)
12. Perante NC, Pajaro MG, Meeuwig JJ, Vincent ACJ. Journal of Fish Biology **60**,821 – 837 (2002)
13. Pratomo A, [Thesis]. Retrieved from Bogor University of Agriculture(2005)
14. Rabiansyah. SebongPereh Maritime University(2005)
15. Rosa IL, Oliveira TPR, Castro AL C, Moraes L, Xavier JHA, Nottingham MC, Dias TLP, Bruto-CostaLV, Araujo ME, Birolo AB, Mai ACG, Monteiro-Neto C. Neotropical Ichthyology **5**,405 – 414 (2007)
16. Santoso. JurnalSaintekPerikanan**2**(1)2006 :83-93(2006)
17. Simanjuntak CF. [Skripsi]. Retrieved from Raja Ali Haji Maritime University. (2015)
18. Vincent ACJ, Marsden AD, Evans KL & Sadler LM. Behaviour 141, 141 – 156(2004)

E3S Web of Conferences **47**, 07002 (2018) https://doi.org/10.1051/e3sconf/20184707002
SCiFiMaS 2018

Make Peace With the Damaged Coast(The Experience of Mentok Coastal Community in Adapting the Offshore Tin Mining)

Ibrahim[1,1,] *Dwi* Haryadi[2] , and *Nanang* Wahyudin[3]

[1]Political Science Department, Bangka Belitung University, Indonesia
[2]Law Department, Bangka Belitung University, Indonesia
[3]Management Department, Bangka Belitung University, Indonesia

Abstract. Tin mining in Mentok coastal area, West Bangka Regency has been the ironic view in the middle of local government effort to progress its territory, including the effort to develop tourism and marine sector. When there are many areas in Bangka Island flare up to refuse the presence of Production Suction Vessel (KIP/KapalIsapProduksi), people around Mentok coastal area spanning from Rubiah gulf to Kalian peninsula received the present of offshore mining. This paper went deep into why KIP could be accepted by Mentok people. It was designed from qualitative research result exploring data through deep interview technique, observation, and documentation. This study result showed that KIP operation in the area run well either as there was compensation modus distributed to every householder in the coastal area in the form of cash money. Provision of cash was managed in terms of securing production process assisted by the administrator to bridge the interest between businessman and society. Local government as the permission issuer and society who didn't give resistance omission sustainability and natural preservation principle; a pragmatic face wrapped with omission principle and justification on the economy aggression.

1 Introduction

Bangka Island is famous worldwide since long time ago with its production, along with the wellknown of pepper as the important world spice with the brand 'Muntok White Pepper' (see Heidhues, 2008: 4-9; Sujitno, 2007: 5-7; Susilo&Maemunah, 2009: 10). Mentok is one of places having important history in this island civilization (see Susilo&Maemunah, 2009: 15-16; Erman, 2009: 20-23). In addition to the entering gate from South Sumatera mainland, Mentok is also an initial place for tin mining occupied by Palembang sultanate. Until then Dutch government made this area into the biggest tin harbor in Bangka Island. No wonder, this area was known as the very important and strategic in this territory.

Along with the expansion of territory, Mentok that was initially a district, in 2003 developed into regency capital as well as West Bangka government center. As a regency capital, Mentok developed as a busy town with various activities; one that ogled was tourism sector development. Coastal area became the important mainstay of Mentok as it has a long and beautiful coastal line, supported by natural sea wealth, such as the fisherman catching, the passenger harbor for ship and ferry, loading and unloading harbor, as well as the harbor for tin ore reserves in the ocean.

In the middle of some local development growth, tin production temptation has become inseparable part of economical business. Tin regulation changing in national level has implicated to the regions since reformation started in 1998 (Rahman *et al.* 2011: 28-29; Yunianto, 2009: 100-101; Ibrahim, 2012: 68-78; Haryadi, 2014; 2015) and provided opportunity for mining to shift from on- to offshore and at the same time giving chance for private parties to develop offshore mining sector. Mentok coast territory spanning kilometers changed into offshore mining area, as if chasing with the development in other sectors, for example, the one in tourism, fishery, and minor fisherman catching.

[1]Correspondingauthor: iim_babel@yahoo.com

Mentok with its coastal wealth changed its face. In land, Mentok taking coastal area as the busy place dressed up as capital regency; however in the ocean that is only 2-3 kilometers away, offshore mining ships were milling about as if competing with fisherman life and freight and passenger vessels. It is interesting to explore on how these two life sides walk side by side.

This writingrevealed the important fact on how the offshore mining ships as production suction vessel could operate and well accepted in Mentok coast area. Coastal people life are getting along well with the ocean peaceful, thatin fact, most of them are fisherman territory. The debate of environmental damage and coastal fisherman anxiety as a result of production suction vessel operation exploring tin employing non-environment friendly method as if to be double standard for the coastal society who were open and accommodative. Is it true if this condition is two sides of coin running each other in support or because there is access inequality wrapped up with pragmatism, this writing provided logical information.

2 Research Method

The method employed in this study was qualitative one. Qualitative method explores the data deepness to explain the study focus (Marsh & Stoker, 2010: 242; see also Chadwick, *et al.*, 1991: 234). Data collection technique was conducted by interview, documentation, and observation. Interview informants were chosen by using purposing sampling that was determining informants by certain considerations. Interview was chosen as major data exploration technique by consideration to give researcher discretion to explore information, view, and value from the field (see Casley& Kumar, 1989: 11). In this regard, the chosen informant criteria were fisherman, housewife, production suction vessel administrator, businessman, and environment activist. The question in the interview was un-structured that was then situationally developed in the field with question lattice that had been already arranged before going to the field. Documentation was performed by searching other source data in the form of photograph taken in the field. Observation was conducted by performing direct observation to the research setting.

3Discussions

3.1 Mentok and Chronic Tin Mining

Tin mining operating in West Bangka Regency coastal area has been generally a Production Suction Vessel (KIP/KapalIsapProduksi). Different with many other coastal areas in Bangka Island that are full with suction pontoon conventional tin mining and TI Rajuk, either legal or illegal (see Irawan*et al.*, 2014), offshore tin mining in Mentok and surrounding has been in suction vessels. Production suction vessel is mostly a former tin mining boat in Phuket Thailand re-operated in Bangka Belitung Islands Province territory.

KIP operation got much controversy as generally it operates in shallow sea. In addition to freely moving outside the area, KIP operation also destroyed environment as it operates big boor/dredger functioning as suction as well as the separator between tin ore and sand in the ship. In many places, KIP received hard rejection from society (Ibrahim *et al.*,2018); however, in Mentok waters, KIP can freely operate.

Below are some pictures of KIP present and operation along the Mentok coast.

Fig 1. KIP in Mentok Coast

Before 2014, KIP operation permission was issued by government with different authority detail from that of after 2014, using the stipulation as follows:

Table 1. Tin regulation and its permission authority

Before Regulation 23/2014	Permission Issue Authority	After Regulation 23/2014
<4 miles	Regency/Municipality Government	-
4-12 miles	Provincial Government	0-12 miles
>12 miles	Central Government	>12 miles

KIP operating in West Bangka Regency territory based on Mentok Head Harbor obtained from RMOL Babel (9/4/2018 edition) showed that there were more than 28 illegal KIP in which 25 operate in Jebus district, 2 KIP in TanjungUlarMentok district, and 1 in DLKP Mentok. All KIPs are owned by 7 different companies.

The tin produced inter location and inter KIP was relative different. In Mentok waters, tin production was more or less 10 drums (equal to 10 tons) in one unloading. This number was considered as low compared to other coastal areas that could reach 45 – 60 drums per unloading (Reza interview, 13/4/2018). Nevertheless, Mentok coast was relative preferred since it could be used as shelter in a bad weather, they could also operate.

KIP operations were basically indicated very destructive toward the environment (see Milieudefensie, 2016; Febriantoet *al.,* 2015; Muslih*et al.*, 2014). Its very flexible cruising range and its big production capacity made KIP as having critical notes for environment conservation movement. Edi Nayo, one of tourism activists (interview 13/4/2018) said that offshore mining in any form is definitely damaging. Coral reef damaged and sea biota decreased is one of impacts resulted from offshore mining. According to him, last time it was easy to find crab; however, recently it is difficult and even the dead fish stranded on the beach as a result of KIP mining. Agus, one of fishermen in the interview (13/4/2018) said that since there was KIP, his catching area was going further into the middle of ocean and the fish quality caught was not as expensive as before.

As the amends for KIP operation in Mentok waters, people around it obtain compensation as donation for every householder. The quantity is various; the average is Rp 300,000/householder monthly. The payment is performed through administrator having important role to bridge the interest between KIP management and coastal people. It's not surprising if KIP operation in Mentok waters is relatively abandoned and there's no meaningful rejection. Indeed, the interview result on site showed that one territory experience with another around KampungTanjung spanning across this Mentok coast was different. However, with the similar spirit, the conflict hushing was performed through compensation modus. This different was on communication pattern of KIP owner with society, compensation distribution pattern, until the compensation quantity.

Until this study was conducted, at least many KIPs leant back or ensconced and operated at once. While people activities also run, such as activities going to the sea for fisherman, material loading and unloading, ship passenger arrival-departure and tourism area development activities by local government in the coastal area: an ironic condition on the fact of coastal people life connected to complicated discussions in terms of exploitation, prosperity, empowerment, and sustainable development.

3.2 Actor Map

KIP was basically a big investment since bringing this kind of ship needed much money. The price for second hand ship was above 20 billion per unit. Besides that, skillful operator who understood with all equipment's used usually operated KIP. They were mostly from Thailand. It isn't surprising ifthe society narrative found a lot on site stated that foreign workers from Thailand dominating this KIP job. It means that mobilizing capital and labor resources is a costly project investment, not even its operation. Therefore, it could be mapped that the key actor in KIP mining business was the capital owner and big businessman. It can be ensured that there's no KIP owned by local people and small scale businessmen. Interestingly, according to Reza interview (13/4/2018), the businessmen do not want directly dealing with society. They preferred to involve certain party to bridge their interest in order not to get rejection from society.

As mining business in coastal area and in large scale, it needs some procedures to get the license to operate. If it's noticed upstream, the permission issuance proceeded with the document accomplishment for analysis regarding environment problem and impact (AMDAL/AnalisisMasalahdanDampakLingkungan), though during implementation this document will often be ignored. Before the issuance of KIP permission as accordance with UU No. 23 2014, KIP mining permission was granted by regency government, in this regard was West Bangka Regency government. Some KIP have operated before this authority shifted. Hence when it shifted to province government, the permission was easier since it didn't have to pass through several other permission gates. If it's noticed here, the first actor involved is local government, either regency or province. It means that the local government involvement as upstream of KIP operation is very much.

In the case of KIP operation in Mentok waters and surrounding, resistance turmoil was relatively limited. The businessmen seems to know better that one of the ways to hush the turmoil was by providing compensation to either directly or indirectly affected society with KIP operation. Compensation choice in the form of either Corporate Social Responsibility (CSR) or the percentage fee has been the main choice to keep secure in operation. How can it connect two different interest in which it is very confrontational in many ways? It certainly needs an actor to bridge both interests. Hence it raised the term 'administrator' in the local language. This administrator is represented by government apparatus and most of them are neighborhood association (RT)/ resident association (RW) or important people respected in the society environment. The term of administrator was then the common term used by local people to describe a figure taking care of KIP and society interest. Therefore, the third actor having important role in the process to connect the two interests is administrator. Selamet (interview 12/4/2018) working as food seller in that coastal area said that the existence of administrator and compensation is to ensure KIP working in secure. Are they legal?Structurally, of course it's not. However, the process in choosing them and their activity are recognized by village and district apparatus. Wahyu, one of administrators since 2012 in an interview (13/3/2018), his election at that moment was attended by government party.

Here is the table of actors and their roles in KIP operation in Mentok waters and around it:

Table 2. Actors and their role

Actor	Role
KIP businessmen	Providing capital and manage mining
Local government	Issue/prolong KIP permission
Administrator	Lobbying, negotiating, and arranging compensation with coastal people

As the costly business unit, during this time KIP is not accessible for coastal people who are very dependent on their life from profession as fisherman or small miners. Supporting by complete permission, KIP is like unstoppable. Nevertheless, there is actually the potential operation rejection arising from society since many other areas in Bangka Island did it. The society rejection makes KIP cannot operate and must move away although they already have complete permission to operate. However, if this rejection can overcome, then KIP operation can gain much profit.

It is interesting to take a close look on general people perspective around KIP location. Most people feel grateful since they get sharing from KIP. The compensation they received making them able to think that KIP is not dangerous and disturb them. Some informants stated that KIP is not more dangerous than Rajuk Nonconventional Mining directly operated by society in the areas closer to the beach. For the people around this area, so far the short term interest is not disturbed, and thus there is no significant reason to refuse KIP, whereas there is direct benefit they gain.

Rohan, one of housewives having profession as salted fish maker, said that they were helped by KIP fund. The fishermen in the place she bought fish to process were actually going far into the middle ocean to catch fish. Therefore according to her, the fishermen were not disturbed by KIP activities. Nevertheless, she admitted that KIP could walk to the beach and very close to their houses.

> *"Nelayan kan jauh ke tengah nyarik ikan e. Kami dak hal dak, karena dapat kompensasi. Per KK 300 ribu. Ngambik e di masjid. Pengurus RT biasa e yang nyuruh ngambik. Tapi asak KIP e bejalan bai. Men dak, dak dapatlah. Untung merik masyarakat, men dek diprotes orang wah".* (Fisherman go far away to the ocean looking for fish. We are not object since we gain 300 thousand rupiah compensation per householder. It is taken in

 https://doi.org/10.1051/e3sconf/20184707002

mosque. Neighborhood association administrator usually asks us to take it. However, it is only if KIP operates. If it doesn't, so we don't get it. Fortunately they give, if not, the society will definitely protest.)

Similar information delivered by Sahrul (interview 13/8/2018) said that in addition to distributing compensation fund, KIP management and administrators involved society in tin sand loading-unloading, and thus in addition to receiving benefit of job opportunity, they also obtained direct fund for every householder. Different with the information delivered by Sahrul and Rohana, around KampungBaru area, Yohan (interview 12/4/2018) said that in their area, the quantity compensation they received depended on the tin produced by KIP. Rp 3,000,- per kilogram became the part from compensation so that the quantity would be different per month. Nevertheless, the same thing from every informant was that they only received compensation if KIP operated. Recently according to Yohana (interview 13/4/2018), it was started a disappointment among people as the compensation they received was getting smaller. Their suspicion was at the village apparatus and administrator dishonesty.

The information above showed that there was successfulness to positively construct KIP presence. Administrator had role to encourage people's trust that KIP gave more actual benefit to the people around KIP operation. Compensation in this regard encouraged the society to receive KIP presence. This short term pragmatism relatively didn't face meaningful resistance, although there were several minor voices of society rejecting it. This rejection, for example, was said by one of fishermen, Rujuk, who was setting a net not far from KIP operation.

"Kami nelayan kecil ni teganggu. Duluk mudah nangkep ikan. Sekarang ni, lah berape kali ku masang jaring, lum adelah dapet e. Dulu ku pacak bejalan di pantai tu agak jauh men surut, sekarang bahaya. Lumpur e sepaha. Lah gati ku tecalep, susah berangket. Orang sine dek hiran, dapet kompensasi, tapi kami teganggu. Men pacak, pegilah jauh-jauh dari pantai ne". (We are small fishermen feeling disturbed. It's easy to catch fish in the past. Nowadays, it's already several times setting net, there's no catching at all. Last time I can walk far to the sea if tidal, now it's dangerous. Mud is as high as thigh. I am often trapped in the mud, difficult to stand up. The people here receive compensation, no wonder, but we feel disturbed. If possible, just go far away from this beach.)

Rujuk opinion was strengthen by Marzuki's (interview 13/4/2018) stating that in the past he was easy to get shrimp as the raw material to make *terasi* (fermented fish); however, since there's KIP around, it became difficult. Zira, Anita, and Jeha, three middle age women interviewed in KampungBaruMentok area, also said that they were surrender with this condition (interview 12/4/2018). In a glance, it seems that the ocean is not damaged; however, actually the catching area for fishermen is getting further since KIP location nowadays is previously their place to catch fish. Being a fisherman at present is not promising anymore, and thus being a tin freight worker becomes one of their choices to keep survive. Yusran, one of fishermen successfully met, stated that fish catching area is getting further and they are aware of it; however, they actually get the compensation distributed (interview 13/4/2018).

Fig 2. The irony of fisherman life and KIP activity

These minor rejections were actually meaningless. Other than outnumbered, this rejection might not also be channeled in the middle community and apparatus pragmatism that have made peace with KIP on behalf of compensation as occurred in some other regions (see Ibrahim, 2015; Ibrahim *et al.*, 2017). One of the citizens, Sahrul (interview 13/4/2018), even said that almost all local citizens received the existence of KIP in the name of solidarity. In the case revealed by Wahyu (interview 13/4/2018), of 700 householders in his region, 6 people stated as refused. The real interest was not balanced, between making a profit in the middle of ocean damage and little compensation; a pragmatism that is not clearly and openly staged.

Below is the pragmatism relation figure between citizen and businessman:

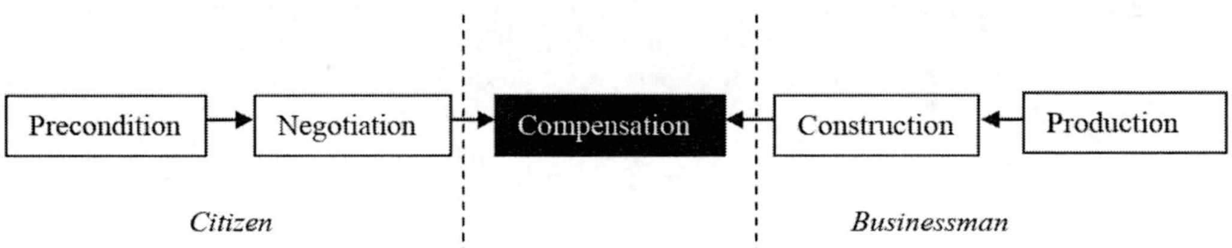

Fig 3. The pragmatism relation figure between citizen and businessman

The figure above showed that there is pre-condition on Mentok tin deposit wealth condition that was not massively exploited except by using dredge owned by PT Timah, Tbk and fisherman community working in marine sector. As KIP entered, it was then performed negotiation through administrator. The compensation became a choice. The businessman who will produce needs positive construction in order to peacefully run KIP operation and the compensation becomes the middle way.

3.4 Apathy: Between Omission and Justification

If we see the rejection wave on KIP operation in many other waters in Bangka Belitung Islands Province, the quiet KIP operation in Mentok coast signed that there's collective apathy. There were several arguments to explain this statement.

First, KIP operating along Mentok coastal line was so obvious. Some of them lean back and take cover; however, some other also operates openly. As the harbor region that becomes the traffic for goods and human being, the peninsula and gulf in Mentok coast are easy to see. Local government and its apparatus would understand well about KIP present. The permission process previously issued placed the government and all apparatus involved as ignoring the resulted impact. In the lowest level, village government and district were involved to supervise the process if there's problem. Wahyu (interview 13/4/2018) said that compensation distribution used neighborhood/resident association' chop. Wahyu and

Sofyan (interview 13/4/208) even stated that apparatus were not only involved, but even also get the bigger share.

Second, society, in the name of compensation eventually felt secure with KIP. Rather than disturbed, KIP even became the vehicle to get earning without working. The way is simple: do not protest on KIP operation. Common people are like performing collective omission wrapped with support on statement of non-disturbing society directly. When there were massive resistances against KIP operation, the people along Mentok gulf and peninsula were enjoying compensation. Iyan, one of fishermen (interview 14/4/2018) clearly said: "We will resist if we don't get compensation." This condition clearly described that KIPs would be secure as far as they were able to act 'fair' by small sharing to the communities around them.

Meanwhile, the third argument was the existence of justification on KIP operation securely constructed, non-destructive, and having potential economy for the surrounding community. Reza and Sofyan, two citizens interviewed (13/4/2018) affirmed that KIP didn't destroy environment. This view was often expressed by informants on field. This construction was campaigned by the involved administrators, supported by passive position of local government on the destruction resulted from KIP. Ecological apathy was then aroused. This condition was called as omission and justification at once. As the community was collectively apathy, hence KIP kept gaining profit from the tin production operation almost without stopping. This is an ironic condition in the middle of effort to safe environment and diminishes the rising crisis.

Uniquely, one of modus used to get support from coastal community was by involving religious figures. Rohana (interview 13/4/2018) said that mosque party was usually involved in distributing compensation fund. It could be clearly seen that there's an effort to legalize KIP existence by using religion instrument as one of its reason. Ilham, one of the Mosque chief stated that "*This mosque was built with many fund resources, but mostly from KIP and PT. Timah. Right now, it is being built fence from KIP.*"

Below does the researcher to describe on how chronic the present occurrence is simplify the apathy scheme:

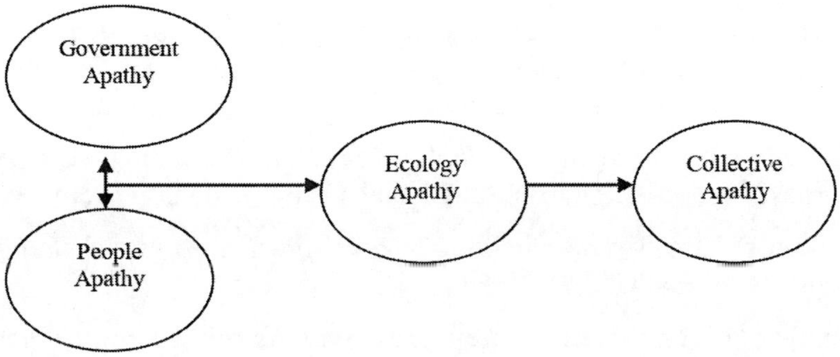

Fig 4.Collective apatism

The scheme above illustrated that collective apathy emerged from government and related stakeholders supported by community apathy toward ecology awareness. This collective apathy will result in omission and justification on the present condition.

4Conclusions

Tin mining in Mentok coast area is basically supported by several key actors; they are businessmen actively looking for offshore mining locations who do not get much resistance from local people, local government issuing operation permission, and administrators in which in the middle of society are known as interest connector. KIP operation in Mentok coast West Bangka regency having relatively good level of acceptance in the middle of society is influenced by the compensation distribution from company to surrounding people in coastal area. By different quantity, cash money distribution is performed through local neighborhood/resident association and the mosque administrator. This compensation is given to ensure that their mining can keep operating. It is interesting that KIP operation is not even constructed disturbing although in reality, KIP destroys environment and marine biota, disturbing fisherman catching, and affecting coastal condition. In the middle of West Bangka regency capital effort to clean up as one of new regencies in Bangka Island as well the effort to push tourism sector, KIP operation feels like anomaly. This condition showed that local people actually has pragmatic attitude, peace with the ocean in which its condition has been damaged, for the sake of short term compensation. Many parties perform omission on the anomaly and as if keep looking for justification on the damage acceleration resulted from KIP, as shown clearly by this paper.

Thank you to the Directorate ofResearch and Community Service that has funded this research in the scheme of 2018 Competency Grant. Thank you also to colleagues, students, and parties who have helped in collecting this study data, especially to Sandy Permana, RidwanDiaguna, Andy Saputra, Fajrina, Nurzaini, Abdul Rahman, and TresnaRahardja.

References

1. Chadwick, Bruce A., et al., *Social Science Research Method*, Translated by Sulistia, et al., IKIP Semarang Press, Semarang (1991)
2. Casley, Dennis J. and Krishna Kumar, *The Collection Analysis, and Use of Monitoring and Evaluation Data*, Second edition, The Johns Hopkins University Press, Maryland (1989)
3. Erman, Erwiza, *From Kampong Formation to Dark Problem, Uncover Tin History in Bangka Belitung*, Ombak, Yogyakarta (2009)
4. Febriyanto, Arief, et al., 2015. *The Impact of Tin Mining on Squid* (Urotheutischinensis) *Fishing Ground in South BangkainInternational Journal of Science : Basic and Applied Research* (IJSBAR), Volume 23, No.1, 283-293 (2015)
5. Haryadi D.2014. *Illegal Tin Mining (Life Time Problem?)*, In: After 14 years (The Journey Note of Bangka Belitung Islands Province). Yogyakarta: Imperium
6. Haryadi D.2015. *Criminogen Factor of Illegal Tin Mining in Bangka Belitung*. Law Problems Journal UNDIP Vol 44 No 1.
7. Heidhues, Mary F. Somers, *Bangka Tin andMentok Pepper, Chinese People Role in 18th to 20th Century Bangka Island Development*, Yayasan Nabil, Jakarta (2008)
8. Ibrahim, *Tin in Economy Pancasila Perspective: A Case Study in Bangka Belitung Islands Province*, Philosophy Doctoral Dissertation inGadjahMada University, Yogyakarta (2012)
9. Ibrahim, Bangka Tin, and The Collapse of The State Power, GSTF Journal of Law and Social Sciences, Vol. 5, Issue 1, pp.1-7 (2016)
10. Ibrahim, et al., *Tin and Authority Contestation : Government, People, and Businessman in Chronic Tin Management*, Proceeding of 135th, The IIER International Conference, Kyoto, Japan, 7-8 November (2017)
11. Ibrahim, et.al., From *Charm to Sorrow : The Dark Portrait of Tin Mining in Bangka Belitung*, Indonesia, PEOPLE : International Journal of Social Sciences, Vol.4, Issue 1, pp. 360-382 (2018)

12. Irawan, R. Rudy, et al., *Strategic Model of Tin Mining Industry in Indonesia (Case Study of Bangka Belitung Province)*in*International Journal of Business and Management Review*, Volume 2, No. 3, 48-58 (2014)
13. Marsh, David & Gerry Stoker, *Theory and Method in Political Science*, Translated byHelmiMahadiandShohifullah, Nusa Media, Bandung (2010)
14. Milieudefensie, *Responsible Mining Tin*, Goodelectronics, Amsterdam (2016)
15. Muslih, Khairul, et al, *Habitat Characteristic andMenduk River Fresh Water Fish Affected from Tin Mining in Bangka Regency*in*AKUATIK-Fishery Resources Journal*, Volume 8, No. 2, 17-23 (2014)
16. Rahman, Bustami, et al, *Tin Problem in Babel (Several Stories and Ideas),* Khomza, Yogyakarta (2011)
17. Sujitno, Sutedjo, *The Impact of Indonesian Tin Throughout History*, PT. GramediaPustakaUtama. Jakarta (2007)
18. Susilo, Joko&SitiMaemunah, *Three Centuries Serving the World, Bangka Belitung Tin Mining Portrait,*JaringanAdvokasi Tambang (JATAM), Jakarta (2009)
19. Yunianto, Bambang, *Tin Mining Problem Review in Bangka Belitung Island Province as the Input for National Tin Policy*, Mineral and Coal Technology Journal Vol. 5, No. 3, July (2009)

Other Sources :

1. Law Number. 23 year 2014 regarding Local Government
2. http://www.rmolbabel.com/read/2018/04/09/8022/Kepala-DKP-Babel-Pasrah-Lawan-28-KIP

Informant List :

1. Agus
2. Anita
3. Edi Nayo
4. Ilham
5. Iyan
6. Jeha
7. Marzuki
8. Reza
9. Rohana
10. Rujuk
11. Sahrul
12. Selamet

13. Sofyan
14. Wahyu
15. Yuhan
16. Yusran
17. Zira

Protection of Traditional Fishermen in The Granting of Fishery Licenses in Indonesia

Muhamad Azhar[1,2], *Suhartoyo* Suhartoyo[1], *Lita Tyesta* ALW[1], *Putut* Suharso[3], *Vivi Endar* Herawati[4]

[1]Faculty of Law, Diponegoro University, Prof. H. Soedarto SH street, Semarang 50275 Indonesia
[2]Center For Administrative Law Studies, Diponegoro University, Prof. H. Soedarto SH street, Semarang 50275 Indonesia
[3]Departement of Library Science, Diponegoro University, Prof. H. Soedarto SH street, Semarang 50275 Indonesia
[4]Department of Aquaculture, Faculty of Fisheries and Marine Science, Diponegoro University, Prof. H. Soedarto SH street, Semarang 50275, Indonesia

Abstract. The granting of fishing licenses for large vessels is very detrimental to the existence of small fishermen. Small fishermen or so-called traditional fishermen are powerless to face the dominance of fishing pens and licenses by large boats. This study aims to determine the rare step in providing protection to traditional Indonesian fishermen on the granting of fishing licenses for large companies in Indonesia. This research is an empirical juridical research using statutory uproach approach, conceptual aproach, and comparative approach. Then analyzed using synthesis analysis. The results show that traditional fishermen tend not to have adequate legal protection in terms of fishing in Indonesian waters. Traditional fishermen tend to be allowed to compete with large vessels that have licenses to catch fish in the Indonesian territory. The granting of fishing licenses is not accompanied by certain obligations associated with traditional fishermen protection. If observed, the permit should have a fishing license charging certain liability obligations for the permit holder.

Keywords: Traditional Fishermen, Licenses, Fisheries. Indonesian Waters.

1 Introduction

The country of Indonesia as an archipelago consisting of large and small islands separated by sea with geographical conditions has a total area of 7,827,087 km² and is located in the equatorial emeralds that have abundant natural resources, especially in marine and coastal areas that reach ¾ Indonesian territory (5.8 million km²). Physically, the sum of the island's 17,504 islands is the largest in the world. Beach near the coast of 81,000 km is the second longest in the world after Canada. The 5.8 million km2 marine area comprises an Exclusive Economic Zone of 2.7 million km², an archipelago of 2.8 km², and a marine area of 0.3 km² [1].

The Unitary State of the Republic of Indonesia (NKRI) is the largest archipelagic country in the world declared in 1957 in the Djuanda Declaration, then confirmed in Law No. 17/1985 on the International Law of the Sea Convention (Law No. 17/1985). There are about 60% of Indonesians living in coastal areas as wide as 50 km from the coastline. In the cities and districts of this region, 80% of Indonesian industries operate to utilize coastal resources and dispose of their waste into coastal areas [2]. Coastal resources are the world's richest tropical marine biodiversity center, where 30% of the world's mangrove forests exist in Indonesia; 30% of the world's coral reefs exist in Indonesia, 60% of protein consumption comes from fishery resources, 90% of fish come from coastal waters within 12 nautical miles of coastline [3]. The large potential of fisheries and marine, causing many Indonesian people who work as fishermen, including traditional fishermen.

Protecting traditional fishermen is a government duty of a country. The existence of traditional fishermen in the fishing industry should be granted access by the government of a country [4]. The task is carried out through adannya government policies in favor of traditional fishermen. One such policy is the government of a country issuing government instruments that protect traditional fishermen, either in the form of *beschikkingen* (decision), *regelingen* (regulation) or *beleidsregel* (rules of wisdom). One of the strategies to protect traditional fishermen is through licensing instruments issued by the government. The type of licensing in the field of fisheries issued by the government (Table 1).

Table 1. Type of Fishery Licenses in Indonesia

* Corresponding author: azharundip@gmail.com

No	Jenis Izin Perikanan	Code	Days
1	Letter of Implementation of Fish auction (TPI)	1.3	14
2	Business License (SIUP)	1.4	14
3	Fishing Permit (SIPI)	1.5	14
4	Letter Fish Breeding (STPbI) in the Sea and Public Waters For Regency / City	1.6	14
5	Permit for Indonesian Fishing Catchers and Transporters (SIKPPII)	1.7	14
6	Permit for Indonesian Fishing Cargo (SIKPII)	1.8	14

Based on the data, it can be seen that licensing instruments that can be used as a means of protection of traditional fishermen are presented as shown in the table. If viewed from the substance, licensing should be able to guarantee that traditional fishermen can keep doing activities, without any barrier from other legal subjects in the form of companies or individuals who get permits.

Companies or individuals who own or obtain licenses from governments tend to ignore the basic rights of traditional fishermen. If explored in depth, according to the experts that the permit (*verguning*), vmerupakan approval from the authorities/government based on legislation to under certain circumstances deviate from the provisions of the prohibition of legislation. So the permit in principle is as a dispensation or release / exemption from a prohibition [5]. The existence of various types of permits obtained by both entrepreneurs and/or individuals tends to harm traditional fishermen. the activities of traditional fishermen became limited and less freely moved to conduct fishing activities of fishery resources.

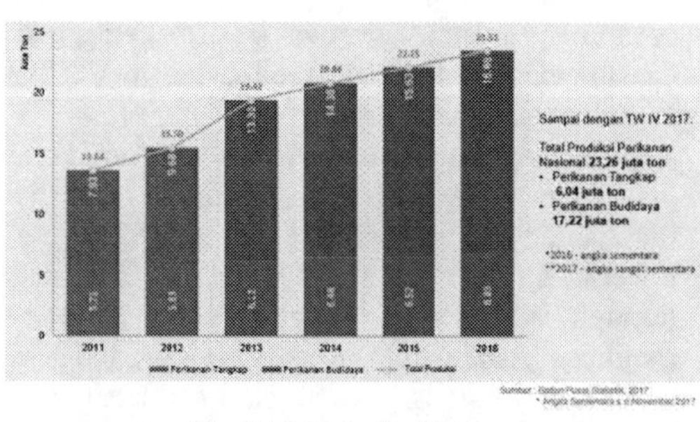

Fig. 1. Fisheries Production

According to research results from the Ministry of Marine Affairs and Fisheries, traditional fishermen tend to be less given access, not at all noticed. Traditional fishermen are not given access, and tend to be teared or underestimated. Although the number of tradional fishermen in Indonesia from year to year has decreased (see BPS data). Traditional fishermen experienced a decrease from the previous 1.6 million to 864 thousand people. In addition to the number of traditional fishermen declining, but the amount of income they jurxu increased. The quality of traditional fishermen's economic income also increased. In 2017, the government managed to lift the growth in the field of fisheries. As in the visible from Fig. 1. while the average exchange rate of traditional fishermen per year has increased as shown in Fig. 2.

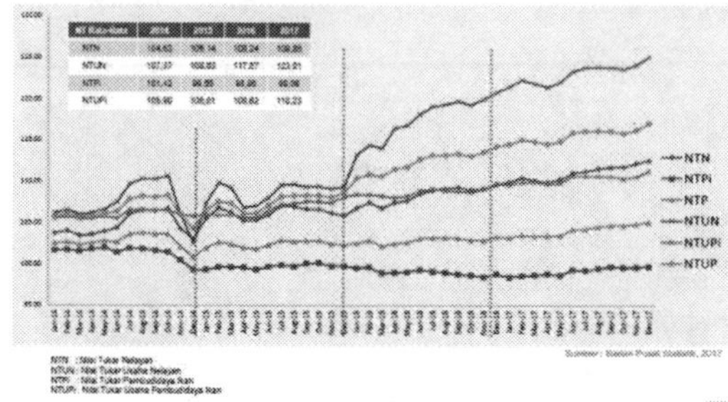

Fig. 2. The average exchange rate of traditional fishermen

Based on the problematic, then the question that will be answered in this article is as a model and strategy of traditional fishermen protection through the granting of fishery business permit in Indoenesia. The question is later described by writing method using empirical juridical research approach using statutory uproach approach, conceptual aproach, and comparative approach. Then analyzed using synthesis analysis.

2 Research Methods

This research is a legal research looking from the aspect of social (non-doctrinal) perspective. The approach used is socio legal but also statutory uproach approach, conceptual aproach, in this case, the law is seen as a space for the process of scientific study in order to seek truth.

Thus writing is not confined by a model of thought alone in looking at a problem to be discussed. The domain of socio-legal studies conceptualizes the law as a norm as well as behavior, the law is seen as merely as concrete, written, sanctioned and issued by authorized institutions which in its work are influenced by various other factors such as economics, politics, culture, religion, etc [6]. The use of relevant legal research would like to understand the law more thoroughly [7]. In this case, socio-legal research leads to a more thorough, whole, not textual understanding of law. Want to know whether the rule is effective, positive, productive, or even disturbing and destructive [8] [9].

4 Result and discussion

In this section of the results and discussion, it will explain three main points related to the protection of traditional fishermen, the justice factor in traditional enforcement, and the granting of fishery permits. The explanation is as follows:

4.1 Protection of traditional fishermen

Based on the provisions of Law no. 45/2009 on Amendment to Law no. 31/2004 on Fisheries affirmed that the utilization of fishery resources should be done in an environmentally friendly manner and fishing gear. This assertiveness is intended as an effort to preserve not only the environment but also the sustainability of fish resources. In support of this policy, the government established a number of policies, among others, defining the type, quantity, and size of fishing gear; type, quantity, size, and placement of fishing aids, and the area, path, and time or season of fishing. This is done in order to maintain the existence of fishery resources remain sustainable and can be enjoyed by all people of Indonesia.

At this time, traditional fishermen are getting high attention from policy policies taken by the government. The policy is in favor or affirmative policy in order to improve the standard of living of traditional fishermen to continue or continue fishing activities [10]. This form of protection of traditional fishermen is poured in the form of easiness to obtain permits for the utilization of fishery resources. The government also encourages local governments to facilitate the existence of traditional fishermen through financial facilities and infrastructure facilities.

Presidential Instruction no. 15 of 2011 on the Protection of Fishermen which reinforces the role and function of all sectors, both government and local government, in protecting and empowering fishermen in Indonesia. Similarly, Law no. 31 of 2004 and its revision No. 45 Year 2009 on Fisheries that mandate to the government also to issue Government Regulation on Empowering Small Fishermen and Small-Scale Fish Cultivators.

In the Presidential Instruction, the President instructs each of the relevant ministers, Governors and Regents/ Mayors, to undertake the necessary steps in accordance with their respective duties, functions and authorities in a coordinated and integrated manner to provide welfare, assurance and protection law for fishermen [11] which operates fishing vessels up to 60 Gross Tonnage (GT). In carrying out these steps, each party coordinates related matters according to their respective fields.

Thus, it can be seen that the protection of traditional fishermen is a mandate of the laws and regulations in the field of fisheries. The mandate is contained in the basic principles of fisheries policy implementation. the necessity to see the sustainability factor of a fishery resource becomes one of the important points of consideration in policy making in the field of fisheries. the principle of sustainability in the utilization of fisheries is one strategy for traditional fishermen to survive and continue.

4.2 Justice factor for fishermen

Nationally, Indonesia has four homework, including homework as described below: First, the high poverty rate. Based on the causes of poverty, the poverty of coastal communities is divided into three types: (i) structural poverty, (ii) cultural poverty, and (iii) natural poverty. Secondly, access to the use of coastal and marine areas by traditional fishermen, and utilize the natural resources in it that are not impartial. Third, the ease of accessing the market for traditional fishermen is classified as an economic (socio-cultural) right expressed specifically in the constitution.

In this case the dependence on the market. At this point, the state must provide maximum service to traditional fishermen. This is because the commodities produced must be immediately sold to meet the needs of everyday life or decay before the sale. Fourth, access to management, where resources are not fully accessible due to the dominance of state law, is sometimes made not in a participatory way and is the result of consideration from the central government without regard to socio-economic aspects of local communities. As a result, the policies issued precisely cause new problems because each party has interests, desires and priorities that exist is the source of triggering the emergence of conflict [12].

The existence of traditional fishermen who generally utilize fishery resources with simple patterns and usually with local coastal wisdom, also competed in the same region. This area should get special protection considering the status of small fishermen using simple fishing gear. In the Fisheries Act there are already several arrangements and on empowering small fishermen and small fish-breeders. This includes playing an important role by utilizing licensing instruments as one of the traditional means of protection of fishermen.

Another emphasis, for example by looking at a number of provisions: (1) Small fishermen are free to catch fish in all fisheries management areas of the Republic of Indonesia; (2) Small fish farmers can cultivate selected fish commodities throughout the fishery management area of the Republic of Indonesia, (3) Small fishermen and small fish farmers are obliged to comply with the conservation provisions. (4) Small fishermen or small fish farmers must participate in preserving the fishery environment and fishery product safety in accordance with applicable provisions [13]; (5) Small fishermen and small fish farmers should register themselves, businesses, and activities to local fisheries agencies, at no charge, for statistical purposes and for small-scale fishermen and fish-breeders.

Such a condition is very important, because it involves a sense of justice. In the context of social psychology, although what they receive is still quite far from the word fair [14]. The distribution of resources and benefits, to rights and remuneration, positions and convenience, which are fair will be judged satisfactory. Conversely, distributions that are considered unfair will cause dissatisfaction. Fair judgment will have an impact on positive social behavior, while an unfair judgment will have a negative impact. One of its forms is the relative deprivation often expressed in protest, anarchy, and rebellious behavior, or even the destruction of coral reefs. The principle of justice in the rule of law, seeks to derive a midpoint between two interests, on a single

https://doi.org/10.1051/e3sconf/20184707003

system of states to exercise power and on the other side society must be protected [15].

From the whole description, it can be seen that the justice factor as one of the drivers or triggers of traditional fishermen protection policy. the existence of traditional fishermen should be able to synergize with the policies issued by the government, so that national development can run as planned. Laying the interests of traditional fishermen on the other interests is one of the efforts to ensure identical poverty in the traditional fishing environment can be pressed.

4.3 The granting of fishery permits

The granting of fishery permits is entirely the authority of the government. However, in the use of such authority, it must always be based on the applicable laws and regulations as well as the general principle of governance [16]. If examined more deeply, the granting of permits must meet the minimum standard suatau permit issued, as the following provisions; 1) Authority: Any legal action by the government, principally within the State of the law, whether in the conduct of the regulatory or service functions, shall be based on the authority granted by applicable laws and regulations. So without clear authority that can never be made concrete decisions juridically. In this case, the granting of a fishery permit can only occur if true authority arises which attribution exists or is known to exist. 2) As a form of stipulation: In a modern State of law, the duties and authorities of the government are not merely to maintain order and order, but also to seek the common good (*bestuurszorg*). The duty and authority of the government to maintain order and security is a classic task that is still maintained today. It is within the framework of this task that the government is granted authority in the field of arrangement, which of this functioning of authority appears to some juridical instrument to face individual and concrete events, this provision is the spearhead of the legal instrument in the administration of government. 3) Government
Institutions: Institutions, theoretically is a rule of the game that knows can take action and determine whether an organization can run efficiently and effectively. Thus, institutional arrangements can be the enablers of success and also if not appropriate in the arrangement, it will be the Contraint of duties including the task of licensing of tea everything that requires the permission of the government/State. 4) Concrete events: It is said that permission is a juridical instrument in the form of statute, which is used by the government in the face of concrete and individual events. Concrete events mean that occurring at a certain time, certain people, certain places, and certain legal facts. Because these concrete events are diverse, permits also vary. Such diverse licenses are made in a process whose procedure depends on the authority of the licensor, the kinds of permits and organizational structure of the institution that issued it. 5) Processes and procedures: Licensing processes and procedures may include licensing service procedures, licensing completion processes which are the licensing completion process undertaken by officers / officers. In each stage of the job, each employee can know their respective roles in the licensing settlement process.

From the description, it can be understood that having policy instruments related to fishery permits, can be used as a means of prevention and control of environmental pollution. Instruments which are solely the authority of the ruler are classified into public legal instruments. Public legal instruments are distinguished in direct regulation and financial arrangements. Direct direct arrangement is done through commands (geboden) and prohibition of prohibition (verboden). The most direct regulatory form of prevention is prohibition (verboden) unless permission is given.

Fishery permit is one form of state administrative decision (beschikking). Decision is the right measuring instrument in relation to fishery management which must be accompanied by the requirements and consideration of fishery area as an effort to prevent the existence of violation in the field of fishery. The authority to regulate fisheries management is based on government authority. Such authority according to the verdict of the Constitutional Court of the Republic of Indonesia is translated as beheersdaad, Beleidsdaad, Berstuursdaad, Regelendsdaad, and toezichthoudensdaad.

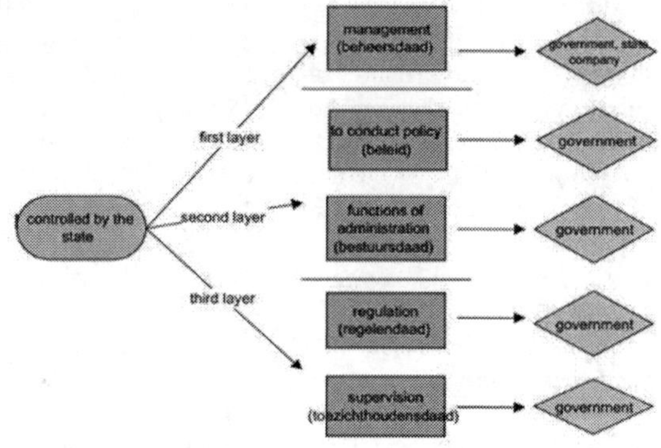

Fig. 3. Contitutional Court interpretation on Controlled by State

In Decision No 002/PUU-I/2003 on the judicial review of the Oil and Gas Law, the Constitutional Court stated that: 'controlled by the state' must be interpreted to include the meaning of a wide-ranging state occupation, as a result of the people's sovereignty concept over 'soil, water, and natural wealth contained therein', comprising also the meaning of public ownership over the aforementioned sources of wealth. The people, collectively constructed by the 1945 Constitution, provide a mandate to the state to conduct policy (*beleid*) and functions of administration (*bestuursdaad*), regulation *(regelendaad)*, management (*beheersdaad*), and supervision (*toezichthoudensdaad*) for the welfare of the people to the utmost. From the Constitutional Court opinion (Fig. 1), it can be construed that the phrase 'controlled by state' can be interpreted widely, not only as ownership in the private law context but also in the context of public ownership. In the context of public ownership, the state has received a mandate from the people over fivem matters: to conduct policy (beleid) and functions of administration (*bestuursdaad*), regulation (*regelendaad*), management (*beheersdaad*) and supervision (*toezichthoudensdaad*).

Fisheries resources as well as other natural resources. Some of the special characteristics are: 1) Resources are generally invisible 2) Resources are common property 3) Exploitation of resources involves high risk 4) The product is very vulnerable (ie, highly perishable).

Furthermore, it can be said that the characteristics that cause more difficult process of utilization of fishery resources compared with other resources. Therefore, the tools of fisheries sciences are needed to enable the utilization of these resources, which include biological, technological, social and economic aspects. The understanding in the Regulation of the Minister of Marine and Fisheries of the Republic of Indonesia No.12 /MEN/2012 can be seen that fishing business is a fishery business based on fishing activities and / or fish transportation activities (Article 1 Paragraph 1). Fishery Company is understood as a company conducting fishery business and conducted by Indonesian citizen or Indonesian legal entity (Article 1, paragraph 3).

Fishing is an activity to obtain fish in waters that are not cultivated by any means or means, including activities that use vessels to load, transport, store, cool, handle, process and/or preserve it (Article 1, paragraph 6) vessels, boats or other floating apparatus used for fishing, supporting fishing operations, fish cultivation, fish transportation, fish processing, fishery exploration and fisheries research / exploration (Article 1, paragraph 8). Fishing vessels are ships specifically used for fishing, including storing, storing, cooling and / or preserving (Article 1, paragraph 9).

In conducting fishery activities, it must be equipped with a fishery business permit. SIUP is a written permit that must be owned by a fishery company to conduct a fishery business using production facilities listed in the license. in addition to the fishing business permit there is also a fishing permit, or known as SIPI is a written permission that must be owned by every fishing vessel to make fishing which is an integral part of SIUP. In addition to these two types, there is also a permit fish carrier ship, hereinafter abbreviated SIKPI. This permit is prestigious as a written permit that every fishing vessel must have to carry fish transportation which is an integral part of SIUP. The existence of permits should be utilized by the government as a means in order to protect traditional fishermen. For example by limiting the existence of permits granted to a legal entity or individual. In fact, various types of permits may be moratorium (suspension).

The regulation also regulates types of licensing concerning fisheries business which include: fishing business, fishery business license and SIPI (Fishing Permit), Fishing Business, SIUP (Fishery Business License) and SIKPI (Surat Izin Fish Fishing and Fishery Business License, Fishery Business License SIUP (Fishery Business License) and Business Processing of SIUP (Fishery Business License). In the Fishing Permit License (SIPI) the main component of capture fisheries is the fishing unit, which consists of boats or ships, fishing gear and labor or fishermen that are interconnected and affect each other.

From the description it can be said that, the government should be able to make efforts in order to limit the existence of legal entities in possession of fishery permits. Such restrictions may be made as long as the objective is within the framework of the protection of traditional fishermen. not only limiting but also can be done penyetopan (moratorium) the existence of fishery permits in order to provide opportunities to traditional fishermen to gain access to existing fishery resources.

5 Conclusion

Based on the results of this study it can be concluded that as follows: Protection of traditional fishermen is a mandate of the laws and regulations in the field of fisheries. The mandate is contained in the basic principles of fisheries policy implementation. Justice factor as one of the triggers or trigger the existence of traditional fishermen protection policy. as well as government participation may limit the existence of legal entities related to fisheries. Such restrictions may be made as long as the objective is within the framework of the protection of traditional fishermen. Not only to limit but also to the moratorium on the existence of fishery permits to provide opportunities for traditional fishermen to gain access to existing fishery resources.

References

1 Ramlan, "*Unlawful Acts of Foreign Investment in the Field of Fisheries in Indonesia* (Perbuatan Melawan Hukum Penanaman Modal Asing Bidang Usaha Perikanan di Indonesia)," *J. Yust.*, vol. 5, no. 1 (Januari-April), pp. 31–40, 2016.

2 D. Yono and I. Permadi, "*Social Justice of Indigenous Peoples in Coastal Management* (Keadilan Sosial Masyarakat Adat dalam Pengelolaan Wilayah Pesisir)," *Semin. Nas. Huk.*, vol. 2, no. 1, pp. 299–322, 2016.

3 M. Azhar, "*Prospect On Recognition Of The Traditional Fishing Rights In The Delimitation Of Marine Waters Between Indonesia And The Philippines* (Prospek Pengakuan Hak Perikanan Tradisional Dalam Penetapan Batas Perairan Wilayah Laut Antara Indonesia dan Filipina)," Yogyakarta: Postgraduate of Gadjah Mada University, 2011.

4 R. Kay and J. Alder, *Coastal Planning and Management*. London: Routledge, 1999.

5 A. Sutedi, *Law of Licensing in the Public Sector* (Hukum Perizinan dalam Sektor Publik). Jakarta: Sinar Grafika, 2011.

6 F. A. Samekto, *Law in the Development of Thought toward Post-Modernism* (Ilmu Hukum dalam Perkembangan Pemikiran Menuju Post-Modernisme). Lampung: Indep Publishing, 2012.

7 S. Halliday and P. Schmidt, *Conducting Law and Society Research: Reflections on Methods and Practices*. New York: Cambrige University Press, 2009.

8 S. Rahardjo, *Building and Renovating Indonesian Law* (Membangun dan Merombak Hukum Indonesia). Yogyakarta: Genta Publishing, 2009.

9 B. Z. Tamanaha, *A General Jurisprudence of Law and Society*. New York: Oxford University Press, 2006.

10 H. Y. Siry, "In search of appropriate approaches to coastal zone management in Indonesia," *Ocean Coast. Manag.*, vol. 54, no. 6, pp. 469–477, 2011.

11 R. Dahuri and I. M. Dutton, "Integrated Coastal and Ocean Management Enters a New Era in Indonesia," *2000*, p. 12, 2000.

12 A. Satria, Legal Assessment of the Protection of Traditional Fishermen in the Management of Marine Resources (Pengkajian Hukum Tentang Perlindungan Nelayan Tradisional dalam Pengelolaan Sumber Daya Kelautan). Jakarta: National Legal Development Board (Badan Pembinaan Hukum Nasional), 2012.

13 I. N. Putra A., A. Hakim, S. H. Pramono, and A. S. Leksono, "The effect of strategic environment change toward Indonesia maritime security: Threat and opportunity," *Int. J. Appl. Eng. Res.*, vol. 12, no. 16, pp. 6037–6044, 2017.

14 A. M. Gai, I. Soewarni, and M. M. Sir, "The concept of community poverty reduction in coastal area of Surabaya based on sustainable livelihood approach," *IOP Conf. Ser. Earth Environ. Sci.*, vol. 137, no. 12099, 2018.

15 Y. J. Utama, " Inauguration of Speech Professor of Law Sciences UNDIP: Building a State Administrative Courts of Authority (Pidato Pengukuhan Guru Besar Ilmu Hukum UNDIP: Membangun Peradilan Tata Usaha Negara Yang Berwibawa)," Diponegoro University, Semarang, 2010.

16 M. Azhar, "Relevance Good Governance Principles In State Administration Implementation System (Relevansi Asas-Asas Umum Pemerintahan Yang Baik Dalam Sistem Penyelenggaraan Administrasi Negara)," *J. Notarius*, vol. 9, no. 2, 2015.

Coastal Community Response to the Movement of Literacy: a study on literacy culture in Demak pesantren's

Putut Suharso[1,2], *Sarbini* Sarbini[3]

[1]Postgraduate Program in Cultural Studies, Sebelas Maret University, Ir. Sutami No. 36A street, Surakarta 57126 Indonesia
[2]Departement of Library Science, Diponegoro University, Prof. H. Soedarto SH Street, Semarang 50275 Indonesia
[3]Communication of Islamic Broadcasting, Faculty of Ushuludin and Da'wah IAIN Surakarta, Pandawa street, Pucangan, Kartasura, Sukoharjo 57168, Indonesia

Abstract. Coastal communities as open societies have historically known the tradition of literacy rather than inland communities. Openness has greatly influenced the teachings of Islam, which later developed into a Coastal Islam. Coastal Islam was born as a consequence of the birth of pesantren along the coast of Java. For pesantren, literacy culture is often interpreted as a routine activity of reading and writing that can not be separated from modern human lifestyle, especially in the world of education (school). In this millennial era, the culture of literacy is declining, along with the rapid use of social media as a new source of information. This study uses a cultural approach that attempts to describe the empirical realities of coastal communities, in the form of social movements, beliefs, social and cultural behaviors related to cultural literacy in pesantren of Demak district. To analyze the event by collecting data through observation, interview, and tracking documentation narrated in the form of qualitative-descriptive with descriptive-interpretative analysis. The results of research include; first Salafiyah-type pesantren view literacy culture is the same as reading al-Quran and tool books (Arabic) to understand the Qur'an and Sunnah. Both pesantrens of combination type, The culture of literacy understood the effort towards the culture of the Qur'an and Sunnah for the personal stabilization of the literate by equipping other readings for life skills. The three pesantren Ashriyah, literacy culture is understood as the refraction of reading and writing of the Qur'an and other literary readings, which are hierarchically started from the Kyai, Asatidz, senior to junior Santri, with collaborative, programmed, synergistic and continuous.

1 Introduction

Indonesia is a maritime country that has a coastline of about 81,000 km, including one of the longest beaches in the world. Indonesia also has a diversity of cultures, religions, and beliefs that live on the coast [1] known to coastal communities. Named coastal communities because they live in coastal areas and the source of economic life depends on the utilization of marine and coastal resources. The term coastal appears in the Mataram Era of the 17th century AD, which distinguishes Java in three cultural areas, Negarigung, Mancanegari, and Coastal. Negarigung is the central area of government that is Solo and Yogya. Located in Mancanegari region outside the territory of the capital but still in the royal power. While the Coast is the northern edge of Java island, which stretches from East to West Cengkal Sewu, Surabaya, Gresik, Sedayu, Tuban, Lasem, Juwana, Pati, Kudus, Jepara, Demak, Kendal, Batang, Pekalongan, Wiradesa, Tegal, Brebes until Cirebon.

Coastal communities, in general, have become part of a pluralistic society, but have a spirit of togetherness. This means that the structure of coastal communities is a combination of urban and rural community characteristics, so as to establish the system and cultural values. Coastal culture is the result of cultural acculturation of each component that forms the structure of coastal communities. His character and attitude tend to be straightforward, spontaneous, words using *Ngoko* language, his art is relatively rough in the sense of uncomplicated, his religious style tends to be puritanical Islam, and his mobility is quite high, lifestyle tends to be extravagant and likes luxury, and like to show off. In the face or solve the problem tend not to like the convoluted. The style of social life tends to be egalitarian. They are more respectful of informal leaders such as Kyai than government officials. Such tendencies are influenced by;
(1) their dwelling environment in a plain or transparent coastal area, unlike the mountainous environment; (2) the Islamic style that emphasizes the "overwhelming".

* Corresponding author: putut.suharso@gmail.com

This egalitarian stance shapes the Islamic coastline often identified as more purist (pure) Islam than the inland Islam that tends to be syncretic. The Islamic style can be seen as the reality of any social entity that describes the phenomenon, that Islam is in the hands of the people experiencing humanization in accordance with its ability to interpret Islam [2]. Interpreting the teachings of Islam greatly influenced the culture of literacy of its adherents. According to Abid al-Jabiri, Islamic tradition is a text tradition; there is no teaching in Islam unless there is a proposition contained in the text of the Qur'an, as well as al-Hadith as the main source of Islam. The emergence of various books of Islamic law from the imams of the madhhab like Imam Hanafi, Imam Maliki, Imam Syafii and Imam Hambali, born from the culture and tradition of literacy in the early Islamic period ie the companions, Tabiin and Tabiit Tabiin very usefully for Muslims until now.

Tradition and culture of literacy pioneered in the period of Tabiin to Muttaqodimin Tabiin inherited by his followers to Indonesia [3] with the emergence of the writings of Muslim poets in Indonesia such as Imam Nawawi al-Bantani which is famous for the book of al-Munir fi Maalimi al-Tanzil, Kiai Bisri Mustofa with his Tafseer, al-Ibriz li Ma'rifah Tafsir al-Qur'an al-Aziz, Kiai Saleh Darat with Majmu'at al-Shariat al-Kafiyat lil Awam and a number of Mufassir in Indonesia today. The culture of literacy among pesantren was born because of cultural, social, political and struggle of thinking problems [4]. The literacy of pesantren is not just a literary tradition that forms discourse, but also idealization, obsession, and historical characteristics.

Pesantren holds the spirit of literacy rooted in the education of pesantren which emphasizes humanization. That is, pesantren prioritizes the teaching of moral ethics [2], and runs the learning function without a narrow paradigm. Hence, unwittingly, literacy has taken root in education in pesantren, although the main literature only deals with religious matters. But in the development of cultural literacy shifted from religious texts to texts with socio-economic themes especially related to life skills. People reading the Qur'an, books like Jurumiyah, Alfiah, Safinah, Fiqih Sunnah are not viewed as a culture of literacy. Cultural literacy is interpreted when reading books that come from library books. In Kabupaten Demak the shift of literacy theme occurred when declared Demak as Literacy District on 29 September 2016 by Demak Regent, HM Natsir together with Kemendikbud, USAID Prioritas, Head of SKPD, and education stakeholders and Demak community in Pendopo of Demak Regency as follow up of Movement Literacy in the form of Literacy Movement (Gerakan Literasi Sekolah).

Demak regency is determined to make the district love to read and write and make reading as an obligation and need. The determination was manifested in the form of a Bupati Instruction Letter requiring every day at 09.00am, all schools and pesantren to stop learning time for a moment to read 10 minutes as a marker of support and commitment to creating a literal Demak district.

The directive aims to facilitate all levels of society, students and parents to jointly make "literacy" as a culture and tradition. This policy as a follow-up "Let's Review (Ayo Mengaji)" previously existed. The last year of facilitation conducted by establishing a reading park in the urban forest, a corner read every SKPD, smart house, Library and book grants to the pesantren. The books donated to pesantren are not all religious books but books with skills and life skills.

From the description above, this study aims to trace the response of Pesantren in Guntur District of Demak Regency to the culture of literacy in response to the Literacy Movement among pesantren.

2 Literature review

The advancement of a nation is not only built with abundant natural wealth, as well as the establishment of established state governance but begins with the existence of the library's role in society through the book civilization [5] or the continuous mastery of literacy from one generation to the next [6]. However, what is happening now, the culture of literacy has been increasingly abandoned by the younger generation, along with the development of knowledge and technology, especially in the digital field. Community activities, especially young people, use the Internet as a means of entertainment.

In fact, literacy-based culture education, including digital literacy, is one of the important aspects that must be applied in educational institutions to foster their deepest interests and talents [7,8]. Nevertheless, high literacy mastery must not ignore the socio-cultural aspect, because literacy is part of human culture or culture. The growth of pesantren education culture is based on the strengthening of writing culture. In some academic studies of literary, it is always illustrated that reading habits correlate significantly with writing habits/skills. From reading intellectual works, including various yellow books, acquired a lot of knowledge that can be pondered, discussed, reflected, processed and finally poured in writing.

UNESCO (The United Nations Educational, Scientific and Cultural Organization) states that literacy is more than just literacy. But more than that, literacy is the ability of the individual to use all the potential and skill possessed in his life [9], with understanding that literacy includes the ability to read the word and read the world. According to UNESCO people's understanding of the meaning of literacy is strongly influenced by academic research, institutions, national context, cultural values, as well as experience. The most common understanding of literacy is a set of real skills-especially the cognitive skills of reading and writing-irrespective of the context in which the skill was acquired and from whom it was acquired. UNESCO explains that literacy skills are the right of everyone and are the basis for lifelong learning [10]. Literacy skills can empower and improve the quality of individuals, families, and communities. Due to its multiplier effects, it can help to eradicate poverty, reduce child mortality, population growth, and ensure sustainable development, and the realization of peace. Illiteracy, however, is an obstacle to a better quality of life. Reading makes human beings full, conferences make human beings ready, and writing makes people certain [11]. One of the certainties is that he will benefit himself and others.

If we explore in the tradition of literature pesantren culture is very thick compared to general education in Indonesia. Pesantren realized that the first revelation that Allah revealed to Prophet Muhammad SAW contains the word *Iqra* (read!) and also the word Qalam (tool for disseminating in writing about what has been read)? (al-'Alaq: 1-5). Even the words 'ilm and 'aql each are not less than 450 and 49 times mentioned in the Qur'an. It also indicates the importance of science and critical thinking in human life, which assumes the task of the Caliphate on earth.

3 Research methods

This study used a qualitative descriptive approach that aims to explain how the pesantren response to the literacy movement in building a culture of literacy. Descriptive qualitative approach used to describe social phenomena occurring in the society [12,13]. Then the data obtained by observation, interview and study dusentasi. From the collected data is selected to be primary data in the form of pesantren activity as the perpetrators of literacy and secondary data from the government as policymakers, literature review and other documents relating to the movement of literacy and culture literacy.

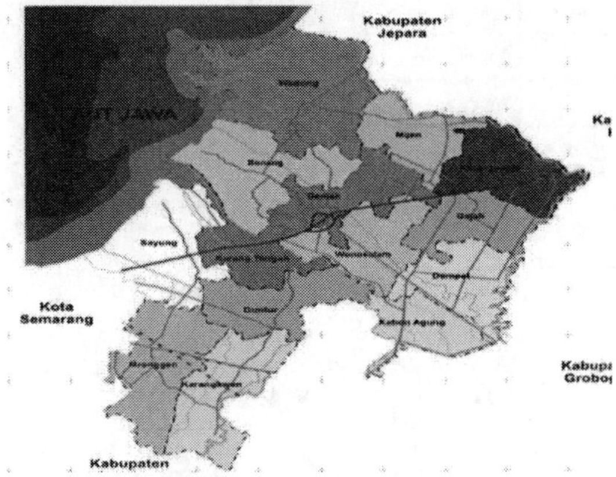

Fig. 1. Research location in Demak district

The research procedure to obtain the data was conducted in-depth interviews with some informants in pesantren in Guntur-Demak sub-district (Fig. 1.) supported by field observation results and juxtaposed with previous research results. Then the results are analyzed through three stages of data reduction, data presentation and conclusion or verification. Methods of data analysis using descriptive qualitative interpretative [14], in the form of interpretation of data through the identification of meaning, categorization, and preparation of structures to disconstruction desired. From the process then report writing.

4 Result and discussion

4.1 Characteristics of pesantren in Guntur, Demak district

Ten pesantren in Guntur has different characteristics in the implementation of educational systems as well as in educational materials (Table 1).

Table 1. Characteristics of pesantren

No	Name of Pesantrean	Address	Number of santri	Characteristics
1	PP. Al 'Arifiyyah	Wonorejo	233	Ashriyah
2	PP. Al Karimah	Temuroso	49	Ashriyah
3	PP. Assa'adah	Genteng	35	Salafiyah
4	PP. Baitut Tholibin	Puritan	60	Salafiyah
5	PP. Bustanut Tholibin	Bogorejo	29	Special
6	PP. Darun Naja	Guntur	50	Special
7	PP. Sabilul Muttaqin	Trimulyo	507	Ashriyah
8	PP. Tanwirul Wafa	Gaji	110	Salafiyah
9	PP. Zahrotul Qur'an	Bogosari	63	Special
10	PP. Putri Al Asnawiyyah	Sindon Trimulyo	63	Special

Source: Processed data, 2018

There are 3 types of pesantren; First pesantren *Salafiyah*, namely the implementation of education in pesantren using the method of Sorogan, Bandongan, and Wetonan. Sorogan is a process of learning that is individual and every student gets the opportunity to learn directly from Kyai or Kyai aide, because Santri required patience, craft, obedience and self-discipline in studying. Bringing themselves to the Sorogan level becomes a pre-requisite to follow the next learning system at the pesantren. While the system of Bandongan (Wetonan) is the learning system of a group of students listening to a teacher/ Kyai who read, translate, and explain the Islamic books in Arabic in the direction and guidance of Kyai consisting of 5 to 500 people Santri. The halaqah is almost the same as the vibe of listening to a teacher or Kyai who read, translate, explain, and review the books in Arabic and Santri each pay attention to his own book and make notes that are considered difficult or important. If kyais are unable to teach in this system, Kyai usually appoints senior Santri to represent him. This method is given to train and test the mental maturity of Santri, so that later become a strong person in religion [15].

Second, pesantren *Ashriyah* (Khalaf/modern) is the education system and educational materials and curriculum boarding school using modern management system. Kyai no longer takes care of pesantren finances but is left entirely to the pesantren treasurer. The existing curriculum gives with modern curriculum pattern and classical learning system. Here is the difference that Ashriyah pesantren use administration system, management, and governance of institution more open than salaf boarding system that is management system, and finance of pesantren always in control of Kyai authority. In the development of pesantren many changes or modifications between Salafiyah with Ashriyah, because Kyai thought and education that develops in facing the development of science and technology while maintaining the basic goals of pesantren education, in spreading and maintaining Islam. In addition, many scholars who have expertise in accordance with the necessary needs. The change of this system that makes pesantren up to now still survive in the middle of society, although there are other factors that make the survive of his pesantren, such as; (2) the tradition of pesantren is not elitist as a valuable capital for the development of humanist education pesantren, (3) the power of pesantren as a bastion of culture and religion for the younger generation, and (4) have strong bonds and intimacy with the surrounding community.

Third, *special* pesantren, namely education system, and educational materials are special. The Santri at the pesantren only read and memorized the Qur'an, or often called the Tanfidzul hut of the Qur'an. The highest level of education Diniyah this never deserted enthusiast. The concept of pesantren is designed as a place for students of the Qur'an memorizing. This is what makes pesantren have a reputation and prestige for the Santri so instilled pride. Most Pesantrens are managed with financial simplicity. The Santri mostly also come from poor families. Pesantren are reluctant to receive supervision from the government because they do not want the curriculum changed. They believe the curriculum and teaching methods that apply are the best. According to the holder of this pesantren policy, it is deliberate that the Santri not demanded many diplomas.

4.2 Culture of pesantren literacy in national literacy movement

Demak community as coastal community has its own characteristics compared to other coastal communities, open communication style, moderate Islamic character on the one hand but on the other hand tend to be Puritan. Even economically have the highest level of poverty in district around, like Jepara, Kudus, Pati and Semarang city.

4.2.1 School literacy policy

Demak as Literacy Regency has been declared since 29 September 2016. The Regent of Demak carries out the declaration. Declaration as District Literacy is intended as a first step to build a literate society Demak literate both in reading and writing. Literacy of a movement, to build, implement the culture of reading and writing as well as maintaining the consistency of its programs. Literacy also means practices and social relationships related to knowledge, language, and culture.

The issuance of Regent's Regulation or other supporting regulations to guarantee and strengthen Demak as Literacy Regency. The commitment of the

Literacy movement is the main factor to encourage and increase the reading interest of the community. Therefore, the above regulation aims to encourage people to make literacy as a culture. Then the provision of adequate reading facilities is an important factor in growing interest in reading. Reading means meant here is reading material, which includes books, magazines, newspapers, and others. However, the fulfillment of the means of reading should be balanced with the availability of librarians who are qualified in the field. This can be accomplished by recruitment of librarians as well as providing technical guidance on library management.

Stakeholder involve in inclusion of Literacy movement of regional leaders (Head and Village Head) and Head of SKPD. As is the policy of the Bupati, the Sub-district Heads to participate in supporting the literacy program by inviting all village heads to revive the village libraries and budgeting village fees to purchase books. village budgets specifically allocated to purchase books or other reading materials. The establishment of a village library and books of collections sufficient to foster a culture of literacy for the general public to obtain the information it wants.

Policies concerning schools where students get character habituation and character development. The availability of school libraries is important in supporting the literacy movement [16]. The library in the school works well in facilitating the students to a literate society (having literacy culture). To support that school must have creative and innovative programs so that children feel the need and necessary with the presence of the library. As the Regent hopes that the school can be as creative as possible it is to compose a reading culture program and provide a list of high quality reading books to students.

4.2.2 Pesantren response

Movement of literacy in the context of this research is understood as the ability to access, understand, and use something intelligently through various activities, such as reading, seeing, listening, writing, and/or speaking. Departing from this concept of literacy declared in Demak Regency in response by pesantren in very varied.

(a) Salafiyah Pesantren, understanding the literacy of schools as an activity for reading, writing and listening, using oral language, as well as written language. Understanding based on the Kyai when following the program of literacy movement in the square Demak and various activities organized by the government of Demak District, which in the literature is often called Early Literacy. (b) Modern Pesantren, provides an understanding that literacy of reading and writing activities through the use of libraries, which is oriented to the development of science and technology. Literacy is always associated with an activity in the library. When only reading the books of pesantren matter has not been named as a culture of literacy. In theory, this is called the Dewey Decimal Classification system as a classification of knowledge that makes it easy to use libraries. (c) Special pesantren provides an understanding that literacy is the ability to listen, speak, read, write, and count, and describes information based on personal understanding and conclusions. In theory, the term is called Basic Literacy.

Of the three responses are based on the tendency of Kyai and Santri submitted based on the results of interviews and observations in the field. Because of the involvement of Kyai and Santri in the Literacy Movement is very minimal even though the Kyai have received a circle from the Government of Demak Kabutan in the form of Habituation. The reading interest growth through the activities 15 minutes to read (Permendikbud No. 23 of 2015). The purpose of literacy activities at the stage of habituation, increase the love of reading outside the hours of study, improving the ability to understand reading, improve self-confidence as a good reader, cultivate the use of various sources of reading.

4.3. Culture of pesantren literacy

Technological developments and information virtualization have caused some pesantren to experience cultural shocks so that among them some forget their identity as a literary house. In the history, pesantren was the forerunner of literary culture that exists in Indonesia. It can be seen that the tradition of Islamic tradition is a text tradition. There is no doctrine in Islam unless there is a proposition contained in the text of the Qur'an, Hadith and the arguments introduced from the two main sources of Islam. The emergence of various books of Islamic law from the Imams of the madhhab such as Imam Hanafi Madzab, Imam Maliki, Imam Syafii and Imam Hambali, teaches us all that the tradition of literacy in early Islam ie the companions, Tabiin, and Tabiit Tabiin very useful for Muslims until now this.

The traditions of the Tabiin Tabiins and their followers to Indonesia can be seen in the writings of Indonesian scholars such as Imam Nawawi al-Bantani who is famous for his Miriki Labid (al-Munir fi Maalimi al-Tanzil) tafsir, Kyai Bisri Mustofa with his Tafsir , al-Ibriz li Ma'rifah Tafsir al-Qur'an al-Aziz, Kyai Saleh Darat with Majmu'at al-Shariat al-Kafiyat lil Awam and a number of Mufassir in Indonesia today. Not many know who actually Kyai Saleh Darat, Kyai Bisri Mustofa (grandfather Gus Mus) with a variety of monumental works because the tradition of literacy among pesantren has not been maximized. Not many pesantren appreciate the writings of the Ulama because it is a tradition that many develop not a tradition of reading and writing. However, a lot of interest by the Indonesian Muslim community is the tradition of listening (*Mustami*).

Literacy is born from a number of problems around it; cultural, social, political, and thought-provoking problems. Literacy is not just a discourse, but also an idealization, an obsession, and a historical setting [17]. Pesantren that have libraries in the coastal areas already exist in one institution, but there are special institutions such as coastal libraries [18]. The library has an important role in the program developing this literacy

Movement [19]. Making pesantren a literacy home that holds a deep-rooted cultural spirit. Education in pesantren based on humanization prioritizes moral ethics teaching and performs didactic functions without narrow adagium. Therefore, without realizing it, literacy is 'directed' in education in pesantren.

The pesantren community has known literature through the literature of the books of the Ulama's work; to know literature through nadham-nadham: Al-Fiyah Ibn Malik, Imrithi or Shorof Science, even recognizing literature through Barzanji's poetry or praises that were echoed after the call to prayer in congregation.

In the literature of pesantren in Indonesia, literature is not only a vehicle for character formation but also plays an important role in the movement of 'new awareness' about Indonesian entity and nation. That was effective because the education in pesantren supports the continuity of its role. Worryingly, in its development, pesantren began to lose its cultural spirit. The pesantren began to experience the emptiness of the elementary figures that guard the identity of an institution called pesantren. Generation of pesantren began to glance at the rides that are considered more promising. The impact of pesantren began to be perceived by the scarcity of the leading figures in the field of religion which also gave birth to monumental works. No more figures of the caliber of Sheikh Nawawi Al-Bantani or the level of Hadratus Shaykh KH. Hasyim Ash'ari.

5 Conclusions

Pesantren is the oldest educational institution in Indonesia. In the history of pesantren has a considerable share in developing the culture of literacy in Indonesia, especially in the era before independence. But in the post-independence development, the role in developing the literacy culture is slowing in comparison with other educational institutions. Pesantren was encouraged literacy movement to change the educational system in Indonesia. One that is expected in change is through a cultural change of literacy.

The culture of community literacy has a correlation to the quality of a nation. The habit of reading, writing someone will greatly affect the insight, mental, and behavior of a person. Habits can be nurtured and developed. Therefore, one of the quality improvements of human resources is determined by culture of literacy. Literacy is concerned with the ability to identify, define, discover, evaluate, create efficiencies and organize, use, and communicate information to address various issues. It will nurture the character of learners through the culture of the pesantren literary ecosystem to become lifelong learners. The National Literacy Movement Program is implemented in stages by considering the readiness of pesantren throughout Indonesia. This readiness includes capacity readiness, preparedness of pesantren stakeholders, and readiness of other support systems.

References

1 Woodward M 2011 *Java, Indonesia and Islam* (New York: Springer)

2 Ahmad I 2011 Democracy and Islam *Philos. Soc. Crit.* **37** 459–70

3 Pohl F 2006 Islamic Education and Civil Society: Reflections on the *Pesantren* Tradition in Contemporary Indonesia *Comp. Educ. Rev.* **50** 389–409

4 Dariah A R, Salleh M S and Shafiai H M 2016 A New Approach for Sustainable Development Goals in Islamic Perspective *Procedia - Soc. Behav. Sci.* **219** 159–66

5 Bradely F 2012 How libraries contribute to sustainable development & the SDGs *IFLA ALP Build. Better Libr. Communities* 40

6 Kress G 2003 *Literacy in the New Media Age* (New York: Routledge)

7 Marsh J and Millard E 2000 *Literacy and Popular Culture: Using Children's Culture in the Classroom* (London: Sage Publication)

8 Horning A S 2007 Defining Literacy and Illiteracy *Read. Matrix* **7** 69–84

9 UNESCO 2006 Literacy: The Core of Education for All *Educ. All Glob. Monit. Rep.* 27–35

10 K.G Saur 2001 *The Public Library Service: IFLA/UNESCO Guidelines for Development* ed C Henry (München: IFLA publications)

11 Guha M and Seale L 2015 Reading maketh a full man *J. Ment. Heal.* **24** 257–60

12 Patton M Q 2015 *Qualitative Research & Evaluation Methods: Integrating Theory and Practice* (California: SAGE Publications)

13 Connaway L S and Radford M L 2017 *Research Methods in Library and Information Science, 6th Edition* (California: Libraries Unlimited)

14 Richey R C and Klein J D 2007 *Design and Development Research: Methods, Strategies, and Issues* (New Jersey: RoutledgeFalmer)

15 Dhofier Z 1994 *Tradisi Pesantren: Studi tentang Pandangan Hidup Kyai (The Tradition of Pesantren: The Study of the Kyai Life View)* (Jakarta: LP3ES)

16 Tierney R J and Rogers T 1986 Functional Literacy in School Settings *Theory Pract.* **25** 124–7

17 Hassan A, Suhid A, Abiddin N Z, Ismail H and Hussin H 2010 The role of Islamic philosophy of education in aspiring holistic learning *Procedia - Soc. Behav. Sci.* **5** 2113–8

18 Suharso P, Sudardi B, Widodo S T and Habsari S K 2018 Library Development Strategy for The Community at Coastal Areas *IOP Conf. Ser. Earth Environ. Sci.* **116** 12002

19 Naibaho K 2007 Menciptakan Generasi Literat Melalui Perpustakaan (Creating Literate Generation Through Library) *Visi Pustaka* **9** 1–8

The Role of Communities in Strengthening Social Capital of Coastal Communities through the Library

Putut Suharso[1,2*], *Bani* Sudardi[1], *Sahid Teguh* Widodo[1], *Sri Kusumo* Habsari[1]

[1]Postgraduate Program in Cultural Studies, Sebelas Maret University, Ir. Sutami No. 36A street, Surakarta 57126 Indonesia
[2]Departement of Library Science, Diponegoro University, Prof. H. Soedarto SH street, Semarang 50275 Indonesia

Abstract. Library is a place that provides information openly and freely for the community. With the principle of openness, libraries can be accessed by all levels of society. The purpose of this study is to provide an analysis of the active role of some communities in establishing private libraries to provide information services to coastal communities. This research is a qualitative research with case study approach in coastal area of Semarang. The results of the research there are several libraries that were established by individual communities. Libraries are established using several terms; as; community reading park, reading house, smart garden. The management of libraries from self-help communities ranging from librarians, collections, buildings and infrastructure facilities. Activities undertaken by librarians have provided creative services to contribute to the literacy capabilities and social capital of coastal communities through a diverse collection of books that support social strengthening and entrepreneurship training.

Keywords: coastal communities; library services; social capital

1 Introduction

The library as a center of knowledge and learning center has a very important role in community development because it provides so much information that can be used by the community. The library provides information openly and freely to the public. With the principle of openness, libraries can be accessed by all levels of society when they intend to have skills that can be used to produce something of economic value, from simple products to sophisticated products [1]. Not only that, the library actually provides so much information and knowledge that can be used to solve the problems of daily life, from family health to information on how to broaden the community in society [2]. The library is a 'storehouse of knowledge', not a book warehouse as it has been perceived by people in general.

Indonesia's country is famous for its rich marine and coastal potential. This is in accordance with the designation of Indonesia as an archipelagic country. The potential of Indonesia's coastal resources is very wide ranging from potential biological resources, regional potential, mineral and energy resources potential, industrial potential, transportation potential and environmental services. With the islands scattered and vast oceans or large waters, then Indonesia has abundant fishery resources. The potential of fishery resources is potential to be developed. This policy orientation also involves strengthening the social capital of coastal communities. Therefore, the development of coastal and marine resources is expected to be an alternative and a source of new growth for sustainable development, especially coastal communities. One of the great potentials of Indonesia's biological resources is fisheries.

Central Bureau of Statistics noted that in March 2017 the number of poor people, the population with per capita expenditure per month under the Poverty Line) in Indonesia reached 27.77 million people (10.64 percent of the total population) [3]. From the results of a study by the Ministry of Fisheries and Marine Affairs (*Kementerian Perikanan dan Kelautan* = KKP) in 2014, the number of poor people in Indonesia is dominated by residents in coastal areas. The number now reaches 7.9 million people or 25 percent of the total number of poor people in Indonesia [4].

Libraries are growing rapidly from time to time adjusting to the development of people's lifestyles, needs, knowledge and information technology. These developments also have an impact on the "grouping" of libraries based on these patterns of life, need, knowledge, and information technology. The terms libraries "swell" become very broad but tend to have a specific specification. From district libraries, village libraries, mosque libraries, private libraries, mobile libraries, and so forth.

* Corresponding author: putut.suharso@gmail.com

2 Literature Review

Management of coastal areas of Semarang city allegedly full of interests, especially from businessmen and institutions located in the coastal region of Semarang. On the other hand, so far, many parties have seen the problem of poverty only on the dose of seemingly external phenomena covering multi dimensions of political dimension, social dimension, economic dimension, and resource dimension or other [5]. The orientation of various poverty reduction programs that focus only on one dimension of the symptoms of poverty basically reflects the partial, sectoral, charity program approach and does not touch the root cause of poverty itself. The programs run are incapable of fostering community self-sufficiency that is finally unable to cultivate sustainable aspects.

Social capital is slightly different from other popular terms, namely human capital. In human capital everything more refers to the individual dimension of power and expertise possessed by an individual. In social capital in a library more emphasis on the potential groups and patterns of relationships between individuals in a group and between groups with space attention on social networks, norms, values, and beliefs between families born from members of the group and become a group norm in society [6,7]. In addition, social capital is also used to understand various other social problems.

Social capital has an important role in the functioning and strengthening of modern life [8]. It is seen from the understanding of social capital is believed to be an important component in mobilizing togetherness, the mobility of ideas, mutual trust and mutual benefit. Andreas Varheim [9] also adds that social capital in libraries is a community's ability to work together to achieve common goals within various groups and organizations. This common goal is related to prosperity, prosperity, success and others in community life on the coast. Some of these things will be easily achieved by a society if each other has a strong belief. The ability to cooperate arises from common beliefs within a society or in the smallest parts of society. Social capital can be institutionalized into the smallest group or in a large community group such as the state.

With regards to community empowerment, social capital can build relationships of individual strengths and abilities with a natural relief system and proactive behavior toward social policy and social change [10]. Community empowerment is oriented towards good improvement interventions, as long as they also have a goal of fixing problems, providing opportunities for their members or individuals to develop knowledge and skills, and engaging professionals who collaborate or cooperate in lieu of experts. Empowerment can not be separated from the participation of outsiders of the group as supporters and facilitators in gaining the power or empowerment of individuals in the community.

3 Research Methods

The purpose of this study is to provide an analysis of the active role of the community in participating in creating and developing libraries to provide information services to coastal communities. Research sites in the coastal city of Semarang in several libraries and parks read (Fig.1).

Source: https://maps.google.com/

Fig. 1. Map of research location in Semarang city

Qualitative research methods with ethnographic research approaches from socio-cultural phenomena [11–13] coastal communities, including observation activities, interviews and ethno-historical research or documents. Source of key informant data consisting of librarian, user and community. In testing validation by using data tranggulasi. Data is processed with descriptive explanation of phenomena and theory.

4 Result and Discussion

4.1 Development of coastal libraries

Coastal region is a dynamic area that is very strategic to develop various business sectors.But unfortunately government empowerment programs by the government for coastal communities have not been worked out proportionally. So far, the supporting facilities in berkehidupan have not been fulfilled, especially related to non-formal education facilities. In this case the library becomes a tool that can be used by the coastal community in assisting the lack of access to education [14]. In the coastal area of Semarang city, now has grown various places with various names that are not using the name of the library, such as; Rumah Baca, Rumah Buku, Taman Baca, Taman Bacaan Masyarakat, Taman Pintar and others. When referring to the function given by the place it can be said as another name of the library.

The existence of various libraries that exist is dependent on the shade institution as a source of funds in developing the library, of all components. This dependence is indeed inevitable because the library holds the principle of being a nonprofit. This principle makes it impossible for libraries to be financially "reciprocated", which is mistakenly perceived as an 'unprofitable' institution and a fund-raising institution, whereas with the concept of regional autonomy, almost all institutions are required to efficiently generate financial revenues.

In terms of the existence of this library, the authors found several libraries that have been successful in running the operations of library services, without any shade institution. As Rumah Baca Seroja in Tambak Lorok, Taman Baca Asthabil is managed by Ninik Ambarwati in Genuk area, Pondok Baca Mortir in Banyumanik and others. The existence of the library has been supported by the government through the library of Semarang city by providing indirect assistance in running its activities. With the limitations that exist in the library level of the city of Semarang in providing services and assistance of the library so much, then the library service programs around the city library to be a solution. Although still less than optimal, because the schedule of visits only 1 time in two weeks, and many areas are still not formed this service.

Looking at the situation is now starting to move some social institutions that focus in the development of education through libraries such as community ASA Edu is a coastal social community, especially in Tambak Lorok region, is one of the community driving motors based on three pillars of education, creativity and health. In addition there is also Foundation Pondok Baca Rotary Indonesia is a foundation that gives attention to the development of reading interest for Indonesian children. The establishment of Pondok Baca is based on "Rotary International Program Focus Area" which one of them is engaged in "Literacy". In implementing the development and continuity of library activities in the city of Semarang, the community must be able to run independently of the library. From the managers there must be people who really moved in high social soul, because without salary must work social in society. So the role of community leaders and village government should be able to provide solutions.

Governmental policies in the field of socio-economic (education, health, economy, infrastructure and institutional) in the development of coastal communities that are "left behind" need to be re-engineered, especially those related to the provision of educational facilities in the form of libraries, because the development of coastal communities living as fishermen in the region the coast is very slow, most of whom are the poor.

4.2 Library as a place of learning

With a wealth of knowledge resources it has, the library is actually one of the main supporting factors for the government when it will implement the development program, both physically and mentally because through the library. Information on development policies can be disseminated to the community in a more effective way and reaching different layers of society, whether social, educational, age, ethnic, or economic. This can happen because in the library there is no distinction to visitors. Everyone is served according to his needs, not his status.

The existence of the library becomes part of civilization and human culture to lead to change in life. One of the goals of the independent Indonesian state to educate the nation's life. One of the learning tools that can be used in the intellectual life of the nation is the library. The role of libraries also bring a positive impact to the community, so it can form an information society. It can be seen how information the way people treat information, respect for information, how people seek information, how people need information. Especially now people should be more careful about the information that is hoax/fake that contains the provocation of news in the community, the development of social media through smartphones and news coverage on television is demanded by the community can select properly and utilize precisely also, now known as the term information literacy.

Fig. 2. Rumah Baca Apung in Tambak Lorok

Coastal communities in the city of Semarang have tried to provide library services innovatively, not only provide services in the form of book lending in the library. Those in the library can increase their knowledge of skills, and the talk of children is needed as a supporter of non-formal learning facilities after returning from school [15]. The existing libraries, in general, provide a space for open space that can be used for group discussions. As in the floating library of houses in Tambak Lorok (Fig.2) has floating buildings established by Ministry of Public Works and People's Housing which consists of 2 floors, first floor for meeting hall and discussion, second floor for book storage space, magazines and others as a library collection.

Other libraries such as Sri Redjeki Hartono Book House in Gayamsari area, which is originally a private library owned by a professor of law at the University of Diponegoro provide access to learning for the community in the surrounding environment. The spirit of library development has begun to be felt by the community. Moved from the lack of learning facilities in the community, so the emergence of a library that without the command can move naturally with the same vision of providing information services that are easy and free in the community.

4.3. Diversity of library activity

The result of the research is that there are several libraries established by individuals and community groups. Libraries are established using several terms; as; community reading park, reading house, smart garden. Management of libraries from non-governmental organizations from librarians, collections, and buildings and their means of prasananya. Activities undertaken by librarians have provided creative services to contribute to literacy skills.

The diversity of library services between one place and another there is no striking difference. A high level of creativity through an activity agenda beyond routine activities should be done. The demand for information from the community, coupled with open access to information in the community, becomes a challenge for library managers in providing services that can be sought by the community. Activities are conducted (Fig.3) in the afternoon until the afternoon after the children come home from school and for adults after they return from their regular work activities.

Fig. 3. Learning activities at Rumah Baca Seroja

Common examples of library activities include: reading together, watching inspirational films, presenting inspiring, storytelling, bulletin community activities, learning to write from authors, retelling books, theater, music, learning from the environment, surgery books, other and so on. Activities conducted by libraries are activities with the principles of the community and for the community.

4.4. Library as social capital

The efforts undertaken by libraries in creating social capital in a coastal society is to build public trust, instill norms and values, and form a social network. The effort is manifested in the form of work programs and routine activities built by libraries, volunteers and other social institutions. In building the trust of the library community organizes activities where the community is involved in the effort to develop the capacity of the community. Starting from children's activities through art training, dancing (Fig.4), storytelling, introduction of environmental management and so forth. For the community where the library becomes a place to unite them through forums of socialization of several programs of activity by village government and community empowerment. The role of libraries as social capital of coastal communities through a diverse collection of books that support social empowerment and entrepreneurship training.

Fig. 4. Exercise dancing in by volunteers in the library

The ASA edu community through its volunteers provides guidance on school subjects, free of charge to school-age children in coastal areas. The accompaniment of Laskar Sedekah is also carried out for the less fortunate and orphans and / or orphans, and some other activities that integrate with existing programs in the library. For now the presence of institutions that can provide full support is expected by this literacy pengiat. When looking at the extent of the existence of the library with the diversity of its activities, the government should pay more attention to provide adequate funding, private sector should also have to enter through the CSR (Corporate Social Responsibility) program for education for these coastal communities.

5 Conclusion

The existence of libraries in the coastal area is now beginning to grow. Slowly but surely, the role of library akn seen by the community will be the importance of library functions. The community has an important role in managing the library independently so that it can survive and run the role of the library as a source of public information. The services provided by the librarian have started creatively to be able to turn this learning tool into social capital.

The authors would like to thank the literacy activists in the city of Semarang who have volunteered to develop libraries in coastal areas, especially for *mas* Udin in Tambak Lorok which has become a pioneer in the community mobilizer. I would also like to thank Prof. Sri Redjeki Hartono a professor of law at Diponegoro University who has made a private library into a library that can be accessed by the general public.

References

1. Durcan T 2011 The future of and for library and information services: a public library view *Libraries and Society: Role, responsibility and future in an age of change* ed D Baker and W Evans (Cambridge: Chandos Publishing)
2. Jochumsen H 2012 The four spaces – a new model for the public library *New Libr. World* **113** 586–97
3. Central Bureau of Statistics 2017 *Persentase Penduduk Miskin September 2017 Mencapai 10,12 persen* (Jakarta: Central Bureau of Statistics)
4. Statistics C B of 2016 Marine and Fisheries Information (Informasi Kelautan dan Perikanan) **Januari**
5. Indarti I 2015 Coastal Resources Development Based Community Development Cooperation in the Coastal Fishermen Semarang *Fokus Ekon.* **10** 44–61

6. Hillenbrand C 2005 Public Libraries As Developers of Social Capital *Australas. Public Libr. Inf. Serv.* **18** 4–12
7. Vårheim A 2014 Trust in Libraries and Trust in Most People; Social Capital Creation in the Public Library *Libr. Q. Information, Community, Policy* **84** 258–77
8. Vårheim A 2008 Theoretical approaches on public libraries as places creating social capital *World Library and Information Congress: 74th IFLA General Conference and Council* (Québec, Canada) pp 1–13
9. Vårheim A, Steinmo S and Ide E 2008 Do libraries matter? Public libraries and the creation of social capital *J. Doc.* **64** 877–92
10. Poeloengasih C D, Bardant T B, Rosyida V T, Maryana R and Wahono S K 2014 Coastal community empowerment in processing Kappaphycus alvarezii: a case study in Ceningan Island, Bali, Indonesia *J. Appl. Phycol.* **26** 1539–46
11. Saukko P 2003 *Doing Research in Cultural Studies: An introduction to classical and new methodological approaches* (London: SAGE Publications)
12. Lincoln Y S 2002 Insights into library services and users from qualitative research *Libr. Inf. Sci. Res.* **24** 3–16
13. Patton M Q 2015 *Qualitative Research & Evaluation Methods: Integrating Theory and Practice* (California: SAGE Publications)
14. Suharso P, Sudardi B, Widodo S T and Habsari S K 2018 Library Development Strategy for The Community at Coastal Areas *IOP Conf. Ser. Earth Environ. Sci.* **116** 12002
15. Nurislaminingsih R 2017 Attitudes of fishermen in tambak lorok semarang towards library *Adv. Sci. Lett.* **23** 10042–4

Coenzyme q10 is potentially usedto improve lipid profile in diabetic hypercholesterolemia-induced periodontitis in the coastal area

Saryono Saryono[1,*], *Hesti* Devinta[2], *AbdulHarisBudi* Widodo[2], and *ArifImam* Hidayat[1]

[1]School of Nursing, Faculty of Health Sciences, University of JenderalSoedirman, Indonesia
[2]Dentistry Department, Faculty of Medicine, University of JenderalSoedirman, Indonesia

Abstract. Contaminated water often affects the occurrence of periodontitis in the coastal area. The diabetic hypercholesterolemia-induced periodontitis in Indonesia is very high. The use of coenzyme Q10 to treat this disease has never been investigated yet. Therefore, this study aimedto analyze the effect of coenzyme Q10 on the lipid profile of diabetic hypercholesterolemia-induced periodontitis.Twenty four rats were randomized into 6 groups (G1-G6). The groups (G1-G3) are healthy, negative and positive control group respectively. The treatment groups (G4-G6) are diabetic hypercholesterolemia-induced rats given coenzyme q10 dose of 13,5; 27 and 54 mg/kg respectively. Rats were induced by periodontitis, hypercholesterolemia and diabetes mellitus. Coenzyme Q10 was administered orally using 2 mL gastric tube once a day for 14 days. Lipid profile including triglycerides, HDL, and atherogenic index (IA) was measured enzymatically by the CHOD-PAP method. Data were analyzed by one-way ANOVA test and followed by the Least Significant Difference (LSD) post hoc test.Coenzyme Q10 with a dose of 54 mg/kgis effective in lowering triglyceride, and atherogenic index and increasing HDL level in diabetic hyperlipidemia-induced periodontitis rats model. This research supports the potential effects of coenzyme Q10 supplementation to improve lipid profile in diabetic hypercholesterolemia-induced periodontitis in the coastal area.

Keywords: coenzyme Q10, lipid profile, hypercholesterolemia, diabetes mellitus

1 Introduction

Water is the source of life. The water content in the body reaches more than 50% in adults. The more densely populated in a region, the water quality gets worse. Contaminated water will have an effect on tooth health[1,2]. Drinking water from surfaces area are generally salty in the coastal areas, so people prefer to dig the wells that contain lots of calcium and phosphate. High levels of calcium, phosphate and other minerals in drinking water in coastal areas play a role in increasing the high prevalence of periodontitis[3].

This water will make the teeth become more easily perforated and damage the periodontal tissue causing periodontitis as well. Periodontal disease in Indonesia is the second number after caries [4]. The prevalence of periodontal disease in Indonesia was 96.58%. While the prevalence of periodontal disease in coastal communities amounted to 10.43% of 2157 elementary school students in Jepara in 2011[5]. Calcium and phosphate

*Corresponding author: sarbiokim@gmail.com

mineral intake as a material of tartar (the main cause of periodontal disease) is very high, especially in coastal communities. Periodontitis disease will be more severe when the patient has diabetes. Prevalence of periodontitis in DM patients at Internal Medicine Polyclinic, Dr. Sardjito General Hospital is 88.24% [6].

Diabetes mellitus is a metabolic disorder with the highest prevalence in Indonesia.It often leads to complications of hypercholesterolemia. Excess cholesterol in the body will increase in the atherogenic index[7]. Lipids are the most potent compounds to be oxidized. Lipid peroxidation products form intermolecular bonds with the amino-terminal group of low-density lipoprotein (LDL) apolipoprotein to form oxidized LDL. This will increase the risk of atherosclerosis formation which is coronary heart disease (CHD) precursor. High oxidative stress will affect in low antioxidant activity levels[8].

Antioxidants couldinfluence themetabolism of cholesterol. Antioxidants can prevent fatty acid oxidation process so that atherosclerosis is not formed. Endogenous antioxidants such as glutathione peroxidase, catalase, and superoxide dismutase under certain conditions are depleted so that the body requires exogenous antioxidants[9]. Antioxidant likes Vitamin E, Vitamin C, beta-carotene, coenzyme q10, quercetin and other bioflavonoids are few examples of exogenous antioxidants. The use of antioxidants, especially coenzyme Q10 to treatdiabetic hypercholesterolemia-induced periodontitishas not been elucidated.Some fish contain more coenzymes Q10 than vegetables.

Coenzyme Q10 has been known to have a high antioxidant capacity and clinical effect[10][11][12]. Besides, coenzyme Q10 is an active redox lipid that serves as a cofactor in the electron transfer chain in the mitochondria and plasma membranes[13]. Coenzyme Q10 300 mg/day is effective antioxidant and anti-inflammatory agent that can reduce TNF-α and IL-6 levels in CHD patients [14]. Previous research has shown that coenzyme Q10 and vitamin E are interrelated, but other studies have shown that is not correlated[15].This study aimed to analyze the effect of coenzyme Q10 on the lipid profile ofdiabetic hypercholesterolemia-induced periodontitis rats.

2 Method

2.1 Research subject

Thisis an experimental research with pretest-posttest with control group design. Rattusnovergicuswas a male rat, Sprague Dawley strain, age 2-3 months, body weight 150-200 grams.Twenty four rats were randomly grouped into 6 groups. A healthy control group (G1) of the rat only fed corn oil. Group 2 (G2) serving as the negative control group washypercholesterolemia and diabetic-induced periodontitis rats, then given corn oil. Group 3 (G3) as the positive control was hypercholesterolemia and diabetic-induced periodontitis rats and then given Vitamin E at the dose of 27 mg/kg. The treatment groups (G4-G6) were hypercholesterolemiaand diabetic-induced periodontitis rat, then given coenzyme Q10 at the dose of 13,5; 27; and 54 mg/kg respectively.Lipid profile examination was conducted before and after the treatment.

2.2 Laboratory experiments

Rats were adapted for one week and placed in separate cages according to the group. Rats were housed in polyethylene cages (5 rats/cage) with stainless steel wire tops and were allowed for commercial standard diet and water ad-libitum. Rats were housed under standard laboratory conditions. Standard care methods were used to keep the animals healthy and free of infections. The principles of the animal care trial have followed the code of conduct of animal research trial. The research was carried out after obtaining an ethical clearance from the Ethics Commission of Faculty of Medicine UNS, Surakarta.

Periodontitis was induced by inoculation of A. actinomycetescomitans bacterial species at a dose of 1 Mcfarland (equivalent to 108 density) of 200 µL or 0.2 mL daily for 7 days in the mandibular incisor.Hypercholesterolemia was induced with a high-fat diet (yellow duck egg 4mL/200g) for 7 days. Rats were induced to be diabeticby streptozotocin (STZ) injection with a single dose of 60 mg/kg. Coenzyme Q10 in the form of a 100 mg blue tablet preparation produced by Nutrimaxwas diluted with corn oil. The coenzyme Q10 at a dose of 13.5; 27 and 54 mg/kg was administered orally using 2 mL gastric tube once a day for 14 days. Lipid profile including triglycerides and HDL were measured enzymatically by using the CHOD-PAP method, while atherogenic index (IA) was calculated using log 10 (TG/HDLc).

2.3 Statistical analysis

The data were expressed as the mean ± standard deviation (SD) and presented in frequency distribution table. The mean difference between groups was examined using one-way analysis of variance (ANOVA) followed by the post Hoc Least Significant Difference (LSD) test for multiple comparisons. The test is considered significant if $p < 0.05$.

3 Results

On the 8th day after induction of periodontitis, one of each group was anesthetized with chloroform and decapitated, then radiographic examination was done in a periapical photograph to ensure bone resorption.The induction of streptozotocin and HFD significantly increased blood glucoseand cholesterol levels compared with healthy controls ($p < 0.05$). Mean triglyceride and atherogenic levels decreased and HDL increased significantly after the treatment period ($p < 0.05$). The results of ANOVA test on triglyceride, HDL and atherogenic index before treatment showed significant differences between group G1 and G2-G6 group ($p < 0.05$) but did not differ significantly between groups G2-G6 ($p > 0.05$). This suggests that cholesterol induction and diabetes mellitus were successfully performed in the G2-G6 group (Figure 1).

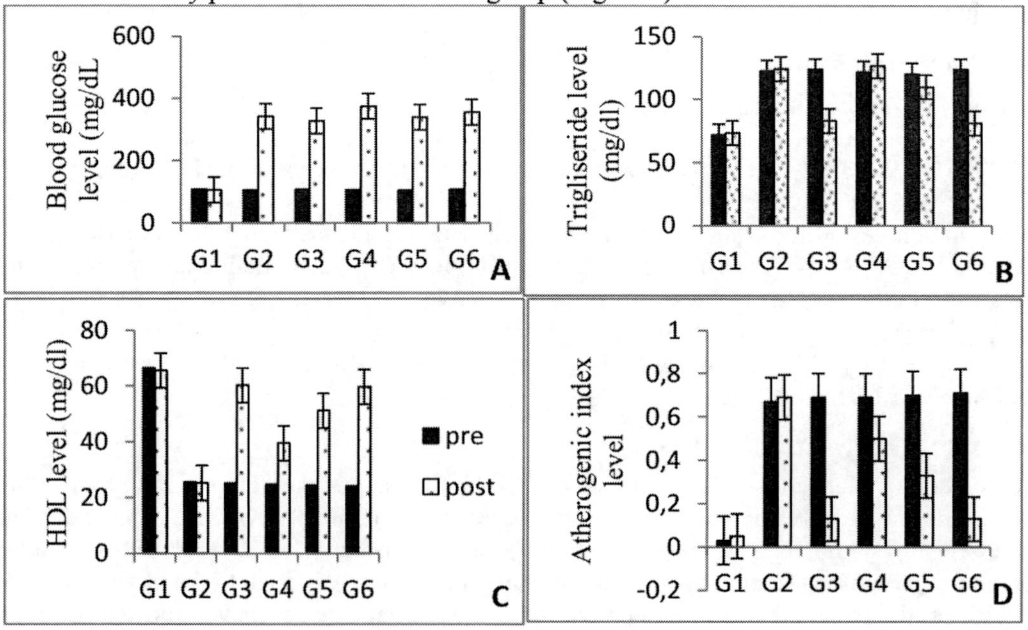

Fig. 1.Blood glucose, HDL, Triglyceride levels andAtherogenic index before and after treatment.

*Corresponding author: sarbiokim@gmail.com

Levels of triglycerides before treatment showed no significant differences between treatment groups G2-G6, but significantly different from healthy control group (G1). After the treatment period, ANOVA test results showed significant differences in triglyceride levels between groups ($p < 0.05$). The G3 and G6 treatment groups had the lowest triglyceride levels (ie 82.94 mg/dl and 81.15 mg/dl) and no significant difference between G3 and G6. This suggests that the administration of coenzyme q10 dose of 54 mg/kg is equal to vitamin E dose of 27 mg/kg. The post hoc LSD test showed differences in all groups ($p<0.05$), except G3, G4, and G6 (Table 1). The groups of G2 with G4 and G3 with G6 are not significantly different.

Table 1. The result of LSD test on triglyceride level after treatment

Groups	G1	G2	G3	G4	G5	G6
G1	-	0.000*	0.000*	0.000*	0.000*	0.003*
G2	0.000*	-	0.000*	0.265	0.000*	0.000*
G3	0.000*	0.000*	-	0.000*	0.000*	0.436
G4	0.000*	0.265	0.000*	-	0.000*	0.000*
G5	0.000*	0.000*	0.000*	0.000*	-	0.000*
G6	0.003*	0.000*	0.436	0.000*	0.000*	-

*= significantly difference

HDL levels before treatment showed no significant differences between groups G2-G6, but significantly different from healthy control group (G1). After the treatment period, there was a significant difference of HDL levels between groups ($p < 0.05$). The mean HDL levels were highest in the G1 group (65.52 mg/dl), but in the treatment group, the highest HDL levels were found G3 (60.15 mg/dl) and G6 (59.57 mg/dl). There was no significant difference in HDL levels between the positive control group (G3) and G6 group given coenzyme Q10 dose 54 mg/kg. Based on the LSD post hoc test, all groups differed significantly ($p<0.05$), except G3 with G6 (Table 2).

Table 2. The result of LSD test on HDL level after treatment

Groups	G1	G2	G3	G4	G5	G6
G1	-	0.000*	0.000*	0.000*	0.000*	0.000*
G2	0.000*	-	0.000*	0.000*	0.000*	0.000*
G3	0.000*	0.000*	-	0.000*	0.000*	0.634
G4	0.000*	0.000*	0.000*	-	0.000*	0.000*
G5	0.000*	0.000*	0.000*	0.000*	-	0.000*
G6	0.000*	0.000*	0.634	0.000*	0.000*	-

*= significantly difference

Atherogenic index before treatment showed no significant difference between treatment groups (G2-G6), but significantly different from healthy control group (G1). After the treatment period, the mean atherogenic index was decreased. The lowest rates of the atherogenic index were found in the G3 (0.138) and G6 (0.135) groups, but there was no significant difference between the groups. This shows that coenzyme Q10 dose of 54 mg/kg is as effective as vitamin E dose of 27 mg/kg. From the post hoc test, it was found that all groups had significant differences in the atherogenic index, except for the G3 and G6 groups (Table 3).

 https://doi.org/10.1051/e3sconf/20184707006

Table 3. The result of LSD test on atherogenic index after treatment

Groups	G1	G2	G3	G4	G5	G6
G1	-	0.000*	0.000*	0.000*	0.000*	0.000*
G2	0.000*	-	0.000*	0.000*	0.000*	0.000*
G3	0.000*	0.000*	-	0.000*	0.000*	0.862
G4	0.000*	0.000*	0.000*	-	0.000*	0.000*
G5	0.000*	0.000*	0.000*	0.000*	-	0.000*
G6	0.000*	0.000*	0.862	0.000*	0.000*	-

*= significantly difference

4 Discussion

Coastal areas are generally synonymous with poor healthy lifestyle patterns, low levels of education and awareness of poor health. The unavailability of qualified drinking water sources has the potential to increase the incidence of periodontitis[3]. Characteristics of infertile soil also cause at least fruit plants so that the consumption of food sources of antioxidants is lower. This is the first study to overcome the problem of periodontitis in coastal areas.

After induction of bacteria in the tooth, rats undergo periodontitis, then STZ-induced to occur diabetes. The mean of blood sugar level before induction was 106,21 mg/dl, but after induction, it increased to 347,75 mg/dl. This suggests that all of the G2-G6 groups had reached the condition of diabetes. The success of hypercholesterolemia and diabetes induction can be seen from the difference of all healthy control group (G1) indicator with the negative control group (G2) significantly ($p < 0.05$). STZ injection works by destroying pancreatic β cells, resulting in DNA changes. STZ inhibits the Krebs cycle and decreases oxygen consumption in the mitochondria, resulting in cell death. STZ also induces the formation of radical oxygen species (ROS) causing oxidative stress. This further results in decreasingof insulin synthesis and leads to diabetes.

Coenzyme Q10 is an important cofactor for the electron transport chain within the mitochondria and also acts as an antioxidant. Coenzyme Q10 works through the mechanism of repair of mitochondrial function in pancreatic cells thus increasing the production of insulin again. In addition, the coenzyme Q10 also works to inhibit lipid and protein peroxidase. Coenzyme Q10 plays a primary role in preventing the formation of lipid peroxidase (LOO) radicals[16], thereby reduce total cholesterol and LDL. Coenzyme Q10 works through microRNA (miRNA) that increases cholesterol efflux and inhibit atherosclerosis[17] so the HDL level increases.

The inflammatory response caused by STZ exposure and a high-fat diet in mice can also be reduced by coenzyme Q10. Free radicals caused by STZ oxidizes the pancreatic beta cells resulting in inflammation. Co-administration of Q10 may work as an anti-inflammatory [10], as evidenced by decreased in TNF-α, CRP and IL-6 in male runners[18]. High LDL after STZ injection can be resulted from an inflammatory reaction due to LDL oxidation. In lipid profile, the antioxidant ability of coenzyme Q10 (at G6) has an equal potency as vitamin E. In another study, coenzyme Q10 was significantly positively correlated with another antioxidant capacity (vitamin E and glutathione peroxidase activity)[19]. Vitamin E is a fat-soluble vitamin that is very effective in protecting cell membranes from free radical oxidation. Vitamin E antioxidants including secondary antioxidants work by capturing and preventing chain reactions, especially on lipid peroxidation by capturing peroxyl radicals.

Administration of coenzyme Q10 with 3 dose variations showed the greatest effect at the highest dose of 54mg/kg (group G6). This suggests that the most effective dose in

*Corresponding author: sarbiokim@gmail.com

improving the lipid profile is a dose of 54 mg/kg. However, the dose of coenzyme Q10 was comparable to that of the positive control group given vitamin E dose 27 mg/kg. Vitamin E and coenzyme q10 work to inhibit lipid peroxidation. Coenzyme Q10 and tocopherol are lipid phase micronutrients that prevent oxidative damage and distributed commonly in peripheral blood plasma[20]. Low levels of coenzyme Q10 in the body will increase mortality due to organ dysfunction[21], and administration of coenzyme Q10 may reduce symptoms of organ dysfunction such as heart failure[22].

Coenzyme Q10 oxidized with reduced ratio was positively correlated with malondialdehyde (MDA) and glutathione peroxidase activity (GPX) ratio[23]. This suggests that the role of coenzyme Q10 as an antioxidant to protect lipids from the free radical oxidation is very important. Coenzyme Q10 can also increase the capacity of antioxidant and cellular energy metabolism in the skin[24]. At a dose of 300 mg/d, coenzyme Q10 supplementation significantly increased antioxidant capacity and reduced oxidative stress and inflammatory response in patients with hepatocellular carcinoma after surgery[25].

Coenzyme Q10 works primarily as an electron carrier during cellular respiration and as an intracellular antioxidant. Reduced coenzyme Q10 is a free radical scavenger that protects LDL against oxidative stress. The reduced coenzyme Q10 works to reduce the lipid peroxyl and tocopheryl radicals and minimize oxidative stress through the dehydrogenation reaction in the oxidized coenzyme Q10. Thus, coenzyme Q10 works to strengthen the performance of vitamin E. The combination of coenzyme Q10 with atorvastatin in hypercholesterolemic rats can improve the biochemical parameters and respiratory function of mitochondrial hepatic rats[14][26][27].

These findings have clinical implications to applied in coastal areas. Coenzyme Q10 can be given to coastal communities to reduce the occurrence of periodontitis. There are many food sources of coenzyme q10 available in addition to vegetables and fruits in coastal areas. Sardines, mackerel, salmon, and tuna fish contains a lot of coenzyme Q10. But still, need further research with larger samples and effective dose in humans.

There are several limitations in this study. The response of adaptation and immunity of rats varies and inflammatory indicators were not measured. Glucose levels after treatment were not measured so it could not explain the effect of coenzyme Q10 on rat blood glucose.

5 Conclusion

The 54 mg/kg dose of coenzyme Q10 is effective in lowering triglyceride and atherogenic index and raising HDL levels in blood model of diabetic hyperlipidemia-induced periodontitis. Antioxidant ability of coenzyme Q10 dose of 54 mg/kg is equal to vitamin E dose 27 mg/kg. Coenzyme Q10 can replace the requirement of vitamin E in the body to increase the body's antioxidant capacity in the coastal area.

Acknowledgment

The researcher would like to extend our deepest gratitude to Dr. Hernayanti for her laboratory assistance.

References

1. S. Kumar, S. Lata, J. Yadav, J.P. Yadav, Appl.. Water. Sci. **7**, 3, 377–84(2017)
2. P. Bhattacharya, A.C. Samal,S. Banerjee,J. Pyne, S.C. Santra, India Environ. Sci. Pollut. Res. **24**, 20300–14 (2017)
3. P. Pandey, A. Sharma, Int. J. Adv. Res. Sci. Eng. **7** 503–9 (2018)
4. Agency for Health Research and Development, Ministry of Health Republic of Indonesia,*Basic Health Research* (Riskesdas) (2013)

5. A. Fitriyanti,A.Susilowati, U.N.A. Darjono, Odonto Dent. J. **1** 6–10 (2014)
6. R. Sari,D. Herawati, R. Nurcahyanti, P.K. Wardani, Maj. Kedokt. Gigi Indones. 3 98–104 (2017)
7. S. Saryono, J. Eliyan,D. Herdiati,A. Khikmatullah, C. Silvana, H. Adi, IOP Conf. Ser. Mater. Sci. Eng. **172** (2017)
8. B. Lee,Y. Huang, S. Chen, P. Lin, Nutrition **28** 250–5 (2012)
9. C. Yen,N. Yang, B. Lee,J. Lin,S. Hsia, P. Lin, Sci. J. **2013** 1–7 (2013)
10. R. Pala,C. Orhan,M. Tuzcu,N. Sahin,S. Ali, V. Cinar,K. Sahin, J. Sport. Sci. Med. **15** 196–203 (2016)
11. L. Peverelli,A. Naini,M. Hirano, M. Catarina, Ann. Clin. Pathol.**2** 1–4 (2014)
12. A. Sharma,G.C. Fonarow,J. Butler, J.A. Ezekowitz, G.M. Felker, Circ. Hear. Fail. **9** 1–9 (2016)
13. M. Potgieter,E. Pretorius,C.F. Van der Merwe, M. Beukes, W.A. Viera, R.E.G. Auer, M. Auer, S. Meyer, Micron. **42** 275–82 (2011)
14. B. Lee, Y. Tseng, C. Yen, P. Lin, Nutr. J. **12** 1–9 (2013)
15. N. Naidoo, R. M. van Dam, W. Koh, C. Chen, Y. Lee, J. Yuan, C. Ong, Chinese J. Nutr.**142** 1046–52 (2012)
16. T. Grenier-larouche, A. Galinier, L. Casteilla, A.C. Carpentier, A. Tchernof, J. Lipid Res. **56** 1985–92 (2015)
17. R.M. Allen, K.C. Vickers, Thromb. Vasc. Biol. **34** 1795–7 (2015)
18. M. Armanfar, A. Jafari, G.R. Dehghan, L. Abdizadeh, Med. J. Islam. Repub. Iran **29**(2015)
19. H. Liu,Y. Huang, S. Cheng,Y. Huang,P. Lin, Nutr. J. **15** 1–9 (2016)
20. A. Franke,J.F. Lai, C.M. Morrison, I. Pagano,X. Li, B.M. Halm, R. Soon, L.J. Custer, Free. Radic. Res.**47** 757–68 (2015)
21. A. Coppadoro,L. Berra,A. Kumar, R. Pinciroli,M. Yamada, U.H. Schmidt, E.A. Bittner, M. Kaneki, J. Crit. Care.**28** 571–6 (2013)
22. A. D. Fotino, A.M. Thompson-Paul, L.A. Bazzano, Am. J. Clin. Nutr.**97** 268–75 (2013)
23. Y. Kaya,A. Çebİ, N. Söylemez, H. Demİr, H.H. Alp, E. Bakan, Int. J. Med. Sci. **9** (2012)
24. A. Knott,C. Smuda,G. Sperling,A. Vogelsang, H. Schwengler, S. Kristof, T. Eisenberg, T. Blatt, BioFactors,**41** 383–90 (2015)
25. H. Liu, S. Cheng,Y. Huang, Y. Huang, Nutrients,**9** 1–9 (2017)
26. M.A. Jiménez-santos, I.E. Juárez-rojop, C.A. Tovilla-zárate, M.T. Espinosa-garcía, M.A. Juárez-oropeza,T. Ramón-frías, D.Y. Bermúdez-ocaña,J.C. Díaz-zagoya, Lipids Health Dis. **13** 1–10 (2014)
27. A. Skarlovnik, M. Janić, M. Lunder, M. Turk, M. Šabovič,Med. Sci. Monit.**20** 2183–8 (2014)

*Corresponding author: sarbiokim@gmail.com

A Development Policy of Networking-Based Creative Marine Small and Medium Enterprises as a Solution for Poverty Alleviation in Indonesia

Slamet Rosyadi[1], *Elpeni* Fitrah[1], *Ayusia* Sabhita Kusuma[1]
[1]Faculty of Social and Political Sciences, University of Jenderal Soedirman, Indonesia

Abstract. Poverty and environmental degradation are issues that have not been resolved in the coastal communities and regions yet. The various policies and development programs have been implemented to address these two main issues. Nevertheless, the poverty rate of coastal communities remains high. To address the problem of poverty in coastal areas, this study tries to offer a policy solution for SME development based on the creative power of network resources. The concept of creative economy is very relevant because the environment and marine fisheries resources provide potential materials for SMEs to manage. Management of marine resources by combining creativity, innovation, and technology can lead to products that have market competitiveness. To support these efforts, the utilization of network resources can provide synergistic power to improve product quality and competitiveness. These networking resources can come from elements of government, business, civil society, financial institutions, professional institutions, universities, and other social communities. The collaboration of these various network resources can be an effective force in poverty alleviation efforts in coastal areas. Studies have supported that poverty alleviation is not effectively implemented if it does not synergize the bureaucratic, business and community forces. The problem of poverty in coastal areas of a complex character cannot be overcome with a linear approach. Moreover, breakthroughs and innovations need to be done to address the complexity of poverty of coastal communities. To that end, the strength of marine SMEs needs to be integrated with the approach of creative economy and networking so that it becomes a driver of the local economy that serves as a solution for poverty alleviation.

1 Introduction

Common characteristics found in coastal areas in developing countries are the low level of community welfare and the damaged natural resources environment [1]. Statistical data in Indonesia shows that 32.4 percent of coastal communities are categorized as poor. Even the number of poor people in the fishing community is twice as many as other communities.

[1]Corresponding author: slamet.rosyadi@unsoed.ac.id

This situation is very paradoxical to see the fishermen's environment that provides a great economic potential for the welfare of fishermen communities.

Some of the causes behind severe poverty rates have been described by many researches. Those causes are such as the lack of access to natural resources, the lack of community participation in policy and program formulation, the lack of human resources, the lack of infrastructure and technology, and the lack of capital [1,2,3]. Generally, the underlying causes of poverty found in coastal communities are low education, highly consumptive lifestyles, dependence on coastal resources, minimal business skills, and limited market access [3].

Various development programs and community empowerment, especially from the government budget has been widely allocated to coastal areas. However, the poverty situation in coastal areas has not changed much. Ironically, it is the government's policy which also increases the reproduction of poverty of coastal communities especially the traditional fishermen. In this context, it is necessary for the strength of social networks to be involved in poverty reduction efforts. This paper is intended to provide thoughts on poverty reduction policy strategies with the strength of social networks that juxtaposed with SMEs. To align with the dynamics of the ever-evolving market changes, the creative economy approach needs to be adopted in the development of SMEs. Various researchers have found that the industrial strength of the creative economy has brought significant changes to the local economy [4]. In the context of coastal areas, various potentials of sea and coastal resources should be managed by combining innovation, technology, social capital in the building of strong industrial posture. Thus, efforts to integrate the power of creative economy and social networking become important to be developed for SMEs in coastal areas have the competitiveness to be a solution to poverty alleviation.

Figure 1 Research Method

To address the problem of poverty in the fishing community, this paper applied the literature study method. Various publications in the form of books, research journals, news and media analysis and statistical reports relevant to the issues are reviewed. Information from the results of the study then aimed to find answers to potential SMEs of creative economy based on network resources as a solution to poverty alleviation.

Figure 2 Results and Discussion

3.1 Problem of Policy Institution

Following institutional theory, this paper argues that policy structures (institutions) for poverty alleviation in coastal community create interaction among coastal communities and government or other stakeholders in making decision and forming judgements about 'good' policy [5].

The issue of poverty alleviation has become a major part of the government's responsibility. Central and local governments have managed various poverty reduction programs with an integrated approach. Efforts to reduce poverty in government perspectives are translated into social assistance programs, community empowerment programs, and empowerment of small businesses. To ensure the effectiveness of this poverty alleviation program, Presidential Regulation no. 15/2010 on the acceleration of poverty alleviation and institutional of National Team of Acceleration of Poverty Alleviation (TNPP2K). With this institution is expected to create coordination across sectors and cross-stakeholders at the center and regional levels in an effort to accelerate poverty reduction (source: http://www.tnp2k.go.id).

With many stakeholders involved in poverty alleviation programs, cross-sector and agency coordination is a formidable challenge. The failure of various policy implementations derived from weak coordination. The more actors and institutions involved, the more difficult it is to bring together diverse interests. As a result, poverty reduction programs aimed at fishing communities are also ineffective in improving their welfare.

Statistics show that the number of poor people with spending size decreased by 1.19 million people or 10.12 percent. Thus, the number of poor people who originally reached 27.77 million people in 2017 was reduced to 26.58 million people [13]. Although the number of poor people declines, the severity of poverty is still high. BPS reports that the index of poverty depth and poverty severity index has increased from March to September 2017. In March 2017, the numbers ranged from 1.74 and 0.44. However, in September 2017 increased to 1.83 and 0.44 (http://www.liputan6.com). Increasing the severity and depth of poverty indicates that the poor are getting harder to get out of economic trouble. Moreover, the poverty found in coastal areas shows that the level of expenditure is very high compared to the mainland. Thus, the poverty rate in coastal areas can reach twice as much than in other areas.

The failures of poverty alleviation policies in the coastal areas are due to weak community capacity in decision-making, lack of community participation in policy and program formulation [1], and lack of government agenda on the issues of empowerment in macro policy [6]. Thus, it is not surprising to find out that the prevention of poverty in the coastal areas is much more difficult than in the mainland. Consequently, the government must spend a much larger state budget to address the problem of poverty in coastal areas. The weak design of poverty alleviation policy especially for coastal area can also be seen from the relation that formed among fishermen community. In his study, Ferry Joko Juliantono [7] found that public policy formulations in coastal areas ignored the dichotomous reality between traditional fishermen and modern fishermen. One example of the case is Ministerial Regulation no. 2 / Permen-KP / 2015 on the prohibition of the use of trawl fishing gear and trawl net. Traditional fishermen oppose the ban because they are relying on the two fishing gear. Ironically, the government does not provide short-term policy solutions to anticipate the impact of ministerial regulations on traditional fishermen whose numbers are larger than modern fishermen. As a result, most traditional fishermen are unable to catch fish and become more entangled with debt to the moneylenders. The poverty of fishermen becomes increasingly difficult to overcome. Thus, policy design in this context is at the root of the problem of poverty.

3.2 SMEs of Creative Economy in Coastal Area

In quantity, amount of SME's in Indonesia develop significantly. Until 2017, the number of SME's are 60 million, it increases 3,1 % from the previous year. The contribution of SME's to the national economy is also quite large at around 61.41% of Gross Domestic Product (GDP) (https://nasional.sindonews.com/read/1278750/18/resolusi-umkm-indonesia-1517531360). Furthermore, employment provided by SME's can accommodate approximately 57.8 million people[8]. Those data show the important role of SME's in local and national development. The development of creative economy sector as a part of SME's also shows good performance. Creative economy sector contributes for about 7,38 % to national economy, and Gross Domestic Product (GDP) of creative economy sector shows 4,38 % in 2015 [9]. Creative industry in Indonesia posted strong and positive growth in the last three years which is contributing 852 IDR to GDP in 2015; 922,58 trillion in GDP in 2016; and 999.4 trillion IDR to GDP in 2017.

Creative economy approach for SME's development is expected to increase the added value of the products. Value added will increase if SME's have the enhancement of knowledge, creativity and technology. In the era of openness market competition, it demands the products that offer value added and benefit which are not provided by the previous market. There are 16 creative economy sectors; 1. Advertising, 2. Fine arts, 3.

Crafts, 4. Fashion, 5. Films, Animations and Videos, 6. Photography, 7. Interior Design, 8. Performing Art, 9. Architecture, 10. Applications and Game Developer, 11. Publishing, 12. Television and radio, 13. Research and Developments, 14. Culinary, 15. Visual Communication Design, 16. Product Design [9]

Development of SME's creative economy in the coastal areas also has excellent prospects. Employment opportunities will be open to coastal communities including fishermen. The coastal region actually holds enormous creative economic potentials to develop. Not only fisheries products, but coastal environment has a range of natural resources, which include coral reefs, sea grass, mangrove and also coastal tourism services. Creative economy sectors that can be develop in coastal areas are seafood culinary; craft from sea shells; film, animations and videos; music; performing arts; research and developments; etc. Those natural resources will provide some potential income to the community if it be managed properly.

The potentials of creative industries in coastal area will have a great effect if the community and fishermen strongly become the main actor in the management of creative economy. In the coastal areas, all the community (fishermen and the villagers) can take advantage of BUMDES (Badan Usaha Milik Desa/ village-owned enterprises) to manage the potential of coastal nature resources collaboratively with local government. BUMDES provides funds and capital to enhance village's potentials economy creative sectors including in coastal areas. Fishermen empowerment also strongly needed in entrepreneurship and human development programs to strengthening institutional management of SME's and creative industry sectors. Moreover, the empowerment of the community will support undeveloped coastal areas to be expanded through a sustainability process in terms of its utilization to nature preservation and tourist destinations. On the other hand, the development of the tourism sector in coastal areas will increase community income of creative products such as souvenirs or culinary. This kind of activities will create the coastal economy into more dynamic progress. Fishermen will not only depend on fishing activity on the high seas. Another impact is coastal ecology will be further preserved and maintained from destructive activities.

3.3 The role of social network

The use of communication and information technology is needed to expand new services and products to potential customers. SME's including creative industries take a huge advantage in the social network era [14]. Technological advances allow SMEs and creative industries to intercept communication with various parties through Facebook, Twitter, Instagram and Wwhatsapp. The study by Franco [10] shows that social networking utilization such as Facebook is related to efforts to reduce costs, improve communication and maintain customer satisfaction.

Social networks help SME's and creative industries to increase innovation and knowledge. Social networks provide important information from external parties such as customers or community institutions. This information is a knowledge acquired by SMEs and creative industries to increase their capacity in production and product management so that the quality of products will be well maintained. Similarly, consumer loyalty can also be maintained with the existence of online social networking.

The use of online social networking media in Indonesia itself according to Report Tetra Pak Index 2017 has reached 132 million users. Interestingly, 40 percent of total online media users are categorized as online media enthusiasts (https://inet.detik.com/cyberlife/d-3659956/132-juta-pengguna-internet-indonesia-40-penggila-medsos). The increase of sophisticated mobile phone technology, lead to the use of social networks as a means to exchange information; marketing ads and online services;

 https://doi.org/10.1051/e3sconf/20184707007

and also to coordinate some various activities. SMEs and creative economy sectors can take advantage of the benefits of social media applications to capture cooperation with various parties and market, so their products and services can expand on a wider scale.

Figure 3 Policy Implication

Future poverty reduction strategies need to consider the development of SMEs based on social networks. As can be seen in Figure 1, social networks can provide resources such as knowledge, experience, and expertise. These resources are needed by SMEs in creative economy to increase value added, expertise and market access so that SMEs have the required competitiveness.

The development of SMEs provides alternative employment opportunities for fishing families. The dependence of fishermen on fishing activities has caused the fishermen not to have time to develop other businesses. The presence of SMEs can provide access to family members of fishermen to process marine resource products. This productive activity provides an alternative source of income for the fishermen's economy. In the long term, SMEs that continue to grow with the power of social networking can improve the creativity of various processed products and will ultimately provide a long-term impact of improving the welfare of coastal communities.

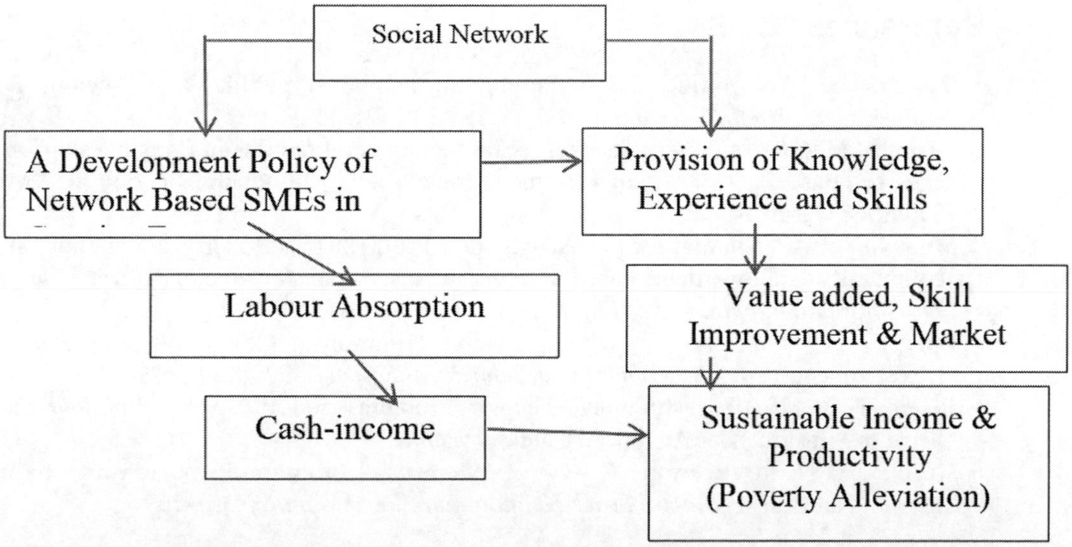

Fig 1. Policy Framework for Development of SMEs of Creative Economy for Poverty Alleviation

To create a conducive climate for the development of SMEs creative economy requires great support and political commitment from the government [11,12]. The case of the ban on the use of traditional fishing gear in the absence of short-term solutions is an important lesson for policymakers. Governments should consider appropriate institutional designs with the sociological context of the group of policy recipients. Therefore, the government should build intensive public communications with recipient groups so that great support from the public will be built on formulation of public policy.

Figure 4 Conclusion

Poverty alleviation for coastal communities requires a new approach. Various program interventions and public policies that have been implemented have proven to have insignificant effect on changes in the socio-economic conditions of better fishermen communities. Poverty and environmental degradation continue to occur in coastal areas. The failure of poverty reduction programs and policies is caused by a less comprehensive policy format in integrating social, economic and cultural issues of coastal communities.

The dependence of coastal communities on external program intervention is increasing. External parties, especially the government, are considered less able to establish effective public communications with coastal communities. As a result, data and information on the roots and poverty solutions received by the government are inconsistent with the actual conditions and needs of the poor in coastal areas. Various program interventions allocated to empower the community create the reproduction of community dependency.

The socio-economic approach through creative economy development with the base of SMEs become an alternative policy solution in poverty eradication of fisherman society. SMEs of creative economy is selected with the consideration that to grow the economic independence of coastal communities needs a strong and creative socio-economic institution. The creative economy model offers solutions that match new challenges at the local and global levels. To cultivate creativity and expertise in coastal communities requires the power of social networks involving elements of government, business, civil society, and college. With the exchange that involves social networking, the socio-economic empowerment process of the fishermen community will develop sustainably and adaptively with the dynamics of environmental change.

References

1 Rosyadi, S., & Tobirin. (2010). Perumusan kebijakan publik dalam menggali akar kemiskinan. *Jurnal Ilmu Administrasi Negara* 10 (2), 114-125.

2 Islam, M.M. (2010). Poverty in small-scale fishing communities in Bangladesh: Contexts and responses. *Dissertation*. Bremen International Graduate School for Marine Sciences, Germany.

3 Muthalib, A.A., Nuryadi, A.M., & Almana. L.O. (2015). Poverty alleviation through community empowerment based farming ponds in coastal areas Southeast Sulawesi. *The International Journal of Engineering and Sciences* (IJES), 4 (10), 32-40.

4 Boccella, N., & Salerno, I. (2016). Creative Economy, Cultural Industries and Local Development. *Procedia - Social and Behavioral Sciences* 223, 291 – 296.

5 Peters, B.G. (2010). Institutional Theory: Problems and Prospects. *Political Science Series* 69. Institute for Advanced Studies, Vienna.

6 Kurniasari, N., & Reswati, E. (2011). Memaknai Program Pemberdayaan Ekonomi Masyarakat Pesisir. *Buletin Sosek Kelautan dan Perikanan* 6 (1), 7-13.

7 Juliantono, F.J. (2015). Dinamika Relasi dan Struktur dalam Reproduksi Kemiskinan Masyarakat Nelayan: Studi atas Masyarakat Nelayan di Desa Teluk, Labuan, Banten. *Dissertation*. FISIP UI, Depok.

8 Putra, A.H. (2016). Peran UMKM dalam Pembangunan dan Kesejahteraan Masyarakat Kabupaten Blora. *Jurnal Analisa Sosiologi* 5 (2), 40-51.

9 Profil Usaha/Perusahaan 16 Subsektor Ekonomi Kreatif (2017). Retrieved May 3, 2018 from http://www.bekraf.go.id/berita/page/17/profil-usahaperusahaan-16-subsektor-ekonomi-kreatif

10 Mário Franco, Heiko Haase, Ana Pereira, (2016) "Empirical study about the role of social networks in SME performance", Journal of Systems and Information Technology, Vol. 18 Issue: 4, pp.383-403, https://doi.org/10.1108/JSIT-06-2016-0036

11 Govori, A. (2013). Factors affecting the growth and development of SMEs: Experinces from Kosovo. *Mediterranean Journal of Social Sciences* 4 (9), 701-708.

12 Lee, J.H., So, Sohn, S.Y., & Ju, Y.H. (2011). How effective is Government Support for Korean Women Enterprenuers in Small and Medium Enterprises. *Journal of Small Business Management* 49 (4), 599-616.

13 BPS: Tingkat Kemiskinan di RI Makin Parah Selama 6 Bulan. Retreived from https://www.liputan6.com/ bisnis/ read/3025513/bps-tingkat-kemiskinan-di-ri-makin-parah-selama-6-bulan

14 Kautsarina. (2013). Electronic marketing through social media (Survey on businessperson of SME in Jambi and Bengkulu). *Jurnal Studi Komunikasi dan Media* 17 (2), 135-147.

Gel aloe vera reduces MMP-9 in diabetic wounds

Yunita Sari[1,*], *Iwan* Purnawan[1], *Dhadhang* Wahyu Kurniawan[2], and *Eman* Sutrisna[3]

[1]*Department of Nursing, Universitas Jenderal Soedirman, Purwokerto, Indonesia
[2] Department of Pharmaceutical Science, Universitas Jenderal Soedirman, Purwokerto, Indonesia
[3] Department of Medical Science, Universitas Jenderal Soedirman, Purwokerto, Indonesia.

Abstract. Gel aloe vera (GAV) is a good candidate for wound treatment for fisheries, since it is cheap, could be grown in coastal areas, and could reduce inflammation, infection and improve reepithelialization in diabetic wound. However, the effect of GAV on reducing matrix metalloproteinase (MMP)-9 is still unknown. High level of MMP-9 is one of the reasons why diabetic wounds frequently fail to heal. Abundant level of MMP-9 will cause degradation of collagen formation. Therefore, the purpose of this study was to investigate the effects of GAV on reducing MMP-9. The induction of diabetes was conducted by using Alloxan Monohydrate. The rats were divided into two groups, GAV group, and control group. The wound appearance, exudate, intensity of fibroblast, and intensity of MMP-9 were compared. Mann-Whitney U test was used to analyze the difference in the intensity of positive cells for MMP-9. The study showed that the wound appearance in the GAV group was smaller compared with control group. In the GAV group, the intensity of positive cells for MMP-9 was significantly lower than in the control group. This study showed that GAV could reduce MMP-9, and therefore GAV could be recommended for wound treatment for patients with diabetic wound.

1 Introduction

Diabetes mellitus (DM) is the condition when the blood sugar above the normal range. The number of patients with DM is increased worldwide. The number of patients with DM in 1980 was about 108 million, and rise becomes 422 million in 2014 [1]. Previous studies showed that the prevalence of DM is higher in the coastal region than other areas [2-3]. The high prevalence of DM in the coastal region among fisheries might occur due to unhealthy lifestyle [3]. In Indonesia, the exact of the number of patients with DM is still unknown, however, the proportion of people with DM among fisheries in Indonesia is high compared with other jobs [4].

The higher number of DM in the coastal regions needs serious attention since the rise of the number of DM also will be followed with a high number of DM complication. One of the complications that most frequently to occur in the DM patients is diabetic wound [5]. Although no study that reported the number of diabetic wounds in the coastal region, however, based on the high prevalence of DM cases in the coastal area, it can be predicted

[1] Corresponding author: sasa.yunita@gmail.com

that the number of diabetic wounds is also high in the coastal region. Moreover, fisheries are susceptible to get wound [6]. Previous study showed that fisherman is susceptible to suffer from suppurative skin lesions due to working on trawlers in the coastal regions [6]. They are also easy to get wound in the marine environment due to coral cut, injury by a knife when filleting the fish, or by penetrating by a hook [6].

Diabetic wound is difficult to treat since it is usually easy to get infected and contain a high level of MMP-9. The bacteria that most frequently present in the wound of fisheries is Staphylococcus aureus [6]. Therefore, fisheries need topical therapy that can reduce wound infection, especially by Staphylococcus aureus. The study by Banu, Sathyanarayana, & Chattannavar (2012) showed that gel aloe vera (GAV) could reduce wound infection by Staphylococcus aureus [7], and therefore, GAV can be a good candidate for wound treatment for fisheries. In addition, aloe vera is cheap, easy to get and can be grown in the coastal region. However, up to present the role of aloe vera in reducing MMP-9 is still unknown.

Gel Aloe vera also has been used for many health purposes since it contains anti laxative agent, antitumor and antiviral agent, antiseptic agent, and anti-aging agent [8]. It is even reported that GAV can also reduce inflammation, infection, and blood glucose level [9]. Many studies have been conducted to assess the effect of GAV on wound healing, either acute wound or diabetic wounds [10-12]. In diabetic wounds, our previous study also showed that GAV could reduce inflammatory cells, and improve the migration of epithelial cells in diabetic wound [12]. We also investigated the effect of different concentration of aloe vera, and we found that Aloe vera 30 % as good as fresh aloe vera in reducing inflammation [13].

As above mentioned, one of the reasons for the failure of healing in the diabetic wound is a high level of matrix metalloproteinases (MMPs) [14]. Although GAV can reduce inflammatory cells and improve reepithelialization, the effect of GAV in reducing MMP-9 in diabetic wound is unknown. Up to the present, no study that investigates the effect of GAV in reducing MMP-9 in diabetic wounds. The study about GAV and MMP-9 is only conducted in *in vitro* study using peripheral mononucleated cells [15].

There are many types of MMPs during wound healing, they are MMPs-1, MMP-2, MMP-3, MMP-7, MMP-8, MMP-9, MMP-10, MMP-11, MMP-13, MMP-14, MMP-15, MMP-16, MMP-17, MMP-19, MMP-21, MMP-23, MMP-24, MMP-25, MMP-26. MMP-18, MT1-MMP, MT2-MMP, MT3-MMP [16]. A study by Liu et.al (2009) showed that the main MMPs that mostly present in diabetic wound is MMP-9 [17]. Previous study using human subject revealed that the level of MMP-9 could predict the delayed healing in diabetic wound. The excessive level of MMP-9 will cause excessive degradation of collagen, and therefore the granulation tissue formation and reepithelialization will be impaired [18]. Impairment of granulation tissue and reepithelialization will cause a delay in the proliferation phase, resulting in delayed wound healing. Since the high level of MMP-9 can delay wound healing, it is needed to treat a wound with topical therapies that can reduce the level of MMP-9.

It is important to assess the effect of GAV on the level of MMPs. If GAV has an effect on reducing MMP-9, it will provide substantial evidence about the use of GAV for wound care treatment. Therefore, the purpose of this study was to investigate the effect of GAV on reducing MMP-9.

2 Method

2.1 Animal

The animal used in this study was male Wistar rats at the age of 12-14 weeks old. The rats were bought from Universitas Muhammadiyah Purwokerto. The animal in the cage has free access to food and water. The induction of hyperglycemic rats was conducted by injecting with alloxan monohydrate. The rats were confirmed as diabetic rats if the blood glucose reaches a value of 200 mg/dl or above.

2.2 Wounding procedure and treatment

The rats were divided into GAV treated group and control group. In this study, we used inner gel aloe vera. The aloe vera was cut and cleaned, and the inner part of GAV was used. Control group received gel that did not contain aloe vera (sodium alginate gel). The making of the sodium alginate gel was based on previous study [13]. Each of groups consisted of 5 rats. The protocols used in the study was approved by ethical research committee, Universitas Jenderal soedirman, Indonesia.

After shaving the rats, one-centimeter of a wound was created in the right side of the abdominal region of the rats. After wounding, the wound was cleansed in both groups by using normal saline 0,9 %. After cleansing, the rats in the GAV group was given GAV while in the control group was given sodium alginat gel. The gel was applied by swabbing method using cotton bud. Finally, both control and GAV group were covered by parafilm dressing, followed by gauze dressing. The wounds were daily observed and photographed for twelve days.

2.3 Tissue staining

Hematoxylin and Eosin staining was conducted to assess tissue damage and fibroblast intensity in the skin tissue. The method of tissue processing was based on previous study [12]. The difference in the fibroblast level was assessed based on the previous study [19].

2.4 Immunohistochemistry

To investigate the intensity level of MMP-9, immunohistochemistry method was used. Polylysine-coated slides was used to place the serial section. The tissue was then deparaffinized and rehydrated through serial graded of ethanol and xylene. To inactivate the endogenous peroxidase , the slide was incubated for 30 minutes in 3% hydrogen peroxide. Antigen retrieval was carried out by immersed in citrate buffer (pH 6.0) for 15 min. The tissue were incubated with primary antibody. The primary antibody used in this study was monoclonal antibody against rat MMP-9 (Abcam, Cambridge, UK) with a dilution of 1:100 The section was then incubated in the biotin-conjugated secondary antibody. In this study, we used streptavidin-peroxidase to visualize the staining. We then used diaminobenzidine (DAB; Sigma, St. Louis, MO, USA) to function as the chromogenic agents and hematoxylin as the counterstaining agent. The images were magnified under a light microscope and recorded on a computer. The difference in the level of positive cells for MMP-9 was assessed using the formula HSCORE. The intensity of labeling were classified as 1, 2, or 3 (weak, moderate, or strong, respectively) [20].

2.5 Statistical analysis

The comparison of body weight and blood glucose level between treatment and control groups were analyzed by independent T-test, while the intensity of fibroblast and intensity of MMP-9 was analyzed by Mann-Whitney U-tests. The p-value less than 0,05 was considered significant. The statistical analysis was conducted by using SPSS version of 20.

3 Result and Disscusion

3.1 Body weight of rats

The average of the body weight of the rats in the GAV group was 190±7 Gram, while the body weight of rats in the control group was 192.5 ± 5 Gram. The result of independent T-test showed that there is that there is no difference in the body weight of the rats in theGAV group and in the control group.

Table 1. The body weight of GAV and control groups

Groups	Body weight (Gram)	p-value
Gel aloe vera	190 ± 7	0,620
Control	192.5 ± 5	

Data as mean ± SD

3.2 Blood glucose injection

The blood glucose after injection of alloxan monohydrate in the GAV group and control group was shown in table 2. The mean of the blood glucose in the GAV group 248.75±27.7 mg/dl, and in the control group was 236.75±21.17 mg/dl. Table 2 showed that the blood glucose level after injection of alloxan monohydrate was increased above 200 mg/dl. Independent T-test showed that there is no significant difference in the blood glucose between two groups.

Table 2. Blood glucose in GAV and control groups

Groups	Blood glucose (mg/dl)	p-value
GAV	248.75±27.7	0,519
Control	236.75±21.17	

https://doi.org/10.1051/e3sconf/20184707008

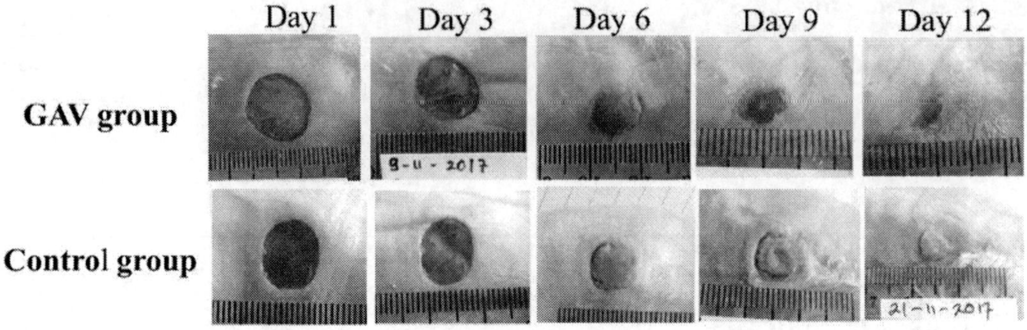

Fig 1. Wound appearance between GAV group and control group

3.3 Wound appearance

The wound appearance can be seen in Figure 1. Figure 1 showed that the appearance of wound on day 1 to day 3 was same between GAV group and control group. There is no necrotic tissue in two groups. In addition, the exudate was still clear in color in both groups. On day 6, the wound in both groups was covered with necrotic tissue. However, the exudate in control group was purulent compared with the exudate in the GAV group. On day 9, the wound in the GAV group was covered with granulation tissue that has a color of red, and the wound size was smaller compared with day 6. Comparing with control, the wound in the control group on day 6 was still covered with necrotic tissue. The necrotic tissue in the control group was thicker on day 9 compared with day 6. The wound size in the control group was almost same compared with on day 9. The exudate in the GAV group was clear in color, while exudate in the control group was purulent. On day 12, the wound size in the GAV group was smaller compared with on day 9. The wound size in the GAV group was also smaller compared with wound size in the control group. Moreover, the wound bed in the control group was still covered with necrotic tissue.

3.4 Intensity of fibroblast

The intensity of the fibroblast in the GAV group was more pronounced compared with control group. Most fibroblast located in the upper part of dermis layer. The difference in the level of fibroblast was shown in table 3.

Table 3. The difference in the fibroblast level

Groups	Fibroblast	p-value
GAV	4*	0.04
Control	3	

Rating scale: 0 = absent, 1 = occasional, 2 = moderate, 3 = abundant, 4 = very abundant
*$p < 0.05$

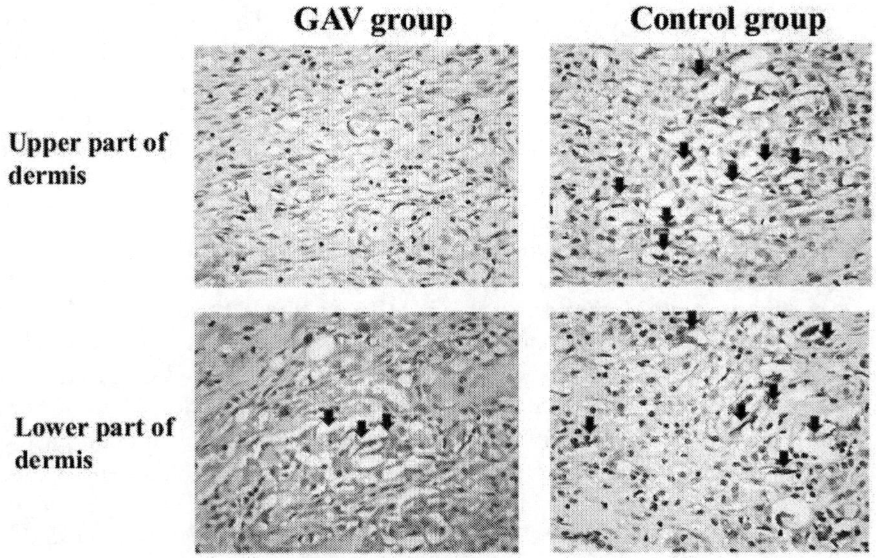

Fig 2. The MMP-9 positive cells (arrow) between GAV and control groups in the upper part of dermis (below epidermis) and lower part of dermis (magnification of 400 times)

The result of MMP-9 staining is shown in figure 2. The blue color means the nucleus cell, while the brown color indicates the positive cells for MMP-9. Figure 2 showed that the positive cells for MMP-9 was more pronounced in the control group compared with the GAV group. The positive cells for MMP-9 were pronounced in the control group, not only in the upper part of the dermis, but also in the lower part of the dermis. The comparison of the intensity of MMP-9 in the dermis layer between GAV group and control group was seen in table 4. From table 4, it could be seen that the intensity of MMP-9 in the GAV group was significantly less than in the control group (P= 0,011).

Table 4. The intensity of MMP-9 in both groups

Groups	Intensity of MMP-9	p-value
GAV	1*	0.011
Control	3	

Values indicated median score
Rating scale: 1 = weak, 2 = moderate, 3 = strong
*p < 0.05

This study is the first study to investigate the role of GAV in reducing MMP-9 in diabetic wound. This study will provide the clinical importance of gel aloe vera in the treatment of diabetic wound. Our study revealed that GAV could reduce MMP-9.

The previous study showed that MMPs have an important role in morphogenesis, wound healing and tissue repair following injury [21]. However, when the MMPs is dysregulated, it can cause many diseases and disorders such as stenosis, heart failure, cancer, arthritis, encephalomyelitis, etc [21]. During wound healing process, MMPs can clean the debris, bacteria, and broken extracellular matrix during early phase of wound healing [16]. It also has a role and affects the differentiation, migration of epithelial, angiogenesis, and apoptosis process [22]. The MMP activity was strictly regulated by the presence of tissue inhibitors of metalloproteinase (TIMP) [23]. The balance between MMP and TIMP will make wound healing progress well. However, the increase of MMPs level will cause excessive degradation of collagen.

There are many MMPs have been studied so far. They are MMPs-1, MMP-2, MMP-3, MMP-7, MMP-8, MMP-9, MMP-10, MMP-11and MMP-13, MMP-14, MMP-15, MMP-16, MMP-17, MMP-19, MMP-21, MMP-23, MMP-24, MMP-25, MMP-26. MMP-18, MT1-MMP, MT2-MMP, MT3-MMP, etc. [16]. Among these MMPs, the MMP that is highly present in the diabetic wound is MMP-9 [24]. The reason why MMP-9 is at a high level in the diabetic wound is unclear. Previous study showed that inflammatory cells are increased in the diabetic wound compared with acute wound [25]. The inflammatory cells such as neutrophil and macrophage will release MMP-9. It might the reason why the MMP-9 is high in the diabetic wound compared with acute wound [26]. Further study is needed to confirm the cause of the increase of MMP-9 in the diabetic wound. In the macroscopical findings, the formation of granulation tissue in the control group is low. The wound bed is also covered with necrotic tissue on day 9 when the wound bed in the GAV group was fully covered with granulation tissue. The lacking of the granulation tissue in the control group might because the high level of MMP-9 as seen in the result of immunohistochemistry. The increased level of MMP-9 might cause prolonged inflammation phase, resulting in the failure of proliferation phase. In our study, the exudate in the control group on day 9 is also purulent compared with the exudate in the GAV group, which is clear and did not contain necrotic tissue. Wound exudates is a wound fluid that is produced during wound healing. The color of exudate reflects the microenvironment of a wound [27]. The reason why the wound exudate in the control group is purulent is due to the presence of thick necrotic tissue.

In the immunohistochemistry result, compared to the control group who does not receive the application of veraGAV, the group that receives GAV showed a reduction of MMP-9. Our study corresponds with the previous study by Vijayalakshmi et al. (2012) that investigate the effect of aloe vera in peripheral blood mononuclear wound [15]. Their research showed that aloe vera can reduce the production of MMP-9. The mechanism involved in the reduction of the production of MMP-9 by gel AV is still unclear. Previous study suggested that down-regulation of MMP-9 might occur due to inhibition of the activation of NF- B and its subsequent binding to MMP-9 promoter or by inhibiting the production of prostaglandins (mediator) which induce MMP-9 expression [15].

Our study showed that GAV could reduce MMP-9. This study provides an important evidence in the use of GAV for wound treatment for patients. Previous study showed that two important factors that can impede healing in the diabetic wound are infection and excessive level of MMP-9. Study by Foote et al.(2017) already showed that GAV could reduce wound infection, including by Staphylococcus Aureus [6]. Based on the result of our study and another researcher, it can be suggested that veraGAV could be a good choice for wound treatment of fisherman in the coastal region.

4 Conclusions

This study is the first to study the role of the GAV in reducing MMP-9 in the diabetic wound. Our study showed that the use of GAV could reduce the level of the intensity of MMP-9 in the diabetic wound. Based on the result of this study, it can be suggested that GAV could be a good candidate for wound treatment for the fisherman in the coastal region.

References

1. NCD Risk Factor Collaboration, *The Lancet* **387**,10027 (2016)
2. C.R.Rao, V.G. Kamath, A. Shetty, & A.Kamath. Int J Diabetes Dev Ctries. **30**, 80–

85(2010)
3. Y. Zhao, E. M.Crimmins, P. Hu, Y. Shen, J. P Smith, J. Strauss, Y. Wang, Y Zhang. Int J Public Health. **61**, 347–356 (2016)
4. Riskesdas. Kemenkes. **90** (2013)
5. K. B. Hobizal, D. K. Wukich. Diabetic Foot & Ankle **3**, 18409 (2012)
6. A. Foote, R. Henderson, A. Lindberg, C. Grigg, C.Greenfield, A.Kirke, K.Auret. Aust Fam Physician. **46**, 923–927 (2017)
7. A.Banu, B. C. Sathyanarayana, G. Chattannavar. Australas Med J. **5**, 305–309 (2012)
8. A.Surjushe, R.Vasani, D. G. Saple. Indian J Dermatol. **53**, 163–166 (2008)
9. M. H. Radha, N. P. Laxmipriya. J Tradit Complement Med. **5**, 21–26 (2015)
10. N.Takzare, M. Hosseini, G. Hasanzadeh, H. Mortazavi, A.Takzare, P. Habibi. Toxicol Mech Methods. **19**, 73–77 (2009)
11. A.Oryan, A.Mohammadalipour ,A. Moshiri, M. R. Tabandeh. Ann Plast Surg. **77**, 37–46 (2016)
12. Y. Sari, I. Purnawan , D.W. Kurniawan, E. Sutrisna. Journal of Evidence-Based Integrative Medicine. **23**
13. Y.Sari, I. Purnawan, E. Sutrisna, D. W. Kurniawan, Nasruddin, 3rd International Conference on Pharmacy and Pharmaceutical Science (Bali, 2018)
14. J. L. Lazaro, V. Izzo, S.Meaume, A. H. Davies, R.Lobmann, L. Uccioli. J Wound Care. **25**, 277–287(2016)
15. D. Vijayalakshmi, R. Dhandapani, S. Jayaveni, P. S. Jithendra, C Rose, A. B Mandal. J Ethnopharmacol. **141**, 542–546(2012)
16. S. M. Ayuk, H.Abrahamse, N. N. Houreld. T. J Diabetes Res. 2897656 (2016)
17. Y.Liu, D.Min, T Bolton, V. Nubé, S. M. Twigg, D. K.Yue, S. V. McLennan. Diabetes Care. **32**, 117–119 (2009)
18. M. J.Reiss, Y. P.Han, E.Garcia, M Goldberg, Y. K Hong, W. L Garner. Surgery. **147**, 295(2010)
19. J.B.Wright, K.Lam, A.G.Buret, , M.E. Olson, R.E. Burrell, Wound Repair Regen **10**: 141–151(2002)
20. A. E. Rizzardi, A. T. Johnson, R. I.Vogel, S. E. Pambuccian , J.Henriksen, A. P. N Skubitz, G.C. Metzger Schmechel. Diagnostic Pathology. **7**, 42 (2012)
21. H.Nagase, R.Visse , G. Murphy.Cardiovascular Research. **69**, 562–573 (2006)
22. Y.Kudo, S. Iizuka, M.Yoshida, T. Tsunematsu, T. Kondo, A. Subarnbhesaj, T.Takata. J Biol Chem. **287**, 38716-38729 (2012)
23. M. P Caley, V. L. C. Martins, E. A.O'Toole, Adv Wound Care, **4(4)**, 225–234 (2015)
24. K. Singh, K. Agrawal, N. K. Gupta, S. K. Mohan, G., S. Chaturvedi The International Journal of Lower Extremity Wounds. **13**, 94–102 (2014)
25. J. J. Salazar, W. J. Ennis, T. J. Koh. J Diabetes Complications, **30,** 746–752 (2016)
26. E. I Deryugina, E Zajac, A. Juncker-Jensen, T. A.Kupriyanova L, Welter, J. P. Quigley. Neoplasia (New York, N.Y.). **16**, 771–788 (2014)
27. Y. Sari, G Nakagami, A Kinoshita, L Huang, K Ueda, S Iizaka, H.Sanada, J.Sugama. International Wound Journal. **5**, 674–680 (2008)

AUTHOR INDEX

Afrizal 408
Agus Mulyadi Hanung 123
Agustriani Fitri 386
Ahda Sumantri Rifki 286
Ahmad Romdoni Taufiq 141
Ahmad Triguardi 170
Alam Fiyesti 170
Amron Amron 170, 273, 281, 312, 329
Andriyani Nuraina 99
Anggoro Sutrisno 149, 261
Apriadi Tri 76
Ardi Idil 42
Arie Prayogo Norman 25, 158, 196, 205
Arsil P. 9
Asyiah B.N. 37
Aulia Zahra 170
Avia Pasha Wanda 312
Ayustina Ratih 170
Azhar Muhamad 396, 427
Basit Abdul 123
Bintoro Rahadi Wasi 348
Budhi Pramono Taufik 25, 286
Budi Wiyanto Dwi 42
Devinta Hesti 450
Dewi Rose 99, 149, 261, 286
Dwirama Putra Risandi 76, 243
Dyah Nur Meinita Maria 141
Ekasanti Anandita 15
Eko Dwi Haryono Florencius 281, 286
Endrawati Hadi 149, 261
Faizal Ibnu 214
Fauziyah 386
Fitrah Elpeni 457
Fitri Mardani Irfina 361
Geraldi Pardede Hanson 182
Gita Tiara Shanti Ainia 182
Gumay Yudha Indra 158
Habsari Sri Kusumo 444
Harada Kazuo 111
Harahap S. Alisyahbana 132
Harini Noor 1
Harisam R. Taufan 323
Harisam Taufan 94, 99, 205
Hartiwiningsih 369
Hartoyo 281
Hary Tjahja Soedibya Petrus 25
Haryadi Dwi 292, 417
Haryanto Tenang 402
Haryono Florencius Eko Dwi 323
Helmi Muflih Azzam 182
Herawati Vivi Endar 396, 427
Hermawan David 1
Heru Prihadi Tri 37, 42, 60
Hidayat Arif Imam 450
Hilmi Endang 329
Hirata Kazumasa 111
Hurly Sefnianti Yessy 273
Ibrahim 292, 417
Idris Fadhliyah 237
Ilham Yuwanda 132
Iqbal Kenedi Muhammad 182
Irawan Henky 76, 243
Joei Koenawan Chandra 76
Juita Sari Ratna 273
Jumsurizal 243
Junaidi Teuku 99, 361
Karlina Ita 237
Karseno 111
Kurniawan Dedi 243
Kurniawan Fery 237
Kurniawan Rika 243
Kusuma Atmadja Wiwin 76
Larasayu Santoso Niken 182
Latifah Emmy 369
Lazuardy Rolian Damar 182, 312
Listiowati Emyliana 25, 158
Maharani Adinda 214
Maharani Karina 182
Mahdiana Arif 99, 117, 196, 361
Maman Suherman Ade 369
Marhaeni Bintang 141, 323
Melda Situmorang Desi 386
Mubarak Helmi 182
Mustakin Haji 170
Nalendra Lazuardy 281
Naufalin R. 9
Nugrayani Dewi 15
Nur Afrizal Satria 182
Nur Cahyo Tri 286
Nur Faizan Rifki 182
Prabowo 182
Pratama Ginanjar 243
Pratomo Arief 408
Pratomo Arif 243
Prayogo Aldo 170
Prayogo Norman Arie 15, 94, 117, 323
Pribadi Rudhi 205
Purba Noir P 214
Purnawan Iwan 464
Purwendah Elly Kristiani 379
Puspaningsih Dewi 37, 68
Raharjo Agus 338, 348, 355
Rasidi 42
Ristiani Ajeng 141
Rizka Safitri Dhananta 182
Rosyadi Slamet 402, 457
Rus Rahayu Diana 226
Sabhita Kusuma Ayusia 457
Saefudin Yusuf 338, 355
Safitri Adini 182

AUTHOR INDEX

Sahri Siregar Asrul 99, 158, 196
Said Raja'I Tengku 76, 243
Salman M.F. 9
Santoso Mahendro 158
Sarbini Sarbini 433
Sari Fitra 170
Sari Siregar Asrul 205
Sari Yunita 464
Saryono Saryono 450
Sekar Cyndana Ajeng 273
Setiadi Eri 50, 60
Setijaningsih Lies 68
Setijanto 141
Sih Piranti Agatha 226
Siregar Asrul Sahri 15, 94
Soedibya Petrus Hary Tjahja 323
Sonny Januar C 196
Subagyo Yusuf 99
Sudardi Bani 444
Sudin Abdul Malik 123
Sudrajat Tedi 338, 348, 355
Sugiyono 205
Suharso Putut 396, 427, 433, 444
Suhartoyo Suhartoyo 396, 427
Suhendra 329
Sukardi Purnama 15, 94
Sulistiawati Linda Yanti 379
Sulistyo Isdy 117
Suryanti Ani 76, 243
Suteja Yuliyanto 386
Sutrisna Eman 464
Taufan Harisam R. 25
Taufik Imam 50, 68
Trenggono Mukti 281, 312
Triatmodjo Marsudi 379
Trihastuti Nanik 396
Triwahyanti Lesa 273
Ulfah Fitria 408
Wahyu A P Trisna 182
Wahyu Kurniawan Dhadhang 464
Wahyudin Nanang 292, 417
Wahyuni Sholehah Tri 158
Wahyuningrat 402
Waluyo Gentur 226
Warkoyo 1
Wasi Bintoro Rahadi 338, 355
Wicaksono R. 9
Widiyati Ani 37, 42
Widodo Abdul Haris Budi 450
Widodo Sahid Teguh 444
Widyaningrum Ria 182
Widyastuti Yohana R. 60
Winanto Tjahjo 149, 182, 261
Winanto Tjahyo 94
Wisudyanti Budi Hastuti Dewi 205
Wisudyanti Dewi 286
Wulandari Yuliarie 1
Yanuar Yogi 132
Yasumasa Bessho Riviani 15
Yulia Ikhsani Idha 123
Yuliantiningsih Aryuni 369
Yulianto Try 76
Yuwono Doddy 170
Zainuri Muhammad 149, 261
Zunika Putri Kharisma 182

EDP Sciences – Web of Conferences
17, Avenue du Hoggar
Parc d'Activité de Courtabouf
BP 112
F-91944 Les Ulis Cedex A
France

ISBN 978-1-5108-6856-4

9 781510 868564

CURRAN ASSOCIATES INC.
proceedings.com